Proceedings of Cycad 2005

The 7th International Conference on Cycad Biology

8–12 January 2005

Jardín Botánico Fco. J. Clavijero
Instituto De Ecología, A.C.
Xalapa, Veracruz, Mexico

Editors: Andrew P. Vovides, Dennis Wm. Stevenson, and Roy Osborne
Graphic design editing by Ciarán Moloney

Memoirs of The New York Botanical Garden, Volume 97

THE NEW YORK BOTANICAL GARDEN PRESS

Published by The New York Botanical Garden Press
200th Street and Kazimiroff Boulevard
Bronx, New York 10458-5126 USA

The paper used in this publication meets the requirements of the National Standards for Information Sciences-Permanence of Paper for Publications and Documents in Libraries and Archives, ANSI/NISO Z39.48-1992.

Printed in the United States of America.

MetLife Foundation is a Leadership Funder of The New York Botanical Garden Press.

Memoirs of The New York Botanical Garden Volume 97 was issued on 28 December 2007.
International Standard Serial Number 0077-8931
ISBN-13: 978-0-89327-490-0
ISBN-10: 0-89327-490-9

Cover: Zamia integrifolia. Section of mature stem that has been cleared to reveal girdling leaf traces.

Contents

Ecology

Ethnobotany

Genomics, Physiology, and Structure

Contributors

Reyneil Cruz-Aguado, Dept. of Experimental Medicine, University of British Columbia, Vancouver, British Columbia, Canada.

Beatriz Rendón Aguilar, Universidad Autónoma Metropolitana Iztapala, Departamento de Ecología.

Georgina Vargas Amado, Instituto de Botánica Centro, Universitario de Ciencias Biológicas y Agropecuarias, Universidad de Guadalajara Zapopan, Jalisco 45110 México.

Jorge González-Astorga, Laboratorio de Genética de Poblaciones, Departamento de Biología Evolutiva, Instituto de Ecología, A. C. km 2.5 Antigua Carretera a Coatepec No. 351, Xalapa 91070, Veracruz, Mexico.

Sandra A. Banack, Department of Biological Science, California State University, Fullerton, CA 92834, USA.

Christina Bárcenas, Dept. de Sistemática Vegetal, Instituto de Ecología, A.C., Apdo Postal 63, Xalapa, Veracruz, 91000 Mexico.

Mark A. Bonta, Division of Social Sciences, Delta State University, Cleveland, OH, USA.

Eric D. Brenner, The New York Botanical Garden, 200th St. and Kazimiroff, Bronx, NY 10458-5126, USA.

Jeff Chemnick, Ganna Walska Lotusland Foundation, 695 Ashley Road, Santa Barbara, CA 93108, USA.

Joanna Chiu, Department of Molecular Biology and Biochemistry, Center for Advanced Biotechnology and Medicine, 679 Hoes Lane, Piscataway, NJ 08854, USA.

Gloria M. Coruzzi, New York University, Department of Biology, 1009 Main Building, New York, NY 10003, USA.

Paul A. Cox, Institute for Ethnomedicine, PO Box 3464, Jackson, WY 83001, USA.

Christopher G. M. Dalzell, Durban Botanic Gardens, PO Box 3740, Durban, 4000 South Africa.

John S. Donaldson, South African National Biodiversity Institute, Private Bag X7, Claremont 7735, South Africa.

Errol Douwes, University of KwaZulu-Natal, School of Biological and Conservation Sciences, King George V Avenue, Durban, 4000 South Africa.

Douglas A. Downie, Department of Zoology and Entomology, Rhodes University, Grahamstown, South Africa.

E. Mark Engleman, Programa de Botánica, Colegio de Postgraduados, CP-56230 Montecillo, Texcoco, México.

Miguel A. Pérez-Farrera, Escuela de Biología, UNICACH Libramiento Norte Poniente S/N Tuxtla Gutiérrez, Chiapas, México, 2903.

Paul I. Forster, Queensland Herbarium, Environmental Protection Agency, Brisbane Botanic Gardens, Mt Coot-tha Road, Toowong, Queensland 4066, Australia.

Alma Delia Chávez González, Universidad Nacional Autónoma de México, Facultad de Estudios Superiores Zaragoza.

Dolores González, Jardín Botánico Fco. J. Clavijero Instituto de Ecología, A.C., Apartado Postal 63, 91000, Xalapa, Veracruz, México.

Root Gorelick, School of Life Sciences, Arizona State University, Tempe, AZ 85287-4501, USA.

Citlalli Castillo-Guevara, Obrero Mundial No. 5 Col. Obrero Campesina, Xalapa, Veracruz, 91020 México.

Erin L. Hawkes, Graduate Program in Neuroscience, University of British Columbia, Vancouver, British Columbia, Canada.

Jody L. Haynes, Montgomery Botanical Center, 11901 Old Cutler Road, Miami, FL 33156, USA.

Ken D. Hill, Royal Botanic Gardens, Mrs. Macquaries Road, Sydney, NSW, 2000 Australia.

Gregory Holzman, P.O. Box 764, Kekaha, HI 96752, USA.

Craig Hull, School of Integrative Biology, The University of Queensland, St. Lucia, Brisbane, 4072 Queensland, Australia.

Carlos Iglesias, Jardín Botánico Fco. J. Clavijero, Instituto de Ecología, A.C., Apdo Postal 63, Xalapa, Veracruz, 91000 Mexico.

Azuncena De Lourdes Luna José, Instituto de Recursos Naturales. Colegio de Posgraduados.

Manpreet S. Katari, New York University, Department of Biology, 1009 Main Building, New York, NY 10003, USA.

Stephen D. Koch, Programa de Botánica, Colegio de Postgraduados, CP-56230 Montecillo, Texcoco, México.

Julio C. Lazcano Lara, Jardín Botánico Nacional Carretera del Rocío Km 3,5, Calabazar, Boyeros C. Habana, Cuba, C.P. 19 230.

Vivian Lee, Dept. of Opthalmology, University of British Columbia, Vancouver, British Columbia, Canada.

Catharine Mannion, University of Florida-IFAS, Tropical Research and Education Center, Homestead, FL 33031, USA.

Thomas Marler, College of Natural and Applied Sciences, University of Guam, Mangilao, Guam 96923.

John K. Marshall, PO Box 552, Denmark, WA 6333, Australia.

Rob A. Martienssen, Cold Spring Harbor Laboratory, 1 Bungtown Road, Cold Spring Harbor, NY 11724, USA.

Richard W. McCombie, Genome Research Center, Cold Spring Harbor Laboratory, 500 Sunnyside Blvd., Woodbury, NY 11797, USA.

Mahinda Martínez, Escuela de Biología, Facultad de Ciencias Naturales Universidad Autónoma de Querétaro, Prolongación Hidalgo S/N, Querétaro, Querétaro.

Alan W. Meerow, USDA-ARS-SHRS, National Germplasm Repository, 13601 Old Cutler Rd., Miami FL 33158, USA.

Jorge Mendieta, Departamento de Botánica, Facultad de Ciencias Naturales, Exactas y Tecnología, Universidad de Panamá, Ciudad Universitaria Octavio Méndez Pereira, Panamá, Panamá.

Chris Moore, Department of Primary Industries and Fisheries, Yeerongpilly, 4105, Queensland, Australia.

Silvia Salas-Morales, Camino Nacional # 80 B, San Sebastián Tutla, Oaxaca, c.p. 71246 Sociedad para el Estudio de los Recursos Bióticos de Oaxaca, Asociación Civil (SERBO, A.C.).

José Pablo Torres Morán, Instituto de Botánica Centro Universitario de Ciencias Biológicas y Agropecuarias Universidad de Guadalajara Zapopan, Jalisco 45110 México.

Fernando Nicolalde-Morejón, Instituto de Ecología, A.C. Km 2.5 Carretera Antigua a Coatepec No. 351, Congregación El Haya, C.P. 91070. Xalapa; Veracruz, México.

Jeremy Moynihan, Department of Biological Sciences, Florida International University, Miami, FL 33199, USA.

Susan J. Murch, Institute for Ethnomedicine, PO Box 3464, Jackson, WY 83001, USA.

Javier Francisco-Ortega, Department of Biological Sciences, Florida International University, Miami, FL 33199, USA.

Roy Osborne, 19 Calhoun Street, McDowall, Queensland 4053, Australia.

Peter M. Palenchar, New York University, Department of Biology, 1009 Main Building, New York, NY 10003, USA.

Jorge Alberto Pérez de la Rosa, Instituto de Botánica Centro, Universitario de Ciencias Biológicas y Agropecuarias Universidad de Guadalajara Zapopan, Jalisco 45110 México.

P. Radha, University School of Basic and Applied Sciences, Guru Gobind Singh Indraprastha University, Kashmere Gate, Delhi 110 006 India.

Stephen A. Rudd, Centre for Biotechnology, Tykistökatu 6, FIN-20521 Turku, Finland.

Suzan J. Runko, The New York Botanical Garden, 200th St. and Kazimiroff, Bronx, NY 10458-5126, USA.

Luis H. Sandoval, Escuela de Biología, Facultad de Ciencias Naturales Universidad Autónoma de Querétaro, Prolongación Hidalgo S/N, Querétaro, Querétaro.

Rita Singh, University School of Basic and Applied Sciences, Guru Gobind Singh Indraprastha University, Kashmere Gate, Delhi 110 006 India.

Swaraj Singh, Dept. of Experimental Medicine, University of British Columbia, Vancouver, British Columbia, Canada.

Prabha Sharma, University School of Basic and Applied Sciences, Guru Gobind Singh Indraprastha University, Kashmere Gate, Delhi 110 006 India.

Christopher A. Shaw, Depts. of Opthalmology and Experimental Medicine, and Graduate Program in Neuroscience, University of British Columbia, Vancouver, British Columbia, Canada.

Dennis Wm. Stevenson, The New York Botanical Garden, 200th St. and Kazimiroff, Bronx, NY 10458-5126, USA.

Mária Y. Sánchez-Tinoco, Instituto de Investigaciones Biológicas, Universidad Veracruzana, Apartado Postal 294, CP-91000 Xalapa, Veracruz, México.

Alberto S. Taylor, Departamento de Botánica, Facultad de Ciencias Naturales, Exactas y Tecnología, Universidad de Panamá, Ciudad Universitaria Octavio Méndez Pereira, Panamá, Panamá.

Irene Terry, University of Utah, Department of Biology, 257 South 1400 East, Salt Lake City, UT 84112, USA.

Juan C. Flores-Vázquez, Camino Nacional # 80 B, San Sebastián Tutla, Oaxaca, c.p. 71246 Sociedad para el Estudio de los Recursos Bióticos de Oaxaca, Asociación Civil (SERBO, A.C.).

Walinda Patricia Reyes Velázquez, Laboratorio de Micotoxinas Centro Universitario de Ciencias Biológicas y Agropecuarias Universidad de Guadalajara Zapopan, Jalisco 45110 México.

Alfredo Saynes-Vásquez, Camino Nacional # 80 B, San Sebastián Tutla, Oaxaca, c.p. 71246 Sociedad para el Estudio de los Recursos Bióticos de Oaxaca, Asociación Civil (SERBO, A.C.).

Piet Vorster, Department of Botany and Zoology, University of Stellenbosch, Private Bag X1, 7602 Matieland, South Africa.

Andrew P. Vovides, Dept. de Biología Evolutiva, Instituto de Ecología, A.C., Apdo Postal 63, Xalapa, Veracruz, 91000 Mexico.

Gimme H. Walter, School of Integrative Biology, The University of Queensland, St. Lucia, Brisbane, 4072 Queensland, Australia.

Christine Wiese, Montgomery Botanical Center, 11901 Old Cutler Rd., Miami, FL 33156-4242, USA.

Jason M. B. Wilson, Graduate Program in Neuroscience, University of British Columbia, Vancouver, British Columbia, Canada.

Acknowledgments

The International Conferences on Cycad Biology are held under the auspices of the Cycad Specialist Group, a section of the Species Survival Commission of the World Conservation Union. The organizers of CYCAD 2005 are particularly grateful to the following persons and organizations for financial, physical, and infrastructural assistance: Instituto de Ecología, A.C., Daniel Piñero, Universidad Veracruzana, Comisión Nacional para el Conocimiento y Uso de la Biodiversidad (CONABIO), The New York Botanical Garden, Tim Gregory, Consejo Estatal del Protección al Ambiente (COEPA), Museo de Antropología, Universidad Veracruzana, Consejo Nacional de Ciencia y Tecnología (CONACyT), Secretaría del Medio Ambiente y Recursos Naturales (SEMARNAT), and Comisión Nacional de Áreas Naturales Protegidas (CONANP).

The following persons kindly assisted in reviewing the papers published in this volume: Michael Balick, Eric Brenner, Paul Cox, Damon Little, John Donaldson, Jack Fisher, Paul Forster, Alejandro Gandolfo, Patrick Griffith, Jorge González-Astorga, Nat Grobbelaar, Michael Hood, Amy Litt, Damon Little, Anders Lindström, Carlos Montaña, Fernando Nicolalde-Morejón, Tim Motley, Rolf Oberprieler, Roy Osborne, John Pate, Miguel Angel Pérez-Farrera, Daniel Pinero, Ken Pringle, Hannes Robbertse, Oliver Sacks, Peter Spencer, Dennis Stevenson, William Tang, Tom Taylor, Irene Terry, Franciso Vergara-Silva, Piet Vorster, Andrew Vovides, Gimme Walter, and Jason Wilson; and Ciarán Moloney and Nate Smith for graphic arts and editing.

Cycad 2005 Organizing Committee and Assistants

Conference Chair	Andrew P. Vovides
Conference Deputy Chair	Dennis Wm. Stevenson
Administrative Assistant	Andrea Cruz
Program Coordinator	Roy Osborne
Nursery Tour Coordinator	Mario Vázquez-Torres
Cultural Activities	Ernesto Rodríguez Luna and Aaron Ojeda Jimeno
Technical Assistants	Phil Brewster, Danae Cabrera, Pablo Carillo, Gómez-García, Orlik Gómez-García, Jorge González-Astorga, Carlos Iglesias, Fernando Nicolalde-Morejón, Pablo Octavio-Aguilar, Victoria Tejeda, and Alejandra Vovides

In Memoriam: Leo Schibli

7 January 1958–8 October 2004

Leo Schibli was born and raised in Zurich, Switzerland. He was educated as a chemist but also loved mapmaking and natural history. As a young man, he traveled extensively, and this reinforced his view of himself as a citizen of the entire world. He adopted Mexico as his home, settling in Oaxaca in 1988. He assisted with various activities in the Graduate Centre of the Technological Institute of Oaxaca without payment and later, with his wife Silvia, was instrumental in founding and operating SERBO (La Sociedad para el Estudio de los Recursos Bióticos de Oaxaca). Through SERBO, Leo drove the use of remote sensing and geographic information systems for mapping and monitoring the condition of the Oaxacan forests. This, coupled with Silvia's direct cataloging of the State's biodiversity, built SERBO into a world class botanical and environmental research organization. In 1998, together with teams from SEMARNAP and SEDENA, Leo produced the cartography that helped control devastating forest fires in Oaxaca. With his extensive physiographic knowledge, he recorded important data on the cloud forests and the "bosques mesofilíticos" (his own phrase), especially in the Sierra Norte and the Chimalapas areas. He was tireless in promoting the importance of physical geography, training many persons in the operation of global positioning equipment and the development of geographic information systems.

A passionate naturalist and conservationist, Leo also loved cycads and exploring for new cycad populations. He admired them for their primeval form, their stubborn persis-

tence through geological time periods, their intimate association with remote and spectacular habitats, and their exemplification of the incredible biodiversity of his beloved Oaxaca. Leo was, above all, a person who loved and embraced every minute of his life. Much respected and dearly missed, he was an inspiration to all who met him.

CHAPTER I

INTRODUCTION

The *International Conferences on Cycad Biology* are held every three years. Since the first conference, held in Nice, France, in April 1987, five more meetings followed in Australia (July 1990), South Africa (July 1993), China (May 1996), the United States (August 1999), and Thailand (July–August 2002). The third and subsequent conferences have been held under the auspices of the Cycad Specialist Group, a section of the Species Survival Commission of the World Conservation Union.

These conferences have now become established as high-level, specialized academic events on this unusual, important, and interesting group of plants, stimulating research and promoting *ex-* and *in-situ* conservation globally.

At the conclusion of the *Sixth Conference* Thailand (2002), Mexico was elected to host the *Seventh Conference* in January 2005. This was partly because Mexico is the richest country in cycad diversity for the neotropics and second richest worldwide. Mexico is also in the vanguard for cycad research and practical conservation. The Jardín Botánico Francisco Javier Clavijero (JBC) holds the living National Cycad Collection and has an important role in the assessment of various nurseries aimed at sustainable management and conservation of cycads in the states of Veracruz, Puebla and Chiapas. This in collaboration with specialists in the Universities of Veracruz (UV) and Sciences and Arts of the State of Chiapas (UNICACH).

A total of 76 foreign and 20 national delegates, and numerous local students, attended the *Seventh International Conference on Cycad Biology* (CYCAD 2005), from 8 to 15 January 2005. After the inauguration in Xalapa's prestigious Anthropology Museum on 8 January, more than 60 presentations were given by Mexican delegates, as well as representatives from 16 foreign countries (Australia, Brazil, Canada, China, Colombia, Cuba, Ecuador, India, Japan, New Zealand, Panama, South Africa, Thailand, United Kingdom, United States, and Vietnam). This conference is held for the first time in Latin América.

Outstanding among the three plenary presentations was a challenging and biomedically important address by Paul Cox (National Tropical Botancial Garden, Hawaii, USA) dealing with the biomagnification of cyanobacterial neurotoxins and neurodegenerative diseases among the Chamarro people of Guam. Dennis Stevenson (New York Botanical

Garden, USA) gave a fascinating presentation on the extensive cycad fossil history and genealogy. Andrew Vovides (Jardín Botánico Francisco Javier Clavijero, Xalapa, Mexico) summarized aspects of cycad conservation and research in Mexico over the past 25 years. The plenary addresses were followed by oral and poster presentations on subjects as diverse as molecular systematics, classical taxonomy, ethnobotany, toxicity, physiology, morphology, genetics, sex expression, ecology, anatomy, conservation, pollination, seed dispersal, genomics, the world cycad list, and a description of a new cycad species from India.

During the workshops and special meetings of the event, Mark Bonta (Delta State University, Mississippi, USA) presented results of his ethnobotanical studies on the edible uses of *Dioon mejiae* by an indigenous group, the Olancho of Honduras, as well as reports of new populations of this cycad. Irene Terry (University of Utah, USA) chaired a session of the cycad insect pollinator working group. Conservation workshops dealt with cycad conservation in Honduras, as well as a world action plan for cycads and an associated global database. An Internet Bulletin Board on "who's who," and what current research topics are being carried out, has been mooted, with the aim of promoting communications and avoiding duplication of effort. John Donaldson (National Botanical Institute, South Africa) reported on the activities of the Cycad Specialist group over the past three years. A travelogue evening was also held where some delegates gave slide shows of their travels and encounters with cycads in habitat and in collections.

The conference ended on 14 January with an excursion to Monte Oscuro and Tlalchinola to visit two cycad nurseries aimed at sustainable management (UMAS) of *Dioon edule* and *Ceratozamia mexicana,* respectively. This was followed in the evening by a specially guided tour of Xalapa's Museum of Anthropology.

I conclude by recognizing the efforts of my colleagues on the CYCAD 2005 Organizing Committee and the many organizations and persons who contributed to the success of this meeting. Please refer to the Acknowledgements section toward the end of this volume for details.

Notice is hereby given that the *Eighth International Conference on Cycad Biology* will be held in Panama City in January 2008. Full details will be made available to the cycad community in due course.

Andrew P. Vovides
Cycad 2005 Conference Chair

CONSERVATION
CHAPTER 2

Recovery Plans for Endangered Cycads: A Model Set of Objectives and Actions Using the Example of Cycas megacarpa from Queensland, Australia

Paul I. Forster

Abstract

The recovery plan process for threatened species is summarized in relation to cycads. A model recovery plan for cycads is presented with the example of *Cycas megacarpa* K. D. Hill from Queensland, Australia. *Cycas megacarpa* is currently listed as "Endangered." The species has a minimum area of occupancy of 25 km^2 in 46 populations with a projected number of 372,964 adult plants, and is locally abundant over a large range in several habitats in southeast Queensland, Australia. A minimum viable population size for *C. megacarpa* is proposed to be in the range of 3500–4500 plants. Only seven populations of *C. megacarpa* are considered viable in the long term because of the presence of large numbers of individuals and a population structure indicating adequate replacement recruitment. Critical life history requirements are the putative pollination mutualism with beetles, dispersal by vertebrates, and the establishment of new individuals. Environmental variables that affect the life history requirements are minimum thresholds of pollination function and efficiency, fire history, dispersal efficiency, and drought. Populations were or are impacted upon by continued land clearing, habitat degradation, and selective eradication. Numbers of individuals have decreased or are still decreasing owing to land clearing, habitat degradation, selective eradication and harvesting, poor long distance dispersal, natural attrition, pollination failure, and/or recruitment failure. *Cycas megacarpa* is present in three national parks, seven state forests and five forest reserves, although the populations are often small and fragmented. Analysis of the data available for *C. megacarpa* indicates that a more appropriate conservation coding is "Vulnerable." Recovery actions to increase the number of individuals and populations for this species and a translocation protocol are outlined.

Resumen

Se resume el proceso del plan de rescate de especies amenazadas en relación a las cícadas. Un plan modelo de rescate para cícadas se presenta con el ejemplo de *Cycas megacarpa* K.D. Hill. Está especie está actualmente enlistada como En Peligro. La especie ocupa una área mínima de 25 km^2 en 46 poblaciones con un número estimado de 372,946 plantas adultas y es localmente abundante sobre un rango de varios habitats en el sureste de Queensland, Australia. Un tamaño poblacional mínimo viable par *C. megacarpa* está propuesto de ser en el rango de 3,500–4,500 plantas. Solo siete poblaciones de *C. megacarpa* son consideradas viables a largo plazo debido a la presencia de un gran número de individuos y una estructura poblacional que indica un reclutamiento de repuesto adecuado. Los requerimientos críticos de la historia de vida son el mutualismo de la polinización putativa por escarabajos, la dispersión por vertebrados y el establecimiento de individuos nuevos. Los variables ambientales que afecta los requerimientos de la historia de vida son umbrales mínimos de la función de la polinización y su eficiencia, historia de incendios, eficiencia de la dispersión y sequía. Las poblaciones fueron y están impactadas por la tala continua, degradación del hábitat y la erradicación selectiva. Números de individuos han disminuido o todavía siguen disminuyendo por las talas, degradación de los habitas, erradicación selectiva y saqueo, dispersión a larga distancia pobre, atrición natural, fracaso de la polinización y/o fracaso en el reclutamiento. *C. megacarpa* está presente en tres Parques Nacionales, siete Bosques Estatales y cinco Reservas Forestales, aunque las poblaciones son a menudo pequeñas y fragmentadas. Análisis de los datos disponibles para *C. megacarpa*, indica que la categoría más apropiada para su conservación es Vulnerable. Acciones de recuperación para incrementar el número de individuos y poblaciones para esta especie y protocolo de su transplante están discutidas.

Introduction

Cycads can be considered as a "flagship species" for conservation biology (Given, 1994; Meffe & Carroll, 1997), as they have unusual life histories that are interesting to many people, are in general restricted in distribution with over 50% of species threatened globally (Donaldson, 2003, 2004), and are of much commercial interest to horticulture or agriculture.

Cycads as a group are considered to be in global decline owing to climate change over time and their inability to out compete most angiosperms (Norstog & Nicholls, 1997). Most species of cycads occur in populations that are often disjunct and with little evidence of genetic flow between them. Incipient inbreeding occurs within many populations (Sharma et al., 1998, 1999, 2004; Keppel et al., 2002; Xiao et al., 2004). It is unlikely that disjunct populations (even those that are geographically adjacent) of most cycad species behave as part of a metapopulation, where the local populations are connected by dispersing individuals or have a flow of genetic material with the potential to recolonize or augment adjacent populations (Hanksi & Gilpin, 1991; Eriksson, 1996; Wakeley, 2004). Rather, cycads fit the "regional ensemble" model of Freckleton and Watkinson (2002) in which systems of essentially unconnected local populations persist in an ill-defined mosaic of suitable and unsuitable habitat. Seed death is high from predation and fire. Shallow burial, rather than no burial, is more likely to result in successful germination (Snow, 2003). Dispersal of seeds is localized (less than 100 m from the parent), rather than long distance (more than 100 m sensu Cain et al., 2000). Many of any resultant cycad seedlings are destroyed by drought, fire, or predation. Cycads are also generally absent from areas of disturbance or succession, where quick establishment of individuals is an advantage (Bond, 1989). Low levels of dispersal and recruitment, slow seedling growth, and specialist pollination requirements mean that cycads are generally restricted in their occurrence, often to areas of periodic dryness and low fertility, as postulated for conifers by Bond (1989).

Norstog and Nicholls (1997: p. 161) stated that "almost all human interaction . . . [with cycads] . . . has been deleterious" and that (p. 168) "the most realistic conservation efforts should involve attempts to provide well-protected cycad habitat reserves . . . and otherwise to ensure that all species of cycads are maintained in viable numbers in cultivation, together with their symbionts (cyanobacteria, mycorrhizae, pollinators), with the hope that some can eventually be returned to their original environments after—as seems inevitable in some cases—they become extinct in nature."

It is difficult to define what comprises a healthy, viable population of cycads, what might be an ideal minimum viable population (MVP) size (Soulé, 1986, 1987; Given, 1994), or how many populations are necessary to maintain evolutionary potential within a species. As noted by Soulé (1987), a population viability analysis (PVA) must be performed for each species, rather than generalizing. In the absence of empirical data, the "magic number" of 500 individuals for dioecious organisms is often cited (Franklin, 1980; Frankel & Soulé, 1981; Shaffer, 1981; Gilpin & Soulé, 1986; Lande & Barrowclough, 1987), based on minimum levels of genetic fitness and resilience to both demographic and environmental stochasticities. At this stage we do not know what is an ideal MVP size for any cycad species. Healthy and viable populations generally have large numbers of individuals (at least 500 reproductive adults), a diversity of individual size classes as occurs in large populations (thousands of plants), more or less equal sex ratios, and obvious seedling recruitment. However, this is not always the case for various reasons.

Individual cycads are long-lived (life span ranges of 60 to 1500 years have been given for *Macrozamia* species; Benson & McDougall, 1993; Pate, 1993), which are resilient to fire and some forms of mechanical disturbance (e.g., road maintenance). Persistent "survivors" present in small numbers are therefore thought to be the last remnants of once healthy populations. All species have an origin, an expansion of range, a decline, and an eventual extinction (Levin, 2000); this also applies to individual populations. Many cycad species may be of recent origin, with species radiations since the Pleistocene (1.75 million to 10,000 years ago) (van der Bank et al., 2001), although it has been postulated that the average length of time that a cycad species exists is about 54 million years (Levin 2000).

In developed nations with government infrastructure and agencies that are devoted to nature conservation, a Recovery Plan process may exist for threatened species. For species that are considered "Endangered" or "Vulnerable" (IUCN, 2001), the current management approach taken with a recovery plan is to prevent the species (or populations) from entering an extinction vortex (Gilpin & Soulé, 1986) and to halt further decline in numbers and area of occupancy. Ideally, the species should be recovered from the threatening processes by a series of actions that result in the species being delisted or downgraded from its threatened status (Burbidge, 1996).

No recovery plan has been previously proposed for any species of cycad (J. S. Donaldson, pers. comm. 2005); hence, the decision of the Australian Government to fund recovery plans (2005–2011) for six endangered species of cycad endemic to Queensland is novel in this respect, especially as the unusual life histories of cycads result in unique problems for many recovery actions.

The current paper outlines recovery actions as they might apply to cycads, using *Cycas megacarpa* K.D.Hill as an example. *Cycas megacarpa* is listed as "Endangered" in the Australian Federal Government *Environment Protection and Biodiversity Conservation* (EPBC) listing (http://www.deh.gov.au/cgi-bin/sprat/public/publicthreatenedlist.pl?wanted=flora) (effective 16 July 2000) and the Queensland State Government *Nature Conservation Act 1992* schedule (*NCA*) and *Nature Conservation Regulation* (*NCR*) (*Amendment* effective 2003). Recovery plans are currently being developed for this species and five other endangered cycads endemic to Queensland (Queensland Herbarium, 2005).

Background on *Cycas megacarpa*

Cycas megacarpa is endemic to southeast Queensland from Bouldercombe in the north, to near Woolooga in the south with a latitudinal-longitudinal range of about 250 × 150 km (Fig. 2-1). *Cycas megacarpa* is the most southerly occurring species of the genus and has an estimated minimum area of occupancy of 2527 ha in 46 populations with a projected number of adult individuals of >372,964 (Table 2-1). The decline in area of occupancy of *C. megacarpa* is not a recent phenomenon, although it is likely that it was accelerated in the twentieth century because of land clearing and habitat degradation for agriculture, and selective poisoning of plants on pastoral land. Keys (1886) noted that the species (as *C. media*) was "gradually disappearing from the [Mt Perry] district," and in 2003, this species was indeed rare and present only in very small populations in that area. Whitelock (2002) stated "without doubt, thousands of plants have been destroyed in past years to protect stock animals from eating the leaves. This fact is easily confirmed when one notes the large numbers of plants along the roadside in their habitat, whereas on the other side of the stock fences there are none to be seen."

Cycas megacarpa occurs within an altitudinal range of 40–680 m, in woodland or open woodland dominated by eucalypts, particularly *Corymbia citriodora* and *Eucalyptus crebra,* but also *Corymbia erythrophloia, C. intermedia, E. melanophloia*, and *Lophostemon confertus.* The substrate is usually rocky and derived from acid volcanics (tuff, rhyolite, trachyte, granodiorite), ironstone, or mudstone, rarely from alluvium.

Although there is no published information, *Cycas megacarpa* appears to have a similar population structure, life history, and ecology to most other Australian species of *Cycas.*

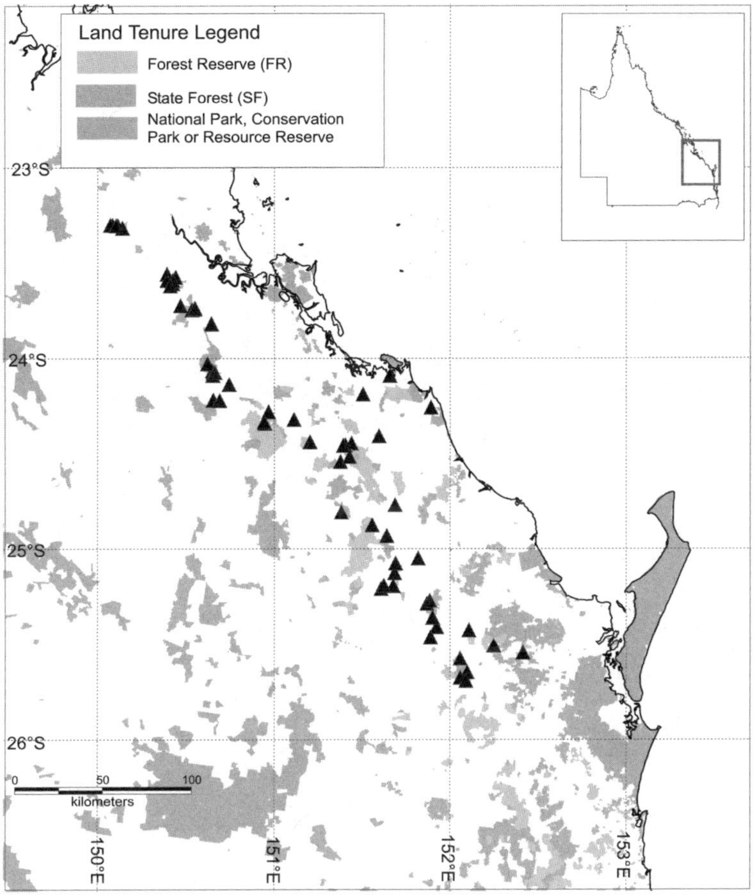

Figure 2-1. Extant distribution of *Cycas megacarpa* (based on herbarium specimen records at the Queensland Herbarium).

Life History and Environmental Constraints

Fire

Cycas megacarpa occurs in habitats that are subjected to periodic fires of varying intensities. Adult plants of *C. megacarpa* are resistant to most fires, although the foliage may be destroyed and some scarring of the stems may occur. Fires probably kill any small seedlings or seed that is either on the plant or locally dispersed.

Table 2-1 Population statistics for *Cycas megacarpa* (populations arranged north to south).

Key: FHT = Free Hold Title; GHPL= Grazing Homestead Perpetual Lease; N.P.= National Park; RR= Road Reserve; S.F.= State Forest; VCL= Vacant Crown Land.

Population number, voucher[a] data: tenure	No. Plants in 50 × 50 m	Projected occupancy of population (ha)	Projected no. plants in population (or actual no.*)	Seedlings present in 50 × 50 m plot
1. Healy RH0108004:GHPL	n/a	n/a	scattered	n/a
2. Smith AQ418803: n/a	n/a	c. 100	thousands	n/a
3. PIF2000: FHT	47	>850	159,800	+
4. R.Healy AQ551377: GHPL	n/a	n/a	hundreds	n/a
5. PIF29707: FHT, RR	14	c. 100	5,600	+
6. PIF29706: RR	n/a	20	28*	+
7. PIF29678: SF	5	1	5*	–
8. PIF29671: SF	36	800	115,200	+
9. PIF29678A: GHPL	49	n/a	49*	+
10. Melzer RM809:VCL	n/a	n/a	n/a	n/a
11. PIF29693: FHT, RR	19	5	19*	–
12. Olsen AQ664738: FHT, RR	4	5	4*	–
13. Antcliff AQ772387: FHT	n/a	n/a	n/a	n/a
14. PIF29774: FHT, RR	18	>200	14,400	+
15. PIF29777: FHT, RR	2	c. 50	35*	+
16. Halford Q3276: NP	n/a	n/a	< 10	n/a
17. PIF29668: FHT, RR	n/a	100	33*	+
18. PIF29667: RR	1	0.1	1*	–
19. PIF16233: FR	77	c. 250	76750	+
20. PIF12247:FHT	n/a	3	119*	+
21. PIF12278: SF	n/a	2	c. 30	+
22. PIF13977: SF	n/a	1.1	90*	–
23. PIF29662:FHT	n/a	4	65*	+ (1)
24. PIF29661:FHT	n/a	4	12*	–
25. PIF12286: FR	n/a	2	54*	–
26. Hind 3279: SF	n/a	n/a	n/a	n/a
27. O'Neill: FHT	n/a	c. 150	c. 300	n/a
28. Melzer RM913: SF	n/a	n/a	n/a	n/a
29. PIF13985: SF	n/a	n/a	< 50	n/a
30. Ritchie AQ555163: SF	n/a	c. 20	"thousands"	n/a
31. Houschildt AQ554865: FHT, RR	7	1	7	–

Table 2-1 (*continued*)

Population number, voucher[a] data: tenure	No. Plants in 50 × 50 m	Projected occupancy of population (ha)	Projected no. plants in population (or actual no.*)	Seedlings present in 50 × 50 m plot
32. PIF29724:FHT	1	0.25	1*	–
33. PIF2448:FHT	n/a	n/a	n/a	n/a
34. PIF2134:FHT	n/a	1	c. 20	n/a
35. PIF29722:VCL	3	0.25	3*	–
36. Machin AQ517088: NP	14	1	14	–
37. PIF29658: NP	25	1	c. 150*	+
38. PIF2264: RR	3	0.1	3	–
39. Young 279: SF	n/a	n/a	n/a	n/a
40. PIF65: FHT	1	0.1	1♂	–
41. Halford Q3504: SF	1	0.0001	1	–
42. Halford Q3413: SF	<10	1	<10	–
43. PIF7470: NP	<20	1	<20	+
44. Halford Q3501: FHT	<20	1	<20	n/a
45. PIF9322: FHT	<20	1	<20	n/a
46. PIF9336: SF	<40	1	<40	n/a

[a]Vouchers at the Queensland Herbarium (BRI).

Pollination Ecology

There is no published information on the pollination ecology of *Cycas megacarpa*, although it is likely that the species is beetle-pollinated. Beetles from the genera *Hapalips* and *Ulomoides* have been recorded from the male cones of *C. megacarpa* (Forster et al., 1994), but were first noted by Keys (1886), who interpreted them as cone destroyers.

Seed Development

Seeds become ripe (sarcotesta is soft and colors pink-purple to orange) around March onward, when they drop from the megasporophylls. The seeds are not ready to germinate for at least 9 months owing to the delayed fertilization unique to cycads (Norstog & Nicholls, 1997).

Dispersal Ecology and Recruitment

There is no information available on dispersal or recruitment for *C. megacarpa*. As with many cycads, the boundaries of populations of *C. megacarpa* can be quite sharp, without a corresponding change in habitat, indicating a dispersal-limited distribution (Primack & Miao, 1992). Snow (2003) demonstrated that the distribution of plants within a population of the cycad *Macrozamia lucida* was strongly clumped and that the bulk of seed dispersal was extremely local and near to the parent plants. A similar situation appears to be the norm for *C. megacarpa*.

Limited dispersal of ripe seeds from cycad species (when the fleshy sarcotesta is pink, purple, red, orange, or yellow) may occur via mammals (e.g., possums, rodents, fruit bats). Evidence from other Australian cycads suggests that seeds are dispersed only a short distance. Dispersal of seeds in *Cycas armstrongii* from the Northern Territory was generally within 1 m of the parent plants (Watkinson & Powell, 1997). Brush-tail possums (*Trichosurus vulpecula*) were able to disperse seeds of *Macrozamia riedlei* up to 24 m from the parent plant (Burbidge & Whelan, 1982), beyond 12 m in *M. communis* (Ballardie & Whelan, 1986), and beyond 4 m in *M. lucida*, although this mammal acted also as a seed predator for the latter species (Snow, 2003). Predation of seeds by bush-rats (*Rattus fuscipes*) from *Macrozamia* species was also recorded (Burbidge & Whelan, 1982; Ballardie & Whelan, 1986; Snow, 2003), with the rodent not considered as an effective dispersal agent. The sarcotesta on seeds of *C. micronesica* is consumed by fruit bats on Guam (Cox & Sacks, 2002; Monson et al., 2003), and van der Pijl (1957) reported fruit bat dispersal of the seeds of *C. rumphii*. It is feasible that a similar interaction with fruit bats may occur for Australian species of *Cycas*. Although the sarcotesta of cycad seeds is usually brightly colored, it is laced with high levels of the toxic chemical BMAA, with the highest levels found in the outer layers of the sarcotesta (Banack & Cox, 2003). Brightly colored flesh on fruit or seed is usually considered as a reward and attractant for seed dispersers (van der Pijl, 1982; Herrera, 1989), although as proposed by Mack (2000), it may have originated as a defensive role for the seed in primitive vascular plants. The high toxicity of seeds, together with the extinction of the Australian megafauna in recent geological time (Choquenot & Bowman, 1998; Roberts et al., 2001; Long et al., 2002) may be responsible for the apparent lack of dispersal from existing populations of this cycad. Few vertebrate dispersers of seeds or fruits of a similar size to cycad seeds now exist in Australia, and are primarily restricted to rainforest communities (Stocker & Irvine, 1983; Theimer, 2001; Dennis, 2002, 2003), where the marsupial Musky rat-kangaroo (*Hypsiprymnodon mochatus*) has been observed to eat and disperse the seeds of the rainforest cycad *Lepidozamia hopei* (Drew & Spencer 1998).

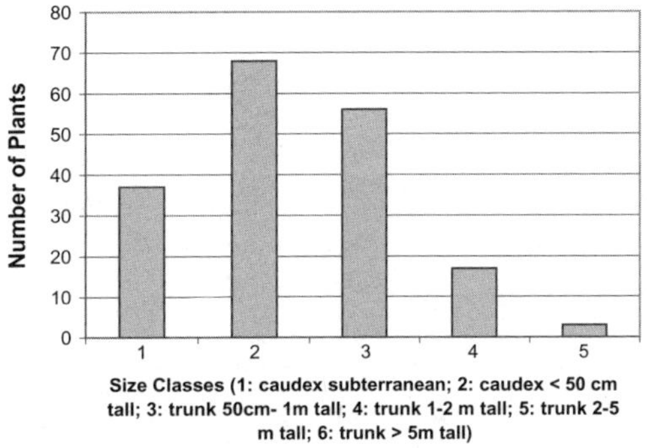

Figure 2-2. Size class distribution in a small population (n=181) of *Cycas megacarpa*. (Population #37 in Table 1).

POPULATION STRUCTURE

Large, and apparently healthy, populations of *C. megacarpa* should have a range of individuals from large adults (5–8 m in height) through to seedlings. As a species, *C. megacarpa* is notable for the occasional presence of very tall individuals (5–8 m high), making it one of the taller growing species of *Cycas*. Detailed survey of a small population (#37) of 181 plants and a large population (#19) of 530 plants (Table 2-1) revealed size classes (Figs. 2-2 and 2-3) ranging from a few large plants over 2 m in height to many small plants (<1m). In the large population, there was a preponderance of small plants, especially juveniles, whereas in the smaller population, the percentage of juveniles was less. The size class structure of Population 19 is comparable to those known for *C. armstrongii* (Ornduff, 1992; Watkinson & Powell, 1997), *C. platyphylla* (Forster, unpubl.) and *C. media* (Ornduff, 1991; Forster, unpubl.) and can be taken as being representative for a long-lived perennial species such as *C. megacarpa* where it occurs in an undisturbed situation and where there is a constant mortality rate within size or age classes that leads to the attrition of individuals through time (Schwartz, 2003). In the small population surveyed, this balance is skewed and there is no longer steady replacement recruitment occurring. Perhaps of greater importance is the actual effective population size (N_e), that is, the number of reproductive-age plants present. Adult plants of *C. megacarpa* are probably greater than 1 m tall (inasmuch as no plants less than 1 m tall have been observed to be reproductive), so in

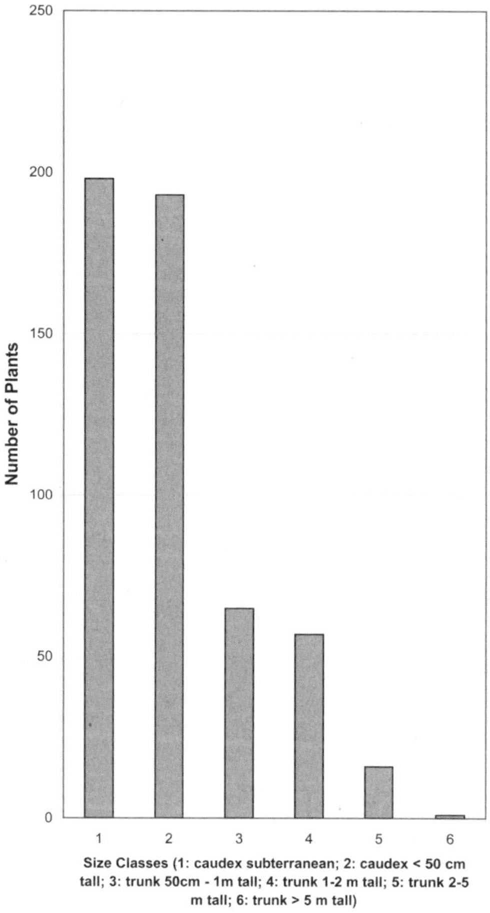

Figure 2-3. Size class distribution in a large population of *Cycas megacarpa* (n = 530) (Population #19 in Table 1).

Population #19 there were 74 adults (14%) in the plot and a projected 10,716 adults in the population. By comparison, in Population #37 there were only 20 adults (11%). If a minimum figure of 500 adults is necessary to avoid long-term loss of genetic variation, then the two surveyed sites with 11–14% adult plants would require between 3570 and 4545 plants to qualify as viable populations. Despite the high number of known populations (46) for *C. megacarpa*, only 11 of these have more than 100 plants and only seven have more than 3500 plants and can be considered as representing viable, dynamic populations of this cycad.

Genetics

A major impediment to any recovery plan for Australian species of *Cycas* is the lack of a species phylogeny based on molecular and morphological data sets. It is likely that the Australian species of *Cycas* are of recent origin, perhaps resulting from speciation events in the last 15 million years (Hill, 1998) during repeated expansion and retreat of different vegetation types (Kemp, 1978), especially the diversification of forests and woodlands where the current species of the genus grow. Until the degree of genetic divergence between and within species in this genus is quantified, it is not possible to prioritize either the species or their component populations in terms of their genetic representativeness (Williams et al., 1991); hence, the degree of genetic loss that might be occurring or that may be inadequately conserved cannot be predicted.

There is no published information on the genetics of *C. megacarpa*. Many populations of this species are very small and greatly fragmented, with only a handful of adult plants. Other species of *Cycas* have high levels of incipient inbreeding and little genetic flow between disjunct populations (Yang & Meerow, 1996; Huang et al., 2001; Keppel, 2002; Keppel et al., 2002). Most genetic studies of cycads have found this to be case, so it is likely that *C. megacarpa* is similar.

The limited dispersal of seeds from populations of *C. megacarpa* may eventuate in genetically similar individuals being most likely to interbreed, thus reinforcing inbreeding and the possibility of inbreeding depression and the accumulation of deleterious genes. It is now possible using genetic techniques to determine not only the amount of variation that is present but also whether individuals within a population originate from that population, or from long-distance dispersal from other populations within the greater metapopulation (Ouberg et al., 1999; Cain et al., 2000; He et al., 2004). As yet these techniques have not been applied to Australian cycads; however, initial studies by Huang et al. (2001) and Xiao et al. (2004) for Asian species of *Cycas* indicate that these methodologies produce worthwhile information.

Predation

The foliage of *C. megacarpa* is commonly predated by insects that use this cycad, and most others, as a host plant. Early records of the foliage being destroyed by marsupials (Keys, 1886) have not been verified, and it is likely that the often, total destruction of the foliage is caused by the lycaenid butterfly (*Theclinesthes onycha*) and the beetle *Lilioceris nigripes* (Forster & Machin, 1994; Forster, unpubl.).

Propagation and Translocation

Propagation

Cycas megacarpa is propagated mainly from seeds. The time from germination to maturity is long, and is likely to be at least 10–20 years; however, this species is considered easy to grow (Whitelock, 2002). If larger plants are damaged they may produce offsets/suckers (Borsboom & Rudd, 2002). These may be removed and rerooted to provide additional plants (Jones, 2002; Whitelock, 2002).

Translocation

Translocation of cycad plants from threatened habitats is one option that has to be considered and is feasible given the biology and form of the plants. Translocation of cycads should be undertaken only for a conservation outcome, where the plants are moved into natural habitats to augment existing populations or to create new ones, preferably within the confines of a metapopulation. Removal of cycads from habitat and placement into regional botanic gardens is not a conservation outcome when natural habitats still exist for the species, although the value of small numbers of plants for educational purposes is not disputed. Removal of cycads from habitat and their dispersal into private hands for purposes of horticulture or landscaping is not a conservation outcome as these individuals will no longer contribute to the genetic pool of the species as it exists in nature or provide potential host plants for the insect pollinator mutualists.

Management Issues

Reasons for Listing

Cycas megacarpa was originally listed as "Endangered" in 2001 by the Queensland Government following ratification in 1995 by the independent scientific committee advising the Minister for Environment. The listing was made on the basis of an apparent widespread reduction in area of occupancy due primarily to land clearing and selective poisoning, and secondarily to collection of plants for horticultural landscaping and collectors. Prior to 1994, there were herbarium collections from 19 of the 46 currently known populations (Table 2-1), but there was no information on the area of occupancy, numbers of plants, or the putative viability of individual populations. The earliest collection of *C. megacarpa* is from the 1880s [*J. Keys* AQ141953 (BRI), see Forster, 1988, for details on this collector]; however, the early herbarium records do not provide sufficient data to document the putative decline in occupancy of the species or whether some of the extant, small populations were more extensive in the last 100 years prior to when the majority of land clearing was undertaken in southeast Queensland. Many populations of *C. megacarpa* continue to be impacted upon by road corridors, quarries, general land clearing, and deliberate eradication by pastoralists. The inclusion of small populations within any conservation status analysis that uses the IUCN (2001) categories may result in an inaccurate assessment because they may be nonviable in the longer term. Two analyses could be presented here, one based only on the seven large populations, and one based on all 46 extant populations. Under the IUCN (2001) criteria for "Endangered" (EN) it is impossible to accurately assess whether there has been a 70% reduction over the last 10 years, or three generations (which is probably 50–75 years). *Cycas megacarpa* is no longer extant south of Population 46 (knowledge current in 2005), despite a recent historical occurrence near Kilkivan. Keys (1886) implied that the species was widespread in the Mt. Perry district, "on the low ground and the hilly country, the chief plants of interest are the cycads [*C. megacarpa*] and macrozamias [*Macrozamia mountperriensis*]"; but is much more restricted today, with many of the populations extremely small and fragmented (e.g., populations 31–34 in Table 2-1). Nevertheless, owing to the number of populations (7 versus 46) (five is the maximum allowed under EN) and the total projected number of individuals (>372,964 versus 2500 under EN), *C. megacarpa* does not fit the criteria to continue being listed as EN. The species does fulfill some of the criteria for listing as "Vulnerable" (VU) given the following assumption: of the 39 populations that are dubiously viable, these were previously viable populations within the last three generations (50–75 years), when a greater number of adult plants were present; however, the numbers of adult plants have been reduced below the minimum viability level owing to the outlined threatening factors.

Cycas megacarpa can be classified as "Vulnerable" based on the following IUCN (2001) categories: VU A2+4; B1ab(i,ii,iii,iv,v)+2a(i,ii,iii,iv,v).

Existing Conservation Measures

There are no specific conservation measures in place for *C. megacarpa* to stabilize or to augment existing populations. Populations of the species are present in four national parks, five forest reserves, seven state forests and one recreation reserve. Four of the populations in forest reserves and state forests are significant for the conservation of the species owing to their large number of plants and natural disposition of size classes (Table 2-1).

Biodiversity Value

All cycads can be considered as "flagship" species (Given, 1994; Meffe & Carroll, 1997) owing to several features, namely, ancient lineage, interesting life histories (including pollination mutualisms), and value to humans (agriculture, horticulture). There are only about 300 species of cycad worldwide (Hill et al., 2004), with several *Encephalartos* species extinct in habitat (Donaldson, 2003). The genus *Cycas* is considered to comprise species that are the most primitive living seed plants (Brenner et al., 2003). *Cycas megacarpa* is the most southerly occurring species of *Cycas* in the world.

Commercial Value

Cycas megacarpa usually does not have "blue" foliage, although some of the northern populations of this species have scattered individuals with foliage that is glaucous-pruinose when young. Hill (1992) initially speculated that these northern populations of *C. megacarpa* showed evidence of introgression from *C. ophiolitica*, but seems to have changed this hypothesis more recently (Hill, 1996, 1998). *Cycas megacarpa* cannot be considered as a "blue cycad of desire" (Forster, 1999), which would be much coveted by cycad collectors for the attractive foliage; however, the species is attractive and easier to grow in subtropical climates than the species originating from the seasonally dry tropics. There is a continuing demand for fertile seeds (value between 50 cents and $2.50 USD per seed) and mature plants for collectors and landscapers (value up to $1500 USD per plant depending on size, and access to overseas markets). Whitelock (2002) stated that "seed is relatively easy to obtain" and that "the awakening of nurseries and landscapers to the usefulness of cycads for decoration will no doubt create added pressure on those colonies [of *C. megacarpa*] remaining in the wild." There has been widespread poaching of plants of this species, particularly in southern Queensland, the most recent and notable being the removal of a 7-m individual from a national park in 2003.

Legislative Context

"Rare or Threatened Plant" includes critically endangered, endangered, or vulnerable plants under the Commonwealth of Australia *Environment Protection and Biodiversity*

Conservation Act 1999. In Queensland, all cycad taxa are regarded as "Restricted" plants under Queensland Government Nature Conservation legislation (Schedule 12 Part 3 *Nature Conservation Amendment Regulation 2003*) in terms of their management for wild harvesting, propagation, and sale. Details on this legislation may be found in Forster (2004a, 2004b). All species of *Cycas* are also included in Appendix II of CITES, which requires a permit from the exporting country to verify that plants are artificially propagated or that trade in wild collected specimens is not detrimental to species survival.

Species Recovery Objectives

- Prevent *C. megacarpa* from entering an extinction vortex, both at the species level and at the individual population level.
- Acquire habitat containing large populations of *C. megacarpa* for conservation reserves.
- Determine the genetic variation both within populations of *C. megacarpa* and between it and adjacent species (e.g., *C. ophiolitica*).
- Determine the integrity of the pollination mutualism in *C. megacarpa* and whether individual populations have lost their pollinators.
- Determine dispersal modes and rates, and whether they are effective for both within-population dispersal and long-distance dispersal.
- Stabilize population numbers (especially for reproductive adults) and augment the number of individuals within populations.
- Create new populations in suitable habitat within the distribution range of the species, preferably within the confines of a genetic metapopulation.

Species Recovery Actions

The following research actions are required or would be seen as ideal:

- Continued survey for additional populations of *C. megacarpa*.
- Survey of many known populations (data not available in Table 2-1) of *C. megacarpa* for numbers of plants, occupancy, recruitment levels, threats, etc.

- Determination of a molecular phylogeny for the Australian species of *Cycas*.
- Genetic study of all or selected populations of *C. megacarpa* to determine the degree of genetic continuity both within and between populations.
- Genetic study of the relationship between *C. megacarpa* and *C. ophiolitica* to determine the degree of genetic pollution to either species.
- Pollination study to determine the dynamics of the mutualism between the cycad and the attending beetles.
- Establishment of permanent monitoring plots to determine rates of attrition and recruitment within populations, particularly in relation to fire.
- Dispersal study to determine whether any is occurring, the vectors involved, and the efficiency rates.

The following land tenure-related actions and management are required:

- Acquisition of land where populations of *C. megacarpa* are present, for conservation in reserves.
- Either management or translocation of populations of *C. megacarpa* that are present on road corridors or in areas being mined/quarried where they are threatened by road maintenance activities, associated land clearing, or quarrying.
- Prevention of poisoning or destruction of *C. megacarpa*. If stock losses as a result of consumption of the plants are a problem, then exclusion of stock by fencing or other means should be undertaken. Alternatively, the plants should be translocated into a conservation reserve within the confines of a metapopulation.
- Maintenance of a fire regime on land where *C. megacarpa* occurs to maximize seed production and recruitment. Current fire management practices should be maintained, unless proposed monitoring research indicates otherwise.
- Management of forestry timber-harvesting operations in state forests where *C. megacarpa* occurs following the guidelines given by Halford (1995).

Details on the recovery plan schedule and costs may be sourced in the draft recovery plan for this species (Queensland Herbarium, 2005).

Species Ability to Recover

Cycas megacarpa is a long-lived plant with apparently poor dispersal, low levels of recruitment, and a putative insect pollination mutualism. *Cycas megacarpa* is locally abundant over a large range in several habitats, one of the "seven forms of rarity" proposed by Rabinowitz (1981) and Kruckeberg and Rabinowitz (1985). It is likely that removal or destruction of adult plants from populations will have an immediate, as well as long-term, disastrous effect on population viability. Raimondo and Donaldson (2003) found that, although removal of seeds from populations of two species of South African *Encephalartos* cycads had little immediate effect on population viability, removal of even a few adult plants was deleterious. It is likely that small, fragmented populations of *C. megacarpa* have little ability to recover to predisturbance levels. For large, and apparently healthy, populations, the ability of the population to tolerate removal or destruction of limited numbers of individuals because of housing development, road maintenance or rerouting, or quarrying or mining may be higher. If the minimum viable population size for *C. megacarpa* is indeed between 3500 and 4500 plants, then none of the smaller populations are recoverable.

Given the causal factors responsible for rarity in this species, together with our ignorance of its general biology, it is more or less impossible to predict whether this species will increase its numbers in habitat without human intervention. Where populations of the cycad are known to have lost the pollinating insects, then either hand-pollination or reintroduction of pollinators from nearby populations to produce seed should be attempted (Kearns et al., 1998). Likewise, as few seeds achieve maturity in habitat, removal of most ripe seed, followed by growth in a nursery for at least 10 years, followed by replanting in habitat, may be the best approach to augmentation of the population. Otherwise, populations of the cycad may remain static or slowly decline, existing only as long as the life span of the individuals within it.

Life History and Environmental Constraint Analysis

In answer to questions raised by Schemske et al. (1994) on species recovery, the following can be stated, although the lack of empirical data for *C. megacarpa* is a confounding problem in all cases.

Question 1. Biological Status of the Species: Are individuals and populations of the species increasing, decreasing, or stable?
Answers:

- In some smaller populations, it is probably fair to say that the number of individuals is decreasing owing to natural attrition and failure of recruitment.
- In other larger populations, the number of individuals is probably stable or even increasing.
- Although the plants are long-lived, the number of populations and the number of individuals are undoubtedly decreasing owing to continued land clearing and habitat degradation for agriculture, water infrastructure, quarries, and mining.

Question 2. Life History: What stages are critical for population growth and species persistence?
Answers:

- Successful maintenance of adults in populations.
- Successful pollination mediated by insect mutualists.
- Successful dispersal of seed.
- Successful germination and subsequent recruitment of individuals to offset or augment natural or anthropogenic attrition of adults.

Question 3. Biological Causes of Variation in Life History: What causes have the most demographic impact?
Answers:

- Attrition of adults due to climate variation, senescence, and "misadventure" (e.g., toppling of large plants).
- Potential for variable pollination success by insect mutualists.
- Variable fire histories resulting in recruitment failure.
- Failure or reduced rate of seed dispersal by potential dispersal vectors (e.g., possums, rodents, fruit bats).

Social and Economic Consequences

There are several current, obvious social or economic consequences from attempting to conserve plants of *C. megacarpa* in habitat, or to restrict or control trade in seeds and live plants.

Economic consequences include the following:

- The need for construction of alternative routes and maintenance procedures on road corridors.
- Restriction of timber harvesting in the vicinity of populations.
- Restriction of mining, quarrying, and other destructive activities in the vicinity of populations.
- The need for fencing or translocation of populations on private land or leasehold land with grazing stock, to prevent stock feeding on the plants.
- Prevention of harvesting or salvage by individuals where the intent is purely "commercial."
- The cost of acquisition for land with populations of *C. megacarpa* at a fair and equitable price.

Social consequences include the following:

- Prevention of traditional harvesting of seeds as a food source by aboriginal peoples (Beaton, 1982; Smith, 1982; Beck, 1992). Traditional harvesting is now unlikely owing to the arduous preparation necessary to remove toxins in the seed. There is no record of the seeds of *C. megacarpa* being utilized in this manner.
- Prevention of cycad enthusiasts or cycad-oriented societies from uncontrolled harvesting of plants or seeds.

Conclusions

Cycas megacarpa is locally abundant over a large range in several habitats, however, it is critically endangered in the southern part of its range, with most known populations considered as nonviable owing to small population sizes. These small populations may have already entered an extinction vortex (Gilpin & Soulé, 1986) and require considerable population management to maintain or augment numbers of individuals. There has been an unknown, but significant reduction of both the number of populations and the number of individuals (especially adults) because of broad-scale land clearing and habitat degradation for cattle grazing, selective eradication by pastoralists, harvesting by commercial landscapers, quarrying, and road building. Lack of recruitment in small populations is also contributing to this decline. Although there are several very large and diverse populations in the northern part of the range, these constitute a minority of populations. While even small populations (40–100 individuals) may appear viable (at least in the short term) because pollinated seeds and seedlings are present, they have an unnatural size class distribution of individuals and little is known or can be inferred about the long-term potential of these populations for the evolutionary diversity of the species. There seems to be little or no dispersal outside of existing populations, nor is it likely that the disjunct populations are acting as part of a metapopulation (Hanksi & Gilpin, 1991; Eriksson, 1996; Wakeley, 2004). Rather they fit the model of populations within a "regional ensemble" (Freckleton & Watkinson, 2002), where there is little genetic connection between populations. The main causal factors that are affecting recruitment in this species (fire, drought, failure of the insect pollinator mutualism) are unlikely to cause the species in its entirety to enter an extinction vortex in the near future; however, long-term monitoring is required using tagged individuals (both adults and seedlings) to determine rates of attrition. Information is still lacking about the genetics, pollination biology, dispersal ecology, and rates of recruitment in this species, all of which are variables critical in management of cycad populations. Under the IUCN (2001) criteria, *C. megacarpa* should be reclassified as "Vulnerable."

Acknowledgments

Thanks to D. A. Halford and P. J. Machin for assistance with fieldwork; P. D. Bostock for the map, and W. Smith for manipulation of included images. Data/information on populations was supplied by J. Healy, G. O'Neill, and T. Ritchie (EPA-QPWS) and P. J. Machin. Development of the translocation protocol was undertaken on the advice of S. Walkley (Plantation 2000). The draft recovery plan for *Cycas megacarpa* (Queensland Herbarium, 2005) on which this paper is largely based, was commented on by A. R. Bean, P. D. Bostock, G. P. Guymer, D. A. Halford, A. E. Holland, M. B. Thomas, and G. W.

Wilson from the Queensland Herbarium. The preparation of the recovery plan for *Cycas megacarpa* was funded by Environment Australia during the first half of 2004. This paper benefited from referee comments provided by J. S. Donaldson.

Literature Cited

Ballardie, R. T. & R. J. Whelan. 1986. Masting, seed dispersal and seed predation in the cycad *Macrozamia communis*. Oecologia 70: 100–105.

Banack, S. A. & P. A. Cox. 2003. Distribution of the neurotoxic nonprotein amino acid BMAA in *Cycas micronesica*. Bot. J. Linn. Soc. 143: 165–168.

Beaton, J. M. 1982. Fire and water: aspects of Australian Aboriginal management of cycads. Arch. Oceania 17: 51–58.

Beck, W. 1992. Aboriginal preparation of *Cycas* seeds in Australia. Econ. Bot. 46: 133–147.

Benson, D. & L. McDougall. 1993. Ecology of Sydney plant species part 1. Ferns, fern-allies, cycads, conifers and dicotyledon families Acanthaceae to Asclepiadaceae. Cunninghamia 3: 257–422.

Bond, W. J. 1989. The tortoise and the hare: ecology of angiosperm dominance and gymnosperm persistence. Biol. J. Linn. Soc. 36: 227–249.

Borsboom, A. & J. Rudd. 2002. *Cycas megacarpa* field assessment and management recommendations. Internal Report. Environmental Protection Agency, Indooroopilly, Australia.

Brenner, E. D., D. W. Stevenson, R. W. McCombie, M. S. Katari, S. A. Rudd, K. F. X. Mayer, P. M. Palenchar, S. J. Runko, R. W. Twigg, G. Dai, R. A. Martienssen, P. N. Benfey & G. M. Coruzzi. 2003. Expressed sequence tag analysis in *Cycas*, the most primitive living seed plant. Genome Biol. 4: R78. http://genomebiology.com/2003/4/12/R78.

Burbidge, A. A. 1996. Essentials of a good recovery plan. Pp. 55–62 *in* S. Stephens & S. Maxwell (eds.), Back from the brink: refining the threatened species recovery process. Surrey Beatty & Sons, Chipping Norton, Australia.

Burbidge, A. H. & R. J. Whelan. 1982. Seed dispersal in a cycad, *Macrozamia riedlei*. Austral. J. Ecol. 7: 63–67.

Cain, M. L., B. G. Milligan & A. E. Strand. 2000. Long-distance seed dispersal in plant populations. Amer. J. Bot. 87: 1217–1227.

Choquenot, D. & D. M. J. S. Bowman. 1998. Marsupial megafauna, Aborigines and the overkill hypothesis: application of predator-prey models to the question of Pleistocene extinction in Australia. Global Ecol. Biogeogr. Lett. 7: 167–180.

Cox, P. A. & O. W. Sacks. 2002. Cycad neurotoxins, consumption of flying foxes, and ALS-PDC disease in Guam. Neurology 58: 956–959.

Dennis, A. J. 2002. The diet of the musky rat-kangaroo *Hypsiprymnodon moschatus*, a rain forest specialist. Wildlife Res. 29: 209–219.

———. 2003. Scatter-hoarding by musky rat-kangaroos, *Hypsiprymnodon moschatus*, a tropical rain-forest marsupial from Australia: implications for seed dispersal. J. Trop. Ecol. 19: 619–627.

Donaldson, J. S. (ed). 2003. Cycads: Status Survey and Conservation Action Plan. IUCN/Cycad Specialist Group. IUCN, Gland and Cambridge.

———. 2004. Cycads, a status survey and action plan. IUCN: The World Conservation Union, Gland, Switzerland.

Drew, A. & H. J. Spencer. 1998. Consumption of *Idiospermum australiense* seeds by the musky rat kangaroo, *Hypsiprymnodon moschatus* (Marsupialia: Potoroidae). Mem. Queensland Mus. 42: 438.

Eriksson, O. 1996. Regional dynamics of plants: a review of evidence for remnant, source-sink and metapopulations. Oikos 77: 248–258.

Forster, P. I. 1988. James Keys F.L.S. and his contribution to early botanical exploration in south-east Queensland. Queensland Naturalist 29: 116–122.

———. 1999. Six blue cycads of desire. Pp. 388–389 *in* K. A. W. Williams, Native Plants of Queensland Vol. 4. K. A. W. Williams, North Ipswich, Australia.

———. 2004a. The cycads of Queensland—diversity and conservation. Palms & Cycads No. 82: 4–28.

———. 2004b. Diversity of the cycads of Queensland, together with an assessment of their conservation status. Pp. 60–72 *in* A. J. Lindstrom (ed.), The biology, structure, and systematics of the Cycadales. Proceedings of the Sixth International Conference on Cycad Biology. Nong Nooch Tropical Botanical Garden, Chonburi, Thailand.

——— & P. J. Machin. 1994. Cycad host plants for *Lilioceris nigripes* (Fabricius) (Coleoptera: Chrysomelidae) and *Theclinesthes onycha* (Hewitson) (Lepidoptera: Lycaenidae). Austral. Entomol. 21: 99–102.

———, ———, L. Mound & G. W. Wilson. 1994. Insects associated with reproductive structures of cycads in Queensland and northeast New South Wales, Australia. Biotropica 26: 217–222.

Frankel, O. H. & M. E. Soulé. 1981. Conservation and evolution. Cambridge University Press, New York.

Franklin, I. R. 1980. Evolutionary change in small populations. Pp. 135–149 *in* M. Soulé & B. Wilcox (eds.), Conservation biology: an evolutionary perspective. Sinauer Associates, Sunderland, Massachusetts.

Freckleton, R. P. & A. R. Watkinson. 2002. Large-scale spatial dynamics of plants: metapopulations, regional ensembles and patchy populations. J. Ecol. 90: 419–434.

Gilpin, M. E. & M. E. Soulé. 1986. Minimum viable populations: processes of species extinction. Pp. 19–34 *in* M. E. Soulé (ed.), Conservation biology: the science of scarcity and diversity. Sinauer Associates, Sunderland, MA.

Given, D. R. 1994. Principles and practice of plant conservation. Timber Press, Portland, OR.

Halford, D. A. 1995. *Cycas megacarpa, C. ophiolitica, Macrozamia conferta, M. cranei, M. fearnsidei, M. lomandroides, M. machinii, M. parcifolia, M. pauli-guilielmi, M. platyrhachis*. Pp. 34–63 *in* R. W. Johnson & D. A. Halford (eds.), Flora and fauna information system—species management manual. Vol. 3. Queensland Department of Natural Resources, Brisbane, Australia.

Hanski, I. & M. Gilpin. 1991. Metapopulation dynamics: brief history and conceptual domain. Biol. J. Linn. Soc. 42: 3–16.

He, T., S. L. Krauss, B. L. Lamont, B. L. Miller & N. J. Enright. 2004. Long-distance seed dispersal in a metapopulation of *Banksia hookeriana* inferred from a population allocation analysis of amplified fragment length polymorphism data. Molec. Ecol. 13: 1099–1109.

Herrera, C. 1989. Seed dispersal by animals: a role in angiosperm diversification? Amer. Naturalist 133: 309–322.

Hill, K. D. 1992. A preliminary account of *Cycas* (Cycadaceae) in Queensland. Telopea 5: 177–206.

———. 1996. A taxonomic revision of the genus *Cycas* (Cycadaceae) in Australia. Telopea 7: 1–64.

———. 1998. Cycadophyta. Pp. 597–661 *in* P. McCarthy (ed.), Flora of Australia. Vol. 48. CSIRO Publications, Melbourne.

Hill, K. D., D. W. Stevenson & R. Osborne. 2004. The world list of cycads. Pp. 219–235 *in* T. Walters & R. Osborne (eds.), Cycad classification: concepts and recommendations. CAB International Publishing, Wallingford, UK.

Huang, S., Y. C. Chiang, B. A. Schaal, C. H. Chou & T. Y. Chiang. 2001. Organelle DNA phylogeography of *Cycas taitungensis*, a relict species from Taiwan. Molec. Ecol. 10: 2669–2681.

IUCN. 2001. IUCN red list categories and criteria version 3.1. IUCN—The World Conservation Union, Gland, Switzerland.

Jones, D. L. 2002. Cycads of the world. 2nd ed. Smithsonian Institution Press, Washington, DC.

Kearns, C. A., D. W. Inouye & N. M. Waser. 1998. Endangered mutualisms: the conservation of plant-pollinator interactions. Annual Rev. Ecol. Syst. 29: 83–112.

Kemp, E. M. 1978. Tertiary climatic evolution and vegetation history in the southeast Indian Ocean region. Palaeogeogr. Palaeoclimatol. Palaeoecol. 24: 169–208.

Keppel, G. 2002. Low genetic variation in a Pacific cycad: conservation concerns for *Cycas seemannii* (Cycadaceae). Oryx 36: 41–49.

———, S.-W. Lee & P. D. Hodgskiss. 2002. Evidence for long isolation among populations of a Pacific cycad: genetic diversity and differentiation in *Cycas seemannii* A.Br. (Cycadaceae). J. Heredity 93: 133–139.

Keys, J. 1886. A contribution to the flora of Mount Perry. Part II. Proc. Roy. Soc. Queensland 2: 41–55.

Kruckeberg, A. R. & D. Rabinowitz. 1985. Biological aspects of endemism in higher plants. Annual Rev. Ecol. Syst. 16: 447–479.

Lande, R. & G. F. Barrowclough. 1987. Effective population size, genetic variation, and their use in population management. Pp. 87–124 *in* M. E. Soulé (ed.), Viable populations for conservation. Cambridge University Press, Cambridge, UK.

Levin, D. A. 2000. The origin, expansion, and demise of plant species. Oxford University Press, Oxford, UK.

Long, J., M. Archer, T. Flannery & S. Hand. 2002. Prehistoric mammals of Australia and New Guinea. UNSW Press, Sydney.

Mack, A. L. 2000. Did fleshy fruit pulp evolve as a defence against seed loss rather than as a dispersal mechanism? J. Biosci. 25: 93–97.

Meffe, G. K. & C. R. Carroll. 1997. Principles of conservation biology. Ed. 2. Sinauer Associates, Sunderland, MA.

Monson, C. S., S. A. Banack & P. A. Cox. 2003. Conservation implications of Chamorro consumption of Flying Foxes as a possible cause of ALS-PDC in Guam. Conservation Biol. 17: 678–686.

Norstog, K. J. & T. J. Nicholls. 1997. The biology of the cycads. Cornell University Press, Ithaca, NY.

Ornduff, R. 1991. Size classes, reproductive behavior, and insect associates of *Cycas media* (Cycadaceae) in Australia. Bot. Gaz. 152: 203–207.

———. 1992. Features of coning and foliar phenology, size classes, and insect associates of *Cycas armstrongii* (Cycadaceae) in the Northern Territory, Australia. Bull. Torrey Bot. Club 119: 39–43.

Ouborg, N., Y. Piquot & J. Groenendael. 1999. Population genetics, molecular markers and the study of dispersal in plants. J. Ecol. 87: 551–568.

Pate, J. S. 1993. Biology of the S.W. Australian cycad *Macrozamia riedlei* (Fisch. ex Gaudich.) C.A.Gardn. Pp. 125–130 *in* D. W. Stevenson & K. J. Norstog (eds.), The biology, structure, and systematics of the Cycadales, Proceedings of CYCAD 90, the Second International Conference on Cycad Biology. Palm & Cycad Societies of Australia, Milton, Australia.

Primack, R. B. & S. L. Miao. 1992. Dispersal can limit local plant distribution. Conservation Biol. 6: 513–519.

Queensland Herbarium. 2005. Recovery plan for *Cycas megacarpa, Cycas ophiolitica, Macrozamia cranei, M. lomandroides, M. pauli-guilielmi* and *M. platyrhachis* 2006–2011. Queensland Herbarium, Environmental Protection Agency, Brisbane.

Rabinowitz, D. 1981. Seven forms of rarity. Pp. 205–217 *in* H. Synge (ed.), The biological aspects of rare plant conservation. Wiley, New York.

Raimondo, D. C. & J. S. Donaldson. 2003. Responses of cycads with different life histories to the impact of plant collecting: simulation models to determine important life history stages and population recovery times. Biol. Conservation 111: 345–358.

Roberts, R. G., T. F. Flannery, L. K. Ayliffe, H. Yoshida, J. M. Olley, G. J. Prideaux, G. M. Laslett, A. Baynes, M. A. Smith, R. Jones & B. L. Smith. 2001. New ages for the last Australian megafauna: continent-wide extinction about 46000 years ago. Science 292: 1888–1892.

Schemske, D. W., B. C. Husband, M. H. Ruckelshaus, C. Goodwillie, I. M. Parker & J. G. Bishop. 1994. Evaluating approaches to the conservation of rare and endangered plants. Ecology 75: 584–606.

Schwartz, M. W. 2003. Assessing population viability in long-lived plants. Pp. 239–266 *in* C. A. Brigham & M. W. Schwartz (eds.), Population viability in plants. Ecological Studies Vol. 165. Springer-Verlag, Berlin.

Shaffer, M. L. 1981. Minimum population sizes for species conservation. BioScience 31: 131–134.

Sharma, I. K., D. L. Jones, P. I. Forster & A. G. Young. 1998. The extent and structure of genetic variation in the *Macrozamia pauli-guilielmi* complex (Zamiaceae). Biochem. Syst. Ecol. 26: 45–54.

———, ———, ——— & ———. 1999. Low isozymic differentiation among five species of the *Macrozamia heteromera* group (Zamiaceae). Biochem. Syst. Ecol. 27: 67–77.

———, ——— & ———. 2004. Genetic differentiation and phenetic relatedness among seven species of the *Macrozamia plurinervia* complex (Zamiaceae). Biochem. Syst. Ecol. 32: 313–327.

Smith, M. 1982. Late Pleistocene Zamia exploitation in southern Western Australia. Arch. Oceania 17: 117–121.

Snow, E. L. 2003. Role of invertebrates in seed dispersal of a locally distributed cycad *Macrozamia lucida* (Cycadales). B.Sc. (Hons.) thesis. University of Queensland, Brisbane.

Soulé, M. E. (ed.) 1986. Conservation biology: the science of scarcity and diversity. Sinauer Associates, Sunderland, MA.

———. (ed.) 1987. Viable populations for conservation. Cambridge University Press, New York.

Stocker, G. C. & A. K. Irvine. 1983. Seed dispersal by cassowaries (*Casuarius casuarius*) in North Queensland rain forests. Biotropica 21: 50–56.

Theimer, T. C. 2001. Seed scatterhoarding by white-tailed rats: consequences for seedling recruitment by an Australian rain forest tree. J. Trop. Ecol. 17: 177–189.

van der Bank, F. H., M. Wink, P. Vorster, J. Treutlein, L. Brand, M. van der Bank & J. Hurter. 2001. Allozyme and DNA sequence comparisons of nine species of *Encephalartos* (Zamiaceae). Biochem. Syst. Ecol. 29: 241–266.

van der Pijl, L. 1957. The dispersal of plants by bats (Chiropterochory). Acta Bot. Neerl. 6: 291–315.

———. 1982. Principles of dispersal in higher plants. Ed. 3. Springer-Verlag, Berlin.

Wakeley, J. 2004. Metapopulation models for historical inference. Molec. Ecol. 13: 865–875.

Watkinson, A. R. & J. C. Powell. 1997. The life history and population structure of *Cycas armstrongii* in monsoonal northern Australia. Oecologia 111: 341–349.

Whitelock, L. 2002. The cycads. Timber Press, Portland, Oregon.

Williams, P. H., C. J. Humphries, & R. I. Vane-Wright. 1991. Measuring biodiversity: taxonomic relatedness for conservation priorities. Austral. Syst. Bot. 4: 665–679.

Xiao, L.-Q., X.-J. Ge, X. Gong, G. Hao & S.-X. Zheng. 2004. ISSR variation in the endemic and endangered plant *Cycas guizhouensis* (Cycadaceae). Ann. Bot. 94: 133–138.

Yang, S.-L. & A.W. Meerow. 1996. The *Cycas pectinata* (Cycadaceae) complex: genetic structure and gene flow. Int. J. Pl. Sci. 157: 468–483.

CONSERVATION

CHAPTER 3

Ex situ Conservation of Microcycas calocoma, Zamia amblyphyllidia, Z. integrifolia, Z. ottonis, and Z. pygmaea germplasm at the National Botanic Garden, Cuba

Julio C. Lazcano Lara

Abstract

Cycads are represented in Cuba by the endemic genus *Microcycas* (Miq.) A. DC., with only one species *Microcycas calocoma* (Miq.) A. DC., and the genus *Zamia* L. with seven species: *Zamia amblyphyllidia* D.W. Stev., *Z. angustifolia* Jacq., *Z. integrifolia* L.f., *Z. pumila* L., *Z. ottonis* Miq., *Z. pygmaea* Sims, and *Z. stricta* Miq., the last three being endemic. In this paper we present the creation of a scientifically documented collection, in which threatened germplasm of *Microcycas calocoma, Z. amblyphyllidia, Z. integrifolia, Z. ottonis*, and *Z. pygmaea* is conserved in ex situ conditions. Seeds, whole individuals, and rhizome fragments were collected from 14 localities, most of them from western Cuba. The germplasm was cultivated under shade house conditions with filtered natural light and good aeration. To develop field gene banks, individuals were planted in three open-air zones of the National Botanic Garden of Cuba. The individuals were distributed randomly in an adequate substrate, and the feasibility for future monitoring was considered. As a result, collections of 156 individuals of *M. calocoma*, 26 of *Z. amblyphyllidia*, 177 of *Z. integrifolia*, 40 of *Z. ottonis*, and 35 of *Z. pygmaea* were established. The collections of *Z. amblyphyllidia, Z. integrifolia*, and *Z. ottonis* were established as field gene banks in the "Mogotes," "Manigua Costera," and "Serpentina" zones of the National Botanic Garden. The relevance of the collection and its efficacy for the conservation of the species are discussed.

Resumen

En Cuba, las cícadas están representadas por el género endémico *Microcycas* (Miq.) A. DC., con una especie *Microcycas calocoma* (Miq.) A. DC. y el género *Zamia* L. con siete especies:

Zamia amblyphyllidia D.W. Stev., *Z. angustifolia* Jacq., *Z. integrifolia* L.f., *Z. pumila* L., *Z. ottonis* Miq., *Z. pygmaea* Sims y *Z. stricta* Miq., las tres últimas, endémicas. En el presente trabajo se muestra el desarrollo de una colección científicamente documentada, en la que se conserva, en condiciones *ex situ*, germoplasma amenazado de: *M. calocoma, Z. amblyphyllidia, Z. integrifolia, Z. ottonis* y *Z. pygmaea*. Se colectaron semillas, individuos completos y fragmentos de rizoma en 14 localidades, la mayoría ubicadas en Cuba occidental. El cultivo del germoplasma se realizó en condiciones de umbráculo con luz natural filtrada y buena aireación. Para el desarrollo de los bancos genéticos de campo se introdujeron los individuos en tres zonas del Jardín Botánico. Los individuos fueron sembrados de manera dispersa. Se tuvo en cuenta la presencia del sustrato adecuado y la factibilidad para el monitoreo de los mismos en el futuro. Como resultado se creó una colección formada por: 156 individuos de *M. calocoma*, 26 individuos de *Z. amblyphyllidia*, 177 individuos de *Z. integrifolia*, 40 individuos de *Z. ottonis* y 35 individuos de *Z. pygmaea*. Las colecciones de *Z. amblyphyllidia, Z. integrifolia* y *Z. ottonis* se desarrollaron como bancos genéticos de campo en las zonas: Mogotes, Manigua Costera y Serpentina del Jardín Botánico. Se caracteriza la colección y se discute la importancia y eficacia de la misma para la conservación de las especies consideradas.

Introduction

Cycads are threatened worldwide; consequently all species are protected under CITES Appendixes I and II. Some Cuban cycads have been reported as threatened by different authors (Borhidi & Muñiz, 1983; World Conservation Monitoring Centre, 1994; Peña et al., 1998; Donaldson, 2003), and thus have been considered a good conservation target.

In Cuba, cycads are represented by the family Zamiaceae, with two genera: the endemic monospecific genus *Microcycas* represented by *Microcycas calocoma* (Miq.) A.DC., restricted to the north-central region of the western Cuban province "Pinar del Río" and the genus *Zamia* L., distributed all around the archipelago and represented by seven species: *Zamia amblyphyllidia* D.W.Stev., *Z. angustifolia* Jacq., *Z. integrifolia* L., *Z. pumila* L., *Z. ottonis* Miq., *Z. pygmaea* Sims, and *Z. stricta* Miq. The last three are endemic species (González Géigel, 2003).

Between 1997 and 2000, field studies of western Cuban cycads were carried out (Lazcano et al., 2001; Peña et al., 2001). As a result, some actions to address the conservation of *M. calocoma, Z. amblyphyllidia, Z. integrifolia, Z. ottonis,* and *Z. pygmaea* were proposed. Ex situ conservation through the creation of germplasm banks was one proposed action.

The effectiveness of ex situ conservation of wild species has often been discussed (Ashton, 1987; IUCN-BGCS & WWF, 1989; Given, 1987, 1994; Wyse Jackson & Sutherland, 2000), especially because of the limitations imposed by the space requirements and maintenance costs of collections large enough to conserve the genetic integrity of species. Nevertheless, ex situ conservation has proved its importance through such actions as the rescue of threatened germplasm, the cultivation of species with recalcitrant seeds that cannot be conserved in seed banks, the availability of material to reduce the pressure from wild collecting, environmental education, and scientific research (Bramwell et al., 1987; Wyse Jackson & Sutherland, 2000).

This study had two objectives: (1) to create well-documented collections of *M. calocoma, Z. amblyphyllidia, Z. integrifolia, Z. ottonis,* and *Z. pygmaea*; and (2) to establish collections of *Z. amblyphyllidia, Z. integrifolia,* and *Z. ottonis* as genetic field banks in zones of the National Botanic Garden. Considering that cycad seeds cannot be stored for a long period, the use of living plants was the best option.

Materials and Methods

Origin of the Germplasm

Eight expeditions were carried out to collect germplasm from localities previously studied (Peña et al., 1996–1997; Lazcano et al., 2001; Peña et al., 2001) (Table 3-1).

Type of Germplasm Collected

Mature seeds, whole individuals of different ages (seedlings, juveniles, adults), and fragments of rhizomes were collected. Collections were made taking into account the population structure of the species in each locality in order to maintain its conservation status (Akeroyd & Wyse Jackson, 1995; Lazcano et al., 2001).

Table 3-1 Localities visited to collect germplasm.

Species	Localities (locality name, county, province)
Microcycas calocoma	Hoyo Grande 1, Viñales, Pinar del Río
	Hoyo Grande 2, Viñales, Pinar del Río
	Sierra de la Caoba, Viñales, Pinar del Río
	Sendero "Maravillas de Viñales", Viñales, Pinar del Río
Zamia amblyphyllidia	Mogote del Mural, Viñales, Pinar del Río
	Mogote del Valle, Viñales, Pinar del Río
	Sierra del Infierno, Viñales, Pinar del Río
	Sierra de la Caoba, Viñales, Pinar del Río
	Escaleras de Jaruco, Madruga, La Habana
	Sierra de Banao, Sancti Spíritus, Sancti Spíritus
	El Yunque, Baracoa, Guantánamo
Zamia integrifolia	Reserva Ecológica Varahicacos, Varadero, Matanzas
	Cayo Sabinal,
Zamia ottonis	Loma La Coca, Habana del Este, La Habana
Zamia pygmaea	Reserva Ecológica Los Indios, Isla de la Juventud
	Laguna Alcatraz Chico, Municipio Sandino, Pinar del Río

Cultivation

The sarcotesta was manually removed, and seeds were germinated in boxes with a mix of coarse sand and humus (3:1).

Whole individuals were cultivated in polyethylene bags with a mix that contained humus or black or red soil and coarse sand in proportion (2:1). Adults of *Zamia ottonis* were cultivated in polyethylene bags with serpentine soil.

Plants were grown under shade house conditions with 50% filtered natural light and good aeration.

Genetic Field Banks of *Zamia amblyphyllidia*, *Z. integrifolia*, and *Z. ottonis*

Characterization of the Zones of the National Botanic Garden Used to Create the Genetic Field Banks

Mogotes (*Zamia amblyphyllidia*): the haystack hill vegetation complex is partially represented in this zone. The vegetation consists of semideciduous forest, and the zone was

Figure 3-1. Zones of the National Botanic Garden that were used to establish the genetic field banks. 1a. "Mogotes" zone (*Zamia amblyphyllidia*). 1b. "Manigua Costera" zones (*Zamia integrifolia*). 1c. "Serpentina" zone (*Zamia ottonis*). (All photos were taken by the author.)

created on the grounds of an abandoned calcareous rock mine. The walls may reach 20 m high (Leiva, 1980) (Fig. 3-1a).

Manigua Costera (*Z. integrifolia*): a coastal xeromorphic shrub vegetation is partially represented in this zone. (Leiva, 1980) (Fig. 3-1b).

Serpentina (*Z. ottonis*): thorny xeromorphic serpentine shrub vegetation on soils derived from ultramafic rocks is represented in this zone (Leiva, 1980) (Fig. 3-1c).

Location and General Characteristics of the Plants Introduced

Individuals were planted randomly but taking into account the presence of an adequate substrate and the feasibility for future monitoring.

All individuals were healthy and more than one year old. With the exception of two exemplars, they had at least one leaf. In the rainy season of 2002, 26 individuals of *Zamia amblyphyllidia*, 31 of *Z. integrifolia*, and 59 of *Z. ottonis* were planted. A second group of 146 individuals of *Z. integrifolia* were planted in the rainy season of 2003.

Monitoring

The genetic field banks were visited every three months, and survival of individuals, presence of reproductive structure, evidence of sexual reproduction, and health of the plants were recorded.

Results

Shade House Collections

Microcycas calocoma

The collection of this species is at the Botanic Garden Conservation Group shade house (Fig. 3-2a), and it contains 156 individuals from four localities. The locality "Hoyo Grande 1" is represented by four individuals from two mother plants, collected as three seedlings, and one juvenile; "Hoyo Grande 2" is represented by nine individuals from five mother plants, collected as two seeds, five seedlings, and two juveniles; "Sierra de La Caoba" is represented by 110 individuals from nine mother plants, 43 seeds, 64 seedlings, and three juveniles; and "Sendero Maravillas de Viñales" is represented by 33 individuals from two mother plants, collected as 21 seeds and 12 seedlings. No individual has yet reached adult size in cultivation.

In the near future, a genetic field bank will be created for this species in the "Mogotes" zone of the National Botanic Garden and in the Pinar del Río Botanic Garden.

Zamia pygmaea

The collection of this species is at the Botanic Garden Conservation Group shade house (Fig. 3-2b), and it contains 35 individuals from two localities. The locality "Reserva

Figure 3-2. Collections in shadow house conditions. **2a.** Collection of *Microcycas calocoma*. **2b.** Collection of *Zamia pygmaea*.

Ecológica Los Indios" is represented by 30 individuals from three mother plants, collected as 23 seeds and seven adults (two females, five males), and "Laguna Alcatraz Chico" is represented by five individuals from one mother plant, all of which were collected as seedlings, and two of them are male plants.

The genetic field bank of this species will be created at Pinar del Río Botanic Garden where a reproduction of white sands ecosystem, natural habitat of the taxon, is going to be created.

Genetic Field Banks

Zamia amblyphyllidia

The bank of this species, placed in the "Mogotes" zone of the Botanic Garden, has 26 individuals from seven localities. The locality "Escaleras de Jaruco" is represented by two individuals from one mother plant, both collected as seedlings. At "Sierra de Banao," one seed was collected. "Sierra de La Caoba" is represented by one juvenile. At "Sierra del Infierno," one juvenile was collected. "Mogote del Mural" is represented by 11 individuals from three mother plants, collected as five seedlings and six adults (two females, four males). "Mogote del Valle" is represented by nine individuals from two mother plants, collected as two seedlings, two juveniles, and five adults (two females, three males), and at "Yunque de Baracoa" one juvenile was collected.

Figure 3-3. Individuals in the genetic field bank of *Zamia amblyphyllidia*. 3a. Nonreproductive adult. 3b. juvenile.

Two years after their introduction, all of the plants are alive and healthy, and most individuals have produced leaves in their new habitat (Fig. 3-3a, b).

Zamia integrifolia

The bank of this species, placed in the "Manigua Costera" zone of the Botanic Garden, has 177 individuals from two localities. The locality "Cayo Sabinal" is represented by six plants collected as seeds from one mother plant; at present three are juveniles and three are adults (two females, one male). "Reserva Ecológica Varahicacos" is represented by 171 individuals from 10 mother plants, 132 collected as seeds and 39 as seedlings (Fig. 3-4a to Fig. 3-4c).

Two years after the introduction of the first group of plants, and one year after the introduction of the second group, all individuals are alive and healthy. Most of the plants have produced leaves in their new habitat (Fig. 3-4), and four have begun to produce cones.

Zamia ottonis

The bank of this species, placed in the "Serpentina" zone of the Botanic Garden, has 40 individuals from "Loma La Coca", from five mother plants, 32 of them collected as seeds, four as seedlings, and four as adults. Two years after plant introduction, 19 of the 59 individuals have died; the remaining plants have produced leaves in their new habitat, and six plants have produced cones (Fig. 3-5a to Fig. 3-5c).

Discussion

Figure 3-4. Individuals in the genetic field bank of *Zamia integrifolia*. 4a. Nonreproductive adult. 4b. advanced stage in juvenile phase. 4c. early stage in juvenile phase.

Developing ex situ collections to adequately represent the genetic diversity contained in all natural populations of a species is extremely difficult, and in many cases is impossible. The main purpose of these collections was to rescue threatened germplasm without affecting the conservation status of the wild populations. Thus, the collection of seeds and seedlings was the best option considering that (1) collection of adults severely affects the reproductive performance of a species; (2) because of their sexual origin, seeds contain a greater genetic diversity than adult fragments, and it is recommended that seeds be used to start new populations (Guerrant, 1996); (3) survival of seedlings in these species in nature is low (Peña et al., 1986; Peña et al., 1996–1997; Lazcano et al., 2001; Lazcano Lara, 2004); and (4) human interference, manifested in habitat destruction and introduction of exotic species, is a real threat to some of the wild populations that were used as a germplasm source (Lazcano et al., 2001; Peña et al., 2001).

Besides the type of material collected, the number of mother plants used as germplasm source is an important issue to consider in the management of genetic diversity. During collection, 10 or fewer seeds or seedlings were collected per mother plant per locality. The number of seeds per mother plant was higher for *M. calocoma* and *Z. integrifolia*, because during collection of these species, many seeds were found in unfavorable sites, for example, lying on rocks without any soil to guarantee survival after germination.

Zamia amblyphyllidia is under-represented because five of the seven localities are represented only by two or one individual. During the collecting period, seeds were

Figure 3-5. Individuals in the genetic field bank of *Zamia ottonis*. 5a. Female adult. 5b. Early stage in juvenile phase. 5c. Early stage in juvenile phase with retarded growth.

found only in one locality ("Sierra de Banao"), five seeds were planted and only one seedling survived.

The ex situ cultivation of a species represents an intrinsic conservation deficiency; nevertheless, the establishment of ex situ collections does allow achievement of good results, and in some cases it is the only available option.

If the two important factors, edaphic conditions and the principal components of the vegetation, are established, and if artificial selection and cultivation are disregarded, then some selection processes that operate in nature may also act under *ex situ* conditions. The development of field gene banks in zones of the National Botanic Garden constitutes a performance test for the germplasm collected, which may result in a collection composed of resistant individuals that could be used for reintroduction or reinforcement with success.

Conclusions

1. The National Botanic Garden has a collection that consists of 156 individuals of *M. calocoma*, 26 individuals of *Z. amblyphyllidia*, 177 individuals of *Z. integrifolia*, 40 individuals of *Z. ottonis*, and 35 individuals of *Z. pygmaea*. The ex situ conservation at national level of these species is supported by this collection.
2. Field genetic banks of *Z. amblyphyllidia*, *Z. integrifolia*, and *Z. ottonis* are a reservoir of resistant material to be used for future conservation actions such as reintroduction and reinforcement.

Recommendations

1. Further expeditions should be undertaken in order to increase the amount of genetic diversity contained in the collections.
2. Monitoring of the collections in order to determine the survival rate and the occurrence of sexual reproduction under ex situ conditions should be continued.
3. A national action plan for the protection of Cuban cycads and their threatened habitats should be implemented.

Acknowledgments

I thank Centro Nacional de Áreas Protegidas, Cuba, for supporting my participation in CYCAD 2005, Leonel del Risco, Luis R. González, and Michel Torres, for helping me to plant the individuals in the zones of the Botanic Garden, Alejandro Palmarola and Luis R. González, for use of digital cameras, and Fabiola Areces for reviewing the manuscript.

Literature Cited

Akeroyd, J. & P. Wyse Jackson. 1995. A handbook for botanic gardens on the reintroduction of plants to the wild. Botanic Gardens Conservation International, UK.

Ashton, P.S. 1987. Biological considerations in situ vs ex situ plant conservation. Pp. 117–130 *in* D. Bramwell, O. Hamann, V. Heywood & H. Synge (eds.), Botanic gardens and the world conservation strategy. Academic Press, London.

Borhidi, A. & O. Muñiz. 1983. Catálogo de plantas cubanas amenazadas o extinguidas. Editorial Academia, La Habana, Cuba.

Bramwell, D., O. Hamann, V. Heywood & H. Synge (eds.). 1987. Botanic gardens and the world conservation strategy. Academic Press, London.

Donaldson, J. S. (ed.). 2003. Cycads. Status survey and conservation action plan. IUCN/SSC Cycad Specialist Group. IUCN, Gland, Switzerland, and Cambridge, UK.

Given, D. R. 1987. What the conservationist requires of ex situ collections. Pp. 103–116 *in* D. Bramwell, O. Hamann, V. Heywood & H. Synge (eds.), Botanic gardens and the world conservation strategy. Academic Press, London.

———. 1994. Principles and practice of plant conservation. Timber Press, Portland, Oregon.

González Géigel, L. 2003. Zamiaceae. Pp. 1–22 *in* Flora de la República de Cuba, serie A, Plantas vasculares. Vol. 8. No. 4. Koeltz Scientific Books, Königstein.

Guerrant, E. O., Jr. 1996. Designing populations: demographic, genetic, and horticultural dimensions. Pp. 171–207 *in* D. A. Falk, C. I. Millar & M. Olwell (eds.), Restoring diversity. Strategies for reintroduction of endangered plants. Island Press, Washington, DC.

World Conservation Union-Botanic Gardens Conservation Strategy [IUCN-BGCS] & WWF. 1989. The Botanic Gardens Conservation Strategy. Gland, Switzerland.

Lazcano, J., E. Peña, P. I. López, L. del Risco & D. Pérez. 2001. Estudios para la conservación de las Zamias cubanas: 2. estado de conservación y factores de riesgo de las especies de Cuba occidental. Revista Jard. Bot. Nac. Univ. Habana 22: 201–207.

Lazcano Lara, J. C. 2004. Biología poblacional y aspectos ecológicos de *Microcycas calocoma* (Miq.) A.DC. en la Sierra de La Caoba, Viñales. M.S. thesis, Jardín Botánico Nacional, Universidad de La Habana.

Leiva, A. 1980. El desarrollo de jardines botánicos en Cuba: aspectos de su organización científica. D.Sc. diss. , Jardín Botánico Nacional, Universidad de La Habana.

Peña, E., L. Díaz & E. Grillo. 1986. *Microcycas calocoma*: caracteres de la semilla y su germinación. Revista Jard. Bot. Nac. Univ. Habana 7: 55–70.

———, P. I. López, J. Lazcano, D. Pérez & Z. Torriente. 1996–1997. La reproducción sexual en *Microcycas*. I, Estudios de monitoreo in situ. Revista Jard. Bot. Nac. Univ. Habana 17–18: 147–158.

———, ———, ———, A. Leiva & U. S. Seal. 1998. Memorias del Primer Taller para la Conservación, análisis y manejo planificado de plantas silvestres cubanas, 13–15 April. IUCN/SSC Conservation and Breeding Specialist Group, Apple Valley, MN.

———, J. Lazcano, P. I. López & D. Pérez. 2001. Estudios para la conservación de las Zamias cubanas: 1. Caracterización de localidades en Cuba occidental. Revista Jard. Bot. Nac. Univ. Habana 22: 195–200.

Wyse Jackson, P. S. & L. Sutherland. 2000. International agenda for botanic gardens in conservation. Botanic Gardens Conservation International, UK.

World Conservation Monitoring Centre [WCMC] . 1994. Plant information at World Conservation Monitoring Centre. Conservation status report as of 30 March 1994, Cuba. WCMC, Cambridge, UK.

CONSERVATION
CHAPTER 4

Taxonomía, Distribución y Estado de Conservación de Zamia en Ecuador

Fernando Nicolalde-Morejón

Resumen

Se reportan cinco especies del género *Zamia* desde Ecuador. *Zamia poeppigiana* Mart. & Eichler, ocurre en hábitats muy variados que van desde los bosques secos en Túmbez (Perú), bosques de transición en la cordillera costanera cerca de Pedernales (Ecuador) hasta los bosques muy húmedos del noroccidente del Ecuador, el tamaño y estado de conservación de las poblaciones de esta especie varían con el estado de conservación de los bosques. *Zamia roezlii* Linden, es la especie con mas amplia distribución en el Chocó biogeográfico Ecuatoriano, donde la destrucción del hábitat (deforestación, agricultura y ganadería) se convierte en la principal amenaza para la conservación de la especie. En la misma región, con distribución muy restringida, ocurren poblaciones de *Z. gentryi* Dodson, especie recientemente descrita, pero la más afectada por la destrucción del hábitat por actividades antrópicas; sin embargo, nuevas poblaciones encontradas al interior de la Reserva Forestal de la Etnia Awá, representa el mejor escenario para la conservación de la especie. La especie *Z. amazonum* Stevenson, relativamente común y ampliamente distribuida en toda la cuenca amazónica, es poco conocida en Ecuador, por lo que determinar su real distribución y estado actual de conservación es muy prematuro y se necesita mas investigación de campo. Situación similar sucede con *Z. ulei* U. Dammer, conocida únicamente desde solo dos poblaciones.

Abstract

Five species in the genus *Zamia* are known to occur in Ecuador. *Zamia poeppigiana* Mart. & Eichler occurs in a variety of habitats, such as dry forest in Tumbez (Peru), transition forest in the coastal cordillera near Pedernales (Ecuador), and moist tropical forest in the northwestern Ecuadorian corner. In this last site population sizes vary with conservation status of the forest. *Zamia roezlii* Linden is the species with the widest range, in the southern end of the biogeographical Chocó, where habitat destruction (long cutting, agriculture, and cattle) is the main threat to that species. Populations of *Z. gentryi* Dodson, a recently described endemic, also occurs in the same region and is the most threatened. However, populations in the Awa Indians Reserve appear to be the most suitable for future biological studies. A relatively common and widely distributed species, *Z. amazonum* Stevenson, is very poorly known in Ecuador, and its actual status and distribution needs to be investigated. The same is true for *Z. ulei* U. Dammer, which is known from only two Ecuadorian populations.

Introducción

Zamia L., representa el género de mayor distribución de Cícadas americanas, en comparación con *Ceratozamia* Brongn., y *Dioon* Lindl., los cuales están mayormente representados en México, con pocas especies aisladas en Guatemala, Belice y Honduras; y *Microcycas* (Miq.) A. DC., endémico a Cuba.

Actualmente se reconocen 59 especies de *Zamia* (Hill et al., 2004), distribuidas desde el sur-este de Florida en los Estados Unidos de Norte América, Centro América,

Antillas Mayores, y Sur América, siendo Bolivia el extremo sur de la distribución para el género.

Además, *Zamia* representa el género con la más alta variación morfológica ecológica y citológica (Norstog, 1980; Vovides, 1983; Moretti & Sabato, 1984; Moretti, 1990; Vovides & Olivares, 1996; Norstog & Nicholls, 1997) de todo el orden Cycadales, únicamente comparable con el género *Cycas* L., en el hemisferio oriental (Norstog & Nicholls, 1997).

El último tratamiento taxonómico completo para *Zamia* fue presentado por Schuster (1932); sin embargo trabajos posteriores (Sabato, 1990; Stevenson, 1987, 1991, 1993, 2001a, 2001b, 2004; Stevenson & Sabato, 1986; Norstog & Nicholls, 1997) han evidenciado muchos problemas taxonómicos y nomenclaturales, debido principalmente al insuficiente trabajo de campo e insuficiente material de herbario observado.

Actualmente, aunque el numero de colecciones a incrementado en comparación a lo que pudo observar Schuster (1932); aun persisten una escasez de colecciones en especial con estructuras reproductivas para ser analizadas, principalmente con las especies de Sur América, donde el número de registros conocidos para muchas especies comprenden apenas dos o tres colecciones botánicas, que son el producto de una baja densidad de poblaciones e individuos registradas para estas especies en condiciones naturales. Esto en gran medida ha obstaculizado la circunscripción de algunos complejos de especies existentes a lo largo de la distribución del género, y por consecuente la elaboración un tratamiento taxonómico completo.

Por lo tanto, este tratamiento de *Zamia* para Ecuador, pretende formar parte de una serie de monografías regionales descritas anteriormente (González Géigel 2004; Stevenson, 1991, 1993, 2001a, 2001b, 2004; Vovides et al., 1983; Vovides, 1999), con miras a completar la revisión taxonómica del género para el neotrópico.

Zamiaceae

ZAMIA AMAZONUM

D. Stev. Fl. Colombia 21: 33–35. figs. 23, 2001. (Fig. 4-1)

TYPE: Brazil. Amazonas, Confluence of Río Tauri and río Içana, *D. Stevenson 886* (Holotype: INPA; Isotype: COL, MO, NY, U).

Figure 4-1. *Zamia amazonum*. **A.** Megastróbilo (inmaduro). **B.** Folíolos.

Tallo subterráneo de 2.5 a 8 cm de diámetro. *Catafilos* con base triangular y ápice linear-lanceolado, 3-7 cm de largo, 1.3-2 cm de ancho. *Hojas* 2–3, ascendentes, generalmente ovadas, 0.6–1.5 m de largo; pecíolo 0.5–0.7 m de largo, armados con aguijones pequeños, fuertes y bifurcados, hojas jóvenes con indumento rojizo; raquis 0.3–0.8 m de largo, con 4–5 pares de folíolos opuesto a subopuestos, no imbricados, con aguijones en el tercio proximal. *Folíolos* cartáceos, oblanceolados a lanceolados, 13–25 cm de largo, 3-7 cm de ancho, sésiles, base atenuada, ápice acuminado, márgenes ligeramente dentados a lo largo del tercio proximal; venas inconspicuas. *Microstróbilos* 2–6, cilíndricos, café-amarillentos, 8 cm de lago, 2 cm de diámetro; pedúnculo 9.5 cm de largo, decumbentes. *Megastróbilo* solitario, cilíndrico, café-rojizo cuando maduros, 14 cm de largo, 5.5 cm de diámetro; pedúnculo 9 cm de largo, erecto. *Semillas* con sarcotesta roja, ovoide 0.9 cm de largo, 0.6 cm de diámetro.

Rasgos característicos: Las diferencias de esta especie con relación a *Z. ulei* radican en que *Z. amazonum* presenta folíolos más cartáceos y más grandes de forma obovados a oblanceolados, frente a los folíolos elípticos y muy papiráceos de *Z. ulei*. No obstante, anteriormente mantenía una confusión con *Z. ulei* por presentar una distribución geográfica similar y apariencia morfológica en especial de individuos infértiles. Nombre vernáculo, "guagrapanga."

Distribución: Esta especie ocurre en las provincias amazónicas de Sucumbíos, Napo y Morona Santiago, en rangos altitudinales de 240–1000 m. Esta especie es la de más amplia distribución en la amazonía ecuatoriana, y de toda la cuenca amazónica de Colombia, Venezuela, Brasil y Perú.

Estado de Conservación: Muy poco se conoce de sus poblaciones en estado silvestre a pesar de tener una amplia distribución en la amazonía ecuatoriana. Existe un bajo número de colecciones botánicas; sin embargo, parece ser que su distribución está mejor representada hacia al sur de la amazonía de acuerdo a los nuevos registros de la provincia de Morona Santiago (*Nicolalde et al. 1397*) donde las poblaciones de esta especie podrían estar mejor representados por ser una de las zonas mejor conservadas de la amazonía ecuatoriana. Debido a la escasez de información disponible, determinar su estado actual de conservación es muy prematuro.

Etimología: Su epíteto específico hace referencia a la presencia de esta especie en el Departamento Amazonas, en Colombia (Stevenson, 2001).

Especimenes observados: Ecuador. Napo: *F. Nicolalde et al. 1397* (QCNE); *D. Neill 9461* (MO). Sucumbios: *W. Palacios et al. 8913* (QCNE, MO).

Especimenes adicionales observados: VENEZUELA. Amazonas: *T. Plowman & W. Thomas 13541* (NY). COLOMBIA. Amazonas: *C. Pañuela et al. 983* (COL), *C. A. Pazmin et al. 020* (COL), *L. E. Urrego 662* (COAH, NY), *L. E. Urrego et al. 623* (NY); Caqueta: *C. Barbosa et al. 8059* (COL), *P. Franco et al. 3840* (COL, NY), *P. Palacios et al. 2806* (COL, MA, NY). PERÚ. Loreto: *R. Simpsom & J. Schunke 800* (F, US, GH), *J. Schunke 14240* (F, NY).

ZAMIA GENTRYI

Dodson, Novon 8: 12–14. Fig. 1 a-f (1998). (Fig. 4-2)

TYPE: Ecuador, Esmeraldas, *C. H. Dodson & H. Gentry 17520A* (Holotype: QCNE; Isotype: AAU, MO, NY, RPSC, SEL, U).

Tallos hasta de 1.2 m de alto, 20 cm de diámetro; algunas veces decumbentes. *Hojas* 5–9 (10), ascendentes, 3 m de largo; pecíolo 0.6–1.1 m de largo, densamente armado con aguijones fuertes; raquis 0.8–1.8 m de largo, con 20–30 pares de folíolos opuesto a subopuestos, no imbricados, con aguijones en el tercio proximal. *Folíolos* coriáceos, linear-lanceolados, ligeramente subfalcados, márgenes enteros, sésiles, base atenuada, ápice agudo,

Figure 4-2. *Zamia gentryi*. **A.** Megastróbilo (imaduro). **B.** Habitat.

20–39 cm de largo, 4-6 cm de ancho, con la superficie abaxial acanalada entre las venas, superficie adaxial con venas no prominentes. *Microstróbilos* 1–7, cilíndricos, vino-rojizos, 30–40 cm de largo, 3-4 cm de diámetro; pedúnculo 20–25 cm de largo, decumbente. *Magastróbilo* solitario, cilíndrico, amarillo a rojizo, 20–30 cm de largo, 8–12 cm de diámetro; pedúnculo corto. *Semillas* con sarcotesta rosada cuando inmadura, roja cuando madura, obovoide, 3.5 cm de largo, 1.7 cm de diámetro.

Rasgos característicos: Esta especie presenta a veces tallos decumbentes al igual que algunos ejemplares de *Z. roezlii*; sin embargo, las diferencias son muy marcadas en cuanto al tamaño de los tallos, que alcanzan hasta 1.20 m de alto en *Z. gentryi* y hasta 2.5 m de alto (7 m en Colombia. Stevenson, 2001) en *Z. roezlii,* además el tamaño y color de los microstróbilos, púrpura-marrón en *Z. gentryi* y crema en *Z. roezlii*, así como la nervaduras poco pronunciadas por el envés de los folíolos, en comparación a los folíolos fuertemente acanalados en *Z. roezlii*. Nombre vernáculo "piña de monte."

Distribución: Z. gentryi ocurre en las provincias de Esmeraldas y Carchi, entre rangos altitudinales de 300–2000 m, en el bosque húmedo tropical y bosque pluvial premontano.

Estado de conservación: Esta especie fue colectada por primera vez en 1988 en una área poco conocida botánicamente, y las subsiguientes colecciones han provenido de esta misma región. Sin embargo, exploraciones recientes durantes los años 2000–2003 demuestran que sus poblaciones naturales están desapareciendo, como producto de una extracción agresiva de madera. Una demanda a la que ha sido sometida esta región en los últimos 10 años, a tal punto que la estimación hecha sobre las poblaciones de esta especie en 1988 determinaban 20 individuos por hectárea (Dodson, 2000), registro que en la actualidad es únicamente referencial.

Observaciones personales demuestran que la distribución de esta especie esta restringida a territorio de la Reserva Awá hacia el oriente de la colección inicial, en un rango altitudinal de 800 a 2000 m, donde se constató la presencia de dos poblaciones muy dispersas y con pocos individuos. Categoría de Lista Roja (Walter & Gillet, 1997) sugerida VU B2a.

Etimología: El epíteto específico honra al Dr. Alwyn H. Gentry botánico explorador de los bosques tropicales de Sur América.

Especimenes examinados: ECUADOR. Carchi: *F. Nicolalde et al. 1395* (QCNE), *G. Tipaz et al. 1878* (QCNE). Esmeraldas: *C. Dodson et al. 17107* (MO, QCNE), *19098* (QCNE), *H. Luther et al. 1235* (SEL), *F. Nicolalde 1396* (QCNE), *B. ;tOllgard et al. 98726* (AUU, QCNE).

Zamia poeppigiana

Martius & Eichler, Fl. Bras. 4(1): 414–416. Tab. 109. 1863. (Fig. 4-3)

Type: PERÚ, Maynas Alto, Toache River, 1830, *Poeppig s.n.* (Lectotype: F *ex Herb. Musei Palat. Vindob.*; Isolectotype: GH *ex Herb. Musei Palat. Vindob*; *fide* Stevenson & Sabato, 1986).

Z. lindenii Regel ex André, Ill. hortic. 22: 23. Planche 195. 1875. Lectotype: Planche 195 in Ill. hortic. 23: 1875. (*fide* Stevenson & Sabato, 1986).

Aulacophyllum lindenii (Regel ex André) Regel, Gartenflora 25: 141. 1876.

Z. baraquiniana Regel, Acta Horti petrop. 4(4): 308–309. 1876. Type: ex Horto Petropolitano, Regel s.n. (Holotype: LE).

Z. wielandii Schuster, Pflanzenr. 99: 149. 1932. *nomen illegit.* superfluous name for *Z. baraquiniana* Regel.

Tallos de hasta 1.5 m de alto, 20 cm de diámetro. *Catafilos* con la base triangular, ápice acuminado, 3.5–4.5 cm de largo, 1.8–3.5 cm de ancho. *Hojas* 8–17, hasta de 3.8 m

Figure 4-3. *Zamia poeppigiana*. A. Megastróbilo. B. Hábito y hojas jóvenes.

de largo, ascendentes, formando una corona abierta, hojas jóvenes verdes; pecíolo 71–80 cm de largo, con aguijones fuertes y pequeños, decreciendo en número y tamaño a lo largo de tercio proximal; raquis 0.7–1.3 m de largo, con 25–40 pares de folíolos, opuestos a subopuetos, no imbricados, con aguijones en el tercio proximal. *Folíolos* coriáceos, linear-lanceolados, 15–42 cm de largo, 2.5-5 cm de ancho, base cuneada, ápice acuminado, venas inconspicuas, márgenes fuertemente dentados a lo largo del tercio distal. *Microstróbilos*, cilíndricos, cremas, 20–27 cm de largo, 3.5-5 cm de diámetro; pedúnculo de 6-8 cm de largo. *Megastróbilos* 2–5, cilíndricos, café-verdoso cuando están maduros, 30–70 cm de largo, 11–15 cm de diámetro. *Semillas*, con sarcotesta rosada cuando están inmaduras, rojas cuando están maduras, 2.8–3.2 cm de largo, 1.2–1.4 cm de diámetro; el promedio de semillas por cono 270. Número cromosómico 2n = 18 (Norstog, 1980).

Rasgos característicos: Los ejemplares de la vertiente del pacífico se reconocen fácilmente por tallos arbóreos de hasta 2.5 m de alto, sus folíolos lisos, lineal-oblongos y fuertemente dentados hacia el ápice. Nombre vernáculo "Palma de Goma".

Distribución: Esta especie ocurre a lo largo de la costa húmeda del Ecuador, desde la provincia del Guayas hasta la confluencia de los ríos Cayapas y Santiago al norte de Esmeraldas en el extremo sur del Chocó. Generalmente ha sido registrada en bosque secundario y en lugares abiertos destinados para cultivos de ciclo corto en suelos bien drenados, en tanto, que al interior de la Reserva Ecológica Manglares Churute ha sido localizada en áreas con bosques bien conservados (Cerón, 2003. Su amplio rango altitudinal oscila entre 0–1200 m en la estribación occidental de los Andes.

Problemas taxonómicos con la especie: Zamia poeppigiana Mart. & Eichler, fue descrita desde material colectado por Poeppig en 1830 a lo largo del río Tocaehe en la Amazonía del Perú. Años después fue publicada *Zamia lindenii* Regel & André, cuya descripción se basó sobre plantas vivas colectadas por Roezl. Debido a la falta de material herborizado Ecuatoriano de Roezl que precediera a la publicación de *Z. lindenii,* Stevenson y Sabato (1986) sinonimizaron *Z. lindenii* bajo *Z. peoppigiana.* No obstante, el tipo y descripción original de *Z. poeppigiana* son muy similares a *Z. lindenii* estos dos taxa son considerados coespecíficas (Sabato 1990; Stevenson & Sabato, 1986, Stevenson, 2004).

En mi opinión, considerando que *Z. poeppigiana* y *Z. lindenii* representan un escenario vicariante muy antiguo, futuras investigaciones deberían considerar un análisis detallado a nivel poblacional. Esto debe incluir los rangos de distribución que estos dos taxa registran en ambos lados de la cordillera de los Andes en los países de Colombia, Ecuador y Perú, en busca de evidencia adicional morfológica y filogenética que confirmen la presencia de una o dos especies.

Estado de conservación: A pesar de su amplio rango de distribución superior a todas las especies que ocurren en el occidente de Ecuador, su principal amenaza es la pérdida de hábitat, debido a que la especie se encuentra registrada en lugares con potencial forestal y agropecuario. Debido a la perdida del hábitat, lo cual resulta en una disminución considerable de las poblaciones a lo largo de la costa de Ecuador, el estado de conservación para esta ésta especie es de alta vulnerabilidad. Categoría de Lista Roja sugerida R, II, V.

Etimología: El epíteto específico es en honor a Eduard Poeppig quien colectó por primera vez esta especie en Perú.

Especimenes examinados: ECUADOR. *E. André 3687* (F, GH, K, NY, P, US), *F. Barclay 709* (US), *H. Eggers 14034* (GH), *C. Horn s. n.* (US), *F. Lehmann 658* (G), *A. Rimbach 85,* (F), *252* (US). Azuay: Huigra, *J. N. Rose & G. Rose 22613* (US); Saraguro, *J. Steyermark 52786* (F, NY, P). Bolivar: Balsapamba, *T. Lockwood 825* (GH), *P. L. Sodiro S. J. 001* (Q). Chimborazo: *S. Lægaard 55253* (QCA). Esmerladas: Borbón, *Finn Borchsenius 235* (QCA); Fila de Bilsa, *L. Richard Abbott 15200* (QCNE), *A. Gentry et al. 72867* (MO), *P. Maas et al. 2920* (U), *F. Nicolalde et al. 375* (MO, QCNE); Eloy Alfaro, *F. Nicolalde et al. 455* (QCNE); Esmapaidas, *P. Maas et al. 2020* (U); Río Mira, *M. G. Patwa s. n.* (US). Guayas: Naranjal, *C. Cerón et al. 17793, 20073* (QAP); *G. Harling & L. Andersson 19417* (GB), *F. Lehmann 5292* (K), *S. Tello Vazquez 002* (QCA). Manabí: *G. Harling & L. Andersson 24751* (GB, QCA); Toache-Las Pampas, *C. Dodson & A. Gentry 9715, 13694* (MO, SEL). El Oro: *Balslev et al. 62526* (AAU, COL, QCA). Pichincha: Congoma Grande, *L. Kvist 40203* (AAU, MO, NY); Bosque Protector La Perla, *V. Zak et al. 5600* (QCA); Santo Domingo, *C. Cerón et al. 29140* (QAP). Los Rios: Jaunche, *C. Dodson & A. Gentry 7991, 12698* (MO, SEL), *C. Dodson et al. 7991* (MO, SEL), *W. W. Thomas & M. Ríos 6724* (QCA); Rio Palenque, *A. Gentry & C. Dodson 18045* (MO, QCA), *K. Mark P. McMahom 4289 (QCA), Norstog 80–5* (FTG).

Especimenes adicionales observados: PERÚ. Amazonas: Bagua, *T. Plowman 5536* (GH). Huanúco: Tingo Maria, *T. Plowman et al. 7575* (F, GH, NY) *11232*, (F, GH, INPA, NY); Loreto: Lower Rio Huallaga, *L. Williams 5373* (F, US); Maynas, *L. Williams 3794* (F). Madre de Dios: Tambopata, *M. Alexiades 1322* (NY); Cocha Cashu Station, *R. Foster 5852* (F). San Martín: Campanilla, *J. Schunke 4200* (F); Tarapoto, *F. Montes* 57 (F, NY); *T. Plowman 7601* (F); *M. Rimachi Y. 5185* (NY), *5356* (MO, NY); Tocache Nuevo, *A. Gentry et al. 25710* (MO, NY); *T. Plowman & J. Schunke 11543* (F). Tumbez: Tumbes, *C. Díaz et al. 5117* (MO); Matapalo, *C. Díaz et al. 7500* (MO); *D. Simpson & J. Schunke 428* (COL, F, GH, NY, US).

Zamia roezlii

Linden, Linden Catologue des plantes nouvelles no. 90, p. 10. 1873. (Fig. 4-4)

NEOTYPE: Planches 133–134 in III. Hortic. 20: 1873. (*fide* Stevenson & Sabato, 1986).

Aulacophyllum roezlli (Linden) Regel, Gartenflora 25: 141.1876.

Figure 4-4. *Zamia roezlii*. **A.** Hábito. **B.** Microstróbilos. **C.** Megastróbilo.

Tallos de hasta 2.5 m de alto, 24 cm de diámetro, algunas veces decumbentes y dicotómicos. *Catáfilos* con la base triangular, ápice largamente acuminado, 3.5-6 cm de ancho, 6–9.5 cm de largo. *Hojas* 4–14 (50), ascendentes, hasta de 3.65 m de largo, formando una corona cerrada, hojas jóvenes con indumento café-amarillento; pecíolo 0.4–0.95 cm de largo, cubierto densamente con aguijones fuertes; raquis 0.9–1.8 (2.3) m, con 20–50 pares de folíolos, opuestos a subopuestos, no imbricados, con aguijones sobre el tercio proximal. *Folíolos* coriáceos, linear lanceolados, ligeramente subfalcados, 50 cm de largo, 4.5 cm de ancho, sésiles, base cuneada, ápice agudo, superfice adaxial y abaxial fuertemente acanaladas entre venas, márgenes enteros. *Microstróbilos* 4–6, cilíndricos, blanco-cremosos, 11–18 cm de largo, 3-5 cm de diámetro; pedúnculo decumbente. *Megastróbilos* 1–2 (3), café, hasta 60 cm de largo, 13 cm de diámetro, cortamente pedunculado. *Semillas* ovoides, 2.8–3.2 cm de largo, 1.2–1.4 cm en diámetro, sarcotesta naranja. Numero cromosómico 2n = 22, 24, 25, 26 (Norstog, 1980).

Rasgos característicos: Se distingue por sus folíolos fuertemente acanalados, lineares o ligeramente falcados, con bordes enteros; por el gran tamaño de los megastróbilos, hasta de 60 cm de largo, y del tallo de hasta 2.5 m de alto (7 m de alto en Colombia, Stevenson, 2001a). Los habitantes del noroccidente del Ecuador le atribuyen los nombres vernáculos "Chigua", "aafachi" y "piña de monte."

Distribución: Esta registrada para las provincias de: Carchi, Imbabura, Esmeraldas y Pichincha, entre 50–850 m de altitud, en zonas de bosque húmedo tropical y bosque pluvial premontano, con temperaturas promedio anuales de 23° y 25° C, y con precipitaciones promedio anuales mayores a 3000 milímetros (Cañadas, 1983).

Estado de conservación: Esta especie es la de más amplia distribución en el noroccidente del Ecuador, específicamente en la provincia de Esmeraldas. Sin embargo, sus poblaciones son apenas representativas en estado silvestres la cuales están amenazadas por procesos de deforestación, por encontrarse en una zona de alto potencial maderero. A ello se suma él avance de la frontera agrícola que transforma los paisajes naturales en extensas zonas de pasturas, área de las cuales las cícadas son extirpadas por presentar cualidades tóxicas perjudiciales para la actividad ganadera.

Estudios poblacionales efectuados entre 1999–2001, muestran evidencia que los individuos encontrados no forman agregados, y que la densidad poblacional puntual para un cuadrante de 5 hectáreas. El cuadrante que registró el mayor número de entre seis de los mismos, es de 5 individuos, de los cuales dos fueron adultos y tres juveniles (Nicolalde,

2000); por lo que, el estado actual de esta especie es de alta vulnerabilidad. Categoría sugerida de Lista Roja es R, II, V.

Etimología: Esta especie fue descrita por Linden en el año de 1873 desde plantas cultivadas en su jardín y el epíteto específico honra a Benedit Roezl (1824–1885) quien colectó por primera vez esta especie a lo largo de la costa del Chocó en la segunda mitad del siglo XIX.

Especimenes examinados: ECUADOR. Esmerladas: Zapallo Grande, *A. Barfot & F. Skov 60116* (AAU, NY, QCA); San Lorenzo, *C. Dodson 19622* (RPSC), *C. Dodson & A. Gentry 19048* (RPSC), C. *Dodson & T. Neudecker 19083* (QCNE, RPSC), *H. E. Luther s.n.* 1992 (MO), *H. E. Luther et al. 2782* (RPSC), *D. Neill et al. 11792* (QCNE, MO); Eloy Alfaro, *F. Nicolalde & J. Palacios 234* (QCNE); San Lorenzo, *F. Nicolalde et al. 443* (QCNE, MO); Reserva Awa, *G. Tipaz et al. 2242* (QCNE, MO); Cotacachi-Cayapas, *G. Tipaz* & *P. Méndez 2255* (QCNE, MO), *M. Tirado et al. 646* (QCNE); Eloy Alfaro, Corriente Grande, *P. Yanez et al. 1413* (QCNE); altitud y colector desconocidos; Referencia Rió Cayapas *48155* (AAU). Imbabura: García Moreno, *C. Cerón et al. 41076* (QAP). Napo: Estación Jatun Sacha, *D. Neill 11160* Cultivada desde semillas de la colección *G. Tipaz 2242* (QCNE, MO).

Especimenes adicionales examinados: COLOMBIA: Chocó: *H. P. Fuchs & L. Zanella 21902* (NY, US); Nariño: *Bernal & Galeano 892* (COL, NY); El Valle del Cauca: *Gentry & Juncosa 53395* (COL, MO, US); *Killip & Cuatrecasas 38967* (COL, US, F), *Killip & García 33337* (BM, COL, F, P, US).

Zamia Ulei

U. Dammer, Vert. Bot. Vereins Prov. Branderburg 47: 117–118. 1907. (Fig. 4-5)

TYPE: Brazil. "Cachoeira obrer Rio Jurá", *E. Ule 5523* (Lectotype, HBG, *file* Stevenson and Sabato, 1986: 142; Isolectotype, F (fragmento), G. K. L. MG).

Zamia cupatiensis Ducke, *Arch. Jard. Bot. Rio de Janeiro 3:20, t. 1. 1922*

TYPE. *Ducke s.n.,* Arch. Jardim Bot. Rió de Janeiro 3: 20, t. 1. 1922 (Lectotype, *file* Stevenson & Sabato, 1986: 136)

Tallo subterráneo y tuberoso, 3-8 cm de diámetro. *Hojas* 2–3(5), 0.9–1.6 m de largo; pecíolo 0.5–0.9 m de largo, armado con aguijones; raquis 0.3–0.6 m de largo, con 3–8

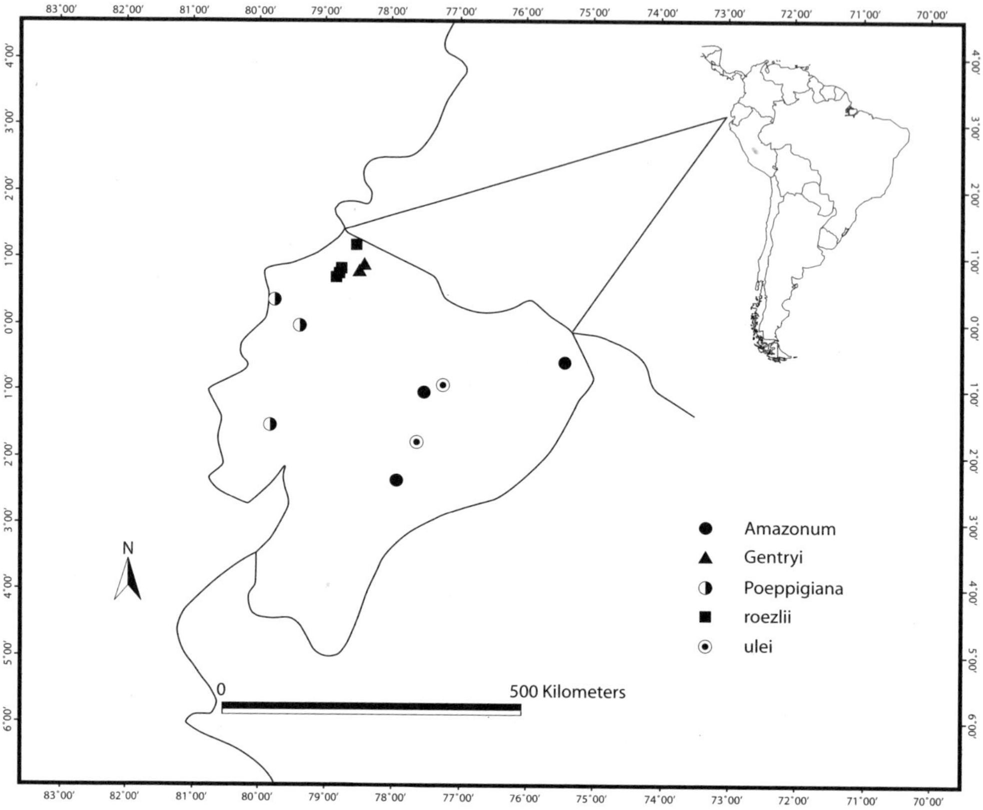

Figure 4-5. Distribución de *Zamia amazonum, Z. gentryi, Z. roezlii, Z. poeppigiana* y *Z. ulei* en Ecuador.

pares de folíolos, subopuestos, con pocas aguijones a lo largo del tercio proximal. *Folíolos* papiráceos, oblongo-elípticos a obovados, 15–26 cm de largo, 6-9 cm de ancho, nervaduras inconspicuas, sésiles, basalmente simétricos, ápice cortamente acuminado, márgenes dentados a lo largo de la mitad distal.

Microstróbilos 2–4, cilíndricos, blanco-cremosos, 6–12 cm de largo, 2–3.5 cm en diámetro de diámetro; pedúnculo 5-7 cm de largo. *Megastróbilo* solitario, cilíndrico, marrón, paredes externas de las megasporófilas de color negro cuando maduro, 12–18 cm de largo, 5-8 cm de diámetro; pedúnculo 7–10.5 cm de largo. *Semillas* ovoides, con sarcotesta roja cuando madura, 1.5 cm de largo, 1 cm de diámetro. Número cromosómico 2n=26 (Caputo et al., 1996).

Rasgos característicos: *Zamia ulei,* se distingue de las otras especies por ser una planta con tallo subterráneo y folíolos oblongo-elípticos, características exclusivas de esta especie con respecto a otras especies de *Zamia* que ocurre en Ecuador. Sin embargo, frecuentemente es confundida con *Z. amazonum,* especie que presenta folíolos oblongo-lanceolados y coriáceos; además, también podría ser confundida con *Z. obliqua* que ocurre en Colombia, pero las diferencias están marcadas por: los folíolos simétricos y no oblicuos en la base y en el caso de esta última en plantas adultas presentan tallos arborescentes.

Distribución: Actualmente muy poco se conoce sobre la distribución de *Z. ulei* en Ecuador, debido a la escasez de colecciones botánicas existentes en los herbarios locales e internacionales. Con anterioridad a este trabajo se consideró a muchos especimenes de *Zamia amazonum* como *Z. ulei*, lo cual hizo pensar erróneamente que la distribución de *Z. ulei* era más amplia.

Estado de conservación: Desconocido debido a la escasez de colecciones. Sin embargo, al considerar el buen estado de conservación de los bosques tropicales en la provincia del Pastaza podríamos asumir la posibilidad de encontrar en el futuro mayores poblaciones.

Etimología: El epíteto específico honra a Ernest Heinrich Georg Ule (1854–1915), botánico alemán que exploró y colectó por primera vez esta especie en la Amazonía durante la segunda mitad del siglo XIX.

Especimenes examinados: Ecuador. Napo: *Maddison et al. 5419* (AAU, F, FTG, K, QCA, SEL, US). Pastaza: *D. Neill et al. 11136* (QCNE).

Especimenes adicionales observados: BRASIL. Acre: *D. C. Daly et al. 6100* (NY), *D.C. Daly et al. 10106, 10061* (NY), *G. T. Prance 12358* (NY). COLOMBIA. Amazonas: *Pipoly et al. 15233* (COL, MO, NY), *T. Plowman 11799, 11800* (F, NY). Putumayo: *Bernal et al. 2049* (COL). PERÚ. Loreto: *J. Schunke V. 14029* (F).

Especies dudosas

El número de especies que ocurren en Ecuador podría incrementarse, debido a la existencia de dos colecciones botánicas (*Nicolalde et al. 234* y *1051*) de las cuales actualmente no se dispone de suficiente información de campo y de herbario para realizar las descripciones y consecuentemente su tratamiento taxonómico, razón por la cual no constan en esta revisión.

Discusión

De las cinco especies actualmente registradas para Ecuador, tres especies ocurren al lado occidental de los Andes, entre rangos altitudinales de 0–2000 m de altitud. *Zamia roezlii*, mantiene poblaciones apenas representativas en estado silvestre, las cuales se ven amenazadas por actividades forestales y agropecuarias. Sin embargo, considerando la extensión que abarca la especie en zonas bajas de la Reserva Ecológica Cotacachi-Cayapas, cabe la posibilidad que al interior de esta área protegida se puedan registrar poblaciones grandes en buen estado de conservación. Nuevas exploraciones son requeridas para establecer su real estado de conservación.

Dadas las condiciones de deforestación a la que ha sido sometida la localidad tipo de *Zamia gentryi* (ver fig. 2B), su endemismo actualmente esta restringido a una reserva Forestal de la etnia Awá en el noroccidente del Ecuador, zona fronteriza con Colombia. Las poblaciones reportadas al interior de esta reserva, representan la mejor alternativa para futuras investigaciones acerca de la biología de esta especie. El estado de conservación para esta especie tendría que ser catalogado como muy crítico.

Zamia poeppigiana representa la especie de más amplia distribución en toda la costa Ecuatoriana, su hábitat es muy variado, va desde el bosque seco y de transición en la cordillera costera cercana a Pedernales (Ecuador), o en Tumbes (Perú), hasta el bosque muy húmedo tropical del Chocó Ecuatoriano. El tamaño de sus poblaciones locales varía de acuerdo al estado de conservación de las masas boscosas. Aunque frecuentemente es reportada desde áreas deforestadas, también se han reportado poblaciones desde la reserva ecológica gubernamental Manglares Churute en la provincia de Guayas y La Estación Biológica y Reserva Bilsa, entidad privada en la provincia de Esmeraldas. Su estado de conservación debe ser considerado como vulnerable.

Z. amazonum a pesar de ser una especie relativamente común y de amplia distribución en la amazonía de Colombia, Venezuela, Perú y Brasil (Stevenson, 2001a), en Ecuador muy poco se conoce sobre su estado actual y potencial distribución. Situación similar sucede con *Z. ulei,* conocida únicamente desde dos poblaciones, por lo que determinar el estado actual de conservación de estas especies sería muy prematuro.

Conclusiones

Más exploraciones de campo son necesarias, tanto al noroccidente del Ecuador como a la región amazónica en busca de nuevas poblaciones, así como de material fértil. Esto nos permitirá esclarecer problemas taxonómicos y eventos de vicarianza entre poblaciones de las especies *Zamia poeppigiana y Z. lindenii* registradas a los dos lados de la cordillera de los andes.

Zamia gentryi, requiere especial atención debido a la disminución drástica de su hábitat. Sin embargo, nuevas exploraciones deben ser planificadas en especial del lado más norte de la reserva Ecológica Cotacachi-Cayapas entre 500 y 2000 m de altitud, y corroborar la presencia de esta especie en esta reserva gubernamental.

Agradecimientos

Expreso mis agradecimientos al Dr. Dennis Wm. Stevenson por su apoyo en mis seis meses de estancia en el Jardín Botánico de Nueva York, en el cual tuve la oportunidad de revisar las colecciones de Zamiaceae para Sur América. Al Dr. Andrew Vovides, Dr. John P. Janovec, Ing. Walter Palacios, Dr. Carlos Cerón, por sus comentarios al documento. Finalmente al proyecto CARE-Subir y CARE-Psur, quienes financiaron gran parte del trabajo de campo en Ecuador, a través de la Fundación Jatun Sacha.

Literatura Citada

Cañadas, L. 1983. Mapa bioclimático y ecológico del Ecuador. Ministerio de Agricultura y Ganadería (MAG). Quito.

Caputo, P., S. Cozzolino, L. Gaudio, A. Moretti & D. Stevenson. 1996. Karyology and phylogeny of some Mesoamerican species of *Zamia* (Zamiaceae). Amer. J. Bot. 83: 1513–1520.

Ceron, A. 2003. Personal communication with author.

Dammer, U. 1907. Cycadaceae. Verh. Bot. Ver. Brandenburg 47: 117–118.

Dodson, C. 1998. A new species of *Zamia* (Zamiaceae) from Ecuador. Novon 8: 12–14.

———. 2000. Personal communication with author.

Ducke, A. 1922. Plantes nouvelles ou peu connus de la région Amazonienne (II Partie). Cycadaceae. Arch. Jara. Bot. Rio de Janeiro 3: 20, tab. 1.

González Géigel, L. 2003. Zamiaceae. Pp. 1–22 *in* Flora de la República de Cuba, serie A, Plantas vasculares. Vol. 8. No. 4. Koeltz Scientific Books, Königstein.

Hill, K., D. Stevenson & R. Osborne. 2004. Appendix 1: The world list of cycads. Pp. 219–235 *in* T. Walters & R. Osborne (eds.), Cycad Classification: Concepts and Recommendations. CABI Publishing, Oxford.

Moretti, A. 1990. Karyotypic Data on North and Central American Zamiaceae (Cycadales) and their phylogenetic implications. Amer. J. Bot. 77: 1016–1029.

——— & S. Sabato. 1984. Karyotype evolution by centromeric fission in *Zamia* (Cycadales). Pl. Syst. Evol. 146: 215–223.

Nicolalde, F. 2000. Estudio de la ecología y biología de *Zamia*, con fines de manejo. Tesis de Grado, Universidad Técnica del Norte, Facultad de Ingeniería en Ciencias Agropecuarias y Ambientales. Ibarra.

Norstog, K. 1980. Chromosome numbers in *Zamia* (Cycadales). Carylogia 33: 419–428.

——— & T. Nicholls. 1997. The biology of the cycads. Cornell Univ. Press. Ithaca, NY.

Regel, E. 1876. Die Cycadeen, deren Gattungen und Arten. Gartenflora 25:140–144.

———. 1876. Cycadearum. Acta Horti Petropolitana 4(2): 308–309.

Sabato, S. 1990. West Indian and South American cycads. Mem. New York Bot. Gard. 57: 173–185.

Schuster, J. 1932. Cycadaceae. *in* A. Engler (ed.), Das Pfanzenreich IV. 1 (Helf 99): 1–168.

Stevenson, D. W. 1987. The West Indian Zamias. Fairchild Trop. Gard. Bull. 42(3): 23–27.

———. 1991. "Flora de Guianas." Ser. A, Fasc. 9: 7–11.

———. 1993. The Zamiaceae in Panamá with comments on phytogeography and species relationships. Brittonia 45: 1–16.

———. 2001a. Cycadales. Flora de Colombia 21: 1–92.

———. 2001b. Zamiaceae. Flora Nicaragua 1: 6–7.

———. 2004. Zamiaceae of Bolivia, Ecuador, and Peru. Pp. 173–194 *in* T. Walters & R. Osborne (eds.), Cycad Classification: Concepts and Recommendations. CABI Publishing, Oxford.

——— & S. Sabato. 1986. Typification of names in *Zamia* L. and *Aulacophyllum* Regel (Zamiaceae). Taxon 35: 134–143.

Vovides, A. P. 1983. Systematic studies on the Mexican Zamiaceae. Chromosome numbers and Karyotypes. Amer. J. Bot. 70: 1002–1006.

———, J. Rees & M. Vázquez-Torres. 1983. Zamiaceae. Flora de Veracruz. 26: 1–31.

———. 1999. Zamiaceae. Flora del Bajío y de Regiones Adyacentes. 71: 1–17.

——— & M. Olivares. 1996. Karyotype polymorphism in the cycad *Zamia loddigesii* (Zamiaceae) of the Yucatan Peninsula, México. Bot. J. Linn. Soc. 120: 77–83.

Walter, K. & H. Gillett (eds). 1997 IUCN Red list of threatened plants. IUCN-The World Conservation Unit, Gland, Switzerland.

CONSERVATION
CHAPTER 5

Managing Cycad Aulacaspis Scale (Aulacaspis yasumatsusi Takagi) at Montgomery Botanical Center, USA

Christine Wiese and Catharine Mannion

Abstract

Since it was first introduced into South Florida in 1995, *Aulacaspis yasumatsui* (cycad aulacaspis scale) has been attacking popular landscape cycads and spreading rapidly. This pest is now found in Florida, Texas, Puerto Rico, U.S. Virgin Islands, and Hawaii. Plant death and damage caused by *A. yasumatsui* is having a marked effect on the ornamental cycad industry and has become a major concern for nursery professionals. The spread of *A. yasumatsui* to other countries through plant sale and trade is also a serious concern as it could threaten native cycad populations. Previous research on cycad aulacaspis scale indicated that maintaining populations at a rate that did not threaten the health of plants required multiple control measures. A series of control measures (monitoring, biological control, and chemical control) were implemented in 2003 and 2004 at Montgomery Botanical Center in order to develop a system of managing this pest as part of a broader integrated pest management program.

Resumen

Desde que fue introducida en el Sur de la Florida en el 1995, *Aulacaspis yasumatsui* (escama de cicadas aulacaspis) ha estado atacando las cicadas populares para la arquitectura de jardines y esparciendose rapidamente. Esta plaga ahora se encuentra en Florida, Texas, Puerto Rico, las Islas Virgenes de Estados Unidos y Hawaii. Las muertes y daños a las plantas causado por *A. yasumatsui* esta haciendo un efecto marcado en la industria de cicadas ornamentales y se ha convertido en una gran preocupación para los profesionales de viveros. La propagación de *A. yasumatsui* hacia otros paises a través de la venta y comercio de las plantas también es una preocupación seria ya que puede amenazar poblaciones nativas de cicadas. Previas investigaciones sobre la escama de cicadas aulacaspis indica que el mantenimiento a una proporción en la cual no amenacen la salud de las poblaciones de las plantas, requiere varias medidas de control. Una serie de medidas de control (vigilación, control biológico y químico) fueron implementadas en el 2003 y 2004 en Montgomery Botanical Center (Centro Botanico Montgomery) con el fin de desarrollar un sistema para controlar esta plaga como parte de un programa más extenso de la IPM.

Introduction

Unusually heavy infestations of a white scale insect were first observed in the area of Montgomery Botanical Center (MBC) and Fairchild Tropical Botanical Garden in 1995. Residents of South Miami were familiar with the pest magnolia white scale (*Pseudaulacaspis cockerelii* Comstock [Coccoidea: Diaspididae]) and assumed these heavy scale infestations to be the same species (Walters et al., 1997). However, these infestations were heavier than those in the past and were concentrated on the undersides of leaves rather than on the upper surfaces, which are the common infestation area of the magnolia white scale (Weissling et al., 1999). Therefore, entomologists were brought in to identify what was then thought to be a new pest species. Dr. Avas B. Hamon, Division of Plant Industry, Florida Department of Agriculture and Consumer Services, Gainesville, identified the scale pest as *Aulacaspis yasumatsui* Takagi (Sternorrhyncha: Diaspididae), cycad aulacaspis scale. The identification was confirmed by Dr. Douglas R. Miller, Systematic Entomology Laboratory, USDA-ARS-SEL, Beltsville, Maryland (Howard et al., 1996).

Cycad aulacaspis scale (CAS) is native to Thailand, where it can be a pest of cycads but is generally kept under control by natural predators (Weissling et al., 1999). CAS is a member of the armored scale group (Hemiptera: Diaspidae), which means that the adult females have a hard outer covering to protect themselves and their eggs (Weissling et al., 1999). CAS has a broad host range among cycads (Howard et al., 1999), but the heaviest infestations at MBC have been observed on species of *Cycas* L. and on *Stangeria eriopus* (Kunze) Baill.

The scale proved to be an extremely difficult pest to control. Once established, it moved rapidly up the state of Florida. Between 1995 and 1999, CAS spread to Hong Kong, Hawaii, and the Cayman Islands, as well as expanding its range of infestation within the state of Florida (Weissling et al., 1999). Most recently, CAS has been reported on the island of Guam—which is of special concern because it has been observed in populations of the native species *C. micronesica* K.D. Hill (Marler, 2004a, 2004b).

Initially, management of this scale was attempted by treating it as if it were magnolia white scale. Both homeowners and horticulturists used pesticides and oils that had been successfully used to treat that pest. However, some pesticides that had been used to control magnolia white scale in the past did not prove to be successful (Howard et al., 1996; Weissling et al., 1999). Horticultural oils, fish oils, and acephate did control CAS effectively when good spray coverage was possible and plants were sprayed frequently (C. Mannion & H. Glenn, unpublished data). Neither insecticidal soap nor thiamethoxam ("Flagship") controlled this scale (C. Mannion & H. Glenn, unpublished data). Dimethoate ("Cygon") and malathion provided adequate control when applied foliarly, but both have been observed to cause phytotoxicity in new growth (Emshousen & Mannion, 2004).

Pyriproxifen ("Distance") was first examined as a potential control for CAS by C. Mannion and H. Glenn at the Tropical Research and Education Center, Homestead,

Florida, in 2002. Results were very promising when pyriproxifen was applied to scale on *C. revoluta* in 1-gallon pots: 100% of eggs and 99% of female scale examined were dead after eight weeks (C. Mannion & H. Glenn, unpublished data). In subsequent tests, Mannion and Glenn found similar results (pers. comm.). Following these tests, we studied the effectiveness of pyriproxifen on heavily and lightly infested leaf samples taken from large cycads in the field at MBC. On six of the nine lightly infested samples, female scale mortality was ≥90%, and eggs of all females were dead On the three other lightly infested samples, female scale mortality was at least 75%, and eggs of all females were dead (Emshousen & Mannion, 2004). On seven of the nine heavily infested samples there was ≥70% female scale mortality and eggs of all females were dead (Emshousen & Mannion, 2004).

In addition to chemical control, biological controls are being evaluated. Two parasitoids, a beetle (*Cybocephalus binotatus* Grouvelle [Colepotera: Nitidulidae]) and a wasp (*Coccobius fulvus* Compere and Annecke [Hymenoptera: Aphelinidae]) were released in 1998 by Baranowski (Hodges et al., 2003). In 2004, USDA entomologists released the wasp at MBC and began an evaluation of its effectiveness.

Objectives

It was necessary for MBC to develop a management plan if CAS was going to be kept under control in the long-term. We had two objectives for our CAS management plan: (1) that it be proactive and (2) that it utilize integrated pest management (IPM) practices.

Beginning in fall 2003, an evaluation was conducted to develop an IPM strategy for managing CAS at MBC. We conducted an examination of each step of the IPM system as it related to CAS management. The steps of the IPM system that we evaluated were as follows:

1. Attend to plant cultural needs.
2. Scout for pests regularly.
3. Correct pest identification.
4. Establish a threshold of tolerance.
5. Evaluate all treatment options.
6. Apply the best treatment option for the situation.
7. Keep detailed records.

For each step in the IPM system, we addressed three questions:

1. What are we doing now to address this step?
2. What else could we do?
3. What changes do we then need to make, if any?

As part of our evaluation of available treatment options, we also further examined the effectiveness of pyriproxifen ("Distance") as a chemical treatment for scale on large cycads in the field. This trial was conducted at MBC in spring 2003.

Materials and Methods

Ten plants were selected for study based on their history of heavy scale infestations. They included one plant each of nine species of *Cycas* and one plant of *Stangeria eriopus*. Each plant was evaluated for its scale population in early May before any treatments for scale were applied. Leaf samples were removed from each plant, and 20 female scale were examined from each sample. The percentage of female mortality and the percentage of females with all their eggs dead were calculated for each sample. One week later, all study plants were treated with a foliar application of pyriproxifen until the point of runoff. The pyriproxifen solution was mixed and applied according to the label. Eight weeks after the pyriproxifen application, the study plants were sampled as before and the percentage of female mortality and percentage of all females with all their eggs dead were calculated.

Results and Discussion

Plant Cultural Needs

Plant cultural needs include appropriate water, light, and soil, as well as other needs. We determined that the water, light, and soil needs were adequately addressed with regard to CAS management and that changes were not necessary. However, we had not adequately addressed the issue of airflow. We observed that when the same species were planted close together, the scale infestation was more intense and plants were more difficult to access for treatment. We also observed that by relocating plants where they had more space, and by removing dense shrubs blocking air flow, CAS infestations seemed to decrease in intensity. In the past, new plants at MBC were often planted too close together. We will now allow more space between plants when constructing new plantings, and relocate some plants to allow them more space. In addition, we will prune particularly dense plants to allow more air to flow through.

Pest Scouting and Identification

Regular scouting and correct identification of the pest were steps that we felt we had adequately addressed in the past and would continue unchanged. A general examination of the collection is conducted once per week, and the collection is examined plant by plant in detail at least once per month. If new pests are found while monitoring the collection, they are sent to entomologists for identification. When routine pests such as CAS are discovered, the intensity and spread of the infestation is evaluated and treatments are applied if necessary.

Threshold of Tolerance

We determined that our threshold of tolerance needed to be at a level that would provide healthy plants for research. Because this scale spreads quickly and can rapidly cause a decline in plant health, MBC's threshold of tolerance for CAS was set low. If first instar CAS, or "crawlers," are seen on more than two plants in the collection, we will begin treatment.

Treatment Options

Beyond the threshold of tolerance, we evaluated the next step, treatment options. We evaluated the pros and cons of several treatment options. According to our observations, horticultural oils had been a successful treatment option for small plants with light scale infestations where good spray coverage was possible. Dimethoate ("Cygon") and malathion in foliar applications had provided control of scale, but these chemicals also kill the natural predators of CAS, which can be used as biological control agents. In addition, dimethoate and malathion have been observed to cause phytotoxicity of new growth. For these reasons, we elected not to use dimethoate or malathion as a treatment option.

Table 5-1 Percentages of female mortality and females with all eggs dead pre- and postapplication of pyriproxifen "Distance."

	Preapplication		Postapplication	
Cycad species	Female mortality (%)	All Females with all eggs dead (%)	Female mortality(%)	All Females with all eggs dead(%)
Stangeria eriopus (Kunze) Baill.	90	25	70	75
Cycas revoluta Thunb.	50	—[1]	100	100
C. taitungensis C.F. Shen, K.D. Hill, C.H. Tsou & C.J. Chen	0	—[1]	100	100
C. angulata R. Br.	100	100	100	100
C. riuminiana Porte ex Regel	100	100	100	100
C. media *R. Br.*	100	100	100	100
C. thouarsii R. Br. ex Gaudich.	95	100	75	—[a]
C. edentata de Laub.	100	100	—[b]	—[b]
C. hongheensis S.Y. Yang & S.L. Yang	50	75	100	100
C. bougainvilleana K.D. Hill	95	100	100	100

[a]No females found with eggs; scale found was primarily immature.
[b]No scale was found on these samples.

In addition to those treatment options we had used in the past, we further examined the effectiveness of pyriproxifen. Results of the evaluation of pyriproxifen were encouraging, although not conclusive. Scale infestations were lighter during the evaluation than they had been in past spring seasons. Many samples showed no live female scale or only crawlers (Table 5-1). After eight weeks, female scale mortality was at least 70% on all samples, and all of the eggs of at least 75% of the females were dead. Of the ten samples, eight had either 100% mortality of female scale and eggs or no scale at all.

One of the most interesting observations from this study was that many of the scale insects were parasitized and that the beetle predator, *Cybocephalus binotatus* Grouvelle, was commonly seen during sampling. This suggested that pyriproxifen was compatible with the use of natural enemies as biological control for CAS. This conclusion was also supported by laboratory testing (C. Mannion & H. Glenn, unpublished).

In addition to these evaluations of pyriproxifen, we also began utilizing the parasitic wasp *Coccobius fulvus* Compere and Annecke, as a biological control agent. The USDA conducted an eight-month release and sampling project at MBC from December 2003 to July 2004. Data from this study are still being evaluated, although preliminary observations indicate that scale was being parasitized by the wasp. We cannot yet say at what rate.

"Best" Treatments

After bringing together all available data on multiple treatment options, we were able to establish a plan that included different treatments, depending on the intensity and spread of the infestations. Horticultural oils were effective for small, low-intensity infestations and where adequate spray coverage was possible. Pyriproxifen provided good control for heavier or more widespread infestations, again where adequate spray coverage was possible. Pyriproxifen was more compatible with biological control than dimethoate or malathion, and did not cause any phytotoxicity. The parasitic wasp could be utilized to assist in controlling scale in areas where spray could not reach (i.e., particularly tall or dense plants).

Record Keeping

Finally, we evaluated our record-keeping system and elected not to make any changes. We keep detailed records of all treatments. It is these detailed records in addition to other data that has made it possible to develop an IPM plan for CAS tailored to our needs and constraints here at MBC.

Conclusion

Eradication of CAS in Florida is no longer possible. CAS has been difficult to control, but as responsible collection managers, we need to make a serious effort to do just that. Not only is it important to maintain healthy plants within our own collection—because of its scientific contribution—but also it is crucial to manage CAS in a way that minimizes spread to new locales. At MBC, we now have a proactive management plan for controlling CAS. We have changed some plant cultural practices to reduce the spread of the scale and to allow better access to plants for treatment. We will continue weekly and monthly scouting so treatments can be applied promptly. We will also utilize multiple available treatments and use the best option for a given type of infestation.

In addition, at MBC we are always seeking new information. We will evaluate new control options as they become available and share this information with those who are affected. Only by working together and sharing all relevant information about CAS control will we be able to effectively manage this pest and stop its spread.

Literature Cited

Emshousen C. & C. Mannion. 2004. Taming Asian cycad scale. Cycad Newslett. 27(1): 8–10.

Hodges, G., F. W. Howard & E. A. Buss. 2003. Update on management methods for cycad aulacaspis scale. U.S. Dept. of Agriculture and Consumer Services, Entomology Section, Washington, DC.

Howard, F. W., A. Hamon, M. Mclaughlin, T. Weissling & S. Yang. 1999. *Aulacaspis yasumatsui* (Hemiptera: Sternorrhyncha: Diaspididae), a scale insect pest of cycads recently introduced into Florida. Florida Entomol. 82(1): 14–27.

———, T. Weissling & D. Hull. 1996. New armored scale insect introduction in Miami area. TropicLine 9(2): 1–2.

Marler, T. 2004a. Fighting back. Pacific Daily News, Hagatna, Guam, 13 March 2004.

———. 2004b. Cycad scale attacking Guam plants. Pacific Daily News, Hagatna, Guam, 25 September 2004.

Walters, T., E. Shroyer & L. Anderson. 1997. Scale and south Florida *Cycas*. Cycad Newslett. 20(1): 13–15.

Weissling, T. J., F. W. Howard & A. B. Hamon. 1999. Cycad aulacaspis scale, *Aulacaspis yasumatsui* Takagi (Insecta: Homoptera: Sternorrhyncha: Diaspididae). Florida Cooperative Extension Service Publication No. EENY-096. Gainesville, FL.

ECOLOGY
CHAPTER 6

Ecología de Ceratozamia aff. robusta en la Sierra Madre del Sur de Oaxaca

Alma Delia Chávez González,

Beatriz Rendón Aguilar,

and Azucena De Lourdes Luna José

Abstract

State of Oaxaca is considered one of the most diversified areas in terms of ecosystems and species. Specifically with genus *Ceratozamia*, Oaxaca corresponds, along with Chiapas, to one of the states where diversification of this genus occurred, so it is considered an area of speciation. Seven species of the genus *Ceratozamia* are reported for Oaxaca, two of which are considered endemic (*Ceratozamia mixeorum* and *C. whitelockiana*). Most of the populations of the different species have been reported from the Sierra Norte and one from the Sierra Sur. Recently, a small population of a putative new species of *Ceratozamia* was discovered at the Sierra Madre del Sur. The population of the putative new species, which is in process of review and description, grows in an area of cloud forest. Forty-five individuals were recorded in a 3250 m^2 plot (65 × 50 m), which showed a congregated distribution ($\chi^2 = 3856.3$, $p < 0.001$, 33 gl, $\alpha = 0.05$). Class structure distribution exhibited type I.

From this pattern, we infer that a recruitment process exists. Even though this is a small population, it is considered established because: a) agricultural activities related with coffee plantations have not impacted this population and b) few indigenous people practice handcraft or horticulture with this species. In all the state, *Ceratozamia* is used as an insecticide, ornamental, and for ritual.

Resumen

El estado de Oaxaca, es considerado como un sitio de gran diversidad de ecosistemas y de especies, particularmente, para el género *Ceratozamia*, este estado representa, junto con Chiapas, uno de los lugares en donde ha ocurrido la diversificación del género por lo que se considera que es un caldero de especiación. Para esta entidad hay siete especies reportadas del género *Ceratozamia*, de las cuales dos de ellas son endémicas (*Ceratozamia mixeorum* y *C. whitelockiana*). La mayoría de las poblaciones se han colectado hacia la Sierra Norte, siendo únicamente dos registros en la Sierra Sur. Dentro de esta entidad ha sido recientemente localizada una nueva especie, que actualmente se encuentra en proceso de revisión y descripción. Esta población se encuentra en un ambiente de bosque mesófilo de montaña. Se registraron 45 individuos en un cuadrante de 3250 m^2 (65 × 50 metros), con una distribución agregada ($\chi^2 = 3856.3$, $p < 0.001$, 33 gl, α 0.05). La curva de la estructura de clase obtenida es de tipo I (Martínez y Alvarez-Buylla, 1995). A partir de esta distribución se puede inferir que dentro de la población existe reclutamiento. Se considera que, a pesar de ser una población pequeña, hasta el momento la población se encuentra estable debido a que: a) las actividades agrícolas relacionadas con la cafeticultura no han incidido en la población y b) a que el uso que las personas realizan de esta especie de *Ceratozamia* es mínimo, ya que son pocos los ejemplares que son usados como ornato en el pueblo. Finalmente, los reportes indican únicamente tres usos del género en el estado de Oaxaca, ornato, insecticida y con fines religiosos.

Introducción

El estado de Oaxaca es uno de los más ricos, tanto en la parte florística, como en otros ámbitos, ya que presenta una gran variedad de ecosistemas que se ubican dentro de las provincias fisiográficas de la Sierra Madre del Sur, el Sistema Montañoso del Norte de Oaxaca y la Planicie Costera Suroriental (Rzedowski, 1988). Hacia el sur de Oaxaca, dentro de la SMS, en la Región Terrestre Prioritaria (RTP) número 129, se presenta un alto grado de conservación y riqueza biológica, ecológica y genética (Arriaga et al., 2000). Esta región alberga por lo menos . . . ecosistemas diferentes (INEGI, 1986; Luna-José, 2001; Aragón, 2004), que van de los bosques de pino, hasta selva baja caducifolia. De éstos, el bosque mesófilo de montaña es de gran importancia al caracterizarse por la presencia de taxones paleoendémicos y neoendémicos, así como especies de distribución restringida o que habitan de manera preferente en estos lugares. Tal es el caso del género *Ceratozamia*, el cual, de acuerdo con González y Vovides (2002) presenta su mayor tasa de especiación en el sur de México.

A nivel mundial, el género *Ceratozamia* está representado por 21 especies, distribuidas a lo largo de regiones templadas y tropicales de América, desde el estado de Tamaulipas, hasta Belize, Guatemala y Honduras (Vovides et al., 2004). En México, la distribución de este género parece coincidir con los refugios florísticos (Toledo 1982; Wendt, 1987; Vovides et al., 2004). Esto corrobora la teoría de los refugios del Pleistoceno, los cuales contienen evidencia fósil de *Ceratozamia* del Mioceno en los refugios florísticos del Cenozoico en Oaxaca (Rzedowski & Palacios, 1977).

Debido a la presencia de este género en el país, así como de los géneros *Dioon* y *Zamia*, pertenecientes a la familia Zamiaceae, México se ubica en el segundo lugar mundial en diversidad de cicadas (Vovides, 2000).

Este género tiene pocos estudios, y sus especies se encuentran en alguna de las categorías que se enlistan en la NOM-059-ECOL-2001 y en la CITES, por lo que es importante realizar estudios precisos, para poder establecer estrategias de manejo y conservación.

Los objetivos de este trabajo fueron: a) Identificar la especie a la que pertenece la población de *Ceratozamia* localizada en la Sierra Madre del Sur de Oaxaca, b) Realizar un diagnóstico del estado de conservación de la población en función de su densidad y estructura poblacional, c) determinar la distribución espacial de los individuos, d) realizar una revisión bibliográfica del género, con mayor interés en su distribución, etnobotánica y características generales, con énfasis en el estado de Oaxaca y e) realizar unja revisión de herbario del género para el estado.

Material y Método

Zona de Estudio

La zona de estudio se localiza en el municipio de San Agustín Loxicha, Oaxaca (Fig. 6-1), a un a altitud de 1760 m. La inclinación de la parcela varía entre 20 y 120%, sin embargo, el sitio donde se localiza la población de *Ceratozamia* tiene una pendiente de 90%, la orientación de la parcela es de 160° Sur. El tipo de vegetación corresponde a un BMM, con un clima de tipo A (INEGI, 1988). La geología del lugar pertenece al complejo Oaxaqueño, el cual esta constituido por rocas metamórficas del Precámbrico (Centeno-García, 2004).

Trabajo de Campo

Se delimitó un cuadrante de 65 × 50 metros, con subdivisiones cada metro, con lo cual se pudo ubicar espacialmente a los individuos, tomándose las coordenadas en los planos *X* y *Y*, al tiempo que se contabilizaban, esto con el fin de poder determinar la densidad poblacional. Se obtuvo un estimado de la estructura por tallas de la población a partir de la altura total, definida como la altura que va de la base del individuo al ápice de la hoja más alta. Se registró el estadio fenológico de los individuos (vegetativo o reproductivo).

Se registraron algunas variables ambientales de la zona de muestreo como: altitud,

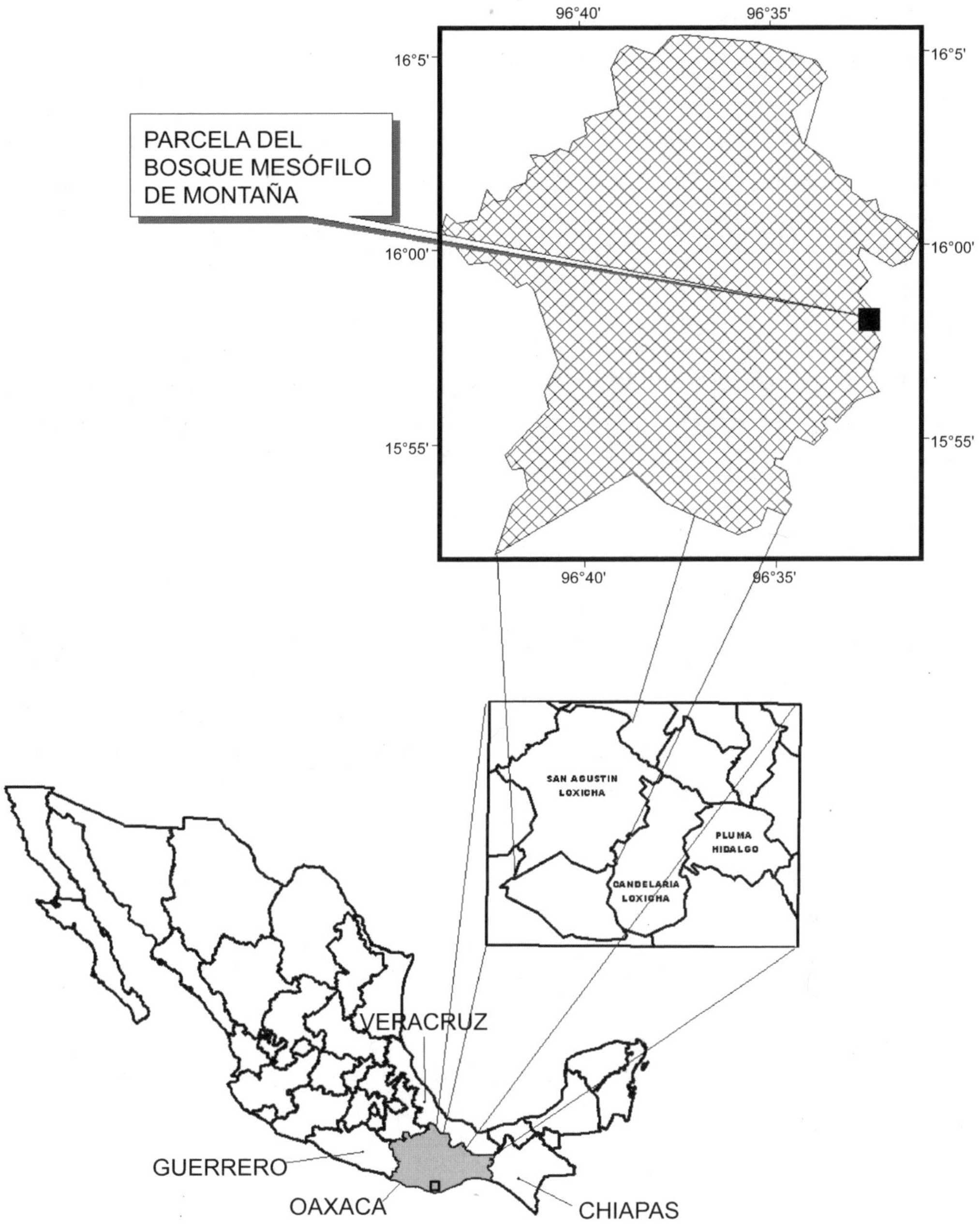

Figura 6-1. Mapa de la zona de estudio.

grado de inclinación de la pendiente y coordenadas geográficas. La composición florística del sitio se obtuvo a partir de los datos obtenidos previamente sobre la estructura y composición florística del BMM (Hernández, 2004). Se consideraron todos los individuos arbóreos que coincidieran con el área muestreada. Se realizó la colecta de hojas de tres individuos, a las cuales se les tomaron datos adicionales, para su identificación y posterior ingreso al herbario.

Trabajo de Gabinete

Se realizó una revisión bibliográfica, tomando en cuenta aspectos de distribución, taxonómicos, biogeográficos, paleontológicos y etnobotánicos. Así mismo, se examinó información sobre características de la zona y estudios previos realizados con cícadas en el estado.

Se realizaron visitas a los Herbarios de la Ciudad de México (MEXU, UAMI, ENCB, FEZA y CHAPINGO) para revisar los ejemplares colectados de la especie. A partir de la información recopilada se elaboró un mapa de las localidades que reportan su presencia al nivel de municipio, esto con la ayuda del Biol. Gilberto Hernández Cárdenas.

El análisis de la estructura poblacional se realizó con base en la estructura vertical, a partir de un histograma de frecuencias de alturas. Para determinar el tipo de distribución de la población se siguió el método descrito por Brower et al. (1989) y Greig-Smith (1983). Para la identificación del ejemplar, se contó con la asesoría del Dr. Andrew P. Vovides, del Instituto de Ecología, A. C., Xalapa, Veracruz.

Resultados

Identificatión de *Ceratozamia* sp.

Al momento, la especie está en proceso de estudio y descripción.

Composición Florística

De acuerdo con Hernández (2004), la composición florística del BMM de San Agustín Loxicha está dada por setenta y nueve especies arbóreas, distribuidas en 15 familias. De éstas, trece especies crecen cerca de *Ceratozamia* (Tabla 6-1).

Estructura de la Población de *Ceratozamia*

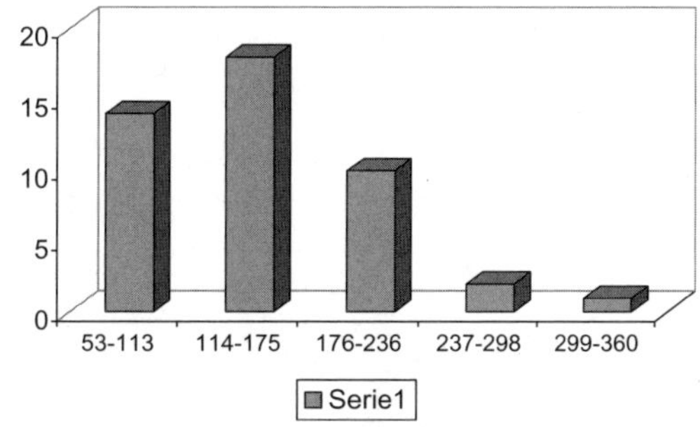

Figura 6-2. Distribución de alturas.

La población consta de 45 individuos con un promedio de altura total con hojas de 86.45 cm. El rango de altura va de 0.53 a 3.56 m.

Para la estructura vertical se obtuvieron cinco intervalos a partir de la altura total (Fig. 6-2), dentro de los cuales, en el primero se encuentran 14 individuos (31.11%), el segundo comprende 18 (40%), el tercero 10 (22.22%), el cuarto 2 (4.44%) y el último 1 (2.22%). La agregación es de tipo I (Martínez & Álvarez Buylla, 1995).

Ubicación Espacial de *Ceratozamia* sp.

Dentro de un cuadrante de 65×50 m, se obtuvo que la población presenta una distribución agregada (χ^2=3856, p<0.001, 33 gl, α 0.05), con un índice de 6.57, una media de

Tabla 6-1 Especies que se localizan cerca de *Ceratozamia*.

Familia	Nombre Científico	Nombre Común
Celastraceae[c]	*Perrottetia longistylis* Rose[a]	Palo de leche
Convolvulaceae	Ipomoea santillanii O'Donell	Desc. 1
Fagaceae[b]	*Quercus* aff. *insignis* M. Martens & Galeotti[a]	Encino
Fagaceae	*Quercus* aff. *paxtalensis* CH Mull[a]	Encino blanco
Fagaceae	*Quercus* sp.[a]	encino delgado
Meliaceae	*Trichillia* sp.	Yegalache
Onagraceae	*Fucsia arborescens* Sims.	Rosado
Rubiaceae[c]	*Rogiera* sp.	Colorado
	Rogiera sp.	Hoja rosada
	Rogiera gratísima Lindes ex Planch	Pimientillo
Thymelaceae	*Daphnopsis* sp.[a]	Palo de siete

Source: Hernández, 2004.

[a]Géneros de plantas vasculares que tienen al menos una especie que habita exclusiva o preferentemente en el BMM (*sensu* Rzedowski, 1996).

[b]Elementos Holárticos.

[c]Elementos Tropicales y Neotropicales (Rzedowski, 1996; Mayorga *et al*, 1998).

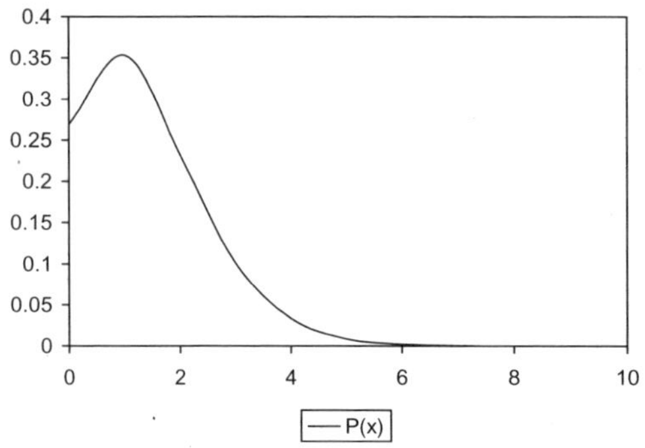

Figura 6-3. Correlación de la probabilidad de localizar número de individuos dentro de un cuadrante.

1.28 y una varianza de 8.45 (Fig. 6-3). El intervalo de confianza es de (74.25–15.74). Dentro de un sub-cuadrante es más probable encontrar un solo individuo, probabilidad que va disminuyendo conforme se incrementa el número de individuos (Fig. 6-4).

Distribución Geográfica de *Ceratozamia* en Oaxaca

De la revisión de Herbario, se obtuvieron 21 registros, los cuales corresponden a siete especies: *Ceratozamia euryphyllidia* Vazquez Torres, Sabato y Stevenson (MEXU, col. Jaime

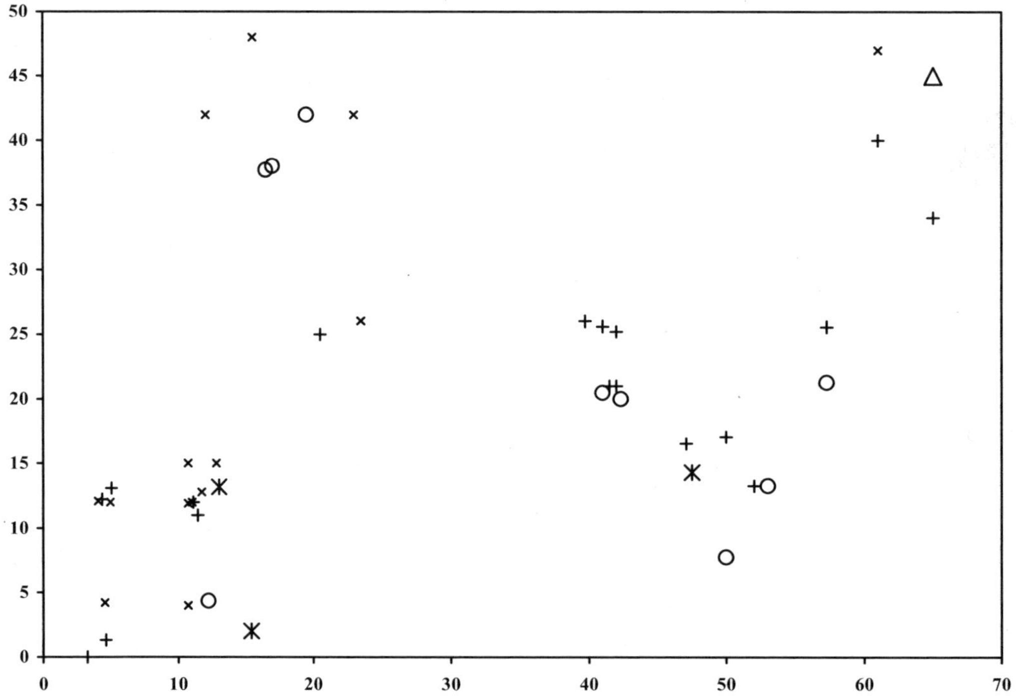

Figura 6-4. Mapa de la ubicación espacial de la población de *Ceratozamia* spp. (×) 53–113 (+) 114–175 (°) 176–236 (*) 237–298 (Δ) 299–360.

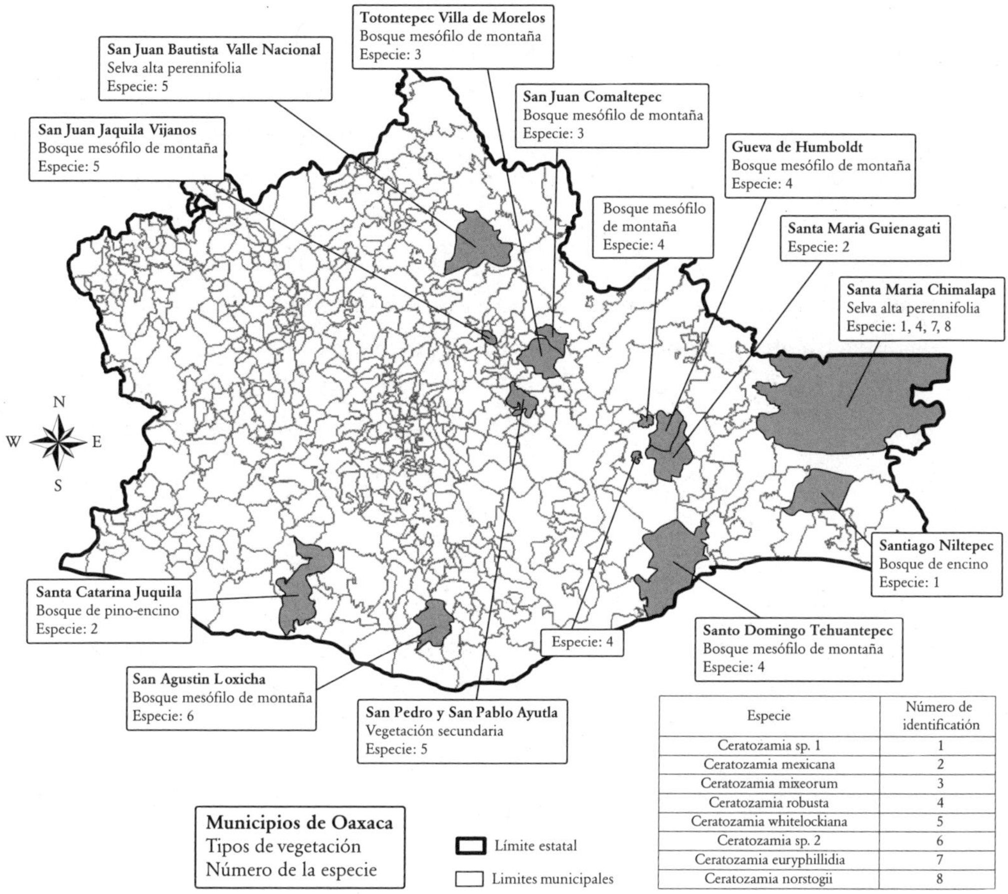

Especie	Número de identificación
Ceratozamia sp. 1	1
Ceratozamia mexicana	2
Ceratozamia mixeorum	3
Ceratozamia robusta	4
Ceratozamia whitelockiana	5
Ceratozamia sp. 2	6
Ceratozamia euryphillidia	7
Ceratozamia norstogii	8

Figura 6-5. Mapa de distribución del género *Ceratozamia* para el estado de Oaxaca.

Rivera; XAL, col. A.P. Vovides), *C. norstogii* DW Stevenson (XAL, S. Salas, C. Torres, H. Morales), *C. matudae* Lundell, Lloydia (MEXU, col. E. Hernández Xolocotzi), *C. robusta* Miquel (MEXU, col. A. Campos, R. Torres), *C. mexicana* Brongn (UAMI; ENCB, col. J. Rzedowski y R. McVaugh; XAL, col. R. García, M. Cerón, S. Salas), *C. mixeorum* Gregory, Chemnick y Salas (MEXU, col. E. Velasco López, E. Vargas Ruiz, R. Torres, R. López Luna, J. Rivera Reyes, W. Ruiz) y *C. whitelockiana* Chemnick y Gregory (XAL, col. X. Munn, R. Del Castillo, R. Rivera), así como dos ejemplares que no se encontraban identificados. De estas especies, dos de ellas son endémicas para la entidad (*Ceratozamia mixeorum* y *C. whitelockiana*).

Cabe mencionar que de las siete especies que se encuentran reportadas para el estado, al menos una, *C. matudae*, se trata de una nueva especie que esta en proceso de descripción, pero en la colecta se encuentra identificada de forma errónea.

Se observa que la gran mayoría de los registros y las especies se localizan al norte y al este de Oaxaca, en los límites con Veracruz y Chiapas (Figura 6-5). Hacia el sur, únicamente se encuentran tres registros de *Ceratozamia*. Con relación a la época en que fueron colectados, la mayoría de los ejemplares datan de los años 30's hasta los 80's.

En cuanto al tipo de vegetación donde se localizan las especies, predomina el bosque mesófilo de montaña, aunque también habitan en bosque de encino, bosque de pino-encino, selva alta perennifolia y vegetación secundaria, en un rango altitudinal que va de 630 a los 2000 m.

Usos Etnobotánicos

Todas las especies son utilizadas ocasionalmente como ornamentales (Jones 1993) y el cono femenino de *Ceratozamia mexicana* es usado como insecticida (www.ine.gob.mx). En específico, para el municipio de San Agustín Loxicha, esta especie llega a ser utilizada en ceremonias religiosas, y principalmente como planta ornamental, ya que algunos campesinos las trasladan al pueblo.

Discusión

El sureste de México es considerado como un importante centro de especiación y diversificación de cicadas, en particular del género *Ceratozamia* (Rzedowski 1996; Rzedowski & Palacios, 1977; González & Vovides, 2002). Estas especies se encuentran comúnmente en sitios considerados paloerefugios, como el BMM, que se caracteriza por reunir taxones paleoendémicos y neoendémicos (Rzedowski & Palacios, 1977). Los resultados obtenidos por Hernández (2004) corroboran este hecho, debido a la presencia de cuatro especies que se distribuyen exclusiva o preferentemente dentro de este tipo de vegetación.

Al parecer, la población de *Ceratozamia* se encuentra conservada, con poco o casi nulo disturbio antropogénico, debido al poco impacto que ha tenido por parte de los pobladores de la localidad, además de que no le dan muchos usos. Esto se correlaciona con el hecho de que la vegetación circundante también se encuentra en buen estado de conservación.

La población presenta una distribución agregada, con evidente formación de manchones, esto probablemente debido que las semillas se dispersan por gravedad y las plántulas se establecen en el fondo de las barrancas, cerca de la planta madre o en la sombra de los árboles. Sin embargo, también pueden existir otros factores, como la competencia entre las plantas circundantes o entre las mismas cicadas, o la profundidad del suelo, los cuales podrían estar jugando un papel importante. Asimismo, las condiciones micro climáticas, de las laderas donde se ubican pueden ser más adecuadas para los individuos. Por otro lado, se ha reportado que algunos mamíferos como el pecarí (*Tayassu tajacu*), ratones (*Peromyscus mexicanus*), el coatí (*Nasua narica*; Pérez-Farrera et al., 2000; Vovides et al., 2003, 2004) y algunas aves (*Geoco coyxvelox*; Chemnick, en prensa) pueden dispersar las semillas, lo que también influye en la distribución de las plantas en la población.

El análisis de la distribución de tallas muestra la existencia un mayor número de plantas jóvenes, esto es, que hay un mayor número de organismos que no exceden de un metro y medio de altura. En este sitio no se observaron plántulas, aunque la mayoría de los organismos presentaron tallas menores a un metro, revelando así una gráfica de tipo I (Martínez & Álvarez-Buylla, 1995).

Para el estado de Oaxaca existen siete registros, con la especie de San Agustín Loxicha, los registros serian ocho, de los cuales tres de ellos se encuentran hacia la vertiente del Pacifico. Por otro lado, cabe mencionar que algunas de estas especies podrían encontrarse mal identificadas, como el caso de *Ceratozamia matudae*, la cual se trata de una nueva especie en proceso de descripción, pero en el herbario de MEXU esta reportada con este nombre; debido a esto, es necesario una revisión más detallada. Así también, se deben realizar más estudios, no solo en esta zona sino en todo el estado.

El género *Ceratozamia*, debido a que no es muy singular por su apariencia y que comparte su hábitat con poblaciones de palmas, sus especies no tienen muchos usos por parte de los pobladores, puesto que estos prefieren las palmas a estas especies, aunque dentro de San Agustín los pobladores prefieren el uso de las hojas de *Ceratozamia* para las ceremonias, debido a que son más duraderas. Pero, por otro lado, para los saqueadores si representan un gran valor económico (Vovides, 1990), por lo que sus poblaciones se ven diezmadas debido a estos saqueos.

En San Agustín Loxicha, la población presenta un alto grado de conservación debido a que se localiza en un sitio que presenta dificultad en su acceso, no es apto para actividades productivas como agricultura o ganadería y, a pesar del huracán Paulina, el área donde se localiza la población no resultó perturbada. Además, esta especie no resulta de interés económico, medicinal o alimenticio para la población. Existen pocos registros de su uso como planta ornamental, lo que redunda en una conservación de la población.

Conclusiones

La Sierra Madre del Sur está considerada como centro de endemismos en muchos grupos (Ramamoorthy et al., 1998), así como zona de paleorrefugios (Rzedowski & Palacios, 1977), esto debido a la historia geológica del estado. En Oaxaca se han reportado dos especies endémicas del género *Ceratozamia*, y, últimamente se han tenido más registros, de los cuales algunos son nuevos. A partir de esto, se puede inferir que aún puede haber más especies que no se han descrito, por lo que esta zona podría ser un centro de endemismos.

La presencia de especies endémicas dentro del BMM de San Agustín Loxicha, así como la población de *Ceratozamia* sp., permite inferir que se trata de una localidad bien conservada, además de ser una potencial zona de paleorrefugio (Vovides, com. pers.).

Debido a los pocos registro que se tienen hacia la SMM, es un indicador de que aún se requieren estudios para esta parte.

Agradecimientos

Agradecemos a CONACYT por el financiamiento recibido mediante el proyecto R-28234, "Evaluación de los Recursos Vegetales de Tres Municipios Zapotecos en la Sierra Madre del Sur de Oaxaca"). A los habitantes del Municipio de San Agustín Loxicha por el apoyo brindado. Al Dr. Andrew Vovides, Dr. Manuel Castillo, Biol. Gilberto Hernández Cárdenas, M. en C. Amelia Cornejo Romero, a los Biol. Ma. del Consuelo Aragón Martínez, Mireya Hernández Hernández, Gabriela Contreras Bernal, Yalina Ventura Aquino, Juana Cardoso Díaz y Humberto Quijano Pérez.

Literatura Citada

Aragón, M. C. 2004. Estructura y composición florística de la asociación selva mediana con pino-encino en Santa María Magdalena, Pluma Hidalgo, Oaxaca. Reporte de Servicio Social. Universidad Autónoma Metropolitana-Iztapalapa. México.

Arriaga, L., J. M. Espinosa, C. Aguilar, E. Martínez, L. Gómez & E. Loa (coord.). 2000. Regiones Terrestres Prioritarias de México. Comisión Nacional para el Conocimiento y Uso de la Biodiversidad. México.

Brower, J., J. Zar & C. von Ende. 1989. Field and laboratory methods for general ecology. 3rd ed. Brown, Dubuque, IA.

Centeno-García, E. 2004. Configuración geológica del estado. Pp. 29–42 *in* A. J. García-Mendoza, M. J. Ordóñez y M. Briones-Salas (eds.), Biodiversidad de Oaxaca. Instituto de Biología, UNAM-Fondo Oaxaqueño para la Conservación de la Naturaleza-World Wildlife Fund, México.

González, D. & A. P. Vovides. 2002. Low interlinage divergence in *Ceratozamia* (Zamiaceae) detected with nuclear ribosomal DNA ITS and chloroplast DNA trnL-F non-coding region. Systematic Botany 27: 654–661.

Greig-Smith, P. 1983. Quantitative plant ecology. 3rd ed. Blackwell Scientific Publications, Oxford.

Hernández Hernández Mireya. 2004. Evaluación y composición florística de un bosque mesófilo de montaña en la sierra madre del sur de Oaxaca. Reporte de Servicio Social. Universidad Autónoma Metropolitana-Iztapalapa. México.

INEGI. 1986. Mapa de uso de suelo vegetación escala 1:1000 000. Instituto Nacional de Estadística, Geografía e Informática.

———. 1988. Carta Hidrológica, edafológica y geológica escala 1:250 000, Oaxaca D14.3

Jones, David L. 1993. Cycads of the world. Smithsonian Institution Press, Washington, D C.

Luna-José, A. L. 2001. Análisis del conocimiento etnobotánico entre los zapotecos de la comunidad Trinidad Buenavista Loxicha, Oaxaca. Reporte de Servicio Social. Universidad Autónoma Metropolitana-Iztapalapa. México.

Pérex-Farrera, M. A., P. F. Quintana-Ascencio & B. Salvatierra Izaba. 2000. Population dynamics of *Ceratozamia matudae* Lundell (Zamiaceae) in el Triunfo Biosphere Reserve, Chiapas, México. Journal of the Torrey Botanical Society 127(4): 291–299.

Ramamoorthy, T. P., R. Bye, A. Lot & J. Fa (comp.). 1998. Diversidad Biológica de México: Orígenes y Distribución. Instituto de Biología, UNAM.

Rzedowski, Jerzy. 1996. Análisis preliminar de la flora vascular de los bosques mesófilos de montaña de México. Acta Botánica Mexicana 35: 25–44.

———. 1988. Vegetación de México. Editorial Limusa. México.

Rzedowski, Jerzy & C. Palacios. 1977. El bosque de *Engelhardtia* (Oreomunnea) *mexicana* en la región de la Chinantla (Oaxaca, México): una reliquia del Cenozoico. Bol. Soc. Bot. Mex. 36: 93–123.

Toledo, V. M. 1982. Pleistocene changes of vegetation in tropical México. Pp. 93–111 *in* G. T. Prance (ed.), Biological Diversification in the tropics. Proceedings of the Fifth International Symposium of the Association for Tropical Biology. Columbia University Press, New York.

Vovides A. P. 1990. Spatial distribution, survival and fecundity of Dioon edule (Zamiaceae) in a tropical deciduous forest in Veracruz, México, with notes on its habitat. American Journal of Botany 77: 1532–1543.

———. 2000. México: segundo lugar mundial en diversidad de cycadas. Biodiversitas XXXI, Año 6, Número 31.

———, M. A. Pérez-Farrera, J. González-Astorga, D. González, T. Gregory, J. Chemnick, C. Iglesias, P. Octavio-Aguilar, S. Avendaño, C. Bárcenas & S. Salas-Morales. 2003. An outline of our current knowledge on Mexican cycads (Zamiaceae, Cycadales). Plant Biology 4: 159–174.

———, ———, D. González & S. Avendaño. 2004. Relationships and phytogeography in *Ceratozamia* (Zamiaceae). Pp. 109–125 *in* T. Walters & R. Osborne (eds.), Cycad classification: concepts and recommendations. CAB International.

Wendt, T. 1987. Las selvas de Uxpanapa, Veracruz-Oaxaca, México: evidencia de refugios florísticos Cenozoicos. Anales del Instituto de Biología UNAM-Serie Botánica 58, 29–54. www.ine .gob.mx

ECOLOGY
CHAPTER 7

Herbivores and Chemical Defenses in Cycads—An Ecological and Evolutionary Approach

Citlalli Castillo-Guevara

Abstract

This project investigates cycad herbivory from separate ecological and evolutionary perspectives. First, I report results of an ecological study of the interactions between the caterpillars *Eumaeus minyas* and *E. debora* (Lepidoptera: Lycaenidae) and their host cycads, *Zamia loddigesii* and *Dioon edule* (Cycadales: Zamiaceae), in an oak forest near Chavarrillo, central Veracruz, México. I describe herbivory levels recorded over a two-year period. Second, I explore the association between herbivory (by caterpillars, weevils, and other insects) and the amount of azoxyglycosides (macrozamin and cycasin) in the Cycadales by using the phylogenetic independent contrasts technique. The azoxyglycosides may play an important ecological role as an antiherbivore defense.

Resumen

Este proyecto investiga la herbivoría en cícadas desde perspectivas separadas ecológica y evolutiva. Primero, se reportan resultados de un estudio ecológico de las interacciones entre orugas, *Eumaeus minyas* y *E. debora* (Lepidoptera: Lycaenidae), y sus cícadas hospederas *Zamia loddigesii* y *Dioon edule* (Cycadales: Zamiaceae), en un bosque de encino cerca de Chavarrillo, en el centro de Veracruz, México. Se describen niveles de herbivoría registrados durante un período de dos años. Segundo, se explora la asociación entre la herbivoría (por orugas, gorgojos y otros insectos) y la cantidad de azoxiglicósidos (macrozaminas y cicasinas) en el grupo Cycadales usando técnicas filogenéticas de contrastes independientes. Los azoxiglicósidos pueden haber jugando un papel ecológico importante como defensas antiherbívoros.

Introduction

Herbivory is an antagonistic interaction in which the herbivore consumes vegetable tissue solely for its own benefit (Dirzo, 1987). This has a potential effect on the survival and reproductive success of the host plants and, consequently, on their population dynamics (Crawley, 1983; Dirzo, 1984; Marquis, 1992). However, studies have demonstrated several levels of plant tolerance to herbivory and compensatory responses to foliage damage, depending on the intensity of the damage and the time elapsed after defoliation (Lennartsson et al., 1997; Mabry & Wayne, 1997; Strauss & Agrawal, 1999). It has been suggested that herbivory is one of the most important selective pressures generating biodiversity of terrestrial plants (Ehrlich & Raven, 1964). Insects are considered the most important of the herbivores in terms of number of taxa, biomass, and transferred mass, as well as for their evolutionary impact on plants (Lovett et al., 2002; Schultz, 2002). Lepidopteran larvae have typically been used as a model of plant-herbivore interactions. Their feeding behavior can have important effects on ecological and/or evolutionary aspects of the host plants, including the production of secondary metabolites (Köpf et al., 1998; Janz et al., 2001). Larvae of *Eumaeus* species (Lycaenidae) feed exclusively on leaves and female reproductive cones of cycads (Clark & Clark, 1991) and can sequester cycasin from these plants (Rothschild et al., 1986; Bowers & Larin, 1989; Bowers & Farley, 1990; Castillo-Guevara, 2002; Castillo-Guevara & Rico-Gray, 2002). This feeding dependence may play an important role in the ecological, demographic, chemical, or evolutionary features of cycads, but studies addressing these relationships are scarce in the literature (DeVries, 1976, 1977, 1983; Clark & Clark, 1991; Beltrán, 1993; Smith, 2000). In this research, I study the ecology of herbivory by *Eumaeus* larvae on two Mexican cycads, *Zamia loddigesii* Miq. and *Dioon edule* Lindl., under natural conditions. Then, in the second part of this research, I examine evolutionary aspects of the association between the herbivores and the amount of azoxyglycosides present in the host cycads (Castillo-Guevara, 2002; Castillo-Guevara & Rico-Gray, 2003).

Part I—Ecological Approach

I studied the ecology of the interactions between the caterpillars *Eumaeus minyas* and *E. debora* and their host cycads, *Z. loddigesii* and *D. edule,* in an oak forest in central Veracruz, México. Here, I describe herbivory levels recorded over two years.

Material and Methods

Study Site

Fieldwork was carried out in a tropical oak forest in central Veracruz, México (19°24'N, 96°48'W, altitude 1000 m a.s.l.). Mean annual precipitation is 1110 mm, and mean annual temperature is 24.5 °C. The climate is mild and humid throughout the year, with a dry and cold season from October to May. This area has been intensively deforested, but remnants still exist of *Quercus oleoides* (Fagaceae), *Brahea dulcis* (Arecaceae), and the cycads *Z. loddigesii* and *D. edule*.

Cycads

A total of 210 cycad plants of both species (*D. edule* = 99 and *Z. loddigesii* = 111) were marked. Plants were sorted into three categories of size according to leaf length. To evaluate the damage by herbivores, I calculated an herbivory index (HI) for each plant, as proposed by Domínguez et al. (1989). This index uses the percentage of foliage area consumed, assigning categories to each frond of the plant as follow: 0 = no damage; 1 = 1–10% damaged; 2 = 11–25%; 3 = 26–50%; 4 = 51–70%; and 5 = 71–100%. Every two weeks over a two-year period I registered the presence of *Eumaeus* or other herbivores observed consuming cycads.

Results

Herbivory

I found significant differences in HI between the cycad species ($t = -3.75$, DF = 208, $P = 0.00$). *Zamia loddigesii* was more attacked by *Eumaeus*. *Eumaeus debora* has only one host plant, *D. edule,* and *E. minyas* feeds on both *D. edule* and *Z. loddigesii*. Adult *Eumaeus* oviposit on immature fronds, but may also oviposit in female reproductive cones.

Dioon edule

I found that 19% of the plants were infested by *Eumaeus* caterpillars, with foliar damage from 26% to 50% per plant. Throughout the study, only two plants were reinfested, and on one host plant two clusters of eggs were recorded in the same month. Among all of the infected plants, 10% were infested by *E. debora*, and 9% by *E. minyas*, but there were no significant differences in HI (t=1.79, DF=17, P=0.09). Plants with medium (0.34–0.67 m) and large (0.68–1.33 m) fronds were more consumed (χ^2_2=10.47, P<0.01). The only other herbivore observed was *Aulacoscelis melanocera* (Chrysomelidae: Aulacoscelinae), which consumed immature cycad leaves and *E. debora* pupae wastes.

Zamia loddigesii

In this species, 56.7% of the plants were infested by *E. minyas* and 75–100% of the foliage area was consumed. Eight plants were reinfested. Plants of small size (0.10–0.33 m) (χ^2_2=9.75, P<0.01) and 1–2 leaves (χ^2_2=10.47, P<0.01) were consumed more that larger plants.

Conclusions

This is the first report of a *E. debora-D. edule* interaction. *Eumaeus debora* has only one host plant, *D. edule,* whereas *E. minyas* uses both *D. edule* and *Z. loddigesii*. There were significant differences in HI between cycad species: *Z. loddigesii* was more infested by *Eumaeus*. Some adult *Eumaeus* oviposit on female reproductive cones, which are also damaged by larvae. Density and survival of *Eumaeus* are low. The impact of herbivory on the plants seemed to be minor, but longer term studies are necessary to be certain of the interaction.

Part II—Evolutionary Approach

Macrozamin and cycasin are toxic azoxyglycosides specific to cycads. Caterpillars (Lepidoptera) and weevils (Coleoptera) feed on cycad root, stem, leaf, and reproductive tissues. Azoxyglycosides may play an important ecological role as antiherbivore defenses. I evaluated the association between herbivory and the amount of azoxyglycosides in cycads by using the phylogenetic independent contrasts technique. I hypothesized that the type of herbivory should be related to the presence of macrozamin and cycasin; thus, herbivory should be lower in species with higher concentrations of azoxyglycosides.

Material and Methods

Distribution of Azoxyglycosides among Cycads

I gathered information available in the literature about the macrozamin and cycasin content, life form, geographic distribution, height, and seed volume of 30 cycad species (De Luca et al., 1980; Moretti et al., 1983; Vovides et al., 1993; Norstog & Nicholls, 1997; Jones, 2000) (Table 7-1). Then, I performed simple linear regressions between plant characteristics (height and seed volume) and azoxyglycosides content to assess possible allometric effects (Table 7-2).

Herbivory Types among Cycads

There is a general lack of information on specific interactions in cycads; most are on herbivory by lepidopteran caterpillars (Appendix I). I found 22 references in the literature about herbivorous insects in ten genera of cycads (Castillo-Guevara & Rico-Gray, 2003).

Cycad Phylogeny

I constructed a phylogenetic tree based on the most parsimonious trees of Bogler and Francisco-Ortega (2004) and Moretti et al. (1993). Owing to the absence of molecular information, some species were left as polytomies (Fig. 7-1).

Association between Azoxyglycosides and Herbivores

To assess the evolutionary association between percent content of azoxyglycosides and herbivory, and to test the hypothesis that higher concentrations of azoxyglycosides in plants result in less damage to the plants, I used a comparative analysis by independent contrasts (CAIC; Purvis & Rambaut, 1995). Contrasts were computed by using Crunch and Brunch algorithms (Purvis & Rambaut, 1995).

Table 7-1 Database used for the comparative analysis: data were obtained from the literature and reviewed for the azoxyglycoside content, plant height, female cone volume, seed volume, and herbivory type of cycad species around the world.

Key: na = not available; – = female *Cycas* plants do not produce true cones but have loosely arranged clusters of megasporophylls.

Species	Macrozamin (% fresh weight)	Cycasin (% fresh weight)	Height (m)	Female cone volume (cm^3)	Seed volume (cm^3)	Herbivory type
Bowenia serrulata	4.33	0.06	1.5	378	6.28	1
Bowenia spectabilis	5.04	0.42	1.5	416	4.81	1
Ceratozamia matudae	1.05	0.08	0.5	510	6.65	1
Ceratozamia mexicana	1.01	0.02	1	603	13.26	3
Cycas basaltica	0.29	0.12	3	–	5.44	2
Cycas cairnsiana	0.2	0.1	3	–	19.35	2
Cycas circinalis	0.42	0.1	5	–	22.08	2
Cycas lane-poolei	0.45	0.72	5	–	58.31	2
Cycas pruinosa	0.33	0.1	2	–	19.35	2
Cycas revoluta	0.26	0.28	3	–	5.65	2
Cycas thouarsii	0.31	0.06	10	–	58.31	2
Dioon califanoi	na	0.04	3	11928	9.41	1
Dioon edule	0.64	0.12	4	3239	7.42	3
Dioon purpusii	na	0.02	5	6953	8.87	1
Encephalartos altensteinii	2.11	0.07	5	18808	8.82	2
Encephalartos ferox	2.1	0.05	1.5	7289	7.61	2
Encephalartos lebomboensis	2.16	0.05	4	11265	7.33	2
Encephalartos umbeluziensis	2.86	0.08	1.5	2624	7.53	2
Encephalartos villosus	2.09	0.08	1.90	2385	5.19	2
Lepidozamia peroffskyana	1.11	0.21	7	23758	30.41	1
Macrozamia diplomera	2.41	0.16	0.9	742	6.62	1
Macrozamia fawcettii	2.49	0.06	0.8	1583	7.28	1
Macrozamia heteromera	na	0.08	0.6	1439	7.15	1
Macrozamia miquelii	3.88	0.09	0.65	1040	4.00	1
Macrozamia moorei	3.72	0.08	8	1225	23.56	1
Microcycas calocoma	0.13	0.07	10	7706	9.94	1
Stangeria eriopus	4.7	0.02	1.5	720	8.61	1
Zamia furfuracea	1.01	0.28	0.6	481	1.33	3
Zamia integrifolia	na	0.38	1.05	32	2.06	3
Zamia latifolia	1.02	0.02	1.1	26	1.10	1

Macrozamin and cycasin (%) in seeds, taken from De Luca et al. (1980) and Moretti et al. (1983); other data are derived from Vovides et al. (1993), Norstog and Nicholls (1997), Jones (2000); nomenclature follows Moretti et al. (1983) and Vovides et al. (1983).

Table 7-2 Regression analyses with and without comparative analysis by independent contrasts (CAIC).

	Comparisons *without* CAIC			Comparisons *with* CAIC		
Variables	r^2	*F*	*P*	r^2	*F*	*P*
Plant height vs. seed volume	0.414	19.793	0.0001	0.433	9.931	0.007
Plant height and seed volume residuals vs. macrozamin	0.003	0.084	0.774	0.006	0.083	0.777
Plant height and seed volume residuals vs. cycasin	0.004	0.100	0.754	0.009	0.141	0.713
Herbivory vs. macrozamin	0.21	6.365	0.018	0.004	0.054	0.819
Herbivory vs. cycasin	0.069	2.076	0.160	0.049	0.779	0.391
Macrozamin vs. cycasin	0.024	0.594	0.448	0.038	0.238	0.642
Plant height vs. seed volume	0.414	19.793	0.0001	0.433	9.931	0.007

Results

Mean amounts of macrozamin are usually significantly larger (mean ± SE = 1.774 ± 0.299, n=26) than mean quantities of cycasin (0.134 ± 0.027, n=30) (t=7.794, DF=54, P=< 0.0001) (Fig. 7-1).

The mean macrozamin content differed significantly (Kolmogorov-Smirnov, d_{max}=0.650, DF=2, P=0.0405) between Old World (2.063 ± 0.363, n=20) and New World (0.810 ± 0.150, n=6) species. However, cycasin content did not differ significantly (Kolmogorov-Smirnov, d_{max}=0.397, DF=2, P=0.2750) between Old World (0.142 ± 0.035, n=21) and New World (0.114 ± 0.043, n=9) species (Fig. 7-1).

Azoxyglycoside content differed significantly (analysis of variance [ANOVA], $F_{1,24}$=14.650, P=0.0008) between plant life forms, macrozamin being higher in shrubby (2.615 ± 0.405, n=13) than in arborescent (0.933 ± 0.300, n=13) life forms. However, cycasin did not differ significantly (ANOVA, $F_{1,28}$=0.137, P=0.7141) between shrubs (0.125 ± 0.033, n=15) and arborescents (0.143 ± 0.045, n=15) (Fig. 7-1).

Regression Analyses with and without CAIC

I found a positive significant association between plant height and seed volume (r^2=0.414) (Table 7-2). To allow for this allometric relationship, I regressed the residuals

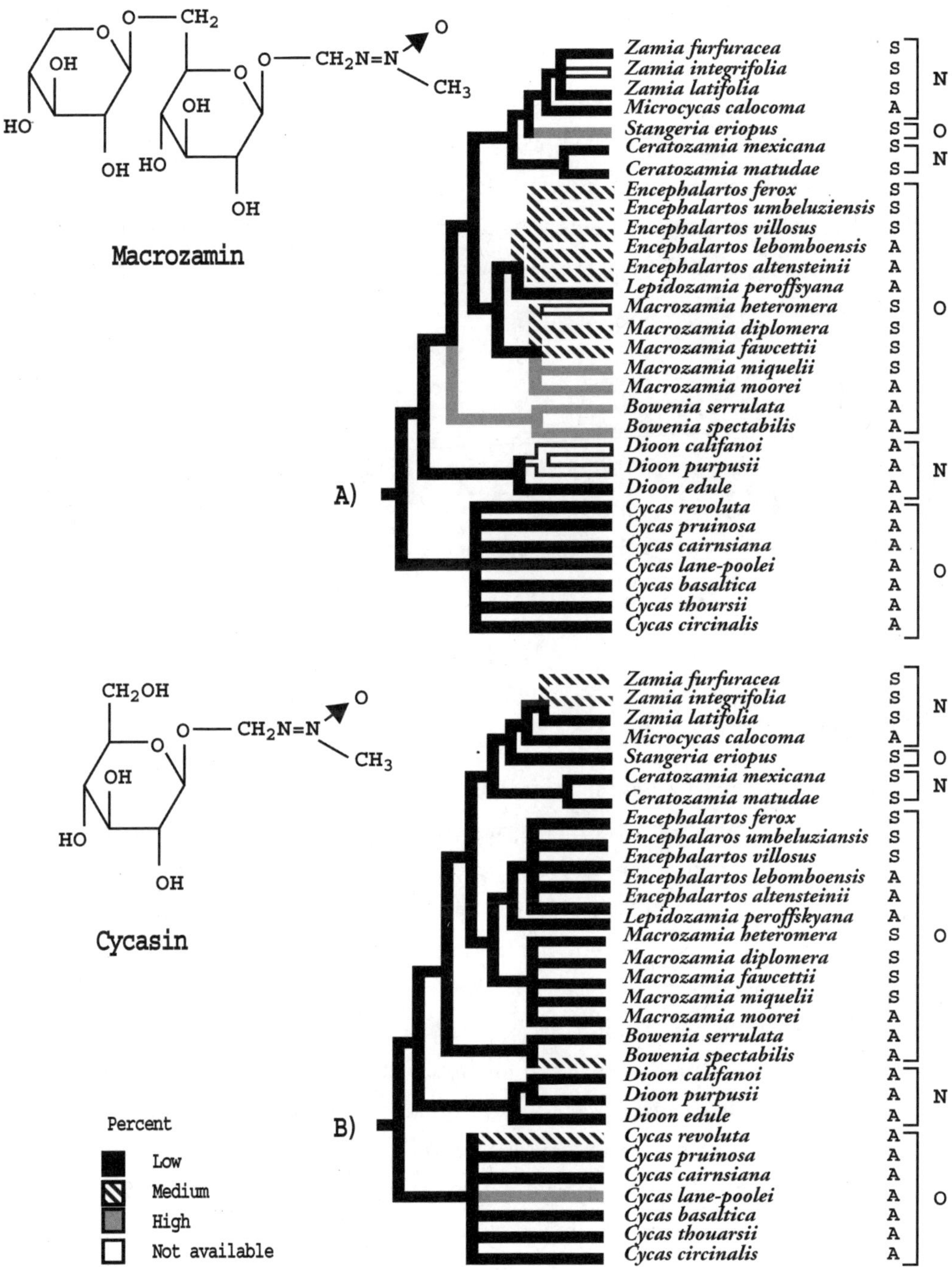

Figure 7-1. Phylogenetic relations among cycad species, based on Bogler and Francisco-Ortega (2002). The figure shows the relations among 30 species for which data on the percentage of azoxyglycosides were available. Life form: S (shrubby), A (arborescent). Distribution: N, New World species; O, Old World species. Structural diagrams show molecules of macrozamin and cycasin. A. Percent macrozamin: ■, low (0.13–1.76); □, medium (1.77–3.40); ■, high (3.41–5.04); □, not available. B. Percent cycasin: ■, low (0.01–0.25); □, medium (0.26–0.50); ■, high (0.51–0.72).

of that association against the percent of macrozamin and cycasin. In both cases, the relationship was not significant ($P>0.05$). I then directly compared herbivory type with percent of cycasin and macrozamin; the former was not significant, but the latter showed a negative significant association. Using CAIC and simple linear regression, I found that the associations between herbivory type and percent of macrozamin and cycasin in cycads were not significant (Table 7-2).

Herbivores and Other Variables

I did not find any significant differences ($H=5.225$, $DF=2$, $P=0.0734$) between plant height and herbivory type (type $1=3.004\pm3.176$, $n=14$; type $2=3.750\pm0.684$, $n=12$; type $3=1.662\pm0.786$, $n=4$). I found that herbivory type is independent of life form (shrubby or arborescent) ($\chi^2_{0.05,2}=5.142$), but not of distribution (Old or New World species) ($\chi^2_{0.05,2}=14.692$).

Conclusions

Herbivory type and macrozamin in this study were negatively correlated, suggesting a possible defensive function of macrozamin against herbivores. However, when analyzed using independent contrasts, thus removing any historical effect, the association disappears. This suggests that the presence of metabolites in plants evolved for some other reason and is possibly maintained by phylogenetic inertia. The presence of macrozamin might thus be considered an exaptation, a characteristic that currently increases fitness but originally evolved for a different purpose, in contrast to an adaptation, a characteristic that increases fitness and that evolved and was selected for to play its current role (Gould & Vrba, 1982; Losos & Miles, 1994). Azoxyglycosides may represent the products of a primitive biochemical pathway designed to store nitrogen (Siniscalco Gigliano,1990), and cycads probably retained these products because of their toxic effects, which may have played an important ecological role during the long evolutionary history of cycads (Siniscalco Gigliano, 1990).

This analysis further shows that there is no relationship between cycasin and geographic distribution, life form, or herbivory type. Macrozamin is high among Old World, shrubby cycads and it may have independently and repeatedly (*Bowenia, Macrozamia, Stangeria*) increased over evolutionary time.

More studies are needed on azoxyglycoside content in order to evaluate and quantify the effect of herbivory on cycad fitness. It is also important to quantify the concentrations of azoxyglycosides in more cycad species and to include environmental factors such as vegetation type and edaphic features.

Concluding Remarks

At the study site, *E. minyas* and *E. debora* butterflies use cycads as host plants on which to lay their eggs, and their larvae are the main herbivores of these plants. These data are consistent with the findings of Clark and Clark (1991) and Castillo-Guevara and Rico-Gray (2002). The growth of cycads and their production of fronds and reproductive cones are related to optimal environmental conditions (Clark & Clark, 1987, 1988, 1991; Ornduff, 1992; Norstog & Nicholls, 1997). The plants can enter prolonged states of repose, and these periods can last several years, during which they produce neither fronds nor cones. Owing to these characteristics, a decrease in the foliage area can affect not only plant structure but also plant survival over the long term. As a result, herbivory of these species might be underestimated, and might be worse than the calculated index. Long-term studies are necessary to evaluate the role of herbivory in cycad fitness.

With regard to the evolutionary approach, these results represent the first comparative analysis of the association between herbivory type and azoxyglycoside content in cycads. They showed a negative relation between herbivory type and percent macrozamin, but this relation disappeared when phylogeny was taken into account. A nonsignificant association was found, however, between cycasin content and herbivory type, whether or not phylogeny was taken into account. An increase in macrozamin content was mainly observed among shrubby, Old World species. The presence of macrozamin may be an exaptation that today plays an important role in the defense against herbivores. More studies are needed on azoxyglycoside content to evaluate and quantify the effect of herbivory on cycad fitness (for example, by guild herbivores). To compare azoxyglycoside content, it is also important to quantify the concentration of azoxyglycosides in more species and take into account environmental factors.

Acknowledgments

This research was supported by the Instituto de Ecología, A.C. (902–16), and a CONACyT (112677) graduate scholarship to CCG. I thank Víctor Rico-Gray and Juan Francisco Ornelas for their help throughout the study, and Carlos Lara for his comments on an earlier version of the manuscript. Comments by Roy Osborne and Andrew Vovides and an anonymous reviewer greatly improved this paper.

Literature Cited

Alejandre, R. J. 1989. Algunos aspectos de la demografía de *Ceratozamia mexicana* Bröng. (Zamiaceae) en un bosque mesófilo de montaña. B.Sc. thesis. Universidad Veracruzana-Xalapa, México.

Beltrán, V. J. M. 1993. Aspectos cualitativos de la herbivoría *Eumaeus debora* Hubner (Lycaenidae) sobre *Ceratozamia mexicana* Brong. (Zamiaceae) en la localidad de Coacoatzintla, Veracruz. B. Sc. thesis, Universidad Veracruzana-Xalapa, México.

Beutelspacher, C. R. B. 1975. Como hacer una colección de mariposas. Universidad Nacional Autónoma de México, México.

———. 1980. Mariposas diurnas del Valle de México. Ediciones Científicas, L.P.M.M. México.

Bogler, D. J. & J. Francisco-Ortega. 2004. Molecular systematic studies in cycads: evidence from trnL intron and ITS2 rDNA sequences. Bot. Rev. 70: 260–273.

Bowers, M. D. & Z. Larin. 1989. Acquired chemical defense in the lycaenid butterfly, *Eumaeus atala*. J. Chem. Ecol. 15: 1133–1146.

——— & S. Farley. 1990. The behaviour of grey jays, *Perisoreus canadensis*, towards palatable and unpalatable Lepidoptera. Anim. Behav. 39: 699–705.

Castillo-Guevara, C. 2002. Herbivoría en cícadas (Cycadales: Zamiaceae) por licénidos del género *Eumaeus* (Lepidoptera: Lycaenidae). Ph.D. diss., Instituto de Ecología A.C., México.

——— & V. Rico-Gray. 2002. Is cycasin in *Eumaeus minyas* (Lepidoptera: Lycaenidae) a predator deterrent? Interciencia 27: 465–470.

——— & ———. 2003. The role of macrozamin and cycasin in cycads (Cycadales) as antiherbivore defenses. J. Torrey Bot. Soc. 130:206–217.

Clark, D. A. & D. B. Clark. 1987. Temporal and environmental patterns of reproduction in *Zamia skinneri*, a tropical rain forest cycad. J. Ecol. 75: 135–149.

Clark, D. B. & D. A. Clark. 1988. Leaf production and the cost of reproduction in the neotropical rain forest cycad, *Zamia skinneri*. J. Ecol. 76: 1153–1163.

——— & ———. 1991. Herbivores, herbivory, and plant phenology: patterns and consequences in a tropical rain-forest cycad. Pp. 209–225 *in* P. W. Price, T. M. Lewinsohn, G. W. Fernandes & W. W. Benson (eds.), Plant-animal interactions: evolutionary ecology in a tropical and temperate regions. John Wiley and Sons, New York.

———, ——— & M. H. Grayum. 1992. Leaf demography of a neotropical cycad, *Zamia skinneri*. Am. J. Bot. 79: 28–33.

Crawley, M. L. 1983. Herbivory. The dynamics of animal-plant interactions. Blackwell, Oxford, UK.

De Luca, P., A. Moretti, S. Sabato & G. Siniscalco Gigliano. 1980. The ubiquity of cycasin in cycads. Phytochemistry 19: 2230–2231.

DeVries, P. J. 1976. Notes on the behavior of *Eumaeus minyas* (Hubner) (Lepidoptera: Lycaenidae) in Costa Rica. Brenesia 8:103.

———. 1977. *Eumaeus minyas* Hubner: an aposematic lycaenid butterfly. Brenesia 12: 269–270.

———. 1983. Checklist of butterflies. Pp. 654–678 *in* D. H. Janzen (ed.), Costa Rican natural history. University of Chicago Press, Chicago.

Dirzo, R. M. 1984. Herbivory: a phytocentric overview. Pp. 141–167 *in* R. Dirzo & J. Sarukhán (eds.), Perspectives on plant population ecology. Sinauer Associates, Sunderland, MA.

———. 1987. Estudios sobre interacciones planta-herbívoro en Los Tuxtlas, Veracruz. Revista Biol. Trop. 35: 119–131.

Domínguez, C. A., R. Dirzo & S. H. Bullock. 1989. On the function of floral nectar in *Croton suberosus* (Euphorbiaceae). Oikos 56: 109–114.

Donaldson, J. S. 1997. Is there a floral parasite mutualism in cycad pollination? The pollination biology of *Encephalartos villosus* (Zamiaceae). Amer. J. Bot. 84: 1389–1406.

Ehrlich, P. R. & D. H. Raven. 1964. Butterflies and plants: a study in coevolution. Evolution 18: 586–608.

Giddy, C. 1978. Cycads of South Africa. Purnell and Sons, Capetown, South Africa.

González, S. E., R. Dirzo & R. C. Vogt. 1997. Historia Natural de Los Tuxtlas. Universidad Autónoma de México, México.

Gould, S. J. & E. Vrba. 1982. Exaptation—a missing term in the science of form. Paleobiology 8: 4–15.

Howard, F. W., A. Hamon, M. McLaughlin, T. Weissling & S. Yang. 1999. *Aulacaspis yasumatsui* (Hemiptera: Sternrrhyncha: Diaspididae), a scale insect pest of cycads recently introduced into Florida. Florida Entomol. 82: 14–27.

Janz, N., K. Nyblom & S. Nylin. 2001. Evolutionary dynamics of host-plant specialization: a case study of the tribe Nymphalini. Evolution 55: 783–796.

Janzen, D. H. 1983. Costa Rican Natural History. University of Chicago Press, Chicago.

Jones, D. L. 2000. Cycads of the world. Smithsonian Institution Press, Washington, DC.

Klots, A. B. 1951. A field guide to the butterflies of North America, east of the Great Plains. Peterson Field Guide Series. Houghton Mifflin, Boston.

Köpf, A., N. E. Rank, H. Roininen, R. Julkunen-Tiitto, J. M. Pasteels & J. Tahvanainen. 1998. The evolution of host-plant use and sequestration in the leaf beetle genus *Phratora* (Coleoptera: Chrysomelidae). Evolution 52: 517–528.

Lennartsson, T., J. Tuomi & P. Nilsson. 1997. Evidence for an evolutionary history of overcompensation in the grassland biennial *Gentianella campestris* (Gentianaceae). Amer. Naturalist 149: 1147–1155.

Losos, J. B. & D. B. Miles. 1994. Adaptation, constraint, and the comparative method: phylogenetic issues and methods. Pp. 60–98 *in* P. C. Wainwright & S. M. Reilly (eds.), Ecological morphology. University of Chicago Press, Chicago.

Lovett, G. M., L. M. Christenson, P. M. Groffman, C. G. Jones, J. E. Hart & M. J. Mitchell. 2002. Insect defoliation and nitrogen cycling in forests. BioScience 25: 335–341.

Mabry, C. M. & P. W. Wayne. 1997. Defoliation of the annual herb *Abutilon theophrasti*: mechanism underling reproductive compensation. Oecologia 111: 225–232.

Marquis R. J. 1992. The selective impact of herbivores. Pp. 301–325 *in* R. S. Fritz & E. L. Simms (eds.), Plant resistance to herbivores and pathogens. Ecology, evolution, and genetics. University of Chicago Press, Chicago.

McDade, L. A, K. S. Bawa, H. A. Hespenheide & G. S. Hartshura. 1994. La selva. Ecology and natural history of a Neoptropical rain forest. University of Chicago Press, Chicago.

Moretti, A., S. Sabato & G. Siniscalco Gigliano. 1983. Taxonomic significance of methilazoxymethanol glycosides in the cycads. Phytochemistry 22: 115–117.

———, P. Caputo, S. Cozzolino, P. De Luca, L. Gaudio, G. Siniscalco Gigliano & D. W. Stevenson. 1993. A phylogenetic analysis of *Dioon* (Zamiaceae). Amer. J. Bot. 80: 204–214.

Nash, R. J., A. Bell & P. R. Ackery. 1992. The protective role of cycasin in cycad-feeding Lepidoptera. Phytochemistry 31: 1955–1957.

Norstog, K. J. & T. J. Nicholls. 1997. The biology of the cycads. Cornell University Press, New York.

Ornduff, R. 1992. Features of coning and foliar phenology, size classes, and insect associates of *Cycas armstrongii* (Cycadaceae) in the Northern Territory, Australia. Bull. Torrey Bot. Club 119: 39–43.

Purvis, A. & A. Rambaut. 1995. Comparative analysis by independent contrasts (CAIC): an Apple Macintosh application for analysing comparative data. Computer Applic. Biosci. 11: 247–251.

Rawson, G. W. 1961. The recent rediscovery of *Eumaeus atala* (Lycaenidae) in southern Florida. Especially for field collectors. J. Lepid. Soc. 15: 237–245.

Ross, G. N. 1964. An annotated list of butterflies collected in British Honduras in 1961. J. Lepid. Soc. 18: 11–26.

Rothschild, M., R. J. Nash & E. A. Bell. 1986. Cycasin in the endangered butterfly *Eumaeus atala florida*. Phytochemistry 25: 1853–1854.

Schultz, J. C. 2002. How plants fight dirty. Nature 416: 267.

Scott, J. A. 1986. The butterflies of North America. Stanford University Press, Stanford, CA.

Siniscalco Gigliano, G. 1990. Chemotaxonomic significance of MAM glycosides and mucilages in cycads. Mem. New York Bot. Gard. 57: 123–131.

Smith, E. M. 2000. A field study and re-establishment of the butterfly *Eumaeus atala* (Lycaenidae) in Miami-Dade County, Florida. M. Sc. thesis, Florida International University. Miami.

Strauss, S. & A. Agrawal. 1999. The ecology and evolution of plant tolerance to herbivory. Trends Ecol. Evol. 14: 179–185.

Teas, H. J. 1967. Cycasin synthesis in *Seirarctia echo* (Lepidoptera) larvae fed methylazoxymethanol. Biochem. Biophys. Res. Commun. 26: 686–690.

Vovides, A. P., K. J. Norstog, P. K. S. Fawcett, M. W. Duncan, R. J. Nash & D. V. Molsen. 1993. Histological changes during maturation in male and female cones of *Zamia furfuracea* L. fil. (Zamiaceae, Cycadales) and their significance in relation to pollination biology. Bot. J. Linn. Soc. 111: 241–252.

Windsor, D., J. Ness, L. D. Gomez & P. H. Jolivet. 1999. Species of *Aulacosceles* Duponchel and Chevrolat (Chrysomelidae) and *Nomotus* Gorham (Languriidae) feed on fronds of Central American cycads. Coleopterists' Bull. 53: 217–231.

Appendix 7-1 Compilation of the animal species reported in the literature as herbivores on cycads.

Common name	Family	Species	Host	Reference
Coontie hairstreak	Lycaenidae	*Eumaeus atala*	*Zamia integrifolia*	4, 5a, 6a, 12, 13,
			Zamia angustifolia	16,17, 18, 19, 20
Florida atala			*Cycas revoluta*	
			Macrozamia lucida	
Zamia butterfly	Lycaenidae		*Zamia loddigesii*	5a, 6a, 6b, 11, 12,
Cycad butterfly		*Eumaeus minyas*	*Zamia skinneri*	14, 18, 20
White-rim cycad hairstreak			*Zamia fairchildiana*	
			Zamia furfuracea	
			Dioon edule	
	Lycaenidae	*Eumaeus debora*	*Ceratozamia mexicana*	1, 2, 13a, 3b, 9
			Dioon edule	
			Cycas revoluta	
			Macrozamia sp.	
	Lycaenidae	*Eumaeus toxea*	*Zamia loddigesii*	5b, 9
			Zamia skinneri	
	Lycaenidae	*Eumaeus childrenae*	*Ceratozamia mexicana*	9
Tiger moth	Artiidae	*Seirarctia echo*	*Zamia integrifolia*	21
Echo moth				
	Nymphalidae	*Taenaris butleri*	*Cycas* sp.	15
	Nymphalidae	*Taenaris catops*	*Cycas* sp.	15
	Nymphalidae	*Taenaris onolaus*	*Cycas* sp.	15
	Nymphalidae	*Theclinesthes onycha*	*Cycas* sp.	15
Miskins blue butterfly	Nymphalidae	*Theclinesthes miskini*	*Cycas* sp.	12
			Macrozamia sp.	
	Nymphalidae	*Luthrodes cleotas*	*Cycas* sp.	15
Leopard moth		*Zeronopsis leopardina*	*Encephalartos altensteinii*	8, 12
			Encephalartos villosus	
			Encephalartos lebomboensis	
			Eencephalartos umbeluziensis	
Emperor moth		*Bunaea alcinoe*	*Cycas thouarsii*	12
Cycad leaf beetle	Chrysomelidae	*Lilioceris nigripes*	*Cycas media*	12
			Bowenia sp.	
	Chrysomelidae	*Aulacoscelis melanocera*	*Dioon edule*	22
		Aulacoscelis costaricensis	*Zamia fairchildiana*	
	Languriidae	*Nomotus* sp.	*Zamia fairchildiana*	22

Appendix 7-1 (*continued*)

Common name	Family	Species	Host	Reference
Cycad seed weevil	Curculionidae	*Antliarhinus zamiae*	*Zamia* sp.	7, 8, 12
			Encephalartos altensteinii	
			Encephalartos villosus	
			Encephalartos horridus	
			Encephalartos lehmannii	
			Encephalartos princeps	
Cycad seed weevil	Curculionidae	*Antliarhinus signatus*	*Encephalartos altensteinii*	12
			Encephalartos villosus	
			Encephalartos horridus	
			Encephalartos lehmannii	
			Encephalartos princeps	
Cycad weevil	Curculionidae	*Tranes internatus*	*Macrozamia* sp.	12
			Cycas sp.	
			Lepidozamia sp.	
			Encephalartos sp.	
Cycad weevil	Curculionidae	*Calendra sommeri*	*Encephalartos altensteinii*	12
Cycad weevil		*Phaceorynus funerarios*	*Encephalartos* sp.	12
Cycad beetle		*Anadastus* sp.	*Cycas* sp.	12
			Macrozamia sp.	
Gall midge fly	Cecidomyiidae		*Encephalartos* sp.	12
Mealy bugs	Pseudococcidae	*Pseudococcus* sp.	*Macrozamia* sp.	12
Flat brown scale		*Eucalymnatus tessellates*	*Macrozamia* sp.	12
			Cycas sp.	
Nigra scale		*Parasaissetia nigra*	*Macrozamia* sp.	12
			Cycas sp.	
Coconut scale		*Pinnaspis aspidistrae*	*Macrozamia* sp.	12
Scale insects	Diaspididae	*Aulacaspis yasumatsui*	*Dioon* sp.	10
			Encephalartos sp.	
			Microcycas sp.	
			Stangeria sp.	
			Cycas sp.	
			Encephalartos sp.	
Aphids	Aphidae		Cycads	12
Termites			Cycads	12
Spider mites			Cycads	12

The list includes caterpillars of lepidoptera, beetles, weevils, sap suckers, scale insects and other insect species.

References: (1) Alejandre, 1989; (2) Beltrán, 1993; (3a) Beutelspacher, 1975; (3b) Beutelspacher, 1980; (4) Bowers & Larin, 1989; (5a) Clark & Clark, 1991; (5b) Clark et al. 1992; (6a) DeVries, 1976; (6b) DeVries, 1977; (7) Donaldson, 1997; (8) Giddy, 1978; (9) González et al., 1997; (10) Howard et al., 1999; (11) Janzen, 1983; (12) Jones, 2000; (13) Klots, 1951; (14) McDade et al., 1994; (15) Nash et al., 1992; (16) Norstog & Nicholls, 1997; (17) Rawson, 1961; (18) Ross, 1964; (19) Rothschild et al. 1986; (20) Scott, 1986; (21) Teas, 1967; (22) Windsor et al., 1999.

ECOLOGY
CHAPTER 8

Seed Dispersal Agents of Two Mexican Cycads

Jeff Chemnick

Abstract

Seed dispersal agents of the world's cycads are poorly known. Animal interactions with cycad seeds are likely responsible, at least in part, for the current distribution of cycad populations. Long-distance seed dispersal and the subsequent speciation of closely related forms among extant cycads are probably the result of avian interactions. A theoretical framework is discussed and anecdotal observations are presented in order to develop a profile for dispersal agents of Mexican cycads.

Resumen

Los agentes dispersores de semillas en el ámbito de las cícadas está pobremente conocido. Las interacciones animales con las semillas de cícadas son probablemente responsables, al menos en parte, de la actual distribución de las poblaciones de cícadas. La dispersión de semillas a largas distancias y la subsecuente especiación de formas cercanamente relacionadas entre las cícadas recientes es posiblemente el resultado de interacciones con aves. En este trabajo se discute un análisis teórico y se presentan observaciones anecdóticas como evidencia para establecer un perfil de los agentes dispersores entre las cícadas mexicanas.

Introduction

The principal mechanisms of cycad seed dispersal are both biotic and abiotic. The current distribution of many of the world's cycad populations are likely due, at least in part, to animal interactions with dehiscing seed. The edible sarcotesta of the seeds, combined with the toxic inner integument, attracts animals that are able to strip off the outer integument or ingest the seed whole, to regurgitate or defecate it later. This enables seed dispersal without seed predation. There is evidence that seeds of some *Macrozamia* species are occasionally dispersed by mammals and birds, for example, *Trichosurus vulpecular* (common brushtail possum, Burbidge & Whelan, 1982). Other vertebrates implicated in seed dispersal of *Macrozamia* species include emus and several other species of birds (Sargent, 1928; Sedgwick, 1940, 1952; Stranger & Stranger, 1970); grey kangaroos, quokkas, brush wallabies (Burbidge & Whelan, 1982), and quolls (Parton, 1952).

While the local distribution of individual plants within a cycad population may be due to the machinations of birds and terrestrial animals, longer distance seed dispersal is probably the result of bird interactions exclusively. In addition, birds are the most likely vector for dispersal of cycad seeds onto extremely steep exposures and cliffs, where mammals would likely have difficulty climbing. Many Mexican cycads occur on steep karst formations and sheer limestone cliffs in exposures that appear inaccessible to terrestrial animals.

Because of their mass, cycad seeds are not likely to be wind dispersed, except, perhaps, in the event of strong tropical storms and hurricanes, which do occur annually along both coasts of Mexico. Many Mexican cycads occur in areas of seasonal rains, which are capable of transporting cycad seeds considerable distances away from the parent plants.

Downhill, downstream seed dispersal due to gravity can expand the range of a population into suitable habitat. In many areas, an abrupt demarcation that has no readily apparent basis exists between the presence and absence of cycads in a habitat. Of the many other plant species that occur within Mexican cycad population ranges, most appear to have a much broader distribution than cycads. The ranges of cycads are almost always the most restricted within any given plant community in Mexico (Vovides et al., 2003).

Whereas the movement of cycad seeds downhill or downstream might result from abiotic factors exclusively, lateral and upward dispersal most likely involves dispersal by animals. Cycad seeds are covered with a fleshy sarcotesta that appears to attract various animals. The seeds apparently offer would-be consumers a meal and may be transported away from the parent plant to a site within the existing population or to a new locality further afield. However, documentation is lacking on long-distance cycad seed dispersal in Mexico. Indeed, few published accounts describe and document animal and cycad seed interactions, even within a population.

Animal interactions are vital to the long-term success of cycad populations. In addition to acting as agents that disperse seed, the other important function that animals perform is pollination. Although insects have been implicated as pollinators, the mere presence of an insect on a cycad cone does not necessarily prove its involvement in the cycad's reproductive life history (Oberprieler, 1995; Chadwick, 1998). The role of an animal in seed dispersal is easier to define. If an animal physically moves a cycad seed, irrespective of distance, it should be regarded as a dispersal agent unless the seeds are consistently damaged so that germination is not possible. The significant questions about dispersal agents are: Which animal is doing the moving, and when, how often, how far, and to where? And, ultimately, does such seed dispersal result in the successful establishment of new individuals and populations of cycads?

Identifying the species involved in cycad-animal interactions and the roles they play is essential to understanding cycad population distributions and dynamics. Such understanding also is necessary to make informed decisions about conservation and management of not only cycad populations but also the animals that are vital to their success (Tang, 1989).

Methods

In investigating the wild cycad populations in Mexico, I have recorded reports of animal interactions with cycad seed through conversations with field assistants and individuals living in close proximity to wild cycad populations. This is a preliminary study of the implications of those reports. Although I am satisfied that the reports are accurate, such reports are not the result of experimentation. Until detailed studies of cycad life histories reveal the animals that are involved with seed dispersal, an important component of cycad population dynamics will not be fully understood. This report contributes to that effort with examples of all three cycad genera that occur in Mexico.

Observations made in the Sierra Norte of Oaxaca are the result of numerous field trips made at various times of the year spanning the period from 1975 to 2005. In May 2004, the Montgomery Botanical Center of Miami, along with the Xalapa Institute of Ecology and Ganna Walska Lotusland of Santa Barbara, coordinated a month-long research field trip to investigate *Dioon* populations along the Pacific coast, from Sonora to Oaxaca (Gregory et al., in prep.). In addition, visits to wild cycad populations by the author have been undertaken throughout Mexico over the last 30 years, usually during the drier months of the year (October-May). During such field trips, cycad populations are surveyed and the data recorded, including details of morphology, phenology, population size, habitat description, edaphic analysis, geographic location, and orographic features. Prior to the recording of the reports of cycad seed dispersal agents contained herein, no attempt has been made to document cycad-animal interactions.

Ceratozamia whitelockiana and *Crax rubra*

Ceratozamia whitelockiana Chemnick and T. J. Greg. occurs in the Sierra Norte of Oaxaca within an elevational range of 340–970 m (Chemnick & Gregory, 1995). During the mid-1970s, the drainage of the Río Valle Nacionál in which it occurred was largely intact (pers. observ.). Large tracts of undisturbed tropical forest extended considerable distances from the road, with only modest evidence of pastures and fields. In the ensuing 30 years, habitat destruction reversed the ratio of undisturbed to disturbed areas. Now, large tracts of altered habitat dominate the landscape, consisting almost entirely of fields and pastures, with only scattered remnant patches of forest. As clearing has progressed, many local populations of *C. whitelockiana* have been extirpated (pers. observ.).

Several years ago a field assistant shot a *faisan*, a Great Curassow *Crax rubra* L., for food and found about a dozen *C. whitelockiana* seeds in the bird's stomach (J. Martinez, pers. comm.). As far as is known, this is the first report of avian dispersal of Mexican cycad seeds in its habitat.

Great Curassows, of the neotropical endemic family Cracidae, are large, turkey-sized birds of humid tropical forests. Males can be as tall as 93 cm and weigh 4.3 to 4.8 kg (del Hoyo et al., 1994). Their diet consists chiefly of fruits, especially figs fallen from trees and taken from the ground. While most curassows have powerful gizzards capable of crushing and assimilating seeds, Great Curassows are thought to be basically or partially frugivorous and are known to regurgitate or defecate the seeds, thus playing an important role in the dispersal of the plants upon which they feed (del Hoyo et al., 1994). In general, curassows consume a remarkable diversity of plants, with different species predominating in comparative importance throughout the year, depending on when different plants are fruiting. Although cracids are not known to migrate, they are known to occupy large ranges within primary forests, perhaps roaming daily within a range of several square kilometers (del Hoyo et al., 1994).

In surveys we conducted of *Ceratozamia whitelockiana* in the 1970s and 1980s, we found numerous clusters of plants distributed throughout the drainage of the Río Valle Nacionál (Fig. 8-1). Populations of *C. whitelockiana*, like those of curassows, have now declined dramatically as a result of habitat destruction (Chemnick & Gregory, 1995). The once widespread distribution of *C. whitelockiana* in the drainage of the Río Valle Nacionál was perhaps due in part to the historic occupation of the drainage by the Great Curassow. Currently, as result of habitat destruction and over collection, both the cycad and at least one of its potential dispersal agents are threatened.

Ceratozamia whitelockiana occurs in the center of the Great Curassow's Mexican range (Fig. 8-1). Because the Great Curassow was found to consume seeds of *C. whitelockiana*, it is possible that the bird serves, or did serve, as a dispersal agent for other species of *Ceratozamia* that also occur within its historic range. Because fruits and seeds make up much of the diets of the other seven Mexican cracids (del Hoyo et al., 1994), it is possible that they (the guans and chachalacas) too serve as dispersal agents for at least some of the cycad species with which they are sympatric.

Figure 8-1. Historic range of the Great Curassow and *Ceratozamia whitelockiana* in Mexico.

Dioon holmgrenii and *Calocitta formosa*

The geographic range of the genus *Dioon* has its northern terminus in the mountains of Sonora state, less than 200 km from the U.S. border. The range of *D. sonorense* (De Luca, Sabato & Vázq. Torres) Chemnick, T. J. Greg. and Salas-Mor. extends southward throughout Sonora and into northern Sinaloa (Chemnick et al., 1997). It is replaced by *D. tomasellii* De Luca, Sabato and Vázq. Torres in the state of Durango to the south. *Dioon tomasellii* likewise has a large range, and consists of various populations that range from Durango southward through Nayarit and Jalisco (Chemnick et al., 1997). Populations of *Dioon* also are known from Michoacan and Guerrero. Along coastal Oaxaca, *D. holmgrenii* De Luca, Sabato and Vázq. Torres predominates (Fig. 8-2).

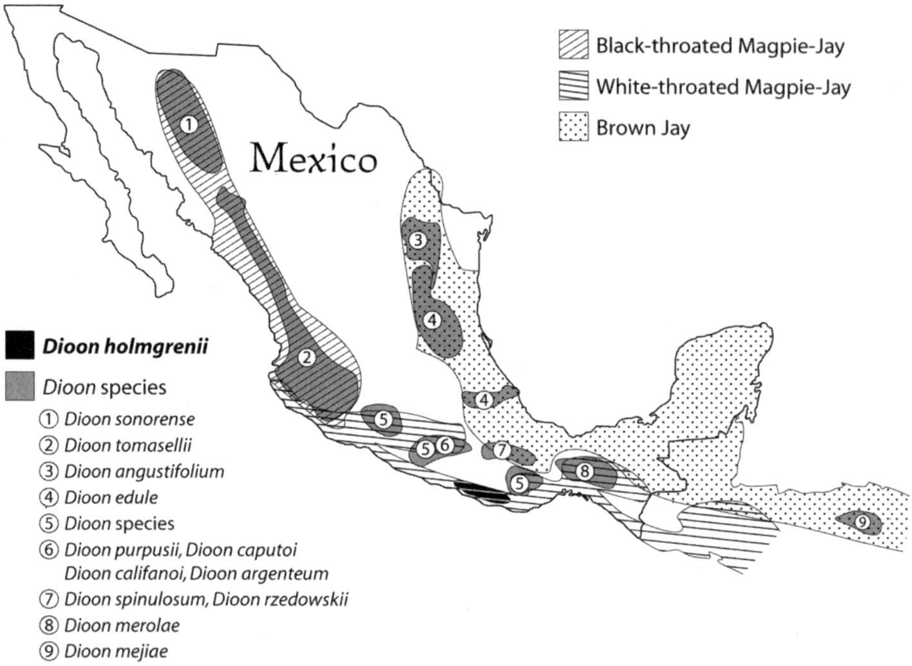

Figure 8-2. Historic range of the White-throated Magpie-jay and *Dioon holmgrenii* in Mexico.

The type locality of *Dioon holmgrenii* is contained almost entirely within a ranch owned and operated by the Reyes-Santiago family, who have rigorously protected the remaining cycads but have removed nearly all the surrounding natural vegetation leaving the remaining individuals of *D. holmgrenii* on terrain completely exposed and vulnerable to erosion. Furthermore, desiccation, due to the loss of plant cover, will likely cause the attrition of nearly all seedlings and result in a significant lack of recruitment.

Field observations have been made of several animal dispersal agents of *D. holmgrenii* seeds at the type locality. Congregations of uracas, that is, White-throated Magpie-jays (*Calocitta formosa* Swainson) often gather to eat and collect the seeds of *D. holmgrenii* as the female cones slowly dehisce over a period of days (F. Reyes-Santiago, pers. comm.). The bright yellow sarcotesta apparently attracts the jays in groups of 4–6 birds, which perch on or nearby dehiscing female plants. Reyes-Santiago has described a resource-sharing scheme whereby each of the jays, screeching loudly, waits in turn to select a seed and then flies either to a nearby perch or over the hill and out of sight. Jays take fallen seed off the ground, occasionally pluck seeds from cones as they dehisce, and fly to a nearby perch. The jays hold the seeds in their talons, clean off the flesh, and drop the naked seeds to the ground. Occasionally, rather than eating the sarcotesta at the site, the jays grasp the seeds with either their talons or bills and fly to some distant perch, presumably to eat the flesh or perhaps feed it to young (F. Reyes-Santiago, pers. comm.). The White-throated Magpie-jay occurs in gregarious family groups along the Pacific slope of southwest Mexico (Fig. 8-2) and as far south as Costa Rica (Howell & Webb, 1995). This species is closely related to the Black-throated Magpie-jay (*Calocitta collie* Vigors), from which it recently split; however, the two taxa are considered conspecific by some authorities (Peterson & Chalif, 1973; Howell & Webb, 1995). The Black-throated Magpie-jay ranges from Sonora to Jalisco (Howell & Webb, 1995), a distribution that corresponds to that of *Dioon sonorense* and *D. tomasellii* (Fig. 8-2). In Mexico, the range of the White-throated Magpie-jay closely overlaps the distributions of *Dioon* species from Michoacan and Guerrero as well as *D. holmgrenii* in Oaxaca and *D. merolae* De Luca, Sabato and Vázq. Torres in southeastern Oaxaca and in western Chiapas (Fig. 8-2). Magpie-jays are omnivorous, taking berries, fruits, invertebrates, and, occasionally, small rodents. The jays often use a foot to hold food while the bill is used to hammer open seeds or pick pieces from fruits and berries. Magpie-jays also cache food, returning later to eat (Langen & Vehrencamp, 1998).

Typically, jays are generalized opportunistic omnivores (Howell & Webb, 1995). Cycad seed-dispersal agents must be opportunistic generalist feeders because the presence of dehiscing female cones is sporadic and unpredictable. The phenology of most Mexican cycad species has yet to be documented, but numerous field trips to wild populations of all three genera over a period of nearly 30 years demonstrate the paucity of available cycad seeds at any point in time (pers. observ.).

The Magpie-jays are among the largest jays in Mexico, but they are rivaled in size by the Brown Jay (*Psilorhinus morio* Wagler), which occurs along the Sierra Madre Oriental of eastern Mexico (Fig. 8-2) from Nuevo Leon south through the Caribbean slope of Central America (Howell & Webb, 1995). The Brown Jay ranges throughout humid lowlands, foothills, forest borders, and open woodlands, neatly overlapping the ranges of *Dioon angustifolium* Miq., *D. edule* Lindl., *D. rzedowskii* De Luca, A. Moretti, Sabato and Vázq. Torres, and *D. spinulosum* Dyer in Mexico, as well as that of *D. mejiae* Standl. and L.O. Williams in Honduras. Like other large-bodied fruit-eating birds such as cracids, trogons, cotingas, and toucans, the Brown Jay is an important long-distance disperser of seeds

(Skutch, 1960; Wheelwright et al., 1984; Bronstein & Hoffmann, 1987). Because its diet consists mainly of the fruits of numerous tropical species, the diet of the Brown Jay may also include cycad seeds. And because of the evidence that at least one Mexican jay species disperses *Dioon* seeds, it is possible that other jays also disperse the seed of cycads that occur within their range.

Zamia integrifolia and *Mimus polyglottus*

Zamia integrifolia L. is the only native cycad in the United States, and there have been several reports of the seed being dispersed by the Northern Mockingbird (*Mimus polyglottus* L.) in southern Florida (T. Broome & W. Tang, pers. comm.). The Northern Mockingbird also has been observed to eat and fly with seed from exotic *Z. furfuracea* L. specimens planted in southern Florida (Eckenwalder, 1980). Likewise, Tang and Broome have observed mockingbirds feeding on seeds from garden plantings of *Z. furfuracea*. The birds take dehiscing *Zamia* seed to remote perches in trees, where they consume the fleshy seed coats and drop the cleaned seeds to the ground (Tang, 1989).

The Northern Mockingbird, of the neotropic endemic family Mimidae, ranges throughout most of Mexico and the Greater Antilles (Howell & Webb, 1995). It is replaced to the south and east by its close congener, the Tropical Mockingbird (*Mimus gilvus* Vieillot), which occurs in Chiapas and the Yucatán peninsula of Mexico as well as in Honduras, the Lesser Antilles, and northern South America. In addition, several other closely related species of mimids occur in Mexico, Central America, and the West Indies (Howell & Webb, 1995). The range of tropical mimids overlaps nearly the entire range of the genus *Zamia*. Because the Northern Mockingbird is known to disperse *Zamia* seed in Florida, it may be a dispersal agent for *Zamia* species elsewhere where they are sympatric. Likewise, other tropical mockingbirds of similar size and diet as the Northern Mockingbird may be dispersal agents for *Zamia* species in Mexico, the Caribbean, Central America, and South America where their ranges overlap.

Other Dispersal Agents

Within the three Mexican cycad genera, there are differences in cone morphology, rate of dehiscence, plant habit, and seed size and color, as well in the habitats in which each genus occurs. A profile can be developed to match particular kinds of birds with particular kinds of cycads.

Most *Dioon* species are dry-adapted plants living in patchy forest, open woodlands, or on cliffs and steep exposures, where dehiscing cones are visible to jays and parrots, which have excellent vision (Sillman, 1973). *Dioon* stems are moderate to large, and, accordingly, the cones are held well above the ground. The female cones dehisce slowly, exposing large, bright yellow seeds. Such large, brightly colored seeds are likely to attract birds in flight, even from considerable distances. It has been shown that diurnal birds have a visual system that is particularly sensitive to yellow and red (Sillman, 1973).

Most *Ceratozamia* and *Zamia* species, in contrast to the majority of *Dioon* species, occur under closed canopy in humid tropical forest. The stems are small to moderate in size, and many are subterranean. Consequently, the cones are held either at or very close to ground level and dehisce quickly, dropping the small seeds into the leaf litter. Though the two genera occupy similar habitats, *Zamia* seeds are quite different in appearance from *Ceratozamia* seeds. Typically, *Zamia* seeds have a bright red sarcotesta and are slightly smaller than those of *Ceratozamia*. In contrast, the sarcotesta of *Ceratozamia* are dull hues of brown and cream color, which suggests that the seeds of each genus attract and are dispersed by different animal vectors. It is likely that the avian dispersal agents of both *Ceratozamia* and *Zamia* are terrestrial forest-dwelling birds that forage in leaf litter, such as cracids, tinamous, ground-doves, and wood-partridges. Within humid tropical forest, avian dispersers might include opportunistic frugivorous birds of the canopy such as trogons, toucans, cotingas, and mockingbirds.

Various mammals also have been implicated as cycad seed-dispersal agents in Mexico. Rodents, such as rats, mice, and squirrels, are closely involved with freshly dehisced cycad seed in Mexico. Faustino Reyes-Santiago (pers. comm.) reported that he has seen squirrels occasionally moving seed of *Dioon holmgrenii* within the vicinity of the parent plants. Vovides et al. (2003) reported that *Peromyscus* sp. disperses seed of *D. edule* in tropical deciduous forest. I have often encountered fresh *Dioon* seeds of various species in the field, stripped clean of their fleshy fruit. Incisor gnaw marks in the dry, remnant bits of sarcotesta may indicate consumption by rodents (Negron-Ortiz & Breckon, 1989). Even as a cone dehisces, rodent activity occurs both among the seeds on the ground and within the cone (pers. observ.). Freshly dehisced seeds are quickly stripped of the sarcotesta. Excavating the ground around female plants often reveals cleaned cycad seeds that have been cached, most likely by rodents for future consumption. Because rodents might occasionally cache more seed than they can eat, or forget the location of all the seed they cache, some cached seeds eventually germinate and grow.

Within weeks or months after female cones dehisce, numerous one-leaf seedlings begin to grow, usually in response to seasonal rains. Typically, such seedlings are shallowly rooted in the moist litter and topsoil close to the parent plant, but personal field observations have shown that desiccation of the first-year seedlings during the ensuing dry season results in almost complete attrition of the seedlings. Studying the location of young plants that have successfully weathered several seasons of drought demonstrates that these plants grew from seeds buried in the ground or within rocky crevices. Only seeds buried sufficiently deep appear to have a chance to withstand the extreme conditions at ground level during the dry season and grow to maturity. Thus, by cleaning and caching cycad seeds, rodents may be necessary and active dispersal agents, at least at the local level. By contributing to population recruitment, rodents are apparently vital to the long-term survival of wild cycad populations.

Larger mammals also are involved with cycad seed dispersal. Faustino Reyes-Santiago (pers. comm.) has observed on occasion the white-nosed coati (*Nasua narica* L.) climb the 2- to 3-m-tall *Dioon holmgrenii* stems and break apart female cones to access the seeds. Reyes-Santiago has described finding scratch marks on the plants' stems as well as numerous tracks on the ground surrounding dehiscing female cycads. The coatis apparently not only eat the sarcotesta but occasionally sample the naked seed as well. Broken and partially eaten seeds are evidence left behind on the ground. The white-lipped peccary (*Tayassu tajacu* L.) has been observed dispersing seed of *Ceratozamia mirandae* Vovides, Pérez-Farr. and Iglesias in Chiapas (Pérez-Farrera et al., 2000).

Other possible mammalian dispersal agents of Mexican cycads include fruit bats. In Mexico, there are 22 species of obligate frugivorous fruit bats and at least 10 more that include some fruit in their diet (R. A. Medellín, pers. comm.). Neotropical fruit bats are the main dispersers of many species of plants (Emmons, 1990). Bats carry large seeds to a night roost, where the seeds are dropped. Small seeds are sometimes swallowed and defecated later, while others are spit out in clumps the size of the bat's mouth. Bat-associated fruits are normally dull-colored, full of sugars, with sweet, pungent smells, and sometimes with juicy pulp, suggesting that *Ceratozamia* seed might be among the fruits consumed by bats (R. A. Medellín, pers. comm.). Furthermore, some *Cycas* seeds are known to be dispersed by megabats in the Old World (van der Pijl, 1957). Therefore, it is possible that the larger fruit bats of the neotropics also are dispersing cycad seed in Mexico.

Threats to Dispersal Agents

Threats to the extant cycad dispersal agents of the world likewise threaten the future success of the world's cycad populations. Conservation of the environment in Mexico and Central America is a major problem (Ramos, 1985). The decline of bird populations in Mexico can be attributed in part to loss of habitat due to human population growth and clearing for agriculture, ranching, and timber, as well as hunting for sport (Howell & Webb, 1995). Loss of nesting sites and habitat locally near wild cycad populations is likely to result in the decline of local bird populations, including those that disperse cycad seed in Mexico.

Great Curassows are rather common in undisturbed areas but are scarce or nonexistent in modified environments. Curassows require the cover of primary forest, and their distribution is seriously affected by any habitat alteration or deforestation (del Hoyo et al., 1994). In addition to habitat destruction, hunting is a serious threat to curassows. The Great Curassow is a prize game bird. In addition to being famously tasty, the strong tail and wing feathers make excellent fletches for arrows. However, current hunting rates are unsustainable. Curassows are particularly vulnerable to hunting because of their large size, terrestrial habit, loud courtship vocalizations, low reproductive capacity, and lack of effective conservation (del Hoyo et al., 1994). In Mexico, the Great Curassow was once widespread with a continuous range throughout the tropical lowlands of the southeastern Atlantic slope (Fig. 8-1). The intense habitat destruction of the last 30 years combined with direct hunting has devastated and fragmented the remaining populations. Now, the Great Curassow has a limited range in a highly transformed environment (del Hoyo et al., 1994).

Some mammalian dispersal agents are hunted as well. In Oaxaca, at the type locality of *D. holmgrenii*, Faustino Reyes-Santiago (pers. comm.) reported that the white-nosed Coati is considered a pest and is occasionally hunted. Similarly, squirrels and other rodents are perceived as pests and frequently hunted.

Summary and Conclusions

By reviewing the common shared characteristics of seed dispersers, a profile emerges of characters that cycad seed-dispersal agents must possess:

- The range of a given cycad species and its dispersal agent(s) must overlap, at least in part.
- The behavior (especially foraging behavior) of any dispersal agent must enable it to encounter cycad seed, at least occasionally.
- Cycad seed-dispersal agents must be opportunistic feeders because the availability of dehiscing female cones is sporadic; thus, cycad seeds constitute an unreliable food source.
- A dispersal agent must be physically capable of transporting cycad seeds away from parent plants while ensuring that at least some seeds remain viable long enough to germinate. Furthermore, such dispersal agents must leave intact, at least some of the time, the inner seed after stripping the edible outer integument. This enables seed dispersal without total seed predation.

Because the cycads are an ancient group of plants with very large seeds, it has been suggested that the original prehistoric dispersal agents are now extinct (Negron-Ortiz & Breckon, 1989). However, in contemplation of the above examples, it is clear that at least some of the extant Mexican cycads currently have dispersal agents, and the seed's edible sarcotesta is the reward for dispersal. Quite possibly, the color and nutritious outer integument of cycad seeds are the result of an ancient coevolved symbiosis between cycads and now-extinct dispersal vectors. As extinction claimed those vectors, as well as other ancient plant groups, the cycads, in a changing world, were apparently favorably preadapted to exploit modern animals as dispersal agents (Bauman & Yokohama, 1976).

When the range maps of various Mexican cycad species are overlaid on those of their putative dispersal agents, geographical overlap is seen. The cycad populations cited herein occur within the ranges of the dispersal agents discussed. Furthermore, *Dioon* and *Ceratozamia* species groups fall within the geographical range of their likely dispersal agents (Figs. 8-1 & 8-2).

Clearly more fieldwork needs to be conducted to correctly assign animal vectors to the cycad seeds that they disperse. One essential surveying tool is a motion-sensitive camera placed near a dehiscing female cone. Another useful technique, marking and recovering dispersed cycad seed, could provide additional data. Such information is essential for understanding both historic and future patterns of distribution as well as predicting the future reproductive success of the cycads. Given the quantity of seed produced by most cycad species, and the much wider distribution of their putative dispersal agents, the question remains as to whether cycad distribution is chiefly the result of dispersal agent behavior or restrictive habitat requirements. Detailed life history and behavioral studies of dispersal agents in the field could help to explain the restricted and discontinuous nature of many cycad populations. Understanding the life histories and interactions of cycads and dispersal agents is critical for making the most effective decisions, for conservation of not only the world's cycads but also the animals upon which the cycads depend for survival.

Acknowledgments

I would like to thank the Ganna Walska Lotusland Foundation, the Montgomery Botanical Center, and SERBO (Sociedad para el Estudio de Recursos Bióticos de Oaxaca). Special thanks go to Satie Airamé, John Donaldson, and Gimme Walter for critical review, Monica Pessino for illustrations, and Timothy Gregory, Silvia Hortensia Salas-Morales, Miguel Angel Pérez Farrera, Andrew Vovides, Roy Osborne, Paul Glenn, Virginia Hayes, Willie Tang, Steven Timbrook, and Terrence Walters, and in memoriam to Leo Schibli. I am especially grateful to Jorge Martínez, Faustino Reyes-Santiago, José Pascual, and all the eyes and ears in the forests of Mexico who make these observations so that we may better understand these enigmatic plants.

Literature Cited

Bauman, H. J. & H. Yokohama. 1976. Seed coat carotenoids of the cycad genera *Dioon, Encephalartos, Macrozamia*, and *Zamia*: evolutionary significance. Biochem. Syst. Ecol. 4: 73–74.

Bronstein, J. L. & K. Hoffman. 1987. Spatial and temporal variation in frugivory at a neotropical fig, *Ficus pertusa*. Oikos 49: 261–268.

Burbidge, A. & R. Whelan. 1982. Seed dispersal in a cycad, *Macrozamia riedlei*. Austr. J. Ecol. 7: 63–67.

Chadwick, C. E. 1998. Some insects and other invertebrates associated with the cycad *Macrozamia communis* (Zamiaceae). Encephalartos 54: 10–18.

Chemnick, J. & T. J. Gregory. 1995. A new species of *Ceratozamia* (Zamiaceae) from Oaxaca, Mexico with comments on distribution, habitat, and relationships. Phytologia 79(1): 51–57.

———, ——— & S. Salas-Morales. 1997. A revision of *Dioon tomasellii* (Zamiaceae) from western Mexico, a range extension of *D. merolae*, and clarification of *D. purpusii*. Phytologia 83: 1–6.

del Hoyo, J., A. Elliott & J. Sargatal (eds.). 1994. Handbook of the birds of the world. Vol. 2. Lynx Editions, Barcelona, Spain, pp. 310–359.

Eckenwalder, J. 1980. Dispersal of the West Indian cycad *Zamia pumila* L. Biotropica 12: 79–80.

Emmons, L. H. 1990. Neotropical rainforest mammals, a field guide. University of Chicago Press, Chicago, pp. 44–79.

Howell, S. N. G. & S. Webb. 1995. The birds of Mexico and northern Central America. Oxford University Press, New York.

Langen, T. A. & S. L. Vehrencamp. 1998. Ecological factors affecting group size and territory size in white-throated magpie-jays. Auk 115(2): 327–339.

Negron-Ortiz, V. & G. J. Breckon. 1989. A note on the dispersal of *Zamia* in Puerto Rico. Caribbean J. Sci. 25(1–2): 86–87.

Oberprieler, R. G. 1995. The weevils (Coleoptera: Curculionoidea) associated with cycads. 1. Classification, relationship, and biology. Pp. 295–334 *in* P. Vorster (ed.), Proceedings of the Third International Conference on Cycad Biology, Cycad Society of South Africa, Stellenbosch, South Africa.

Parton, G. F. 1952. Encounter with native cat. W. Austral. Naturalist 3: 93.

Pérez-Farrera, M. A., P. F. Quintana-Ascencio, B. Salvatierra-Izaba & A. P. Vovides. 2000. Population dynamics of Ceratozamia matudai Lundell (Zamiaceae) in El Triunfo Biosphere Reserve, Chiapas, Mexico. J. Torrey Bot. Club 127: 291–299.

Peterson, R. T. & E. L. Chalif. 1973. A field guide to Mexican birds. National Audubon Society, Boston.

Ramos, M. A. 1985. Problems hindering the conservation of tropical forest birds in Mexico and Central America, and steps towards and conservation strategy. International Council for Bird Preservation Tech. Publ. 4: 67–76.

Sargent, O. H. 1928. Reaction between birds and plants. Emu 27: 185–192.

Sedgwick, E. H. 1940. Birds of the Rockingham district. Emu 40: 129–152.

———. 1952. Birds and *Zamia* seeds. W. Austral. Naturalist 3: 117.

Sillman A. J. 1973. Avian vision. Pp. 349–387 in D. S. Farner & J. R. King (eds.), Avian biology. Vol. 3. Academic Press, New York.

Skutch, A. F. 1960. Life histories of Central American birds II. Pacific Coast Avifauna No. 34. Cooper Ornithological Society, Bend, OR.

Stranger, M. J. & R. H. Stranger. 1970. Grey Butcher-bird feeding on *Macrozamia* fruit. W. Austral. Naturalist 11: 145.

Tang, W. 1989. Seed dispersal in the cycad *Zamia pumila* in Florida. Canad. J. Bot. 67: 2066–2070.

van der Pijl, L. 1957. The dispersal of plants by bats. Acta. Bot. Neerl. 6: 291–315.

Vovides A. P., M. A. Pérez-Farrera & C. Iglesias. 2001. Another new species of *Ceratozamia* (Zamiaceae) from Chiapas, Mexico. Bot. J. Linn. Soc. 137(1): 81–85.

———, ———, J. González-Astorga, D. González, T. Gregory, J. Chemnick, C. Iglesias, P. Octavio-Aguilar, S. Avendaño, C. Bárcenas & S. Salas-Morales. 2003. An outline of our current knowledge on Mexican cycads (Zamiaceae, Cycadales). Curr. Topics Pl. Biol. 4: 23–25.

Wheelwright, N. T., W. A. Haber, K. G. Murray & C. Guindon. 1984. Tropical fruit-eating birds and their foodplants: a survey of Costa Rican lower montane forest. Biotropica 16: 173–192.

ETHNOBOTANY
CHAPTER 9

Ethnobotany of Honduran Cycads

Mark A. Bonta

Abstract

Ethnobotanical studies of Cycadales, though scarce, are crucial to our understanding of the order. The current study illustrates the range and depth of human relationships with cycads, describing uses for five species in three genera (*Ceratozamia*, *Dioon*, and *Zamia*) of Zamiaceae in Honduras. My intent is not only to document and outline a specific regional cycad ethnobotany but also to provide an example of the type of ethnobotanical data that need to be systematically recorded, analyzed, and published for cycads in other regions, and to supplement current knowledge, update historical records, and contribute to cycad protection and conservation strategies. Specifically, the current study records historical and contemporary alimentary, decorative, medicinal, and other uses, vernacular nomenclature, folk beliefs, and functions within human-affected ecosystems, including de facto and de jure traditional protective measures as well as intercropping of cycads and other food plants.

Resumen

Estudios etnobotánicos de los Cycadales son escasos pero centrales para nuestro conocimiento del orden. El presente estudio ilustra el rango y profundidad de las relaciones entre el ser humano y Cycadales; se describen los usos para cinco especies de tres géneros de Zamiaceae (*Ceratozamia*, *Dioon* y *Zamia*) en Honduras. El intento del autor no es solamente documentar y así formular el croquis de un tratamiento etnobótanico de una región específico, sino también proveer un ejemplo de los tipos de datos que necesitan ser reportados, analizados, y publicados en forma sistemática para Cycadales en otras regiones, para suplementar el conocimiento actual, actualizar datos históricos, y contribuir a un cuerpo de literatura que puede informar al proceso de protección y conservación de Cycadales. Específicamente, el presente estudio se enfoca en los usos alimentarios, decorativos, medicinales, y otros, sean históricos o actuales; nomenclatura vulgar; creencias; y sus funciones dentro de los ecosistemas antropogénicos, incluyendo medidas tradicionales de protección tanto como el "intercropping" de Cycadales con otras plantas alimenticias.

Introduction

Ethnobotany—the study of the relationships between people and plants—has not been a prime consideration in recent studies of cycads. Though Thieret's classic article in *Economic Botany* appeared nearly 50 years ago (Thieret, 1958), in the interim, despite an explosion in the number of cycad species known to science, very little specific treatments of cycad ethnobotany have been published in scientific journals (e.g., Beck, 1992; Osborne et al., 1994; Vázquez Torres, 1990 Whiting, 1963). The remainder of the original, post-1958 work on cycad ethnobotany occurs as minor sections within published and gray literature on cycads and within broader botanical treatments as in Duke (1975) (i.e., of regional flora, regional ethnobotany, or ecosystems; see "Literature cited" in Bonta & Osborne, this volume, for an extensive list). Though substantive treatments of ethnobotany do occur in a few popular works on cycads (e.g., Jones, 2002; Whitelock, 2002), these largely recycle Thieret's work, which in itself was left incomplete (*Zamia* was never treated). The literature on *Cycas micronesica*'s possible connection to human neurogenerative diseases on Guam is an exception (e.g., Brenner et al., 2003; Whiting, 1963, but even in this case little has been done to follow up with studies on cycad toxicity in the numerous other human cultures in Asia, Australia, Africa, and the Americas where cycads are consumed.)

Complemented by Bonta et al. (in prep., an in-depth study of *Dioon mejiae*) and informed by five years of fieldwork in Honduras, this article provides a comprehensive and systematic overview of cycad ethnobotany in one country, in the hopes of inspiring a renewed and sustained focus on the "human side" of cycads in general. After all, in ethnobotany we can uncover clues that allow us better to protect cycads in their native habitats and perhaps achieve sustainable conservation of cycads, a key goal of the IUCN's Cycad Specialist Group (Donaldson, 2003).

Cycads in Honduras are used much as they are across the world—as sources of starchy food (seeds and stems), particularly in times of scarcity, and as religious icons (-whole plants and leaves). Other uses also follow from their typical characteristics and are similar to uses elsewhere—as rodenticides, purgatives, decoratives, and so forth. Though none of the types of uses of Honduran cycads discussed below is "new" in terms of what Thieret (1958) reports for other countries, we must emphasize that several of the major uses we have investigated are perhaps unique in the sense that they are still widespread in local culture—indeed, *Dioon mejiae* Standl. and L. O. Williams (*tiusinte*) is one of the most important food sources for 33,000 rural Hondurans. Because Thieret (1958) draws from centuries of literature on cycads, it is hard to know which of the practices he describes are still in use. Parts of rural Honduras are, then, living laboratories for cycad ethnobotany, and therein lies the impetus for the current research.

At present, Honduras is known to contain seven species of Zamiaceae in the genera *Ceratozamia* Brong. (1), *Dioon* Lindl. (1), and *Zamia* L. (5) (for detailed taxonomic treat-

Table 9-1 Nomenclature of Honduran cycads in Spanish, Pech, and Tol.

Species	Spanish names	Pech names	Tol names (Tolupan ethnic group)
Ceratozamia "*hondurensis*" (in process of description)	*camotillo* (little sweet potato, *Ipomoea batatas*)	—	—
Dioon mejiae Standl. & L. O. Williams	*tiusinte*—often written *teosinte* or *teocinte*; *palma teosinte*; *palmera*	—	*tiñúc*; *tiusinte*
Zamia "Atlántida" (in process of description)	*camotillo*, *yuca de ratón*	—	—
Z. "Cortés" (in process of description)	*camotillo*; *cica* (commercial nursery term only)	—	—
Z. herrerae Calderón & Standl.	(unknown)	—	—
Z. "*microchigua*" (in process of description)	(unknown)	—	—
Z. standleyi Schutzman	*camotillo*; *yuca deratón*"	*yojrá unicej* (venomous manioc)	*yuca de ratón* (mouse manioc)

ment see Table 9-1 and Haynes, 2007. Six are endemic to Honduras, though cultivated examples occur in neighboring countries and overseas. Thus far, the focus of cycad ethnobotany research in Honduras has been on *D. mejiae,* and that work will appear in detail in Bonta et al. (in preparation); a summary is incorporated herein (see also Haynes & Bonta, 2007).

The present author, along with Jody Haynes and others, has been investigating *D. mejiae* (henceforth *tiusinte*) since 1999; it is a common species in its restricted range, well-regarded, and easy to investigate in the field. The case of the other cycads is distinct. On the one hand, with the exception of *Zamia standleyi* Schutzman, they are range-restricted, highly endangered (and thus in undivulged locations); on the other hand, they are all considered by local ethnic groups to be harmful, even "evil" plants (*plantas malvadas*), and as a result are extremely difficult to investigate in the field. Nevertheless, we have achieved moderate success through culturally sensitive questioning employing open-ended interviews and protection of subjects' identities. Field research often occurred within the context of botanical collecting expeditions, including the combined Montgomery Botanical Center/Honduras National Herbarium (TEFH) trip of July/August 2003 (Haynes & Bonta, 2003). The Honduran cycad data in the present article draw from the above-mentioned field research (published in preliminary form in Bonta, 2003b and Haynes & Bonta, 2003) and are also informed by the scant published and archival material on the subject by Honduran authors (Aguilar Paz, 1999; Hardy Nelson S., 1986; Landa, 1935, 1940; Monografía de Yoro, 1926).

Honduras (112,090 km^2) is 80% mountainous and oriented along a west-east axis; it contains cycads only on its northern fringes, in the Caribbean slope lowlands, mostly below 700 m above sea level (*Z. herrerae* Calderón & Standl. possibly occurs more widely, but this has not been confirmed). Only *Z. standleyi* and *D. mejiae* are found inland beyond the coastal humid forests, and even these species have never colonized most of the available habitat. In terms of ethnobotany, this means that while local residents of long habitation in or near native populations of Honduran cycads know their uses well, these same cycads have little cultural visibility elsewhere, especially across central, western and southern Honduras, where the majority of the Honduras's seven million inhabitants are concentrated. Honduran cycads tend to occur in out-of-the-way places and as a result have been as ignored by most Hondurans, as they have been by botanical collectors until very recently.

Ceratozamia "hondurensis," recently discovered and currently being described, is the southernmost and easternmost representative of its genus. In Honduras, this species is barely known, as it is restricted to the few remaining forests in an area where recent migrants comprise the bulk of the population (Honduran cycads are generally known well only to autochthonous groups who have occupied the same landscape for several generations). The *Ceratozamia* is a *camotillo* and thus equated with the *Zamia* (see below), but thus far we have not discovered any of the types of use recorded for Mexican ceratozamias (Vázquez Torres, 1990.

Zamia standleyi, a common species, is the main focus of the present *Zamia* investigation—it is known as either *yuca de ratón* or *camotillo*. The latter name, and associated uses, are also applied to all other Honduran zamias for which ethnobotanical data have been recorded and appears to have an autochthonous Honduran origin, as this name is not applied elsewhere to cycads (though Roy Osborne records *palma de camote* for *C. robusta* Miq. in Mexico: see Bonta & Osborne, 2007). A variant of *yuca de ratón*, on the other hand, is also found as a dominant common name for cycads in Cuba (Bonta and Osborne, in prep.) and points to a cultural-historical connection between Honduras and that country, as described below. Overall, *Zamia* uses are similar or identical to those found, at least historically, in Puerto Rico, Florida (Small, 1921), Cuba (Roig-Mesa, 1965), and elsewhere throughout the Caribbean and the southern Caribbean slope of Central America and into South America (Pérez-Arbelaez 1956; Schultes & Raffauf, 1990).

The entire ethnobotanical profile of *tiusinte* is nearly identical to what we know of Mexican *Dioon* species, particularly those of Oaxaca, in terms of its alimentary uses (e.g., Chamberlain, 1919). Its nomenclature is Classical Nahuatl and links it to *Ceratozamia*, *Dioon,* and *Zamia* uses on the Gulf slope of Mexico and particularly among the Nahuatl and Teenek Maya in the Huasteca (Alcorn, 1984; B. Schutzman, pers. comm.). Though *tiusinte* is endemic to northeastern Honduras, there is virtually nothing in its ethnobotany that cannot be ascribed on one hand to the influx of settlers from pre-Columbian Mexico into northeastern Honduras from AD 1000 to 1600 (and most likely during the latter part of this period) and on the other hand to the introduction of Roman Catholicism from the early 1500s onward (for detailed historical treatments of northeastern Honduras, see Bonta, 2001, and Newson, 1986).

Vernacular Nomenclature

CAMOTILLO

All Honduran zamias and *Ceratozamia* are called *camotillo*, the diminutive for *camote*, sweet potato (*Ipomoea batatas* [L.] Lam.). *Camote* derives from the Classical Nahuatl *camotli* (tuber, edible root). The reference is morphological and evokes the subterranean stem, thought of locally as a *raíz* (root). According to Hardy Nelson S. (1986), the following Honduran plants are also called *camotillo*: *Curcuma longa* L. (Zingiberaceae), turmeric, a cultivated species; *Xiphidium caeruleum* Aubl. (Hemodoraceae), a palm-like species also called *camotillo de palma* or *lirio*; *Operculina pteripes* (G. Don) O'Donell (Convolvulaceae), a vine; *Ipomoea meyeri* (Spreng.) G. Don and *I. minutiflora* (Mart. & Gal.) House (Convolvulaceae), vines also known as *bejuquillo* and *campanilla*; *Dioscorea macrostachya* Benth. (Dioscoraceae), a vine also called *cabeza de negro*, *cuculmeca*, and *yampi*. Hardy Nelson S. identifies the latter species, as well as "*Zamia loddigesii*" (misidentification of a Honduran zamia in this case?), as highly poisonous. Depending on the cultural context and location, *camotillo* can inspire fear or simply invoke the sweet potato; in some cases, the same name applies to distinct species in the same village. For example, in one community in San Esteban, Olancho (eastern Honduras), recent settlers from western Honduras referred to a local *Diplazium* fern as *camotillo*, while long-time inhabitants applied the same name to *Zamia standleyi*; both groups thought of theirs as the "true" (and thus deadly) *camotillo* but appeared ignorant of the properties of the other plant.

YUCA

Zamia standleyi and *Z.* "Atlántida" are known as *yuca de ratón*, meaning "mouse manioc" and referring to their use as a rodenticide. The allusion is to *Manihot esculenta* Crantz (Euphorbiaceae), a poisonous tuber with high starchy content made edible through a process similar to that used for zamias. However, *yuca* is also a generic Honduran term for a tuber. The Pech indigenous group call *Z. standleyi, yojrá unicej* ("venomous yuca").

While *yuquilla de ratón* (little mouse manioc) is recorded for most species of Cuban cycads (Bonta & Osborne, 2007), in Honduras *yuca de ratón* is used only in the municipality of Olanchito, which was in colonial times a *ciudad* (administrative town inhabited by

Spaniards and *criollos*). Presumably, settlers brought the name from Cuba (or, possibly, vice versa) at an unrecorded date; the name has been recorded nowhere else in Central America or the Caribbean.

Hardy Nelson S. (1986) lists only one other Honduran plant with a name related to *Manihot*: the ubiquitous *yuquilla*, *Maranta arundinaceae* L. (Marantaceae), also called *macuxe* or *sagú* ("sago"). The uses for this ubiquitous "sago" starch-producing species are identical to those of the zamias but much more widespread and well-known.

Tiusinte

Though there are several spelling and pronunciation variants of *tiusinte* in both Spanish and Tol, there appears to be only one Nahuatl-derived root term for *D. mejiae*, despite the fact that it is found within the territory of several distinct indigenous languages (extant as well as extinct). The Classical Nahuatl prefix *teo* derives from *teotl*, deity; *teo* was commonly used to denote the sacred, as in *teocalli*, "sacred houses" (temples). A few plants and animals were thus named: *teozanatl* was today's *zanate*, the Great-tailed Grackle, *Quiscalus mexicanus*. *Teocintli*, most commonly associated with the various *Zea mays* subspecies that are evolutionary ancestors of maize, means "sacred ear of maize" (*cintli* = "maize ear"); the name is also applied in Mexico to several species of *Zamia* in the Huasteca (B. Schutzman, pers. comm.). Teenek Maya inhabitants of this region refer to these same zamias as "shepherds of maize" (Alcorn, 1984), while contemporary Nahuatl and mestizo names in the Huasteca also make a clear connection between maize and cycads (Bonta & Osborne, 2007; Vázquez Torres, 1990). Dioons are thought of as "old-time" maize in Oaxaca (Bonta & Osborne, 2007), and though in Honduras the actual meaning of *tiusinte* has been lost, a cultural connection between maize and *D. mejiae* remains (see below). Though Standley and Williams (1950), who originally described *D. mejiae*, found it "futile to speculate" on why the name *teocinte* was used for this species, it has turned out to be an important clue linking northeastern Honduran prehistory to Mexican influence, as well helping reveal a symbolic, if not evolutionary, relationship with maize.

Muddying the water significantly, *teosinte* and variants are applied to Poaceae species across central, western, and southern Honduras, as well as in Nicaragua, El Salvador, and Guatemala. Hardy Nelson S. (1986) mentions the following species: *Tripsacum latifolium* Hitchc. (Gramineae), also known as *caña brava, hierba prodigiosa, pasto de Guatemala, pasto prodigioso*, *taucinte*, *teocinte,* and *tepecinte*; *Euchlaena mexicana* Schrad., also called *maíz café* and *Setaria paniculifera* (Steud.) Fourn. (Gramineae). Hardy Nelson S. (1986) also claims that *Cycas revoluta* Thunb. in Honduras is known as *taucinte* and variants (but does not mention *D. mejiae* at all in the work); he also ascribes the name *taucinte* to *Zamia*. Finally, Hardy Nelson S. reports (in Haynes & Bonta, 2003) that a species of *teosinte* that he collected in 2003 in western Honduras was recognized by Hugh Iltis as *Zea luxurians* × *Tripsacum fasciculatum*, or *T. andersonii* J. R. Gray.

Clarification of the Poaceae-Zamiaceae nomenclatural homology has been important for the purposes of botanical exploration, as the locative *-al* is often used to refer to places where the plants grow (or grew). In northeastern Honduras, *Tiusintal* signifies a *tiusinte* population; elsewhere, it refers to a population of maize-like Poaceae.

Though there is a widespread casual assumption that the *tiusinte* is a species of palm (*palmera*), most people with intimate experience of the plant are aware that the resemblances are superficial. No link between zamias and *tiusinte* is widely known, but one informant with intimate knowledge of both species guessed, without prompting, that *D. mejiae* and *Z. standleyi* were related.

Alimentary Uses

We have only one report of *Z. standleyi* consumption as a regular part of the diet. Among the Pech in El Carbón, Olancho, where the plant is generally feared and hated, only one man (now deceased) was known to eat them regularly for food. After extracting the poison through a series of washings, he fried the *raices* (subterranean stems) in strips (*tajadas*). All other willful consumption of zamias reported to this researcher was for medicinal purposes.

Tiusinte is consumed regularly by approximately 33,000 people, including both mestizo ("Ladino") and indigenous (various ethnic groups) inhabitants of the departments of Olancho, Yoro and Colón. Year-old female cones (*cabezas*) are harvested during the *verano* (dry season, March to May) at numerous wild populations that together comprise upward of 600,000 adult plants (Haynes, 2007; Haynes & Bonta, 2003). The demand is so high that most accessible female *tiusinte* cones are harvested in a given year (though natural production of cones fluctuates). The harvest coincides with the time of scarcity when maize, the staple food, runs out, but in several municipalities tiusintes are highly regarded as food items in their own right and thus are also consumed by those who are not suffering from hunger. The cones are broken apart and the *cumbitos* (seeds) extracted and brought home in sacks, where they are split apart and processed into a variety of foods with names and shapes nearly identical to their maize counterparts. The most common and highly regarded product is the *tamal de tiusinte*, the cycad tamale; second in importance, though not regarded as that tasty, is the *tortilla de tiusinte* (cycad cake). Other products are the *atol de tiusinte*, a sweetened slurry sometimes prepared with milk; the *rosquilla*, an unleavened, ring-shaped pastry containing cheese; the *rosquete*, a leavened, ring-shaped pastry; and *pan de tiusinte*, cycad bread, made using baker's yeast and wheat flour. The latter two products are rarely prepared today.

Techniques of preparation vary widely, as each product has different desired qualities and there is also the issue of toxin removal. The *tamale* and the *tortilla* and usually the *atol*, are prepared from a *nixtamal* mixture that results from the grinding of seed kernels cooked three to six or more times in *Quercus* ash (other ashes are also used occasionally). The wet *nixtamal* is shaped into elongated or, occasionally, spherical *tamales*, which are wrapped in any of a number of types of leaves (the preferred is *jucuyul*, the *uva de montaña* (wild grape), an unidentified Myrsinaceae tree, stacked in a pot and cooked for 30 minutes to an hour. The ruddy-colored *tamales* that result are extremely filling, and their quality is measured by their rubberiness (*hulosidad*). *Tortillas* are cakes made from the same *nixtamal*, cooked on the griddle; cycad *tortillas* have also been reported from the Chinanteca in Oaxaca, Mexico, made with *D. spinulosum* seeds (Chamberlain, 1919).

The *rosquillas*, *rosquetes*, and *pan de tiusinte* (collectively, *mascaduras*), and sometimes the *atol*, are prepared following a distinct process. Though all cycad seeds can be stored with the sclerotesta intact, the outer flesh is generally removed and the kernels dried in the sun for better storage. After sun-drying, the seed kernels are ground and then suspended in a cloth in water overnight, thus leaching the toxins. The detoxified paste may then be used immediately, or dried again for storage. It is drained through a sieve to produce the *atol*, or combined, dry, with various ingredients to make the three baked products.

Toxicity

Zamia

As *camotillo*, *Z. standleyi* and probably other zamias are widely feared by mestizos in the culturally autochthonous department of Olancho. Among the indigenous Pech of eastern

Olancho department, their fear of the plant is mingled with their hatred for it, because mestizos generally believe that the Pech use *camotillo* as their primary means of assassinating their enemies. However, the documented practice of poisoning one's enemies with *camotillo* (by slipping it into drink) has been practiced widely throughout mestizo Olancho as well, and, indeed, people living in northern Honduras and elsewhere associate closely assassination by *camotillo* with Olanchano culture in general. In Olanchito municipality (Yoro department), which borders Olancho to the northwest, *camotillo* is recognized as the name of a semimythical plant that the Olanchanos use to kill their enemies; Olanchito residents insist it is distinct from their *yuca de ratón*, poisonous to mice but not to people.

The Pech are extremely reticent to discuss the plant, which probably explains why the ethnobotanical literature (e.g., Conzemius, 1928; Lentz, 1993) does not mention it. Nevertheless, Pech leaders in El Carbón in 2003 discussed *yojrá unicej* and Pech perceptions of it with botanists (Haynes & Bonta, 2003). The plant is often found growing in maize and bean fields, primarily because it is believed to be dangerous to touch or get near to, because if harmed the plant gives off a *polvillo tóxico* (poisonous dust). Pech have traditionally been able to eliminate the plant after repeated burning of fields.

Death by *camotillo* poisoning involves hours, days, or weeks of bloody diarrhea—the universal popular belief in Honduras is that *si tiene [x número de] días de cortado, se muere en [x número de] días* (if it was cut *x* number of days before, it will take the same number of days to die). This suggests that freshly prepared *camotillo* is far more lethal than *camotillo* that has been stored. (One informant, who believes that his father's death was caused by *camotillo*, relates that it took six months for him to die.) Even in areas where zamia *camotillo* is unknown, the belief that the number of days it takes to die equals the number of days since the *camotillo* has been prepared is still known (though it is uncertain in many cases which of the several Honduran *camotillos* is actually meant). As to the method of preparation, it is said that the mashed, uncooked stem is mixed in food or beverage.

Very few people believe that there is a remedy, but according to Ciro Navarro, director of the Lancetilla Botanical Garden in Tela, two Olanchanos once visited him looking for the *contrahierba* (homeopathic remedy) for their brother, whom someone had given *camotillo*. They were unfamiliar with the plant, but believed that *camotillo* itself could be used to combat poisoning.

Standley (1931) reported the practice of *camotillo* poisoning on the Caribbean coast near Tela, Honduras, and referred to it repeatedly in later publications on Belize,

Guatemala, and Costa Rica (Standley and Record, 1936; Standley, 1937; Standley and Steyermark, 1958): "In the coast of Honduras there is a species . . . (*Z. furfuracea* L. f.), which is known by the name Camotillo. Its root is highly poisonous, and has been employed at times for criminal poisoning, as well as for poisoning noxious animals. There is a popular belief that the root, if out of the ground two days, kills its human victim in two days; if dug a week before, it kills in a week, and so on. I do not know whether the Costa Rican species is used in the same manner, or whether the same belief is held there regarding it. It goes without saying that, although the root is a dangerous poison, it does not kill in accordance with the popular schedule" (Standley, 1931: pp. 63–64).

Dioon mejiae

Only those informants who had had direct experience with acute *tiusinte* poisoning were aware of its toxicity. There was no local awareness of long-term, progressive effects of *tiusinte* consumption. Acute poisoning was associated solely with poor preparation and with consumption of either raw seeds or the *apaguaste* (waste water that is produced during the process of *almidón* [starch; also *harina*, flour] preparation). The *apaguaste* is known to cause death in piglets and violent purgative episodes in humans. Though we observed cattle and pigs chewing on leaves, we encountered no concern about livestock poisoning among local ranchers (quite the contrast from *D. edule* in Mexico and the situation of the "zamia staggers" in Australia [Jones, 2002]).

How are *tiusinte* seeds or seed kernels detoxified? It appears that either washing in ash (*tamales*, *tortillas*, some *atoles*) or sun-drying combined with leaching in water (*rosquillas*, *rosquetes*, *pan*, some *atoles*) neutralizes the toxins (C. Shaw, pers. comm.). In one case, geography researcher Daniel Graham, who was filming *tiusinte* consumption (Bonta et al., in prep.), was served apparently insufficiently processed cycad foods and suffered nausea, though not vomiting, as a result. It appears that his hosts, under pressures of his time constraint, used shortcuts in preparation. Despite this episode and many accounts of families near starvation who were "rescued" by the *tiusinte* harvest, systematic or widespread poisoning by desperate people has not been reported to us.

In relation to the Guam cases (Whiting, 1963), there has never been unequivocal evidence of traditional cycad users systematically poisoning themselves elsewhere, lending credence to the idea that Guam was an anomaly, probably due to conditions on the island during and immediately after WWII (when islanders consumed a vast quantity of cycad food, apparently with minimal processing). Thieret (1958) reports cycad consumption in numerous Australian aboriginal, Asian, and African societies, and in no case does it appear that the societies were poisoning themselves in ignorance; "famous" cases of mass cycad poisoning involves outsiders with little knowledge and patience (and great need) who have not followed all the necessary steps.

Medicinal Uses

An *atol* (slurry) prepared from well-washed and cooked *yuca de ratón* (*Z. standleyi*) in Olanchito is used as an effective remedy for children's diarrhea, particularly bloody diarrhea, among indigenous people (in Agalteca and among the Tolupan of Tribu Zapotal). Lemon is added to the *atol* in Agalteca. This use appears to be related to the use of *camotillo* as a homeopathic remedy.

The *brea* (tar or mucilage) on the female *tiusinte* cones is used for staunching wounds. Formerly, *brea* was also employed as a fixative for a cloth bandage soaked in *danto* (Baird's tapir, *Tapirus bairdii*) fat, which, when placed on the forehead, aided decongestion. Mucilage has also been mentioned as effective for helping *curar algunos granos* (to cure skin lesions).

Decorative Uses

In Religious Context

Tiusinte leaves have long been associated with death and rebirth, and this association is common to cycads across the world (Jones, 2002). The probable explanation is the plant's well-known longevity combined with its phoenix-like capacity to be "born again" after fire (indeed, several *Cycas* species bear "phoenix" in their vernacular epithets in China: Bonta and Osborne, in prep.). In Honduras, the reasons *tiusinte* leaves are preferred for several Roman Catholic rites of death and rebirth are no longer popularly known (just as the Nahuatl translation of its name has also been forgotten); Mexico *Dioon* leaves are similarly and widely used in culturally depauperate circumstances. However, cycad leaves carry deeper, apparently pre-Christian meanings among the Chiapanec of Tuxtla Gutierrez, Chiapas, who perform a six-day pilgrimage to carry leaves of *D. merolae* De Luca, Sabato & Vázq. Torres (*nimalari*=feather-leaf, or *espadaña*=church steeple) to household shrines as part of Día de la Cruz (Day of the Holy Cross, May 3) celebrations (Gómez-Pompa et al., 1997). This fully intact syncretized rite, which includes a "cycad prayer" intoned in the extinct Chiapanec language, provides strong evidence that cycad leaves had symbolic significance to Mesoamericans before the coming of Catholicism.

In El Paraíso department, Honduras (directly south of Olancho), *tiusinte* leaves are fashioned into wreaths marking the graves of children. In southern Olancho, wreaths (*coronas*), decorated with silver or gold paint and glitter, are carried to the cemetery to mark graves on *Día de los Difuntos* (Day of the Dead, All Souls' Day, Nov. 2), though the use of *tiusinte* leaves for this purpose has largely given way to paper and plastic alternatives. *Tiusinte* leaves are also used to decorate household shrines erected at the anniversaries of the deaths of family members. In northeastern Olancho, *tiusinte* leaves are used to decorate churches during *Semana Santa* (Holy Week). In northwestern Olancho and neighboring Olanchito, Yoro, *tiusinte* leaves are associated primarily with church facade decorations on Christmas Eve and with nativity scenes.

Arcades decorated with *tiusinte* leaves were formerly used widely to honor the visits of bishops to remote communities and for weddings. Finally, *tiusintes*, some of great age, are found growing in cemeteries in several communities.

For Secular Purposes

Tiusintes have no secular decorative uses within the species' native range. However, where they are found only cultivated, they have a variety of functions. *Para adorno* (for ornament) is a reason private owners of cultivated *tiusintes* often give for why they have the plant. Leaves have been used to decorate parade floats. Plants are highly regarded as a decorative species for public parks (*parques centrales*) and in courtyards of public buildings across Honduras; prime examples are in Parque Herrera and the Ministerio de Gobernación in Tegucigalpa and at the edge of the city park in La Ceiba.

Most domestic cycads in Honduras are cultivated in a handful of commercial nurseries, though zamias have never been recorded as decorative plants in traditional settings. However, urban Hondurans (perhaps unaware of the reputation of *camotillo*) have recently begun buying and cultivating them. Two nonnative species, *Z. furfuracea* L. *f.* and *Cycas revoluta*, however, appear to be the most popular.

Honduran species are also sold internationally, with *D. mejiae* the preferred species.

Miscellaneous Uses

As Toys

Dry *tiusinte* seeds, hollowed out, have been used as whistles (*cachinflín, chiflador, silbador,* or *pito*) much in the form of an ocarina; a similar use is reported in Mexico for *Dioon* seed kernels (Chamberlain, 1919). They were formerly heard sounding celebratory notes to mark the beginning of the *tiusinte* harvest. Another common toy was the *enchute* (pin-and-target)—hollowed-out seed kernels were perforated with numerous holes, then attached by an *Agave* string to a sharpened wooden pin. The object of a game between two players was to make the seed hop onto the pin, while grasping the pin. The winner would then announce *te puse una panza* (I impregnated you).

As Housing Material

The Tolupan of Tribu Zapotal, Yoro, occasionally use *tiusinte* leaves as roof thatch for wall-less shelters called *galeras*. Various palm leaves are used widely among rural people in Honduras, and indeed across the world, for the identical purpose.

As Rodenticide

Yuca de ratón, primarily *Z. standleyi*, gets its name from its utility as a preparation to poison mice that are damaging *frijolares* (bean fields). The Tolupan of Tribu Zapotal mash the *raíz* to a watery pulp and use it as bait, placing it not in the relevant *frijolar* but "on the other side of the hill," to ensure that the mice are lured away and die elsewhere.

As Domestic Animal Feed

Tiusinte seeds, cooked, can be ground up and given to chickens. However, *tiusinte* in Olancho is generally seen as a solely human food item, thus too valuable to waste on domestic animals.

As Laundry Starch

Previously, *yuca de ratón* was used among indigenous and mestizo inhabitants of Olanchito municipality as a fine-quality *almidón* (starch); this practice has been widely recorded for zamias across the Caribbean and at one time was a major industry in Florida (Small, 1921). The method related to us in Agalteca village was to grate the *raíz* and roast it in the oven before grinding it to powder.

Though the *harina* (flour) of *tiusinte* is also called *almidón*, we have recorded no utilization of the species for this purpose. Nevertheless, a place-name may indicate that this was once the case: the name of the largest single population of *tiusintes* in existence (about 168,000 plants at present and historically twice that number) is "Saguay" (in Gualaco municipality, Olancho), and it is the only place in Honduras with that name. The locative *-ay* (place of) is derived from a Misumalpan language, leaving *sagu* ("sago"), formerly a generic word for starch-producing plants. Unfortunately, the fragmentary Honduran archival record for Olancho is silent on this subject.

As Trade Items

Tiusinte has a minor exchange value; no traditional exchange value of other Honduran cycads has been reported. *Tiusinte* was once bartered for coffee in Gualaco municipality. In Olanchito, and to a lesser extent across Olancho, *tiusinteros* gather seeds and sell them in communities, providing an income supplement, primarily during the months of March, April, and May. *Tiusinteros* also have dug up and sold young plants on demand (leaves for ceremonial purposes are not sold but rather given away). In Olanchito, *tiusinteros* sell seeds to a variety of consumers, including to *tiusinteras* in the villages, who prepare tamales and sell them in the Olanchito city market, or to private homes.

Folk Beliefs and Sayings

Camotillo is found euphemistically in the expression *dar camotillo*, "to give *camotillo*," as in *parece que le dieron camotillo* ("it appears as if they slipped him *camotillo*"), in reference to a particularly sudden and inexplicable death. The primary belief associated with the actual consumption of *camotillo*, that number of days cut=number of days to die, as mentioned above, is found across Honduras, even in places where the actual identity of the plant is not known.

Very few folk beliefs are associated with *tiusinte*, which in itself is intriguing, as most plants in Honduras that are well-known and oft-utilized are associated with numerous beliefs. When pressed, several informants related that the main belief is that *tiusinte* is *maravilloso,* a magnificent food fit only for human consumption—indeed, the supremely human food and superior to maize. In the town and municipal center of Esquipulas del Norte, it is often said that "one who eats *tiusinte* will never leave" or "one who eats *tiusinte* will soon return to Esquipulas." Though the saying does not appear to be taken seriously, it is nevertheless often mentioned to the outsider; coincidentally, Esquipulas is the most *tiusinte*-centered muncipality, with little to no shame associated with consumption of the plant. Quite the opposite situation is found in neighboring Guata municipality, where *tiusinte* consumption is equated with extreme poverty and need, and, though engaged in quite regularly, is often not openly or pridefully admitted to.

Relationships to Agroecosystems and the Human Landscape

Intercropping

We have no contemporary or historical evidence of *tiusinte* being planted as a crop by itself, but in effect it is often "intercropped" by being protected—rather than removed—within fields planted to *Zea mays* L., *Phaseolus vulgaris* L., and musaceous plants. In one village in Esquipulas del Norte municipality, farmers who are also *tiusinteros* have traditionally planted maize after preparing fields through burning; the numerous tiusintes growing in the fields have never intentionally been removed, because harvest of the cones has provided a dietary as well as a cash supplement (maize is the primary subsistence crop for household consumption). Fire was observed to aid leaf flush (*vestimento*), but removal of *tiusinte* leaves that shaded out maize was observed to damage cycad growth and cone production; farmers adapted a method to balance maize production and *tiusinte* production as a result.

Camotillo/yuca de ratón often grows in agricultural fields only because farmers are loathe to disturb it and repeated burning does not destroy it as rapidly as other vegetation is disturbed. We have encountered no evidence of intentional "shepherding"—zamias planted or protected in maize fields because of symbolic importance as maize "lords" or shepherds—as has been recorded by Alcorn (1984) among the Huastec Maya of San Luis Potosí, Mexico. However, we cannot discount the possibility that zamias are left in Honduran fields traditionally, for reasons that have been forgotten by contemporary farmers, extending the idea forwarded herein that Honduras's *Dioon* and *Zamia* ethnobotany, largely derivative of historical and pre-Columbian Mexican sources, is consequently culturally depauperate. It should be added emphatically, however, that this paucity of deeper meaning does not denigrate the extensive and intricate local factual knowledge of cycads that has developed in situ, often in the context of one farmer or one family, independent of and largely unrelated to traditions, symbolism, and/or history. These local situations are, indeed, no doubt more relevant and important for those involved with schemes for human-centered protection of cycads; the larger connections drawn in this article are meant to inform the development of wider and generalized conclusions on the relationships between humans and cycads.

Integration into Pastoral Systems

Few traditional pastoral systems can be encountered in Honduras owing to the predominance of "modernization" in pasture preparation and cattle breeding and management (though see Johannessen, 1963). *Tiusintes* for the most part and zamias entirely are unintended but inevitable victims of the complete conversion of forest to pasture that takes place over the course of several years. However, in the case of *tiusinte*, ranchers who are aware of the local importance of *tiusinte* often leave the trees alone during the removal of scrub and forest and the planting of African pasture grasses. Though ranchers have not expressed any overt dislike of *tiusinte* (e.g., because of its toxicity, an issue that has never been raised at all by them), the mechanics of pasture preparation do eventually dictate the erosion and extirpation of *tiusinte* populations in any case. Pastures that are fired twice or thrice annually, within five years progress from dense *tiusinte* groves to open fields devoid of cycads. Zamias fare similarly.

Traditional pastoral systems, however, do not noticeably degrade *tiusinte* populations. These generally involve lower intensity burning in more marginal, hillside areas. The best example is the Tiusintalito population in Gualaco, which became cattle ranch land in the late 1700s ("Saguay," 1779) and remains so; today, it contains the largest *tiusinte* trees known and has healthy cone and seedling production (Haynes & Bonta, 2003).

Some ranchers who use modern, intensive methods nevertheless make an overt effort to protect *tiusintes,* and in one case in Gualaco limit fire to "sustainable" levels, achieving a balance of dense *tiusinte* growth and lush grass (peons are instructed to employ machetes, not just fire, to remove nonpalatable growth).

Integration into Silvicultural Systems

Zamias grow well in a wide variety of forest types, from dense, unburned rain forest, to occasionally burned gallery forest, to annually or biannually fired pine forest and pine/oak forest. They fall within the myriad silvicultural regimes and schemes applied to surrounding woods, without ever being singled out for special treatment.

Tiusintes are rarely found in pine forest (which is generally subject to artisanal and, increasingly, modern commercial logging), but are perceived as part of the natural tropical dry forest in places such as Gualaco, which can have many management schemes applied to it (as extension of kitchen gardens, shelter for livestock, orchards, municipal watersheds, private or public woodlots, etc.). The term used for Gualaco populations is *finca,* signifying in this case a tree plantation much like a shaded coffee *finca.* Inhabitants of Gualaco insist that all destruction of the Saguay forest in which the *tiusinte* traditionally grows is degradation of the *finca.* Informants in other municipalities are not so insistent upon this point.

Land Tenure Issues

Where zamias and *tiusintes* occur within village, town, and city watersheds, they are protected along with the rest of the forest vegetation. Watershed protection in Honduras has become an extremely important part of environmental protection in recent years, but has also been a part of legislation and tradition in Honduras at least since colonial times (Bonta, 2003a). More recent public land designations—national forest, national park, biological reserve, national wildlife refuge—may by accident contain zamias (more commonly) and *tiusintes* (infrequently), but in most case cycads occur on either private land or on *tierra nacional*—in either case, some sort of land use is generally occurring. State land is either under concession or "invaded" for agricultural, pastoral, silvicultural (i.e., timber), or other purposes; private land may be owned by individuals or by cooperative entities. Commonly, land rights are murky and conflictive. Within the extraordinarily complex de jure and de facto land tenure panorama, *Ceratozamia* and *Zamia* have until now been ignored, while *D. mejiae* is a clearly defined common property resource (it is *patrimonio*) that is customarily owned by the community and the municipality, not the individual or corporate entity that may hold title or usufruct rights to the land on which it grows. In Gualaco until the 1980s, all wild *tiusinte* was protected by municipal decree (as *el embargo*), and the annual harvest was regulated by the municipal government. Today, *tiusinteros* still claim completely open and unfettered access to all wild *tiusinte* trees, even those growing literally in the front yards of landowners. Some ranchers, on the other hand, think of themselves as the "owners" of all the *tiusintes* on their land and of the *tiusinteros* as trespassers and thieves.

Other municipalities have not decreed specific protection of *tiusintes* per se, but *tiusinteros* and their families still consider tiusintes to be common property everywhere the species is found.

Conclusion

This article has summarized relationships between humans and cycads in Honduras without entering into exhaustive detail but nevertheless portraying the importance that systematic ethnobotanical data gathering can have for our understanding of cycads as part of the human landscape. To elaborate, much as some might wish that cycads were free of human influence, in a realistic attempt to describe and understand cycads as we actually encounter them, we are led to the study of human interactions. By way of example, though at present we lack substantive data (at least from Latin America) we cannot rule out the possibility that humans have been in part responsible for cycad dispersal—a hypothesis that is only possible following a study of human uses and that, it appears to me, is crucial in our understanding of how, why, and where cycads thrive. Overall, my intent in the current article is to stimulate the framing of questions about the role cycads have played and continue to play in human culture and what roles humans play in the biology of cycads.

Acknowledgments

At time of writing, the author is a Montgomery Botanical Center Research Fellow; MBC generously supported the 2003 Honduras cycad expedition; I would like to thank Jody Haynes, Terrence Walters, and the MBC board for that seminal event, as well as for MBC's subsequent Honduras-focused research and promotion. The Louisiana State University Board of Regents supported the initial phase of this research; Delta State University has also supported the author's cycad research since 2003. In no particular order, I would like to thank the following people and institutions who have been integral to this research: Oscar Flores Pinot, German Sandoval, Ricardo Steiner, Cyril Nelson, Marta Moreno and AFE-COHDEFOR, Ciro Navarro, Pepe Herrero, Manuel and Mirza Rey, Isidro Zúniga, and communities and community members of Santa María del Carbón, Gualaco, Guata, Jano, Esquipulas del Norte, Juticalpa, and Olanchito.

Literature Cited

Aguilar Paz, J. 1999. Flora tradicional de Honduras. 2nd ed. Francisco Salvador Fundación Cultural, Ediciones Kukulcán, Tegucigalpa, Honduras.

Alcorn, J. B. 1984. Huastec Mayan ethnobotany. University of Texas Press, Austin, TX.

Beck, W. 1992. Aboriginal preparation of *Cycas* seeds in Australia. Econ. Bot. 46: 133–147.

Bonta, M. 2001. Mapping enredos of complex spaces: a regional geography of Olancho, Honduras. Ph.D. diss., Louisiana State University, Baton Rouge.

———. 2003a. Seven names for the bellbird: conservation geography in Honduras. Texas A&M University Press, College Station, TX.

———. 2003b. Teocinte, 'Ear of God'. Cycad Newslett. 26(1): 7–12.

——— & R. Osborne. 2007. Cycads in the vernacular—A compendium of local names. Mem. New York Bot. Gard. 97: 182–193.

Bonta, M., O. Flores Pinot, D. Graham, J. Haynes & G. Sandoval. In prep. Ethnobotany and conservation of tiusinte (*Dioon mejiae* Standl. & L. O. Williams, Zamiaceae) in northeastern Honduras.

Brenner, E. D., D. W. Stevenson & R. W. Twigg. 2003. Cycads: evolutionary innovations and the role of plant-derived neurotoxins. Trends Pl. Sci. 8(9): 446–452.

Chamberlain, C. J. 1919. The living cycads. Hafner Publishing, New York.

Conzemius, E. 1928. Los indios Payas de Honduras: estudio geográfico, histórico, etnográfico, y lingüístico. Société des Américanistes de Paris, Paris.

Donaldson, J. S. (ed.). 2003. Cycads: Status Survey and Conservation Action Plan. IUCN/Cycad Specialist Group. IUCN, Gland and Cambridge.

Duke, J. A. 1975. Ethnobotanical observations on the Cuna Indians. Econ. Bot. 29: 278–293.

Gómez-Pompa, A., A. P. Vovides, N. Ogata, R. Castro-Cortés, J. A. González García & A. Corona López. 1997. Cycads, the endangered living fossils/Cycadas, fósiles vivientes en peligro de extinción. Q'Taxa CD-ROM, Plant Resources Information Laboratory, Department of Botany and Plant Sciences, University of California, Riverside.

Hardy Nelson, S. C. 1986. Plantas comunes de Honduras. Editorial Universitario, Tegucigalpa, Honduras.

Haynes, J. 2007. Cycad flora of Honduras: species, taxonomy, population structure and conservation status. Mem. New York Bot. Gard. 98: 000–000.

Haynes, J. & M. Bonta. 2003. Montgomery Botanical Center Honduras 2003 cycad expedition final report. Submitted to AFE-COHDEFOR, Tegucigalpa, Honduras. Montgomery Botanical Center, Miami, FL.

——— & ———. 2007. An emended description of *Dioon Mejiae* Standl. & L. O. Williams (Zamiaceae). Mem. New York Bot. Gard. 97: 418–443.

Johannessen, C. 1963. Savannas of interior Honduras. Ibero-Americana 46. University of California Press, Berkeley.

Jones, D. L. 2002. Cycads of the world: ancient plants in today's landscape. Ed. 2. Reed New Holland, Sydney, Australia.

Landa, L. 1935. Gimnospermas. Revista Arch. Bibliot. Nac. Honduras, Bot. Sist. series, 13(11): 715–718.

———. 1940. Botánica. Ariston, Tegucigalpa, Honduras.

Lentz, D. 1993. Medicinal and other economic plants of the Paya of Honduras. Econ. Bot. 47(4): 358–370.

Monografía de Yoro. 1926. Sociedad Pedagógica de Yoro, Olanchito, Honduras.

Newson, L. 1986. The cost of conquest: Indian decline in Honduras under Spanish rule. Westview, Boulder, CO.

Osborne, R., A. Grove, P. Oh, T. J. Mabry, J. Ng & A. A. Seawright. 1994. The magical and medicinal usage of *Stangeria eriopus* in South Africa. J. Ethnopharmacol. 43: 67–72.

Pérez-Arbelaez, E. 1956. Plantas útiles de Colombia. Ed. 3. Librería Colombiana, Bogotá, Colombia.

Roig-Mesa, J. T. 1965. Diccionario bótanico de nombres vulgares cubanos. Editorial Nacional de Cuba, Editora del Consejo Nacional de Universidades, La Habana, Cuba.

"Saguay." 1779. Archivo Nacional de Tierras/Olancho. Land title #166. In the Archivo Nacional Histórico, Tegucigalpa, Honduras .

Schultes, R. E. & R. F. Raffauf. 1990. The healing forest: medicinal and toxic plants of northwest Amazonia. Vol. 2. Dioscorides Press, Portland, OR.

Small, J. K. 1921. Seminole bread: the conti, a history of the genus *Zamia* in Florida. J. New York Bot. Gard. 22: 121–129.

Standley, P. C. 1931. Flora of the Lancetilla Valley, Honduras. Publ. Field Mus. Nat. Hist., Bot. Ser. 12: 66–67.

Standley, P. C. & S. J. Record. 1936. The Forests and Flora of British Honduras. Publ. Field Mus. Nat. Hist., Bot. Ser. 17: 1–432.

———. 1937. Flora of Costa Rica. Part 1. Publ. Field Mus. Nat. Hist., Bot. Ser. 18 (1): 63–64.

——— & L. O. Williams. 1950. *Dioon mejiae*, a new cycad from Honduras. Ceiba 1: 36–38.

——— & J. A. Steyermark. 1958. Flora of Guatemala, Part I. Publ. Field Mus. Nat. Hist., Bot. Ser. 24 (1): 11–20.

Thieret, J. W. 1958. Economic botany of cycads. Econ. Bot. 12: 4–41.

Vázquez Torres, M. 1990. Algunos datos etnobotánicos de las cicadas en Mexico. Mem. New York Bot. Gard. 57: 144–147.

Whitelock, L. M. 2002. The cycads. Timber Press, Portland, OR.

Whiting, M. G. 1963. Toxicity of cycads. Econ. Bot. 17: 270–302.

ETHNOBOTANY
CHAPTER 10

Cycads in the Vernacular—A Compendium of Local Names

Mark A. Bonta and Roy Osborne

Abstract

We present the first ever worldwide compilation of traditional cycad names and discuss the significance of our findings. Using all available cycad literature and much personal consultation, we have compiled a table of scientific names, localities, languages, vernacular names, and, where known, translations into English. We have uncovered at least one vernacular name for 146 of the 303 presently known cycad species and a total of 495 vernacular names in 129 languages. Patterns of meaning revealed in our compendium include cycad-palm and cycad-fern homologies, a strong maize analogy (in Mesoamerican countries), toponymic references, allusions to morphological features, non-Western binomial taxonomies, references to local folklore, and references to economic uses of cycad products.

Resumen

Se presenta el primer compendio mundial de nombres tradicionales de Cycadales; se discute su significado. Haciendo uso de toda literatura disponible además de la consulta personal, hemos compilado una tabla de nombres científicos, localidades, idiomas, nombres vernaculares y traducciones al inglés (donde se sabe). Hemos descubierto por lo menos un nombre vernacular por 146 de las 303 especies de Cycadales reconocidas al presente, por un total de 495 nombres vernaculares en 129 idiomas. Nuestro compendio revela homologías entre Cycadales y palmas y entre Cycadales y helechos, una fuerte analogía con el maíz (en paises mesoamericanos), referencias toponímicas, alusiones a rasgos morfológicos, taxonomías binomiales no occidentales, referencias al folclore, y referencias a los usos económicos de productos de Cycadales.

Introduction

The last few years have seen much activity in the field of cycad systematics, including many new species circumscriptions and a watershed meeting of cycad systematists in April 2002, in Miami, Florida, United States, with the consequent publication of the text *Cycad Classification: Concepts and Recommendations* (Osborne & Walters, 2004. These efforts address fundamental questions relating to the biological species concept vis-à-vis cycads and the relationships within and between cycad taxa. To complement this much-needed systematics work, it is also crucial that we not overlook the importance of cycads to local communities throughout the world. This importance is often demonstrated by a plethora of vernacular names and accompanying beliefs, and in some regions (e.g., China), the evolution of elaborate folk cycad taxonomies.

Why are the names and taxonomies of cycads in non-Western scientific contexts important to scientists? Although there are several reasons, perhaps paramount is the significance that traditional, ethnobotanical knowledge systems have for conservation in the broadest sense. A name, a meaning, a story, a toponym—where cycads are concerned, these indicate local awareness of populations of what are in many cases globally threatened species. The better these plants are known and appreciated locally, the easier it may be for conservation programs to argue for the type of village-level protection that botanists such as Willie Tang, Andrew Vovides, and colleagues have pioneered in China and Mexico (Donaldson et al., 2003, Vovides et al., 2002).

Local names are more than mere curiosities or outdated folklore—rather, they are portals of entry into worlds of meaning that complement Western scientific insight. As cycad systematists labor to establish biological certainty, we offer this compilation as a parallel effort to inspire cycad aficionados to search out the cultural meanings that will allow a common language to be spoken by outsider and local person alike.

Another benefit of a cycad vernacular compendium is that it enhances the ability of scientists to locate current and historic populations of the plants. Recently, Bonta (2007), upon discovering that the term *tiusinte* referred to *Dioon mejiae* in northeastern Honduras, located several populations during interviews with local cycad cone harvesters. In toponyms, the root *tiusinte* was found, for example, in *El Tiusintal* (place of cycads, a hamlet), *Cerro del Tiusinte* (cycad hill), and *Quebrada del Tiusinte* (cycad stream). Previously, outside researchers had assumed that the term (often spelled "teocinte") referred to *Zea mays*, thence overlooking *D. mejiae*. Without this knowledge of names, Bonta would never have been able to track down remote populations (Bonta, 2003). In a similar sense, Osborne used the key word *imFingo* as a prompt to communication with Zulu medicine-men in kwaZulu-Natal, South Africa, leading to valuable revelations about the magical and medicinal uses of *Stangeria eriopus* (Osborne et al., 1994).

Note: For clarity of presentation, we have omitted the authorities to species names in this text; all such authorities are shown in Table 10-1.

Methods

To summarize cycad folk taxonomies, we searched all the cycad literature known to us, and condensed the details into a table (Table 10-1) showing scientific names, localities, languages, vernacular names, and translations into English (where known).

Table 10-1 is organized into regional areas on the basis of broad cultural delineations—for instance South Asia (the Indian subcontinent) is recognized as a cultural region distinct from East Asia. To our knowledge, other attempts to compile such a list

Table 10-1 Scientific names, localities, and vernacular names of African, American, Asian, and Oceanian cycads.

Key: Ab. = Aboriginal; Aust. = Australian; Ch. = Chinese; dimin. = diminutive; Eng. = English; Fr. = French; Jap. = Japanese; lang. = language; lit. = literal; Náh. = Náhuatl; pref. = preferred; ref. = reference; Sp. = Spanish; translit. = transliteration; unident. = unidentified; Viet. = Vietnamese.

(A) Cycads of Africa and adjacent Indian Ocean islands.

Scientific name	Locality	Vernacular name	References
Cycas thouarsii R.Br. ex Gaudich.	Comoros, Madagascar, Mozambique, Tanzania	Betsileo—*fahou, fahu, voafaho*; Fr.—*faux sagoutier* (false sago), *samble*; Merina—*batsimisaraka* (man-eating plant), *fato, fatra, fatzon, voafako*; Swahili—*mpapindi* (also used for the palm *Dypsis pembanus*), *mtapo, mtapu*; unident. lang. —*tsambou*	Lamarque & Razafimahery, 1950; Schuster, 1932; Vorster & Vorster, 1973; Williams, 1949; H. Beentje, pers. comm.
Encephalartos altensteinii Lehm.	S. Africa (E. Cape)	Xhosa—*iSundu, umGuza, umPhanga*	Dold & Cocks, 1999; Pooley 1993; von Breitenbach & von Breitenbach, 1992
Encephalartos barteri Carruth. ex Miq.	Benin, Ghana, Nigeria, Togo	Krobo—*kpadei-atah, pardi attar* (both ghost palm); Twi-akapim—*mamponmere*	Dalziel, 1937; Prain, 1917; Schuster, 1932
Encephalartos ferox Bertol. f.	Mozambique, S. Africa (kwaZulu-Natal)	Ronga—*chipissana*; Ronga & Shona—*untopani* (palm-like plant); Tsonga—*chihanga* (vegetable coal); Zulu—*umThobane, uThobani*	da Graça Silva, 1983; de Koning, 1993; Pooley, 1993; von Breitenbach & von Breitenbach, 1992
Encephalartos friderici-guilielmi Lehm.	S. Africa (E. Cape, kwaZulu-Natal)	Xhosa—*umGuza, umPhanga*	Pooley, 1993; von Breitenbach & von Breitenbach, 1992
Encephalartos ghellinckii Lem.	S. Africa (E. Cape, kwaZulu-Natal)	Xhosa—*umGuza, umPhanga*; Zulu—*isiDawu, isiGqikisomkhovu* (zombie's pillow)	Pooley, 1993; von Breitenbach & von Breitenbach, 1992
Encephalartos gratus Prain	Malawi, Mozambique	Chichewa(?)—*gwerekwe*; Yoa—*lichaungwe*	Binns, 1972
Encephalartos heenanii R.A. Dyer; *E. paucidentatus* Stapf & Burtt Davy	S. Africa (Mpumalanga), Swaziland	Swazi—*siJekwane*	von Breitenbach & von Breitenbach, 1992

Table 10-1 (A) (*continued*)

Scientific name	Locality	Vernacular name	References
Encephalartos hildebrandtii A. Braun & C.D. Bouché	Kenya, Tanzania	Boni—*icheli, tielle*; Giriyama—*kisapo*; Sanya—*balacha, muka*; Swahili—*mapo, mkamwa, mkwanga, msapo*	Beentje, 1994; Dharani, 2002; Weiss, 1979; H. Beentje, pers. comm.
Encephalartos kisambo Faden & Beentje	Kenya	Taita—*kisambo*	Faden & Beentje, 1989; Hurter, 1994
Encephalartos laurentianus De Wild.	Angola, Dem. Rep. Congo	Kiyaka—*malele, bundu-nguma*	Engler, 1908; Lebrun, 1930; Prain, 1917; Robyns, 1948
Encephalartos macrostrobilus S. Jones & Wynants	N. Uganda	Madi—*ci-cia* (like a date palm)	Jones & Wynants, 1997
Encephalartos manikensis (Gilliland) Gilliland and allied species incl. *E. chimanimaniensis* R.A. Dyer & I.Verd., *E. munchii* R.A. Dyer & I.Verd. and *E. turneri* Lavranos & D.L. Goode	Mozambique, Zimbabwe	Ronga—*chipissana*; Ronga & Shona—*untopani* (palm-like plant); Shona—*mu Tata, mu Tete, unguarara* (dwarf plant); Tsonga—*chihanga* (vegetable coal)	da Graça Silva, 1983; deKoning, 1993; Gilliland, 1939; P. Capela, pers. comm.
Encephalartos marunguensis Devred	Dem. Rep. Congo	Kitabwa—*kavunduvundu, tusu*	Malaisse, 1969
Encephalartos middelburgensis Vorster and allied species incl. *E. dolomiticus* Lavranos & D.L. Goode, *E. dyerianus* Lavranos & D.L. Goode, *E. eugene-maraisii* I.Verd., *E. inopinus* R.A. Dyer and *E. lanatus* Stapf & Burtt Davy	S. Africa (Mpumalanga, with allied species also in Limpopo)	North Sotho—*mofaka* (singular), *mefaka* (plural)	von Breitenbach & von Breitenbach, 1992; J. de Wet Bösenberg, pers. comm.
Encephalartos natalensis R.A. Dyer & I.Verd., and allied species incl. *E. aemulans* Vorster, *E. lebomboensis* I.Verd. and *E. senticosus* Vorster	Mozambique, S. Africa (kwaZulu-Natal), Swaziland	Swazi—*siJekwane*; Xhosa—*umGuza, umPhanga*; Zulu—*isiGqikisomkhovu* (zombie's pillow), *uJobane, uJubane, umHlungulo, umNgquabe*	Pooley, 1993; von Breitenbach & von Breitenbach, 1992; R. Osborne, pers. observ.
Encephalartos poggei Asch.	Dem. Rep. Congo	Baluba—*biondo*; Kabwe—*katende*; Kanioka—*tchiondo, tchiunda*; Lulua—*kalaba, kalaba kabo, kalala kabo, lulondo*	Gentil, 1904; Prain, 1917; Robyns, 1948; Schuster, 1932; Whitelock, 2002

(*continued*)

Table 10-1 (A) (*continued*)

Scientific name	Locality	Vernacular name	References
Encephalartos princeps R.A. Dyer	S. Africa (E. Cape)	Xhosa—*umGuza, umPhanga*	Pooley, 1993; von Breitenbach & von Breitenbach, 1992
Encephalartos schaijesii Malaisse, Sclavo & Crosiers	Dem. Rep. Congo	Ndembo—*makindu ya dilungu*	Malaisse et al., 1993
Encephalartos schmitzii Malaisse	Dem. Rep. Congo, Zambia	Unident. lang.—*kanchindu*	Gono & Phiri, 2001
Encephalartos septentrionalis Schweinf.	Sudan, Uganda	Azande—*bopotele*; Bongo—*kagga-kunda, kaggo kondo*; Kredy-Ndugu lang.—*kotto* (alcoholic beverage from trunk); Niam-Niam lang.—*mevul-pia, mwue piah*	Prain, 1917; Robyns, 1948; Schuster, 1932; Schweinfurth, 1871
Encephalartos tegulaneus Melville	Kenya	Samburu—*lpision*	Beentje, 1994
Encephalartos transvenosus Stapf & Burtt Davy and allied species	S. Africa (Limpopo)	North Sotho—*mofaka* (singular), *mefaka* (plural); Venda—*mutondolo, tshifhanga, tshitondolo*	von Breitenbach & von Breitenbach, 1992
Encephalartos umbeluziensis R.A. Dyer	Mozambique, Swaziland	Tsonga—*muanga*	de Koning, 1993
Encephalartos villosus Lem.	S. Africa (E. Cape, kwaZulu-Natal), Swaziland	Xhosa—*umGuza, umPhanga*	Dold & Cocks, 1999; von Breitenbach & von Breitenbach, 1992
Encephalartos spp. (generic)	S. Africa	Afrikaans—*broodboom* (bread tree) (often compounded with morphological or geographic adjectives)	common knowledge
Stangeria eriopus (Kunze) Baill.	S. Africa (E. Cape, kwaZulu-Natal)	Afrikaans—*bobbejaankos* (baboon food); Pondo(?)—*finguane, juma*; Xhosa—*umFingwane* (whole plant), *umNcuma* (cone); Zulu—*imFingo, inFingo*	Osborne et al., 1994; Vorster & Vorster, 1974

(B) Cycads of the Americas.

Scientific name	Locality	Vernacular name	References
Ceratozamia hildae G.P. Landy & M.C. Wilson	Mexico (Queretaro, San Luis Potosi)	Sp.—*pata de gallo* (cock's foot)	A. Vovides, pers. comm.
Ceratozamia mexicana Brongn.	Mexico (Hidalgo, Puebla, Veracruz)	Huastec Maya (also known as Teenek)—*t zalam-thipac* (Thipaak's shade—see *Zamia fischeri*, below); Sp.—*costilla de león* (puma's rib), *palma imperial* (imperial palm), *piña del monte* (wild pineapple)	Alcorn, 1984; Vovides et al., 1992
Ceratozamia microstrobila Vovides & J.D. Rees	Mexico (San Luis Potosi)	Sp. dimin. of Náh.—*chamalillo* (little *chamal* or little *Dioon edule*)	Vázquez Torres, 1990
Ceratozamia miqueliana H. Wendl.	Mexico (Chiapas, Tabasco, Veracruz)	Sp.—*palmita* (little palm)	Vovides et al., 1992
Ceratozamia mixeorum Chemnick, T.J. Greg. & Salas-Mor.	Mexico (Oaxaca)	Sp.—*carrete* (ox cart, as toy)	Chemnick et al., 1997
Ceratozamia morettii Vázq. Torres & Vovides	Mexico (Veracruz)	Náh.—*tepetmaizte*, *tepemaizte* (hill or mountain maize)	Vázquez Torres & Vovides, 1998
Ceratozamia robusta Miq.	Belize, Guatemala, Mexico (Chiapas, Oaxaca, Veracruz)	Sp.—*costilla de león* (puma's rib), *palma de camote* (sweet potato [*Ipomoea batatas*] palm); *palma imperial* (imperial palm), *palmilla* (little palm), *piña del monte* (wild pineapple); Tzeltal—*tushku*	Vázquez Torres, 1990; Vovides et al., 1992; R. Osborne, pers. observ.
Ceratozamia Brongn. (generic, also including *Zamia* spp.)	Mexico (Chiapas)	Unident. langs.—*amendaui*, *amendu*, *amenduai*; Sp.—*palma de jangó*	Patiño, 1989; M. A. Pérez-Farrera, pers. comm.
Chigua bernalii D.W. Stev. & *C. restrepoi* D.W. Stev.	Colombia	Sp.—*chigua* (ref. to both cycads and ferns)	Jones, 2002
Dioon caputoi De Luca, Sabato & Vázq. Torres	Mexico (Puebla)	Sp.—*palma real* (royal palm)	Vázquez Torres, 1990; Whitelock, 2002
Dioon edule Lindl.	Mexico (Gulf of Mexico coast)	Náh.—*chamal*, *quiotamal*, *sotol*, *tiotamal* (sacred tamale), *tzamal*; Sp.—*palma de Dolores* (Our Lady of Sorrows palm), *palma de macetas* (flowerpot palm), *palma de Santa Teresita* (Saint Teresa's palm), *palma de serrucho* (handsaw palm), *palma de la Virgen* (Virgin's palm), *palmita* (little palm), *tullidora* (crippler)	Martinez, 1979; Schuster, 1932; Vázquez Torres, 1990; Vovides et al., 1992; Whitelock, 2002; J. Chemnick, pers. comm.

(continued)

Table 10-1 (B) (*continued*)

Scientific name	Locality	Vernacular name	References
Dioon holmgrenii De Luca, Sabato & Vázq. Torres	Mexico (Oaxaca)	Sp.—*palma del sol* (sun palm), *plumilla* (little feather)	Whitelock, 2002
Dioon mejiae Standl. & L.O. Williams	N.E. Honduras	Sp. translit. of Náh.—*tiusinte* (pref. to *teocinte, teosinte*) (sacred ear of maize, *Zea mays*); Tol—*tiñuc*	Bonta, 2003, 2007
Dioon merolae De Luca, Sabato & Vázq. Torres	Mexico (Chiapas, Oaxaca)	Chiapaneco—*nimalari* (feather leaf); Sp.—*espadaña* (church steeple), *maíz viejo* (old time maize), *morrito* (small tree gourd with spherical fruit, *Crescentia* sp.), *palma espinuda* (spiny palm), *palmilla* (little palm)	de Luca et al., 1981; Warnholtz, 2005; A. Vovides, pers. comm.
Dioon purpusii Rose	Mexico (Oaxaca)	Chontal—*la-fane-tejua, lan-zi-le, lan-zi-li*; Náh.—*chamal*; Sp.—*palma real* (royal palm)	de Luca et al., 1978; Martinez, 1979; Schuster, 1932
Dioon rzedowskii De Luca, Moretti, Sabato & Vázq. Torres	Mexico (Oaxaca)	Mazatec—*tush-kju*	Whitelock, 2002
Dioon sonorense De Luca, Sabato & Vázq. Torres	Mexico (N.W. coast)	Nah.—*peyote*; Sp.—*palma de la Virgen* (Virgin's palm), *peine* (comb, ref. to the leaf),	de Luca, et al. 1984; A. Vovides, pers. comm.
Dioon spinulosum Dyer	Mexico (Oaxaca, Veracruz)	Sp.—*coyolito de cerro* (little hill coyol [*Acrocomia*] palm), *palma de chicle* (rubber palm), *palma de chicalite* (ref. to the seed preparation for foodstuff), *palma de Dolores* (Our Lady of Sorrows palm), *chicalito* (the seed, used to make bull-roarers or *zumbadores*), *coyolillo*	Schuster, 1932; Vázquez Torres, 1990; Whitelock, 2002; A. Vovides, pers. comm.
Dioon tomasellii De Luca, Sabato & Vázq. Torres	Mexico (S.W. coast)	Sp.—*palma, palma de la Virgen* (Virgin's palm), *palma del diablo* (Devil's palm), *palmita* (little palm), *peinetillo* (little comb)	de Luca et al., 1984; Gregory et al., 2007
Dioon Lindl. (generic)	Mexico (Oaxaca)	Sp.—*hueso de pescado* (fishbone), *maíz viejo* (old time maize), *palma de oro* (golden palm), *palma espinada* (spiny palm), *palmilla* (little palm)	Chemnick et al., 1997; Martinez, 1979; A. Vovides, pers. comm.

Table 10-1 (B) (*continued*)

Scientific name	Locality	Vernacular name	References
Microcycas calocoma (Miq.) A.DC.	Cuba	Sp.—*corcho, palma corcho, palma de corcho* (all ref. to cork palm)	Schuster, 1932
Zamia amazonum D.W. Stev.	Ecuador (Napo)	Quichua—*guagrapanga*; Secoya—*ñanami jicó*; Sp.—*rabo de raya* (ray [type of fish] tail)	F. Nicolalde-Morejón, in prep.
Zamia angustifolia Jacq.	Bahamas, Cuba	Sp. (Cuba)—*guáyara, yuca cimarrona* (wild manioc, *Manihot esculenta*), *yuca guáyara, yuquilla de paredón* (little cliff-dwelling manioc), *yuquilla de paredón de hoja menuda* (little small-leaved cliff-dwelling manioc), *yuquilla de ratón* (little mouse manioc)	Roig y Mesa, 1965
Zamia boliviana (Brong.) A. DC.	Bolivia, Brazil	Sp. (Bolivia)—*mauro*; Port. (Brazil)—*sagueiro da terra*	Patiño, 1989
Zamia chigua Seem.	Colombia (Choco), Panama (Chiriquí)	Sp.—*chigua macho* (male *chigua*, ref. to both cycads and ferns), *helecho* (fern)	Norstog & Nicholls, 1997; Whitelock, 2002
Zamia cunaria Dressler & D.W. Stev.	Panama	Cuna—*obset* (plant), *obset e sana* (female cone)	Stevenson, 1993
Zamia encephalartoides D.W. Stev.	Colombia (Santander)	Sp.—*cacao del indio* (Indian cocoa, *Theobroma cacao*)	Stevenson, 2001; Whitelock, 2002
Zamia fairchildiana L.D. Gómez	Costa Rica, Panama	Sp.—*palma de cerro* (hill palm), *palma de pegamento* (glue palm)	J. Chemnick, pers. comm., C. Lopez-Gallego, pers. comm.
Zamia fischeri Miq. and *Zamia vazquezii* D.W. Stev., Sabato, A. Moretti & De Luca	Mexico (Querétaro, San Luis Potosi, Veracruz)	Huastec Maya (also known as Teenek)—*ahaatik a eem* (maize lord), *Thipaak* (mythical boy who brought maize to the Huastec), *tsakam way* (small ear of maize), *tsakam Thipaak* (little Thipaak), *tsalaam Thipaak* (Thipaak's shade), *ts'een thipaak* (mountain Thipaak); Náh.—*chamalillo* (little *chamal* or dwarf *Dioon edule*), *teocinte* (pref. to *teocintle*) (sacred ear of maize, *Zea mays*), Sp.—*amigo del maíz* (friend of maize), *mazorca* (maize ear, ref. to cone)	Alcorn, 1984; Vovides et al., 1992; B. Schutzman, pers. comm.

(*continued*)

Table 10-1 (B) (*continued*)

Scientific name	Locality	Vernacular name	References
Zamia furfuracea L. f.	Mexico (Veracruz)	Sp.—*palma bola* (drunk palm)	A. Vovides, pers. comm.
Zamia gentryi Dodson	Ecuador (Esmeraldas)	Sp.—*piña de monte* (wild pineapple)	F. Nicolalde-Morejón, in prep.
Zamia inermis Vovides, J.D. Rees & Vázq. Torres	Mexico (Veracruz)	Sp.—*palmita* (little palm)	Whitelock, 2002
Zamia integrifolia L. f.	Bahamas, Cayman Islands, Cuba, USA (Florida, Georgia)	Eng. (Bahamas)—*bay-rush*; Eng. (USA)—*comfort root, Florida arrowroot, wild sago*; Seminole—*coontie* (pref. to *compte, comptie, comtie, conti, Conti Hateka* [white root, white bread-plant], *coonti, coontia, koontee, konti, koontie, kunti*)	Correll & Correll, 1982; Sleight, 1953; Small, 1921, 1933
Zamia kickxii Miq.	Cuba	Sp. (Cuba)—*yábuna, yuquilla* (little manioc, *Manihot esculenta*), *yuquilla de ratón* (little mouse manioc), *yuquilla de sabana* (little savanna manioc)	Roig y Mesa, 1965
Zamia lecointei Ducke	Brazil, Colombia (Amazonas, Antioquia), Peru (Loreto), Venezuela	Taiwano—*gaw*	Schultes & Raffauf, 1990
Zamia loddigesii Miq.	Guatemala (Petén), Mexico (Hidalgo, Oaxaca, Veracruz)	Náh.—*teocinte, teocintle, teosinte* (sacred ear of maize, *Zea mays*) (plants), *mazorca* (ear of maize); Sp.—*amigo del maíz* (friend of maize), *camotillo, cocalito, palmiche* (palmetto), *palmilla* (little palm); Zoque-poluca—*cahua, chacuhua, maíz de coshca, poua*; unident. langs. —*guachumpoyó; tzompollo*	Patiño, 1989; Standley & Steyermark, 1958; Vázquez Torres, 1990; Vovides et al., 1992; B. Schutzman, pers. comm.; A. Vovides, pers. comm.
Zamia melanorrhachis D.W. Stev.	Colombia (Amazonas, Cordoba, Santander)	Sp.—*corocito* (little cohune palm); unident. lang.—(*muí*) *ibaracú*;	Stevenson, 2001
Zamia muricata Willd.	Colombia, Venezuela	Unident. lang.—*acesiva*	Patiño, 1989
Zamia neurophyllidia D.W. Stev.	Costa Rica, S. Nicaragua, N. Panama	Ngobere—*tadowa*; Afro-Panamanian- —*guade teet*	Holzman and Haynes, 2004
Zamia obliqua A. Braun	Colombia (Antioquia, Chocó, Valle), S. Panama	Sp.—*chigua macho* (male *chigua*, ref. to both cycads and ferns), *maicito* (little maize)	Bernal and Restrepo, 1991; Stevenson, 2001; F. Nicolalde-Morejón, in prep.

Table 10-1 (B) (*continued*)

Scientific name	Locality	Vernacular name	References
Zamia poeppigiana Mart. & Eichler	Ecuador	Sp.—*palma de goma* (glue palm), *palma tabaquillo* (little tobacco palm); Tsa'fiqui—*pichi richi*	MacBride, 1936; F. Nicolalde-Morejón, in prep.; Schultes & Raffauf, 1990
Zamia portoricensis Urban	Puerto Rico	Sp. (Puerto Rico)—*marunguey*	Schuster, 1932
Zamia pumila L.	Cuba, Dominican Republic, Puerto Rico	Sp.—*guáyara, guáyiga, guáyuya, yábuna, yuquilla de paredón* (little cliff-dwelling manioc, *Manihot esculenta*), *yuquilla de ratón* (little mouse manioc), *yuquilla de sabana* (little savanna manioc)	Grey, 1927; Roig y Mesa, 1965; Zanoni, 1982
Zamia pygmaea Sims	Cuba (W. Cuba, Isla de la Juventud)	Sp. (Cuba)—*yuquilla de ratón* (little mouse manioc, *Manihot esculenta*), *yuquilla de sabana* (little savanna manioc)	González Géigel, 2003; Roig y Mesa, 1965
Zamia roezlii Linden	Colombia (Amazonas, Chocó, Nariño, Valle), Ecuador	Chá palaa—*aafachi*; Sp.—*chigua* (ref. to both cycads and ferns), *piña de monte*; Tsa'fiqui—*pichi-richi*	Bernal and Restrepo, 1991; F. Nicolalde—Morejón, in prep.
Zamia skinneri Warsz. ex A. Dietrich	Central & N.W. Panama	Cuna—*elkia*; Sp.—*cebolla* (onion), *cebolla roja* (red onion), *cebollina* (little onion)	Duke, 1975; J. Haynes, pers. comm.; D. Stevenson, pers. comm.
Zamia standleyi Schutzman	N. Honduras	Pech—*yojrá unicej* (venomous manioc, *Manihot esculenta*); Sp.—*camotillo* (little camote or sweet potato, *Ipomoea batatas*), *yuca de ratón* (mouse manioc)	Bonta, 2007
Zamia ulei Dammer	W. Brazil, Colombia, Ecuador, Peru	Karijona—*ma-re-ká-mê*; Yukuna—*koo-roó-chee*; Tukano—*ku*	Bernal and Restrepo, 1991; Schultes & Raffauf, 1990; Whitelock, 2002
Zamia variegata Warsz.	Guatemala, Mexico (Chiapas)	Sp.—*camotillo*	Standley & Steyermark, 1958
Zamia wallisii A. Braun	Colombia (Antioquia)	Sp.—*chigua* (ref. to both cycads and ferns); unident. lang.—*acesiva*	Pérez-Arbelaez, 1956
Zamia L. spp. from the northwest Amazon (generic)	Colombia	Karapaná—*koo*; Kubuyarí—*maw'-ti-ree-na*; Tanimuka—*koo*	Schultes & Raffauf, 1990

(C) Cycads of Oceania: Australia, Eastern Indonesia, Papua New Guinea and S. Pacific islands.

Scientific name	Locality	Vernacular name	References
Bowenia spectabilis Hook. ex Hook. f.	Australia (Qld)	Yalanji Ab.—*bungkay*; unident. Ab. langs. —*chiroo, gunyoo, jayur*	Wilson, 2002; R. Hill, pers. comm.
Cycas angulata R.Br.	Australia (NT, Qld)	Aust. Ab.—*munbuwa* (seed), *ngathu* (plant)	Beck, 1992, 1993
Cycas apoa K.D. Hill	Indonesia (Maluku, Papua), Papua New Guinea	Ekor village lang. (Halmahera I.)—*bico*; Kaka—*apoa*; Sentani—*handambo*	Hill, 1994a
Cycas armstrongii Miq.	Australia (NT)	Iwaidja Ab.—*manggugbu, wurarrngarr*; Malakmalak Ab.—*walangkirrik*; Matngala Ab.—*turtur*; Tiwi Ab.—*kwaka* (seeds), *minta* (female plant),	Blake et al., 1998; Lindsay et al., 2001; Puruntatameri et al., 2001
Cycas arnhemica subsp. *muninga* Chirwin & K.D. Hill	Australia (NT)	Anindilyakwa Ab.—*muninga*	Hill, 1994b
Cycas arnhemica subsp. *natja* K.D. Hill	Australia (NT)	Burarra Ab.—*natja*; Djambarrpuyngu Ab.—*warraga* (plant), *laluk* (seeds)	Hill, 1994b; Wightman & Smith, 1989
Cycas bougainvilleana K.D. Hill	Papua New Guinea (Bougainville, New Britain), Solomon Is.	Ambungi village lang. (New Britain)—di di; Matong and Pomio village langs. (New Britain)—*pago*; Rennel & Baga Island langs. (Solomon Is.)—*bae bae, bai bai*; Suvai village lang. (Bougainville)—*diriuna*; Tonolei village lang. (New Britain)—*diduna*	Hill, 1994a
Cycas calcicola Maconochie	Australia (NT)	Malakmalak Ab.—*walangkirrik*; Matngala Ab.—*turtur*; Ngan'gikurunggurr and Ngan'giwumirri Ab.—*yerrsya* (plant), *misya* (seeds); Wardaman Ab.—*ngewerrin*	Lindsay et al., 2001; Marrfurra et al., 1995; Raymond et al., 1999
Cycas canalis K.D. Hill	Australia (NT)	Malakmalak Ab.—*walangkirrik*; Matngala Ab.—*turtur*	Lindsay et al., 2001
Cycas maconochiei subsp. *lanata* K.D. Hill	Australia (NT)	Malakmalak Ab.—*walangkirrik*; Matngala Ab.—*turtur*; Ngan'gikurunggurr and Ngan'giwumirri Ab.—*yerrsya* (plant), *misya* (seeds)	Lindsay et al., 2001; Marrfurra et al., 1995
Cycas maconochiei Chirgwin & K.D. Hill subsp. *maconochiei*	Australia (NT)	Batjamal Ab.—*tjuntuju*; Emi Ab.—*marra*; Tiwi Ab.—*minta* (female plant), *kwaka* (seeds)	Puruntatameri et al., 2001; Smith & Wightman, 1990

Table 10-1 (C) (*continued*)

Scientific name	Locality	Vernacular name	References
Cycas media R.Br. subsp. *banksii* K.D. Hill	Australia (Qld)	Yalanji Ab.—*mara, marra;* unident. Ab. langs.—*badur, bodell, kamamma, kimalo, mununa* (seeds), *nartu* (seeds), *nijar*	Bailey, 1902; Hill & Baird, 2003; Roth, 1901; Specht, 1958
Cycas ophiolitica K. D. Hill	Australia (Qld)	Bayali Ab.(?)—*baveu*	Bailey, 1902; Roth, 1901
Cycas orientis K.D. Hill	Australia (NT)	Rirratjinu Ab.—*nathu*	Yunupingu et al., 1995
Cycas papuana F. Muell.	Papua New Guinea	Unident. lang.—*warnara*	Bailey, 1909
Cycas scratchleyana F. Muell.	Indonesia (Ambon, Maluku, Papua), Papua New Guinea, Australia (Qld)	Dsimakani lang.—*kataki*; Mekeo lang.—*enge-enge*; Medino village lang.—*notuweh*	Hill, 1994a
Cycas seemannii A. Braun	S.W. Pacific islands (Fiji, New Caledonia, Tonga, Vanuatu)	Bauan (Eastern Fijian)—*logologo* (pref. to variants *lagalaga,langalanga, langolango, longolongo, rongorongo*); Bismala (Vanuatu)—*namele;* Nadroga (Western Fijian)—*sisila;* Sie (Eromanga Island, Vanuatu)—*no-moll;* unident. langs. (Fiji)—*roro, tuawawa niu* (devil's coconut); unident. langs. (Vanuatu)—*mwele* (Pentecost I.), *namail* (Maskelynes Is.), *nemol* (Espiritu Santo); unident. langs.—*mandrai* (bread), *wiro*	Degener, 1949; Hill, 1994a; Keppel & Osborne, 2004; Parham, 1943; Smith, 1979
Lepidozamia hopei Regel	Australia (Qld)	Yalanji Ab.—*julbin, miray*; other Aust. Ab.—*binggira, bingir, ngarumba, wunu*	Forster, 1996; R. Hill, pers. comm.
Macrozamia communis L.A.S. Johnson and allied spp. (also sometimes incl. *Lepidozamia peroffskyana* Regel)	Australia (NSW)	Dhurak Ab.—*burrawang* (pref. to *burrawan, burrawong*)	Kennedy et al., 2001
Macrozamia douglasii W. Hill ex F.M. Bailey	Australia (Qld)	Badtjala Ab.—*coobine, goulbine*	Osborne, 2003
Macrozamia macdonnellii (F. Muell. ex Miq.) A. DC.	Australia (NT)	Eastern Arrerente Ab.—*atyikwarle, atywekekwerle*; Western Arrerente Ab.—*tywekekwerle*	Latz, 1995; Osborne, 1999
Macrozamia miquelii (F. Muell.) A. DC.	Australia (Qld)	Bayali Ab.(?)—*banga*	Thozet, 1866

(*continued*)

Table 10-1 (C) (*continued*)

Scientific name	Locality	Vernacular name	References
Macrozamia riedlei (Gaudich.) C.A. Gardner	Australia (WA)	Aust. Ab.—*booyo*; Noongar Ab.—*baian, budjan, djiriji, dyergee, gigijee, jeerajee, jeerja, jeerli, koondagoor, kundagur, quinine, quinning*	Abbott, 1983; Bindon, 1996
Macrozamia Miq. (generic)	Australia	Eng. (Aust.)—*wild pineapple, zamia fern, zamia palm*	White, 1928; common knowledge

(D) Cycads of S. Asia (the Indian subcontinent).

Scientific name	Locality	Vernacular name	References
Cycas beddomei Dyer	India (Andra Pradesh)	Telugu—*madhana—kamakshi* (lust-eyed one, possible ref. to aphrodisiac properties), *per ita*; unident. lang.—*konda itha*	Schuster, 1932; Whitelock, 2002
Cycas circinalis L.	India (Andhra Pradesh, Karnataka, Kerala, Maharashtra, Tamil Nadu)	Hindi—*jangli-mada-mast-ka-phul* (forest flower of sweet desire); Kanarese—*mundi, mundicalu, mundi-ichlumara*; Malayalam—*chananga, eandu, indu, indum podi, intalapanna, inthu, mouta panna* (egg plant), *todda panna* (palm-like plant), *tutapanna* (palm-like plant); Oriya—*odasa-mari, orguna*; Port.—*armatoria das igresias, palma d'igresia*; Sanskrit—*varaguna*; Tamil—*alangarakonnai, alangarasolai, canningay* (convulsions of the hand), *eanchy, madanagama*; Telugu—*kamkshi, per ita*	Miquel, 1840, 1844; Pant, 1962; R. Singh & P. Radha, pers. comm.; Srivastava & Chauhan, 2004; van Rheede, 1682.
Cycas nathorstii J. Schust.	Sri Lanka	Sinhalese—*madu* (nectar)	Schuster, 1932
Cycas pectinata Buch.-Ham.: see under (E), below			
Cycas zeylanica (Schuster) A. Lindstr. & K. D. Hill	India (Andaman Is.), Sri Lanka	Sinhalese—*maha-madu* (great nectar)	Lindstrom & Hill, 2002

(E) Cycads of East Asia.

Scientific name	Locality	Vernacular name	References
Cycas bifida (Dyer) K.D. Hill	China (Guangxi, Yunnan), N. Vietnam	Ch.—*cha-ye su-tie* (fork-leafed cycad), *long-kou su-tie* (Longzhou cycad)	Chen et al., 1995; Walters & Yang, 1994
Cycas chamaoensis K.D. Hill	Thailand	Thai—*prong* (cycad)	Hill & Vatcharakorn, 1998
Cycas chevalieri Leandri	N. Vietnam	Viet.—*cáy ngen*	Whitelock, 2002
Cycas clivicola K.D. Hill	Malaysia, Thailand	Thai—*prong* (cycad), *prong khao* (mountain cycad)	Hill & Vatcharakorn, 1998
Cycas curranii (J. Schust.) K.D. Hill	Philippines (Palawan)	Tagalog—*bayit* (pref. to *bait, vait*), *oliva* (plant, pref. to *oliba, uliba, uliva*), *pitógo* (seed, pref. to *bitogo, patubo, pitugo, potago*)	Amoroso, 1986; Schuster, 1932; Zamora & Co, 1986; K. Hill, pers. comm.
Cycas edentata de Laub.	Philippines (Sulu)	Tagalog—*bayit* (pref. to *bait, vait*), *oliva* (plant, pref. to *olíba, uliba, uliva*), *pitógo* (seed, pref. to *bitogo, patubo, pitugo, potago*)	Amoroso, 1986; Schuster, 1932; Zamora & Co, 1986; K. Hill, pers. comm.
Cycas elongata (Leandri) D. Yue Wang	S. Vietnam	Viet.—*cáy-xuong-te*	Poilane, 1924
Cycas guizhouensis K.M. Lan & R.F. Zou	China (Guanxi, Guizhou, Yunnan)	Ch.—*feng-wei-cao, guan-yin-lian, shan-bo-lu, su-tie* (cycad)	Walters & Yang, 1994
Cycas hainanensis C.J. Chen	China (Hainan)	Ch.—*ci-bing su-tie* (spiny petioled cycad), *fenghuangdan* (phoenix egg), *hainan su-tie* (Hainan cycad)	Chen et al., 1995; Walters & Yang, 1994
Cycas hongheensis S.Y. Yang & S.L. Yang	China (Yunnan)	Ch.—*ba-he su-tie, hong-he su-tie* (Honghe [Red River] cycad)	Walters & Yang, 1994
Cycas litoralis K.D. Hill	Malaysia, Myanmar, Indonesia (Sumatra), Malaysia, Thailand, S. Vietnam	Malay—*kwale pahang* (woods on seashore); Thai—*prong thale* (sea cycad)	Hill & Vatcharakorn, 1998; Ridley, 1893
Cycas macrocarpa Griff.	Malaysia, Thailand	Malay—pakoo galowe; Thai—*prong* (cycad), *pha kood ha yee* (Pohnpei State, FSM)—*manuatababa;* Sonsorolese (Palau)—*faretoul*; unident. langs. (Palau)—*klemia, kokeal, remiang, rumiyan*; unident. langs. (Yap State, FSM)—*fallutier, fratel, frotel*	Griffith, 1854; Hill & Vatcharakorn, 1998

(continued)

Table 10-1 (E) (*continued*)

Scientific name	Locality	Vernacular name	References
Cycas multipinnata C.J. Chen & S.Y. Yang	China (Yunnan)	Ch.—*dujuetie* (single leaf cycad), *duoqi su-tie* (multipinnate cycad), Ha-ni—*lei-ze-ze-bo* (ghost's palm)	Chen et al., 1995; Yang, 1995
Cycas nongnoochiae K.D. Hill	Thailand	Thai—*prong* (cycad)	Hill & Vatcharakorn, 1998
Cycas pachypoda K.D. Hill		Viet.— *thiên tu chân voi-chân voi* (elephant leg cycad)	Hill et al., in prep.
Cycas panzhihuaensis L. Zhou & S.Y. Yang	China (Sichuan, Yunnan)	Ch.—*e-bao-gong, e-boa-chi, panzhihua su-tie* (Panzhihua cycad)	Walters & Yang, 1994
Cycas pectinata Buch.-Ham.	Bhutan, China (Yunnan), India, Laos, Myanmar, Nepal, Thailand, Vietnam	Assamese—*thaljimura*; Burmese—*môndaing*; Ch.—*bi-chi su-tie* (cycad with comb—like megasporophylls), *feng-wei-jiao, feng-huang-dan*; Khasi (India)—*dieng-sia-goda*; Nepali—*thakal*; Thai—*boka, plong, prong khao, prong pa* (forest or field cycad)	Chen et al., 1995; Hill & Vatcharakorn, 1998; Pant, 1962; Pant et al., 1994; Smitinand, 1972; Walters & Yang, 1994
Cycas pranburiensis S.L. Yang, K.D. Hill, W. Tang & Vatcharakorn	Thailand	Thai—*ma phrao si da*	Hill & Vatcharakorn, 1998
Cycas revoluta Thunb.	Japan	Ch.—*tieshu* (iron tree), *feng-wei-jiao-ye* (phoenix tail banana), *su-tie* (cycad); Jap. (main islands)—*sotetsu* (preferred) (coming back from the dead by iron), *ban shou* (barbarian's banana), *hou bi* (bird's tail feathers), *hou bi shou* (broken banana leaf), *sha ka shou* (firebreak banana), *tesshou, tessio* (iron banana), *tosso*; Jap. (Ryukyu Islands)—*hichichi, hitichi, satetsu, shichichi, sichi, sichidzi, sidzichi, sidzidzui, sihittu, sirichi, sitechi, sitichi, sitidzi, sitochi, sitsuchi, sitsudzu, situchi, suchichi, suidzu, susitykuki, sutachi, suticha, sutichi, sutta, sutuku, suutichi, syutta, syutto, tsudzu* (all ref. to whole plants); *kyungama, mii, nadzu, nari, sutitsi-nari, yanabu* (ref. to seeds only)	Osborne & Tomiyama, 1995

Table 10-1 (E) (*continued*)

Scientific name	Locality	Vernacular name	References
Cycas riuminiana Porte ex Regel	Philippines (Luzon)	Ilocano—*sawang* (pref. to *sauang*); Tagalog—*bayit* (pref. to *bait, vait*), *oliva* (plant, pref. to *oliba, uliba, uliva*), *pitógo* (seed, pref. to *bitogo, patubo, pitugo, potago*), *tamok*	Amoroso, 1986; Schuster, 1932; Zamora & Co, 1986; K. Hill, pers. comm.; R. Villas, pers. comm.
Cycas rumphii Miq.	Indonesia (S. Borneo, N.E. Java, Maluku, Papua, Sulawesi), Papua New Guinea	Ambonese Malay (Maluku)—*utta niwel, utta nuer*; Biak (Amdoei village, Papua)—*nufuès*; Javanese—*pakis rady*; Malay—*haji, hajo, pakoe laut, paku laut* (both sea fern), *paku rajah* (Rajah's fern), *sajor calappa utam*; Ternate (Maluku)—*djudjara, djoedjaroet, madjong utu*; Wapi (Yakoi village, Papua)—*sumo*; Weda (Halmahera I.)—*ftoi*; unident. lang. (Mussau Is., PNG)—*otou*; unident. lang.—*bakutu*	Hill, 1994a; Lepofsky, 1992; Miquel, 1844; Pant, 1962; Schuster, 1932
Cycas saxatilis K.D. Hill & A. Lindström	Philippines (Palawan)	Tagalog—*bayit* (pref. to *bait, vait*), *oliva* (plant, pref. to *oliba, uliba, uliva*), *pitógo* (seed, pref. to *bitogo, patubo, pitugo, potago*)	Amoroso, 1986; Schuster, 1932; Zamora & Co, 1986; K. Hill, pers. comm.
Cycas siamensis Miq.	Cambodia, Laos, Myanmar, Thailand, S. Vietnam	Burmese—*môndaing*; Lao—*ma prou tou*; Thai—*maphrao tao, phrao tao, prong pa* (forest or field cycad), *talapat ruesi* (pref. to *talapat rusi*)	Hill & Vatcharakorn, 1998; Pant, 1962; Schuster, 1932; Smitinand, 1972
Cycas simplicipinna (Smitinand) K.D. Hill	China (Yunnan), Laos, Myanmar, Thailand, N. Vietnam	Ch.—*kong-que-bao-dan, shan-ba-buo, shen-xian-mi, xiang-wei-cai*; Thai—*maphrao tao, phrao tao, plong, prong pa* (forest or field cycad), *talapat ruesi*	Hill & Vatcharakorn, 1998; Smitinand, 1972; Walters & Yang, 1994
Cycas szechuanenis C.Y. Cheng, W.C. Cheng & L.K. Fu	China (Fujian, Guangdong)	Ch.—*feng-wei su-tie, feng-wie-tie, guizhou su-tie* (Guizhou cycad), *nanpan jiang su-tie* (Nanpan River cycad), *(qiannian)-feng-wei cao* (thousand-year phoenix grass), *sichuan su-tie* (Sichuan cycad)	Chen et al., 1995; Walters & Yang, 1994; Zhong & Zhou, 1999

(*continued*)

Table 10-1 (E) (*continued*)

Scientific name	Locality	Vernacular name	References
Cycas taitungensis C.F. Shen, K.D. Hill, C.H. Tsou & C.J. Chen	China (Taiwan)	Ch.—*tai-dong su-tie* (Taidong cycad); *taiwan-feng-wei-jiao*	Chen et al., 1995; Walters & Yang, 1994
Cycas taiwaniana Carruth.	China (Guangdong)	Ch.—*guangdong su-tie* (Guangdong cycad), *hai-te-koe* (sea-iron fowl)	Shen et al., 1994
Cycas tansachana K.D. Hill & S.L. Yang	Thailand	Thai—*prong* (cycad)	Hill & Vatcharakorn, 1998
Cycas tropophylla K.D. Hill & P.K. Loc	N. Vietnam	Viet.— *thiên tuê ha long* (Ha Long Bay cycad)	Hiêp & Kiew, 2000
Cycas wadei Merrill	Philippines (Culion)	Tagalog—*bayit* (pref. to *bait, vait*), *oliva* (plant, pref. to *oliba, uliba, uliva*), *pitógo* (seed, pref. to *bitogo, patubo, pitugo, potago*)	Amoroso, 1986; Schuster, 1932; Zamora & Co, 1986; K. Hill, pers. comm.

have been somewhat cursory and fraught with inaccuracies (e.g., Whiting, 1963). The sole comprehensive review on the economic botany of cycads (Thieret, 1958) makes scant mention of vernacular names, and, distressingly, few of the recently published "popular" cycad books (e.g., Osborne et al., 2007) include details of vernacular names (an exception is Whitelock, 2002).

The scientific names and plant localities in Table 10-1 are taken from the current World List of Cycads (Hill et al., 2007). Where there have been changes to the taxonomy, we have transferred the vernacular names to the new species names. For example, all except the Indian names used in the earlier literature for *Cycas circinalis* have been reassigned to *C. rumphii, C. seemannii,* and allied species, in accordance with the revised circumscription for the first species (Hill, 1995).

We have generally retained variants of a common theme. Where there is strong evidence for a dominant spelling or pronunciation, we have drawn attention to the apparent preferred term. For example, among the Tagalog (Filipino) words *bitogo, patubo, potago, pitógo,* and *pitugo* used in reference to the seeds of Philippine *Cycas* species, we have highlighted *pitógo*, as it is most commonly cited in the literature. Similarly, from the Náhuatl-derived variants *teocinte, teosinte,* and *tiusinte*, as used for *D. mejiae*, we prefer *tiusinte* because it is consistent with the pronunciation most often encountered among local users in its native range. We caution that in cases such as these, one may hear numerous pronunciations and see several spellings, all with the same meaning, but varying from village to village and province to province. A third example is the nomination of the better known term *coontie* from among the Seminole variants *conti, coonti, coontie, contihateka, koonti, koontie*, and so on, for the native Florida *Zamia*.

Our data-gathering process was followed by extensive email correspondence with cycad specialists, and our final listing has been further scrutinized and refined by authorities on the different genera (see Acknowledgments). Finally, we have erred on the side of caution, eliminating names of doubtful authenticity. Here, we are reminded of an anecdote where an intrepid botanist had faithfully recorded the name *rahampitso maraina* for a Madagascan palm, only to find out later that the harassed guide had used the phrase for "tomorrow morning" in answer to the botanist's persistent demands to locate the plant (H. Beentje, pers. comm.).

In the interest of remaining faithful to vernacular nomenclature, we have excluded Western horticultural terms altogether, because these are often of recent invention (for the convenience of the cycad trade) and rarely reflect local meanings. Indeed, widespread use of horticultural terms and their translations into local languages serve only to obscure, confuse, and in some cases eventually replace traditional concepts. (We do, however, include all known vernacular names in English, French, Spanish, and Portuguese.)

Discussion

All 11 cycad genera are represented in Table 10-1, and we have located at least one vernacular name for 146 (48%) of the 303 cycad taxa currently recognized (Hill et al., 2007). The compendium reveals a rich cultural diversity involving a total of 495 vernacular names in 129 different languages, broken down as follows: Africa, 81 names for 38 spp. in 35 languages; Americas, 140 names for 50 spp. in 27 langs.; Oceania, 91 names for 22 spp. in 38 langs.; South Asia, 30 names for 5 spp. in 8 langs.; and East Asia, 153 names for 31 spp. in 23 langs. (English, French, and Portuguese names occur in more than region). Not surprisingly, the genus *Cycas*, with the largest number of recognized species (100), contains the greatest number of vernacular names. Within that genus, the greatest taxonomic diversity is recorded for the Japanese species *Cycas revoluta* (41 names for the plant alone). The rich documentation of *Cycas* folk taxonomy owes much to the work of Ken Hill and coworkers, who have taken pains to include details of vernacular names in their work. Furthermore, Hill has taken the laudable step of assigning two such names as subspecific epithets for *C. arnhemica*.

By contrast, relatively few names have been documented for the 41 species in another large genus, *Macrozamia*, in Australia. Sadly, this does not imply a lack of indigenous names but the fact that Australian ethnobotanists have given limited attention to local traditional cultures in areas where *Macrozamia* cycads are found. It also reflects the fact that Westernized generations of aboriginal people have lost much of the cultural knowledge of their forebears (Osborne, 1999).

The vernacular names we have recorded in Table 10-1 embody a variety of concepts. Because cycads occur across broad cultural regions, we have separated the species into broad geographic groups to facilitate understanding of the conceptual linkages in folk knowledge. We discuss some of the more prominent themes in vernacular names below.

Cycad-Palm and Cycad-Fern Homologies

A unifying theme in Latin American Spanish names is the morphological similarity of palms to the four Mesoamerican cycad genera, as evidenced in names such as *palma bola, palma corcho, palma de chicle, palma de Dolores, palma de macetas, palma de pegamento, palma de Santa Teresita, palma de serrucho, palma del sol, palma de la Virgen, palma espinada, palma imperial, palma real, palmilla, palmita,* and *palmiche* (in total, we have recorded 26 names with the "palma" root). The lumping of cycads with palms in folk taxonomies (many people in Latin American countries think of cycads as odd types of palms) may be "wrong" in scientific terms, but it is "right" in cultural context. Cycad leaves, particularly of *Dioon*, often complement or are used in place of palm fronds for Catholic ceremonies such as the Day of the Cross (*Día de la Cruz*) and the Day of the Dead (*Día de los Difuntos*), when they are fashioned into decorations such as wreathes, crosses, arcades, and altars.

The palm homology is not confined to the New World. In African languages one finds the Krobo terms *kpadei-atah* and *pardi attar* (both meaning "ghost palm," *Encephalartos barteri*, Nigeria), the Madi word *ci-cia* ("akin to a date palm," *E. macrostrobilus*, Uganda), and the Ronga and Shona name *untopani* ("palm-like plant," *E. manikensis* and allied species, Mozambique). In Australia, the term *zamia palm* is frequently used by farmers in generic reference to *Macrozamia*.

Much less common than the allusion to palms is the occasional reference to a morphological similarity between cycads and ferns, which we ascribe to the shared features of circinate venation and pinnately compound leaves. The genus name *Chigua* and the species *Zamia chigua* both derive from a Colombian term embracing both cycads and ferns and used to refer to at least six *Zamia* and *Chigua* species. Similarly, the Spanish term *helecho* (fern) is applied to *Z. chigua*. In Southeast Asia, we find the Malay phrase *pakoe laut* (sea fern) used for *Cycas rumphii*, while in Australia, *Macrozamia* is sometimes known as "zamia fern."

Maize Analogy

A unifying feature of Mesoamerican cycads that emerges from this compendium is the analogy to maize seen in the comparisons to *Zea mays* subspecies such as *teosinte* ("sacred ear of maize") and the references to "old-time maize," "maize lord," "maize shepherd," and similar terms. Across the Mesoamerican region (Mexico and northern Central America), *Ceratozamia*, *Dioon,* and *Zamia* appear to have been linked to maize even in pre-Columbian times, presumably because of the morphological similarity of their "ears," as well as their dietary importance. Starch from cycad cones as well as from cycad stems is consumed widely in the region, and cycad food products, for example from *D. mejiae*, are given the same names as maize food products: most notably, "tamales," "tortillas," and "atol" (Bonta, 2007). In Mexico and Honduras, *Dioon* seeds have traditionally supplemented local diets in times of scarcity, particularly when maize stores became depleted. There is, in other words, a yet-unexplored complementarity of maize and cycads that is all the more intriguing because it pairs a rather unusual wild food with one of the world's preeminent domesticated crops, and with that crop's genetic ancestor.

References to Starch and Other Household Uses

Several names refer to cycads as a source of starch. For instance, *yuca* and *yuquilla* (referring to manioc [cassava], *Manihot esculenta*) is used for Cuban and Honduras *Zamia* cycads, while *Z. integrifolia* has been called the Florida arrowroot or Conti Hateka, the "white bread plant." The Afrikaans word for *Encephalartos* is *broodboom* (bread tree), and Chamorro names *federico* and *fadang* both refer to the flour obtained from starch in *Cycas micronesica* seeds.

Many cycad names are inspired by household uses. For example, the Spanish names *palma de chicle* (rubber palm) and *palma de chicalite* (referring to the seed cooking and preservation process) are used for *D. spinulosum* in Mexico. Also in Mexico, *carrete* (ox cart) refers to a toy fashioned from the male cone of *Ceratozamia mixeorum*, while *chicalito* derives from the use of the stony sclerotesta of *Dioon* seeds to make bull-roarers (*zumbadores*), ocarina-type whistles that hum when tied to strings and whirled around. In Panama, the name *palma de pegamento* (glue palm, *Z. fairchildiana*) alludes to the use of the plant's resinous gum as a sort of office paste. In Mozambique, the Tsonga name *chihanga* (vegetable coal) refers to the practice of roasting *Encephalartos* trunks to provide a convenient source of charcoal.

References to Localities and Habitats

As with many other plants, habitats or areas of occurrence feature commonly in cycad vernacular names and local place names. We have already mentioned the toponyms *El Tiusintal* (place of cycads, a village), *Cerro del Tiusinte* (cycad hill), and *Quebrada del Tiusinte* (cycad stream) in reference to *D. mejiae* in Honduras. Also in the New World, habitat descriptors are found in *coyolito de cerro* (little mountain coyol palm, *D. spinulosum*), *piña del monte* (wild pineapple, *Ceratozamia mexicana*, *C. robusta*, *Zamia gentryi* and *Z. roezlii*), *tepetmaizte* or *tepemaizte* (hill or mountain maize, *C. morettii*), *yuquilla de paredón* (little cliff-dwelling manioc, *Z. angustifolia* and *Z. pumila*) and *yuquilla de sabana* (little savanna manioc, *Z. kickxii* and *Z. pumila*). In Asia, we encounter the Thai names *prong khao* (mountain cycad, *Cycas clivicola*), *prong pa* (forest or field cycad, *Cycas pectinata*, *C. siamensis*, and *C. simplicipinna*); the Chinese name *nanpan jiang su-tie* (Nanpan River cycad, *C. szechuanenis*); and the Vietnamese *thiên tuê ha long* (Ha Long Bay cycad, *C. tropophylla*).

References to Morphology

Direct references to plant morphology are often incorporated in cycad names. In the New World, we encounter the Spanish names *pata de gallo* (cock's foot, *Ceratozamia hildae*), *costilla de león* (puma's rib, *C. mexicana* and *C. robusta*), the Chiapaneco term *nimalari* (feather leaf, *D. merolae*) and the Spanish *espadaña* (church steeple) and *morrito* (small tree gourd with round fruit, *Crescentia* sp.), both also for *D. merolae*; as also the terms *palma de serrucho* (handsaw palm, *D. edule*), *peine* (comb, the comb-like leaf of *D. sonorense*), and *camotillo* (small sweet potato, *Zamia standleyi*). The Colombian name *cacao del indio* (Indian cocoa, *Z. encephalartoides*) is apparently based on the similarity of its female cone to the fruit of *Theobroma cacao*. The Chinese use the words *cha-ye su-tie* (fork-leafed cycad, *Cycas bifida*), *ci-bin su-tie* (spiny petioled cycad, *C. hainanensis*), *duoqi su-tie* (multipinnate cycad, *C. multipinnata*), and *bi-chi su-tie* (cycad with comb-like megasporophylls, *C. pectinata*), while the Japanese use *hou bi* (bird's tail feathers) and *hou bi shou* (broken banana leaf) for *C. revoluta*. In Vietnam, native cycads are generically referred to as *thiên tu* (thousand-year-old cycad) with extensions for species, for example, *thiên tu chân voi-chân voi* (elephant leg cycad, *Cycas pachypoda*).

Specialized Cycad Vocabularies

Cycads that are the subject of intensive subsistence regimes inspire rich vocabularies describing morphological features, harvest techniques, pest control, indigenous conservation, folklore, food preparation, and other ethnobotanical aspects. For *D. mejiae*, Bonta (2007) has identified over 60 specialized terms, including *churute* (pollen cone), *cabeza* (female cone), *vestimento* (flush of young leaves), *alma* or *espíritu* (sprout), *tiusintero* (male cycad harvester), *tiusintera* (female cycad food preparer), *pan de tiusinte* (cycad bread), *cumbito* (seed), and *embargo* (season when cycad harvesting is prohibited by municipal decree) (Bonta et al., in prep.).

A similar situation is seen in some Australian aboriginal languages. The Rirratjinu people in eastern Arnhem Land use terms for *Cycas orientis* which include *banbuy* or *nayakandi* (immature plants), *mulurrukun* (seeds lying on the ground), and *meltjuntjun* and *melnathu* (the white sarcotesta within the seeds) from which the foodstuff *guyalpadi* is prepared (Yunupingu et al., 1995).

Non-Western Taxonomic Schemes

Cycad names in several Asian countries are often nomenclatural binomials or polynomials, analogous to the Western scientific genus name and specific epithet, and founded on a similar basis of hierarchical classification. In China the generic root *su-tie*, "cycad," is seen in the names *ci-bing su-tie* (*Cycas hainanensis*), *ba-he su-tie* (*C. hongheensis*), *long-kou su-tie* (*C. micholitzi*), *bi-chi su-tie* (*C. pectinata*), and *feng-wie su-tie* (*C. szechuanensis*), among others. In Thailand, the generic *prong* occurs in the names *prong khao* (*C. clivicola*), *prong thale* (*C. litoralis*), and *prong pa* (collectively for *C. pectinata*, *C. siamensis,* and *C. simplicipinna*). A similar situation exists in Vietnam, where all cycads are collectively known as *tu* and native cycads are called *thiên tu* (one-thousand-year-old cycad), as seen in *thiên tu chân voi-chân voi* (elephant leg cycad, *Cycas pachypoda*) and *thiên tu H Long* (Ha Long cycad, *C. tropophylla*).

Folkloric Associations

The least investigated vernacular names are those apparently derived from folklore. The Merina (Madagascar) name *batsimisaraka* (man-eating plant, *Cycas thouarsii*) appears to be founded on a legend in which the leaves are blood-sucking tentacles (Schuster, 1932). Another African example is the Zulu *isiGqikisomkhovu*, zombie's pillow (*Encephalartos natalensis*).

For *Cycas circinalis*, we encounter the Hindi *jangli-madan-mast-ka-phul*, "forest flower of sweet desire," and for *C. beddomei* the Telugu name is *madhanakamakshi*, "nectar of [goddess] Kamakshi" (identified with sexual desire). These terms are thought to be based on a concept of cycads as aphrodisiacs on the Indian subcontinent (Nadkarni, 1954; S. Bonta, pers. comm.). A similar situation is found in the names *guade teet* and *taudowa*, which both refer to the aphrodisiac properties attributed to a "tea" prepared from the stem of *Zamia neurophyllidia* in Panama's Isla de Bastimentos and Isla de Popa, respectively (Holzman & Haynes, 2004).

Conclusion

We hope this article generates a greater awareness of vernacular cycad taxonomies and inspires sustained scientific engagement with the "cultural side" of cycads. We urge cycad taxonomists to include in their publications details of indigenous names, languages, meanings, and any relevant ethnobotanical notes. This follows exactly one of the stated "guidelines for taxonomic descriptions" specified in Chapter 15 of *Cycad Classification: Concepts and Recommendations* (Osborne & Walters , 2004).

This compendium is also a window onto a little-explored field of cycad research. For example, Bonta's work on Honduran cycads has revealed not only names but also an intricate and hitherto unexplored local knowledge system that, in turn, is becoming the base for conservation intertwining local "vernacular" experts and outside "Western scientific" experts. We suggest that numerous rewarding opportunities exist for in-depth studies of cycad knowledge systems, and we hope that this compendium inspires not only biological scientists but also social scientists to explore these avenues.

Acknowledgments

We are indebted to many members of the cycad community for their useful input to our investigation. In this respect, we wish to record our appreciation to Henk Beentje, Pedro Capela, Ramiro Chaves, Jeff Chemnick, De Wet Bösenberg, Dale Dixon, Paul Forster, Bill Gaynor, Jody Haynes, Ken Hill, Rosemary Hill, Fernando Nicolalde-Morejón, Miguel Angel Pérez-Farrera, Loc Phan Ke, P. Radha, Lou Randall, Bart Schutzman, Rita Singh, Dennis Stevenson, Willie Tang, Alberto Taylor, Romeo Villas, Piet Vorster, Andrew Vovides, Terrence Walters, and Glenn Wightman. We also wish to thank Michael Balick for his useful comments in reviewing this text. Steve Bonta kindly provided translations of some South Asian languages.

Literature Cited

Abbott, I. 1983. Aboriginal names for plant species in Western Australia. Forest Department of Western Australia Technical Paper No. 5, Perth, Western Australia.

Alcorn, J. B. 1984. Huastec Mayan ethnobotany. University of Texas Press, Austin.

Amoroso, V. B. 1986. Morphological study of the sporophylls of Philippine *Cycas*. Philipp. J. Sci. 155: 177–198.

Bailey, F. M. 1902. Cycadaceae. The Queensland flora. Brisbane, Australia.

———. 1909. Contribution to the flora of British New Guinea. Queensland Agric. J. 22(3): 147–150; cited by Read & Solt, 1986.

Beck, W. 1992. Aboriginal preparation of *Cycas* seeds in Australia. Econ. Bot. 46: 133–147.

———. 1993. Cycad use by Australian aboriginal people. Pp. 8–15 *in* Proceedings of CYCAD 90, the Second International Conference on Cycad Biology: the biology, structure, and systematics of the Cycadales. D. W. Stevenson & K. J. Norstog, eds. Palm & Cycad Societies of Australia Ltd., Milton, Queensland, Australia.

Beentje, H. 1994. Kenya trees, shrubs and lianas. National Museums of Nairobi, Kenya.

Bernal, D. & D. Restrepo. 1991. Las cicadáceas fósiles vivientes: Peligran dinosaurios vegetales. Ecologica (Colombia) 3: 10–15.

Bindon, P. 1996. Useful bush plants. Western Australian Museum, Perth, Australia.

Binns, B. 1972. Dictionary of plant names in Malawi. The Government Printer, Zomba, Malawi.

Blake, N.M., G. Wightman & L. Williams. 1998. Iwaidja ethnobotany: aboriginal plant knowledge for the Gurig National Park, Northern Australia. Northern Territory Botanical Bulletin No. 23. Parks and Wildlife Commission of the Northern Territory, Darwin, Australia.

Bonta, M. 2003. Teocinte, 'ear of God'. Cycad Newslett. 26(1): 7–12.

Bonta, M. 2007. Ethnobotany of Honduran cycads. Mem. New York Bot. Gard. 97: 120–142.

———, O. Flores Pinot, D. Graham, J. Haynes & G. Sandoval. In prep. Ethnobotany and conservation of tiusinte (*Dioon mejiae* Standl. and Williams).

Chemnick, J., T. J. Gregory & S. Salas-Morales. 1997. *Ceratozamia mixeorum* (Zamiaceae), a new species from Oaxaca, Mexico, with comments on distribution, habitat and species relationships. Phytologia 83: 47–52.

Chen, C. J., L. Zhou, S. Y. Yang & Z. S. Han. 1995. *Cycas* in China, with notes on its conservation status. Pp. 177–292 *in* P. Vorster (ed.), Proceedings of the Third International Conference on Cycad Biology. Cycad Society of South Africa, Stellenbosch, South Africa.

Correll, D. S. & H. B. Correll. 1982. Flora of the Bahama Archipelago. J. Kramer, Vaduz, Liechtenstein.

da Graça Silva, M. 1983. 1. Cycadaceae. Pp. 1–33 *in* E. J. Mendes, (ed.), Flora de Moçambique. Instituto de Investigaçáo Cientifica Tropical, Lisbon.

Dalziel, J. M. 1937. The useful plants of tropical West Africa. Appendix *in* J. Hutchinson & J. M. Dalziel, Flora of tropical West Africa. Crown Agents, London.

de Koning, J. 1993. Checklist of vernacular names in Mozambique. Wageningen Agric. Univ. Pap. 93(2).

Degener, O. 1949. Naturalist's South Pacific expedition, Fiji and Honolulu, pp. 258–259; cited by Read & Solt, 1986.

de Luca, P., S. Sabato & M. Vázquez Torres. 1978. *Dioon purpusii* Rose (Zamiaceae), a misknown species. Delpinoa 20: 31–35.

———, ——— & ———. 1981. *Dioon merolae* (Zamiaceae), a new species from Mexico. Brittonia 33: 179–185.

———, ——— & ———. 1984. *Dion tomasellii* (Zamiaceae), a new species with two varieties from western Mexico. Brittonia 36: 223–227.

Dharani, N. 2002. Field guide to common trees and shrubs of East Africa. Struik, Cape Town, South Africa.

Dold, A. P. & M. L. Cocks. 1999. Preliminary list of Xhosa plant names from Eastern Cape, South Africa. Bothalia 29: 267–292.

Donaldson, J. S., B. Dehgan, A. P. Vovides & W. Tang. 2003. Cycads in trade and sustainable use of cycad populations. Pp. 39–47 *in* J. S. Donaldson (ed.), Cycads. Status survey and conservation action plan, IUCN/SSC Cycad Specialist Group, IUCN, Gland, Switzerland and Cambridge, UK.

Duke, J. A. 1975. Ethnobotanical observations on the Cuna Indians. Econ. Bot. 29: 278–293.

Engler, A. 1908. *E. septentrionalis*. Natürlichen Pflanzenfamilien Vol. 1. Leipzig, Germany.

Faden, R. B. & H. J. Beentje. 1989. *Encephalartos kisambo*. A new cycad from Kenya, with a note on *E. tegulaneus*. Utafiti 2(1): 7–10.

Forster, P. I. 1996. *Lepidozamia hopei* (Zamiaceae), the world's tallest cycad. Encephalartos 48: 12–15.

Fosberg, F. R. & M. H. Sachet. 1975. Flora of Micronesia Part I: Gymnospermae. Smithsonian Contr. Bot. (20): 1–15.

Gentil, L. 1904. *Encephalartos* from the Congo. The Gardeners' Chronicle, June 11 1904, pp. 370–371: cited by Read & Solt, 1986.

Gilliland, H. B. 1939. A note on the Rhodesian cycad. Proc. Rhodesia Sci. Assoc. 37: 133–134.

Gono, A. & P.S.M. Phiri. 2001. In search of a rare cycad in Zambia. *Encephalartos* 66: 21–23.

González Géigel, L. 2003. Zamiaceae. Flora de la República de Cuba 8(4). A. R. Gantner Verlag KG, Liechtenstein.

Grey, R. M. 1927. Report of the Harvard Botanical Gardens. Cycadaceae and Palmae, pp. 64–66; cited by Read & Solt, 1986.

Griffith, W. 1854. Notulae ad plantas asiaticas. Calcutta 4: 1–17. Reprinted 1976, Bisher Singh, Mahendra Pal Singh & Periodical Experts, Delhi.

Hiêp, N. T. & R. Kiew. 2000. Wild plants of Ha Long Bay. Thanh Nien Publishing House, Hanoi, Vietnam.

Hill, K. D. 1994a. The *Cycas rumphii* complex (Cycadaceae) in New Guinea and the western Pacific. Austral. Syst. Bot. 7: 543–567.

———. 1994b. A taxonomic revision of the genus *Cycas* (Cycadaceae) in Australia. Telopea 7: 1–64.

———. 1995. The genus *Cycas* (Cycadaceae) in the Indian region, with notes on the application and typification of the name *Cycas circinalis*. Taxon 44: 1–9.

———, D. W. Stevenson & R. Osborne. 2007. The world list of cycads/[La Lista Mundial de Cícadas]. Mem. New York Bot. Gard. 97: 454–483.

——— & P. Vatcharakorn. 1998. The cycads of Thailand. Pp. 178–190 *in* D. R. Hodel (ed.), The palms and cycads of Thailand. Kampon Tansacha, Nong Nooch Tropical Garden, Thailand.

Hill, R. & A. Baird. 2003. Kuku-Yalanji rainforest Aboriginal people and carbohydrate resource management in the wet tropics of Queensland, Australia. Human Ecol. 31: 27–52.

Holzman, G. & J. Haynes. 2004. Cycads of the sand: the beach-dwelling zamias of Bocas del Toro, Panama. Cycad Newslett. 27(4): 3–6.

Hurter, J. 1994. Focus on . . . *Encephalartos kisambo*. Encephalartos 39: 4–8.

Jones, D. L. 2002. Cycads of the world—ancient plants in today's landscape. Ed. 2. Reed New Holland, Sydney, Australia.

Jones, S. & J. Wynants. 1997. *Encephalartos macrostrobilus* (Zamiaceae), a new cycad species from northern Uganda. Encephalartos 50: 13–17.

Kennedy, P., C. Thompson & R. Osborne. 2001. Focus on . . . *Macrozamia communis*. Encephalartos 68: 13–22.

Keppel, G. & R. Osborne. 2004. Focus on . . . *Cycas seemannii*. Encephalartos 78: 12–20.

Lamarque & Razafimahery. 1950. Le 'tsambou' des Comores et de Madagascar (Cycas thouarsii R. Br.). Agron. Trop. (Nogent-sur-Marne) 5: 397–401;cited by Read & Solt, 1986. [No first initials given.]

Latz, P. 1995. Bushfires and bushtucker: aboriginal plant use in central Australia. IAD Press, Alice Springs, Australia.

Lebrun, J. 1930. Note sur l'*Encephalartos laurentianus* De Wild., de la Vallée du Kwango. Rev. Zool. Bot. Africaines 19(3–4): 384–390.

Lepofsky, D. 1992. Arboriculture in the Mussau Islands, Bismarck Archipelago. 1992. Econ. Bot. 46(2): 192–211.

Lindsay, B.Y., K. Waliwararra, F. Mijlat, H. Kuwarda, R. Pirak, A. Muyung, E. Pambany, J. Marruridj, P. Marrfurra & G. Wightman. 2001. Malakmalak and Matngala plants and animals: aboriginal flora and fauna knowledge from the Daly River area, Northern Australia. Northern Territory Botanical Bulletin No. 26. Parks and Wildlife Commission of the Northern Territory, Darwin, Australia.

Lindstrom, A. & K. D. Hill. 2002. Notes on the species of *Cycas* (Cycadaceae) from Sri Lanka and Islands of the Andaman Sea. Novon 12: 237–240.

MacBride, J. F. 1936. Flora of Peru. Publ. Field Mus. Nat. Hist., Bot. Ser. 13(1): 81–82.

Malaisse, F. 1969. *Encephalartos schmitzii* Malaisse, Cycadacée nouvelle du Cong-Kinshasa. Bull. Jard. Bot. Natl. Belg. 39: 401–406.

———, J. P. Sclavo, & C. Crosiers. 1993. Recherches sur les *Encephalartos* Lehm. (Zamiaceae) d'Afrique centrale. 2. Apport de la morphologie foliare dans le différation spécifique. Bull. Jard. Natl. Belg. 62: 205–219.

Marrfurra, P., M. Akanburru, M. Wawul, T. Kumunerrin, H. Adya, K. Kamarrama, M. Kanintyanyu, T. Waya, M. Kannyi, G. Wightman & L. Williams. 1995. Ngan'gikurunggurr and Ngan'giwumirri ethnobotany: aboriginal plant use for the Daly River area, Northern Australia. Northern Territory Botanical Bulletin No. 22. Darwin, Australia.

Martinez, M. 1979. Catálogo de nombres vulgares y científicos de plantas mexicanas. Fondo de Cultura Económica, Mexico City.

Miquel, F. A. W. 1840. Commentarii phytographici. Leyden.

———. 1844. Genera et species cycadearum viventium. Linnaea 17: 675–744.

Nadkarni, K. M. 1954. Indian materia medica. Ed. 3, Vol. 1, Bombay, India; cited by Whiting, 1963.

Nicolalde-Morejón, F. In prep. Ethnobotany of *Zamia* in Ecuador.

Norstog, K. & T. J. Nicholls. 1997. The biology of the cycads. Cornell University Press, Ithaca, NY.

Osborne, R. 1999. Focus on . . . *Macrozamia macdonnellii*. Encephalartos 59: 4–7, 17–18.

———. 2003. Focus on . . . *Macrozamia douglasii*. Encephalartos 74: 6–11.

———, A. Grove, P. Oh, T. J. Mabry, J. Ng & A. A. Seawright. 1994. The magical and medicinal usage of *Stangeria eriopus* in South Africa. Journ. of Ethnopharmacology 43: 67–72.

———, A. Grove, P. Oh, T. J. Mabry, J. Ng & A. A. Seawright. 1994. The magical and medicinal usage of *Stangeria eriopus* in South Africa. J. Ethnopharmacol. 43: 67–72.

———, K. D. Hill, H. T. Nguyen & L. Phan Ke. 2007. Cycads of Vietnam. Osborne & van Eeden, Brisbane & Cape Town.

——— & H. Tomiyama. 1995. Focus on . . . *Cycas revoluta* Thunberg. Encephalartos 41: 5–15.

——— & T. Walters. 2004. In search of the true tree: guidelines for classification. Pp. 195–217 *in* T. Walters & R. Osborne (eds.), Cycad classification: concepts and recommendations. CAB International, Wallingford, UK.

Pant, D. D. 1962. *Cycas* and the Cycadales. Central Book Depot, Allahabad.

———, R. Singh & D. K. Chauhan. 1994. On *Cycas pectinata* Hamilton from north-east India. Encephalartos 38: 17–30.

Parham, H. B. R. 1943. Fiji native plants with their medicinal and other uses. Polynesian Society Memoirs 16: 42; cited by Read & Solt, 1986.

Patiño, V. M. 1989. Notas preliminares sobre el uso de las Zamiaceas por los pueblos primitivos y aculturados del intertrópico americano. Perez-Arbelaezia 2(8): 429–442 (Colombia).

Pérez-Arbelaez, E. 1956. Plantas útiles de Colombia. Ed. 3. Librería Colombiana, Bogotá.

Poilane, M. 1924. Les *Cycas* d'Indochine. Rev. Int. Bot. Appl. Agric. Trop. 4: 472–473;cited by Read & Solt, 1986.

Pooley, E. 1993. Trees of Natal, Zululand and Transkei. Natal Flora Publications Trust.

Prain, D. 1917. Cycadaceae. Pp. 344–354 *in* D. Oliver (ed.), Flora of tropical East Africa. Vol. 6, Sect. 2, Reeve, London, UK.

Puruntatameri, J., R. Puruntatameri, A. Puruntatameri, L. Burak, C. Tipuamantymirri, M. Tipakalippa, J. Puruntatameri, P. Puruntatameri, J. B. Pupangamirri, R. Kerinaiua, D. Tiploura, M.-M. Orsto, B. Kantilla, M. Kurrupuwu, P. F. Puruntatameri, T. D.

Puruntatameri, L. Puruntatameri, K. Kantilla, J. Wilson, J. Cusack, D. Jackson & G. Wightman. 2001. Tiwi plants and animals: aboriginal flora and fauna knowledge from Bathurst and Melville Islands, Northern Australia. Northern Territory Botanical Bulletin No. 24. Parks and Wildlife Commission of the Northern Territory, Darwin, Australia.

Raymond, E., J. Blutja, L. Gingina, M. Raymond, O. Raymond, L. Raymond, J. Brown, Q. Morgan, D. Jackson, N. Smith & G. Wightman. 1999. Wardaman ethnobiology: aboriginal plant and animal knowledge for the Flora River and south-west Katherine Region, Northern Australia. Northern Territory Botanical Bulletin No. 25. Parks and Wildlife Commission of the Northern Territory, Darwin, Australia.

Read, R. W. & M. L. Solt. 1986. Bibliography of the living cycads. Lyonia 2(4): 33–200.

Ridley, H. N. 1893. On the flora of the eastern coast of the Malay Peninsula. Trans. Linn. Soc. London, Bot., Series 2, 3: 267–408; cited by Read & Solt, 1986.

Robyns, W. 1948. Cycadaceae. Flore du Congo Belge et du Ruanda-Urundi. I.N.E.A.C., Brussels.

Roig y Mesa, J. T. 1965. Diccionario bótanico de nombres vulgares cubanos. Editorial Nacional de Cuba, Editora del Consejo Nacional de Universidades, Havana.

Roth, W. E. 1901. Food: its search, capture & preparation. N. Queensland Ethnogr. Bull. 3: 31.

Schultes, R. E. & R. F. Raffauf. 1990. The healing forest—medicinal and toxic plants of northwest Amazonia. Vol. 2. Dioscorides Press, Portland, OR.

Schuster, J. 1932. Cycadaceae. Pp. 1–168 *in* A. Engler & L. Diels (eds.), Das Pflanzenreich, Vol. 4, Part 1.

Schweinfurth, G. 1871. Cycadaceae. Bot. Zeitung (Berlin) 29: 333.

Shen, C. F., Hill, K. D., Tsou, C. H. & Chen, C. J. 1994. Cycadaceae, a new name for the widely known cycad species endemic in Taiwan. Bot. Bull. Acad. Sin. 35: 133–140.

Sleight, F. W. 1953. Kunti, a food staple of Florida Indians. Florida Anthropol. 6(2): 46; cited by Read & Solt, 1986.

Small, J. K. 1921. Seminole bread—the conti, a history of the genus *Zamia* in Florida. J. New York Bot. Gard. 22: 121–137.

———. 1933. Manual of the southeastern flora. New York.

Smith, A. C. 1979. Flora vitiensis nova: a new flora of Fiji. Tropical Botanic Garden, Lawai, Hawaii.

Smith, N. M. & G. M. Wightman. 1990. Ethnobotanical notes from Belyuen, Northern Territory,

Australia. Northern Territory Botanical Bulletin No. 10. Conservation Commission of the Northern Territory, Darwin, Australia.

Smitinand, T. 1972. Cycadaceae. Flora of Thailand. Vol. 2. Part 2. pp. 185–192.

Specht, R. L. 1958. An introduction to the ethno-botany of Arnhem Land. Rec. Amer.-Austral. Exped. Arnhem Land 3: 479–503; cited by Read & Solt, 1986.

Srivastava, P. C. & D. K. Chauhan. 2004. A strange branching cycad: *Cycas circinalis* var. *swamyii* Pant. Cycad Newslett. 27(1): 5–7.

Standley, P. C. & J. A. Steyermark. 1958. Flora of Guatemala, Part I. Publ. Field Mus. Nat. Hist., Bot. Ser. 24(1): 11–20.

Stevenson, D. W. 1993. The Zamiaceae in Panama with comments on phytogeography and species relationships. Brittonia 45: 1–16.

———. 2001. Orden Cycadales, Monografia 21, *in* R. Bernal & E. Forero (eds), Flora de Colombia. Instituto de Ciencias Naturales, Facultad de Ciencias, Universidad Nacional de Colombia.

Thieret, J. W. 1958. Economic botany of cycads. Econ. Bot. 12: 4–41.

Thozet, A. 1866. Notes on some of the roots, tubers, bulbs, and fruits, used as vegetable food by the Aboriginals of northern Queensland, Australia. W. H. Buzacott, Rockhampton, Queensland, Australia.

van Rheede, H. A. 1682. Horti Indici Malabarici III, Amsterdam, Holland.

Vázquez Torres, M. 1990. Algunos datos etnobotánicos de las Cícadas en Mexico. Mem. New York Bot. Gard. 57: 144–147.

——— & A. P. Vovides. 1998. A new species of *Ceratozamia* (Zamiaceae) from Veracruz, Mexico. Novon 8: 87–90.

von Breitenbach, F. & J. von Breitenbach. 1992. Tree atlas of southern Africa. Dendrological Foundation, Pretoria, South Africa.

Vorster, P. & E. Vorster. 1973. The African *Cycas*. Excelsa 3: 14–26.

——— & ———. 1974. *Stangeria eriopus*. Excelsa 4: 79–89.

Vovides, A. P., C. Iglesias, M. A. Pérez-Farrera, J. C. Astorga & U. Schippman. 2002. Peasant nurseries: an effort to conserve cycads in Mexico. Cycad Newslett. 25(4): 3–7.

———, P. Resslar & R. Pérez–López. 1992. The cycads of Veracruz, Mexico. El Eco de Virginia Bilingual Publishers, Norfolk, Virginia.

Walters, T. & S. L. Yang. 1994. The cycads of China—findings from the Montgomery Foundation/Fairchild Tropical Garden 1992 Expedition. Pp. 6–11 *in* W. Tang (ed.), The cycads of China. J. Cycad Soc. (1).

Warnholtz, G. S. (ed.). 2005. Catálogo de especies vulnerables al aprovechamiento forestal en bosques templados del estado de Oaxaca. Available at: http://www.semarnat.gob.mx/vida/dioonmer.htm (accessed February 2005).

Weiss, E. A. 1979. Some indigenous plants used domestically by East African coastal fishermen. Econ. Bot. 33: 35–51.

White, C. T. 1928. *Macrozamia*. Queensland Agric. J. 30: 608.

Whitelock, L. M. 2002. The cycads. Timber Press, Portland, OR.

Whiting, M. G. 1963. Toxicity of cycads. Econ. Bot. 17: 270–302.

Wightman, G. M. & N. M. Smith. 1989. Ethnobotany, vegetation and floristics of Milingimbi, Northern Australia. Northern Territory Botanical Bulletin No. 6. Conservation Commission of the Northern Territory, Darwin, Australia.

Williams, R. O. 1949. The useful and ornamental plants in Zanzibar and Pemba. Cheshire, UK.

Wilson, G. W. 2002. Focus on . . . *Bowenia spectabilis* Hook ex J.D. Hook. Encephalartos 70: 10–16.

Yang, S.-L. 1995. Adventures in Asia. Cycad Newslett. 18(11): 2–3.

Yunupingu, B., L. Yunupingu-Marika, D. Marika, B. Marika, B. Marika, R. Marika & G. Wightman. 1995. Rirratjinu ethnobotany: aboriginal plant use from Yirrkala, Arnhem Land, Australia. Northern Territory Botanical Bulletin No. 21. Parks and Wildlife Commission of the Northern Territory, Darwin, Australia.

Zamora, P. M. & L. Co. 1986. Guide to Philippine flora and fauna, Vol. 2. Natural Resources Management Center, Philippines.

Zanoni, T. A. 1982. Guayica (*Zamia*) in Hispaniola. Cycad Newslett. 5(4): 5–12.

Zhong, Y. C. & D. A. Zhou. 1999. The distribution of *Cycas szechuanensis* in Guangxi Province. Pp. 70–72 *in* C. J. Chen (ed.), Proceedings of the Fourth International Conference on Cycad Biology. International Academic Publishers, Beijing.

ETHNOBOTANY
CHAPTER 11

The Ethnomedicinal Use of Cycads in South Africa

Errol Douwes and Christopher G. M. Dalzell

Abstract

The utilization of cycads in South Africa is apparently not limited to only a few species, because of the generic use of these plants for medicinal/magical purposes. Plant gatherers collect cycad material or even whole plants from any population they can access easily, often with devastating results for the plants and populations. Utilization is primarily for medicinal and magical purposes, and it is continuing at a rate considered unsustainable by the majority of conservation authorities. Studies aimed at understanding the extent of harvesting and trade of these plants have benefited the drafting and implementation of improved legislation, and have also boosted conservation efforts to protect wild cycad populations. However, owing to insufficient manpower and resources required for the implementation of efficient, proactive conservation programs by the authorities, unsustainable use of cycad plants is likely to continue.

Resumen

Poblaciones de *Encephalartos* y *Stangeria* están amenazados severamente dada la sobrecolecta por parte de sanadores tradicionales en Sudáfrica. Un ejemplo anterior relata las razones por lo cual fue transferido el último ejemplar silvestre de *Encephalartos woodii,* del Bosque Ngoye al Jardín Botánico de Durban en los años 1903 y 1907, decisión que fue tomada para preservar la especie y por consiguiente el resguardo de los ejemplares remanentes que estaban amenazados por los sanadores tradicionales. Existen muchos ejemplos subsecuentes del uso medicinal de la corteza de *Encephalartos* por médicos tradicionales en Sudáfrica y en particular en kwaZulu-Natal, ahora estamos frente a una etapa crítica en la cual podríamos perder muchas especies como resultado de esta actividad extractiva. Un caso ejemplar es la destrucción de una de las colonias más grandes de *E. natalensis* en las afueras de Durban. Éste trabajo describe una de las poblaciones más grandes, la cual estamos tratando de salvar con el apoyo de las autoridades CITES en kwaZulu-Natal, también se evalúa la factibilidad del programa. Un segundo proyecto llevado a cabo durante los últimos cuatro años es la colección de material de *Stangeria,* desde poblaciones en kwaZulu-Natal, proyecto con mucho éxito, debido a que se colectó exitosamente material de *Stangeria* desde 28 poblaciones y la información biológica fue registrada en una base de datos específica. Éste trabajo subraya las amenazas para dos géneros de cícadas Sudafricanas y que el Jardín Botánico de Durban ha logrado preservar sus poblaciones. También se discute el desarrollo y éxito de la base de datos.

Introduction

There are 66 cycad species in Africa, including the adjacent Indian Ocean islands (Hill et al., 2007). The genus *Encephalartos* Lehm. (with 65 species) and the genus *Stangeria* T. Moore (one species) are both endemic to Africa, while *Cycas thouarsii* R. Br. ex Gaudich. is the only representative of the widespread *Cycas* L genus (Donaldson, 2003).

Southern African indigenous people have been utilizing cycads for many centuries (Dyer, 1965). In the mid- to late-twentieth century, it became apparent that such utilization, particularly for use in medicinal/magical preparations was escalating (Osborne et al., 1994). Such escalation was due to increased human populations, destruction of large areas of natural cycad habitats through forestry and agriculture, and easier access to remote populations. The exponential increase of the illegal international trade of South Africa's cycads in the 1970s also had a major impact on in situ populations (Tang, 1985). In response to this worldwide decline in cycad populations, a number of studies were initiated to help us gain insights into the extent of the problem and how it could be monitored. Proposals for various means of stemming the decline in cycad populations have been numerous, many of which have positively influenced the drafting and implementation of legislation for the protection and prohibition of trade in cycads.

Donaldson and Bösenberg (1999) presented a method of matching photographs of in situ cycad populations at various time intervals. Photographs were analyzed in order to identify changes in habitat and population sizes/structures. Other studies include surveys conducted at medicinal plant markets, such as the *Stangeria eriopus* (Kunze) Baill. lignotuber trade assessment (Osborne et al., 1994). This study revealed that the combined number of lignotubers sold in two prominent medicinal plant markets in Durban, South Africa, was 3410 during July 1992 alone. Other ethnomedicinal surveys have also incorporated assessments of the cycad trade. Mander (1998) ranked *Encephalartos* species as the 28th most frequently demanded plant by consumers at traditional medicine markets in kwaZulu-Natal, South Africa, and *S. eriopus* as 63rd in the same study. While *Stangeria* may not be regarded as highly sought after, it is certainly widely traded. A survey of plants traded in Witwatersrand traditional medicine shops in Gauteng Province, South Africa (Williams et al., 2001), revealed that only two of 50 shops stocked pieces of *Encephalartos*, compared with 15 that stocked *Stangeria*. Of note is the fact that *S. eriopus* does not grow naturally in the Gauteng province—illustrating the extent of demand for these plants far from their natural habitat.

Ethnomedicinal Uses of Cycads in South Africa

STANGERIA ERIOPUS

This monotypic cycad genus, although utilized by several ethnic groups in South Africa, is most extensively used by the Zulu and Xhosa people. This is unsurprising, as *Stangeria* is limited to the eastern seaboard of southern Africa, an area inhabited mainly by Zulu, Xhosa, and Swazi peoples.

Chopped *Stangeria* lignotubers, known locally as *imfingo* are traded most frequently (Fig. 11-1), but whole plants, particularly young specimens, are also sold. Once purchased, the lignotubers are usually chopped into much smaller pieces and mixed with similarly prepared portions of other plants. Approximately 75% of *Stangeria* plants used are incorporated into magical preparations known as *intelezi*. A further 20% of the plants are used medicinally as an emetic, which reportedly achieves an internal cleansing of the body. Other uses include the warding off of lightning from homesteads, as a snuff to relieve congestion in infants, and as an agent to reduce blood pressure (Osborne et al., 1994).

Figure 11-1. Lignotubers of *Stangeria eriopus* offered for sale at a herbal street market in Durban, kwaZulu-Natal. (Photo: R. Osborne.)

Encephalartos Species

There are approximately 39 species of *Encephalartos* in South Africa, of which 15 occur in kwaZulu-Natal (South Africa National Biodiversity Institute, 2004). The Zulu name *isigqiki-somkovu* is used to refer to various *Encephalartos* species and translates as "seat of the familiar" (Doke et al., 1958). It is believed that when plants are placed around the homestead, any roaming evil spirits (witches' familiars) wanting to cause mischief will be thwarted through having to sit on the plant outside (Crouch et al., 2003). Other uses of *Encephalartos* include incorporation into necklaces, to be worn as charms to ensure longevity (Batten & Bokelmann, 1966), and consumption of rootstock infusions as love charm emetics (Gerstner, 1939). The use of entire stems to make starchy flour for diet supplementation is also known to occur during times of famine. *Encephalartos* species are used rather generically for medicinal/magical purposes, and any species that is available will be used.

While whole plants and cone sections are known to be traded, it is the sections of stem-bark that are most commonly harvested and sold, as can be observed in many medicinal plant markets in southern Africa. The removal of stem sections of about 15 cm in width from arborescent species (usually by means of a bush knife) has been recorded

Figure 11-2. A tall female specimen of *Encephalartos natalensis* at Elandskop, kwaZulu-Natal, showing the destructive effects of extensive bark stripping. (Photo: R. Osborne.)

from both the Eastern Cape and kwaZulu-Natal. Such bark collection done on a small scale is not likely to have a significant impact on wild cycad populations, because plants do recover after the wound has healed. However, if collecting is intensive the plants soon die. A number of cycad populations are known to have declined significantly as a direct result of bark collecting (e.g., *E. friderici-guilielmi* Lehm. at Tsolo in the Eastern Cape, and several *E. natalensis* R.A. Dyer & I. Verd. populations in kwaZulu-Natal) (Fig. 11-2). Results of the earlier-mentioned matched photograph study indicate that bark collecting has resulted in a significant decline in *Encephalartos* populations at about 5% of sites. Of concern is the observation that bark collecting is occurring even from the "Critically Endangered" *E. latifrons* Lehm. plants, with less than 100 remaining (Donaldson, 2003).

Literature Cited

Batten, A. & A. Bokelmann. 1966. Wild flowers of the eastern Cape. Books of Africa, Cape Town, South Africa.

Crouch, N. R., G. F. Smith, M. Lotter & R. Symmonds. 2003. *Encephalartos woodii*—the first historically documented ethnomedicinal plant casualty in southern Africa. Encephalartos 74: 25–28.

Doke, C. M., D. M. K. Malcolm & J. M. A. Sikakana. 1958. English and Zulu dictionary. Witwatersrand University Press, Johannesburg, South Africa.

Donaldson, J. S. 2003. Regional overview: Africa. Pp. 9–19 *in* J.S. Donaldson (ed.), Cycads. Status survey and conservation action plan. IUCN/SSC Cycad Specialist Group. IUCN, Gland, Switzerland and Cambridge, UK.

——— & J. D. Bösenberg. 1999. Changes in the abundance of South African cycads during the 20th century: preliminary data from the study of matched photographs. Pp. 240–247 *in* C. J. Chen (ed.), Proceedings of the Fourth International Conference on Cycad Biology. International Academic Publishers, Beijing, China.

Dyer, R. A. 1965. The cycads of southern Africa. Bothalia 8: 405–516.

Gerstner, J. 1939. A preliminary checklist of Zulu names of plants with short notes. Bantu Studies 13: 131–149, 307–326.

Hill, K.D., D.W. Stevenson & R. Osborne. 2007. The world list of cycads. /[La lista mundial de cícades]. Mem New York Bot. Gard. 97: 454–483.

Mander, M. 1998. Marketing of indigenous medicinal plants in South Africa. A case study in kwaZulu-Natal. Food and Agriculture Organisation (FAO) of the United Nations, Rome, Italy.

Osborne, R., A. Grove, P. Oh, T. J. Mabry, J. C. Ng & A. A. Seawright. 1994. The magical and medicinal usage of *Stangeria eriopus* in South Africa. J. Ethnopharmacol. 43: 67–72.

South Africa National Biodiversity Institute. 2004. PRECIS (Pretoria Computerized Information System) databank. SANBI, Pretoria, South Africa.

Tang, W. 1985. Cycad trade and conservation. Encephalartos 4: 22.

Williams, V. L., K. Balkwill & E. F. T. Witkowski. 2001. A lexicon of plants traded in the Witwatersrand umuthi shops, South Africa. Bothalia 31: 71–98.

GENOMICS, PHYSIOLOGY, AND STRUCTURE
CHAPTER 12

Genomic Studies in Cycas rumphii Miquel

Eric D. Brenner, Dennis Wm. Stevenson, Manpreet S. Katari, Richard W. McCombie, Stephen A. Rudd, Suzan J. Runko, Rob A. Martienssen, and Gloria M. Coruzzi

Abstract

Although cycads are the oldest living seed plants, little work has been performed on cycads at the molecular level. Here we report the first genomic treatment of cycads (Brenner et al., 2003b) in which 4210 expressed sequence tags (ESTs) were created from RNA harvested from the young leaves of *Cycas rumphii*. These ESTs were clustered into 2458 contigs and compared to the genes in other plants using BLAST analysis. Of these *C. rumphii* contigs, 1764 were similar to other plant genes: 1718 to angiosperms, 1310 to gymnosperms, and 734 to lower (nonseed) plant genes. A smaller subset, 46 contigs, was found that matched genes in lower plants and gymnosperms but not angiosperm genes. The complete sequence was gathered from the cDNAs of these 46 contigs and compared again to angiosperms, from which 14 still matched only gymnosperms. Conversely, a surprisingly large number of cycad contigs had homology to genes in angiosperms, with roles in development and signaling pathways.

Resumen

Aún cuando las cícadas son las plantas más antiguas con semillas, muy pocos estudios se han realizado en cícadas a nivel molecular. En este estudio presentamos el primer análisis genómico de cícadas (Brenner et al., 2003b). 4.210 ESTs (expressed sequence tags) fueron creados de ARN cosechado de hojas jóvenes de *Cycas rumphii*. Estos ESTs fueron agrupados en 2458 contigs y se compararon con los genes de otras plantas mediante el análisis de BLAST. De los 2458 contigs de *C. rumphii* disponibles, 1764 mostraron similitud con genes de otras plantas, de los cuales 1718 fueron similares a angiospermas, 1310 a gimnospermas y 734 con genes de plantas inferiores (sin semillas). Encontramos un grupo pequeño de 46 contigs que sólo mostraron similitud con genes en plantas inferiores y gimnospermas pero sin similitud a genes en angiospermas. La secuencia completa de estos 46 contigs se obtuvo de los correspondientes cDNA y se comparó nuevamente a genes de angiospermas, con 14 contigs todavía mostrando similitud sólo con gimnospermas. En contraste, un número sorprendente de contigs de cícadas presentó homología con genes en angiospermas con funciones en desarrollo y vías de transducción de señales.

Introduction

Cycads have survived for more than 270 to 280 million years since their origins in the Lower Permian (Mamay, 1969; Gao & Thomas, 1989). Of the gymnosperms, the Cycadales are considered to be the most ancestral compared to Ginkgoales, Gnetales, and Coniferales and thus are at a key node in plant evolution (Nixon et al., 1994; Magollón & Sanderson, 2002; Soltis et al., 2002). Cycads, which are nonflowering seed plants, exhibit a variety of characteristics evolutionarily intermediate between ferns and flowering plants, which make cycads an important species for molecular and genomic developmental studies (Brenner et al., 2003a). Plesiomorphic characteristics include motile male gametes, dichotomous branching, and ovules and seeds borne on the margins of leaflike megasporophylls—a trait that is particularly apparent in the genus *Cycas* (for review see Chamberlain, 1935; Stevenson, 1990; Norstog & Nicholls, 1997).

Despite the importance of cycads in the study of plant evolution, virtually nothing is known about the genes responsible for development in cycads. As a genetic system, cycads are not amenable to manipulation. Cycads are dioecious (male and female cones on separate plants), can take up to 30 years or more to become reproductive, and may require up to two and one-half years to produce viable seed after pollination (Norstog & Nicholls, 1997). Fortunately, recent advances in plant genomics can overcome many of these challenges. The availability of completely sequenced and annotated angiosperm genomes makes the study of genes from evolutionarily important plants possible by comparing the expressed genes of cycads (expressed sequence tags [ESTs]) to the complete genomes of higher plants.

Cycas rumphii Miq. was chosen for an EST survey because this genus is the sister taxon to the rest of the Cycadales (Treutlein & Wink, 2002; Hermsen et al, 2006). Described here is the construction of a cycad EST library from immature, developing *C. rumphii* leaves based on a previous report (Brenner et al., 2003b). In this study, cycad ESTs were discovered with similarity to genes involved in development in angiosperms. We also describe a subset of cycad clones with similarity to other gymnosperm genes but with no similarity to known angiosperm genes. Together these analyses pave the way for future genomic studies in cycads.

Methods

Tissue Collection, cDNA Library Construction, and DNA Purification

A cycad cDNA library was constructed as described (Brenner et al. 2003b). Briefly, recently emerged leaves (approximately 5 to 30 cm long) from the main apex of a female *C. rumphii* plant, accession 808/59 A (The New York Botanical Garden Conservatory), were frozen in liquid nitrogen. RNA was extracted from pulverized, frozen tissue using the RNeasy Maxi Kit (Qiagen). As a further cleanup step, *C. rumphii* RNA was precipitated in 2 m LiCl. The RNA pellet was washed twice with 70% ethanol and resuspended in water. PolyA RNA was purified with the Oligotex Maxi Kit (Qiagen) to make a cDNA library, which was constructed with the Lambda ZAP-CMV cDNA synthesis kit (Stratagene) from 10 μg of polyA RNA. During the cloning procedure, the cDNA was size fractionated over a Sepharose CL-6b column to select for high-molecular-weight fragments. Approximately 100 ng of high-molecular-weight cDNA was collected from the first five 100 μL fractions, precipitated, and resuspended in 3.5 μL water. Then 0.5 μL cDNA was directionally subcloned into the vector at the *Eco*RI and *Xho*I sites.

To confirm that the gymnosperm-specific cycad ESTs were indeed genuine cycad genes, polymerase chain reaction (PCR) analysis was performed according to standard protocols and DNA was collected using the DNeasy purification kit (Qiagen) from *C. rumphii* sporophylls approximately 2 to 4 weeks prior to emergence.

EST Sequencing

Plasmid DNA was collected as described in the manual (Stratagene), catalog number 200450 in the in vivo mass excision section. Sequencing was performed with either the M13 forward and/or reverse primer at the Cold Spring Harbor Laboratory using an ABI 3700 capillary sequencer. All sequencing reactions used the 1/16 Big Dye Terminator.

EST Clustering

Cluster analysis on the ESTs was performed with the HarvESTer application (Biomax Informatics AG, Martinsried, Germany) (http://www.ncbi.nlm.nih.gov/VecScreen/

UniVec.html). The default HarvESTer settings were optimized to screen for vector against the UniVec nonredundant database of vector and polylinker sequences. The hashed position tree (HPT) clustering used a similarity link threshold of 0.7. A maximum distance of six steps was required to define a cluster from the similarity network, thus encouraging the separation of likely paralogs. Cluster consensus sequences and concomitant alignments were also derived from the HPT clusters using the CAP3 application with default settings (Huang & Madan, 1999). The HarvESTer assemblies and coordinate alignments were imported into the Sputnik EST and cluster analysis application (Rudd et al., 2003).

BLAST Comparison of Cycad ESTS to Plant Gene Databases

All cycad contigs sequences were aligned against the PlantEST database using TBLASTX and BLASTX (Altschul et al., 1990) against the NR(aa) database. Todd Wood from Clemson University provided the PERL script that creates the PlantEST databases as described above. All PlantESTs were downloaded from GenBank and segregated into the following three gene sets: angiosperms, gymnosperms, and nonseed plants (ferns, mosses, and algae). Matches with an expect value < 1e 10–5 were considered significant.

Results

A cDNA Library from *Cycas Rumphii*

In this study, 10- to 40-cm immature leaves protruding from the crown were collected shortly after emergence (Fig. 12-1) and used to construct as cDNA library to generate 4210 ESTs. The majority of these reads (3917) were generated from the 5' end of the cDNA, and a small subgroup (293) were sequenced from the 3' end. Cluster analysis performed produced a total set of 2458 contigs (unigene) consisting of 1917 singletons and 541 assemblies. Additional cycad EST sequences are currently being generated from leaves, ovules, and megasporophylls from this cycad specimen (Brenner et. al., unpublished data). The entire unigene set can be viewed and downloaded at http://nypgenomics.org. Sequence and peptide annotations can be viewed at http://sputnik.btk.fi/.

Cycad Contig Matches to Genes in Other Plants

Using TBLASTX (Altschul et al., 1990), the *C. rumphii* unigene set was compared to all available ESTs from GenBank and all predicted genes from the *Arabidopsis* and rice genomes from TIGR at http://www.tigr.org/. This combined EST and predicted gene dataset were separated into three files: angiosperms, gymnosperms, and nonseed plants. Of the total genes used in this analysis against the cycad contig dataset, the angiosperm subgroup consisted of 84.5%, the gymnosperms 6.5%, and the lower plants 9.0%.

Figure 12-2 displays the total number of cycad contigs shared between one or more of the plant gene datasets. Using BLAST stringency values of expect < 1eX10–5, a majority of cycad contigs (1764/2458) have matches in other plants; however, 694 have no detectable counterpart to other plant genes. Most *Cycas rumphii* matches (1718) are to angiosperms—because of the predominant number of flowering plant accessions available to the scientific community. A large number (1416) of cycad matches to angiosperms are also found in gymnosperms and/or lower plants. Of these, 1310 cycad contigs match gymnosperm genes, and 734 match genes from lower plants.

Full-Length Sequencing of Cycad Clones That Match Only Gymnosperm Genes

In Figure 12-2, 46 *C. rumphii* ESTs specifically match genes only in the gymnosperms and lower plants, but not in angiosperms. The full-length cDNAs of these 46 contigs that were sequenced as ESTs represent only a portion of the actual gene. Thirty-seven of the 46 *C. rumphii* cDNAs were sequenced in their entirety (nine clones were not successfully recovered for sequencing), and 14 clones still had no detectable similarity to any known angiosperm genes. None of these 14 *C. rumphii* cDNA clones have homology to any known genes outside of the plant kingdom. Gene-specific primers designed to each of the 14 genes were able to amplify a fragment from genomic DNA isolated from a different *C. rumphii* specimen and different tissue (sporophyll) than the source tissue of the cDNA library (data not shown). This distinct *C. rumphii* specimen was cultivated in a geographically separate location (Florida) from the cDNA source *C. rumphii* specimen used for cDNA library construction (New York).

Figure 12-1. *Cycas rumphii* leaves that were used for expressed sequence tag (EST) analysis. RNA was extracted from a flush of young, emergent leaves (arrow) from *C. rumphii* growing in the Conservatory of the New York Botanical Garden. PolyA RNA was purified and used to generate a cDNA library for EST sequencing.

Cycad Genes Similar to Developmental Regulators

Cycad ESTs matched a surprisingly large number of genes (TBLASTX score $<$ e-5) with previously characterized roles in growth and development (Table 12-1) in angiosperms including the transcription factors *CONSTANS* (Suarez-Lopez et al., 2001), two distinct

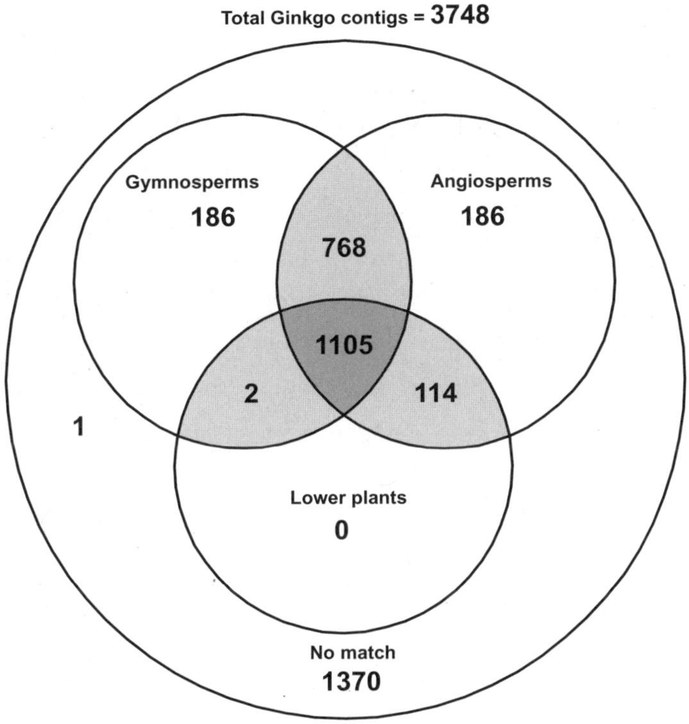

Figure 12-2. Shared gene sets between cycads versus lower plants, gymnosperms, and/or angiosperms. BLASTX (cutoff E value>e 10–5) was used to compare the cycad contigs against expressed sequence tags (ESTs) from all angiosperms, annotated genes from the full *Arabidopsis* and rice genome sequence from GenBank, and all available ESTs from gymnosperms and lower plants available in GenBank. Common genes between cycads and other plant subgroups are shown in the shaded intersecting areas.

homeobox genes (Chan et al., 1998), and a *YABBY* gene (Eshed et al., 2001). Other cycad ESTs have similarity to post-transcriptional and post-translational regulators of *Arabidopsis* development, including *ARGONAUT* (Bohmert et al., 1998) and *COP9* (Chamovitz & Glickman, 2002), respectively.

Conclusions

Because of their key evolutionary position, molecular studies on cycads will contribute vital input into our understanding of plant evolution as a whole. However, despite their historical importance in the field of terrestrial plant evolution, little is known regarding gene expression during development in the Cycadales. To correct this deficiency, a genomic treatment on *C. rumphii* is presented. In this work, a cDNA library was made from young, developing *C. rumphii* leaves to produce a cycad EST database.

We subsequently took a comparative genomics approach to analyze this dataset and subsequently discovered cycad genes similar to those found in higher plants, as well as cycad genes that appear unique to gymnosperms. In this analysis, we found 46 cycad contigs that matched only gymnosperms and/or lower plant ESTs but were not found in the genomes of higher plants or in nonplants. Complete sequencing of 37 of these cycad cDNA clones showed that 14 clones were still found only in other gymnosperms. This absence of homology to the completely sequenced genomes of angiosperms suggests that these 14 genes are found only in gymnosperms or lower plants in which genomic studies now underway. However, because ESTs, even when they are clustered into contigs, represent only a portion of the full-length gene, assignment of these genes as specific to gymnosperms is preliminary. As more ESTs are sequenced, preferably from normalized libraries, greater representation of the cycad transcriptome will result. Thus, future sequencing of cycad ESTs will increase support of clade specificity of genes that subsequently can be correlated with evolutionary innovations.

Our comparative genomics approach has detected genes in cycads that are likely likely to be important in development in development. One of these cycad ESTs one showed similarity to the *YABBY* gene family in angiosperms. *YABBY* genes are transcription factors expressed on the abaxial side of all lateral organs that promote abaxial cell fate (Eshed et al., 2001). In *Arabidopsis*, mutations in the *YABBY* gene, *INO* (*INNER-NO-OUTER*), lead to the loss of the outer integument (Villanueva et al., 1999) reminiscent of gymnosperm (and cycad) unitegmy (single integument). Unitegmy is considered to be the ancestral condition in seed plants (Villanueva et al., 1999). An analysis of *YABBY* gene expression in cycads may help explain the origin of the integument in gymnosperms and/or possibly the second integument in angiosperms.

Another transcription factor potentially involved in cycad development has highest

Table 12-1 Expressed sequence tags (ESTs) from *Cycas rumphii* with matches to genes involved in plant development.

GenBnk #	Subject description	e-value	% identified	Score
Development				
CB092871	Argonaute-like protein Argonaute (AGO1)-like protein *Arabidopsis thaliana*	8.00e-10	>0.85	34
CB090033	YABBY2 *A. thaliana*	2.00e-36	0.58	151
CB089539	Multisubunit regulator protein COP9—spinach COP9; *Spinacia oleracea*	2.00e-31	0.62	98
CB092157	CONSTANS B-box zinc finger family protein; *A. thaliana*	1.00e-47	0.48	221
CB092462	CRHB3 homeoprotein; *Ceratopteris richardii*	3.00e-44	0.70	131
CB089344	Homeodomain protein HB2; *Picea abies*	3.00e-29	0.62	117

C. rumphii ESTs were compared to GenBank with a BLASTP score < e-5. The top match produced from the BLAST search to the cycad EST is listed under the subject description.

similarity to the *CONSTANS* gene family, which are regulators of flowering time that follow internal and external (environmental) inputs in *Arabidopsis* (Putterill et al., 1995). Because cycads predate the evolution of flowering plants, it would be of interest to determine if *CONSTANS* genes in cycads or other gymnosperms temporally regulate seasonality of reproductive growth.

Acknowledgments

We would like to thank Francesco Coelho, Javier Francisco Ortega, Terrence Walters, and Jody Haynes of the Montgomery Botanical Center, Florida, for providing plant tissue and advice on cycad phenology; Vivekanand Balija for sequence generation; and Alex Clark for technical help. Funding for this work comes from the Plant Genomics Consortium. The Plant Genomics Consortium itself is funded by a grant from the National Science Foundation (DBI-0421604) as well as generous contributions from the Altria Group, Inc.; The Mary Flagler Cary Charitable Trust; The Eppley Foundation for Research, Inc.; The Leon Lowenstein Foundation, Inc.; The Ambrose Monell Foundation; and The Wallace Genetic Foundation, Inc.

Literature Cited

Altschul, S. F., W. Gish, W. Miller, E. W. Myers & D. J. Lipman. 1990. Basic local alignment search tool. J. Mol. Biol. 215: 403–410.

Bohmert, K., I. Camus, C. Bellini, D. Bouchez, M. Caboche & C. Benning. 1998. *AGO1* defines a novel locus of *Arabidopsis* controlling leaf development. EMBO J. 17: 170–180.

Brenner, E. D., D. W. Stevenson & R. W. Twigg. 2003a. Cycads: evolutionary innovations and the role of plant-derived neurotoxins. Trends Pl. Sci. 8: 446–452.

———, D. W. Stevenson, R. W. McCombie, M. S. Katari, S. A. Rudd, K. F. Mayer, P. M. Palenchar, S. J. Runko, R. W. Twigg, G. Dai, R. A. Martienssen, P. N. Benfey & G. M. Coruzzi. 2003b. Expressed sequence tag analysis in *Cycas,* the most primitive living seed plant. Genome Biol. 4: R78.

Chamberlain, C. J. 1935. Gymnosperms: Structure and Evolution. University of Chicago Press, Chicago, IL.

Chamovitz, D. A. & M. Glickman. 2002. The COP9 signalosome. Curr. Biol. 12: R232.

Chan, R. L., G. M. Gago, C. M. Palena & D. H. Gonzalez. 1998. Homeoboxes in plant development. Biochim. Biophys. Acta. 1442: 1–19.

Eshed, Y., S. F. Baum, J. V. Perea & J. L. Bowman. 2001. Establishment of polarity in lateral organs of plants. Curr. Biol. 11: 1251–160.

Gao, Z. & B. A. Thomas. 1989. A review of fossil cycad megasporophylls, with new evidence of *Crossozamia pomel* and its associated leaves from the lower Permian of Taiyuan, China. Rev. Palaeobot. Palynol. 60: 205–223.

Hermsen, E., E. Taylor, T. Taylor, & D. Stevenson. 2006. Cataphylls of the middle Triassic cycad *Antarcticycas schopfii* and new insights into cycad evolution. Amer. J. Bot. 93: 724–738. 2006.

Huang, X. & A. Madan. 1999. CAP3: a DNA sequence assembly program. Genome Res. 9: 868–877.

Magollón, S. & M. J. Sanderson. 2002. Relationships among seed plants inferred from highly conserved genes: sorting conflicting phylogenetic signals among ancient lineages. Amer. J. Bot. 89: 1991–2006.

Mamay, S. H. 1969. Cycads: fossil evidence of late paleozoic origin. Science 164: 295–296.

Nixon, K., W. Crepet, D. W. Stevenson & E. Friis. 1994. A reevaluation of seed plant phylogeny. Ann. Missouri Bot. Gard. 81: 484–583.

Norstog, K. J. & T. J. Nicholls. 1997. The biology of the cycads. Cornell University Press, Ithaca, NY.

Putterill, J., F. Robson, K. Lee, R. Simon & G. Coupland. 1995. The *CONSTANS* gene of *Arabidopsis* promotes flowering and encodes a protein showing similarities to zinc finger transcription factors. Cell 80: 847–857.

Rudd, S., H. W. Mewes & K. F. Mayer. 2003. Sputnik: a database platform for comparative plant genomics. Nucleic Acids Res. 31: 128–132.

Soltis, D. E., P. S. Soltis & M. J. Zanis. 2002. Phylogeny of seed plants based on evidence from eight genes. Amer. J. Bot. 89: 1670–1681.

Stevenson, D. 1990. Morphology and systematics of the Cycadales. Mem. New York Bot. Gard. 57: 8–55.

Suarez-Lopez, P., K. Wheatley, F. Robson, H. Onouchi, F. Valverde & G. Coupland. 2001. *CONSTANS* mediates between the circadian clock and the control of flowering in *Arabidopsis*. Nature 410: 1116–1120.

Treutlein, J. & M. Wink. 2002. Molecular phylogeny of cycads inferred from rbcL sequences. Naturwissenschaften 89: 221–225.

Villanueva, J. M., J. Broadhvest, B. A. Hauser, R. J. Meister, K. Schneitz & C. S. Gasser. 1999. *INNER NO OUTER* regulates abaxial-adaxial patterning in *Arabidopsis* ovules. Genes Dev. 13: 3160–3169.

GENOMICS, PHYSIOLOGY, AND STRUCTURE
CHAPTER 13

Age Influences Photosynthetic Capacity of Cycas micronesica Leaves

Thomas E. Marler

Abstract

The photosynthetic characteristics of *Cycas micronesica* K.D. Hill leaves aged 2 to 24 months were determined in cultivated plants. Net photosynthesis (A_{CO2}) was measured as net carbon dioxide exchange. Light-saturated A_{CO2} declined as leaves aged; that for the oldest leaves was approximately 75% of that for the youngest leaves in the study. The photosynthetic photon flux at which A_{CO2} reached capacity was about 1550 μmol m^{-2} s^{-1} for 2-month-old leaves and 1350 μmol m^{-2} s^{-1} for 22-month-old leaves. These results indicate that photosynthetic capacity of *C. micronesica* leaves remains high for at least two years after leaf emergence. However, these results retain relevance only at the landscape level in the absence of any changes in severity of leaf damage by cycad leaf consumers already on Guam and the absence of new alien leaf consumers becoming established on Guam. The number of hours during the photoperiod that the sustained photosynthetic capacity is achievable also declines with leaf age for several reasons. The contribution of these older leaves to the plant's carbon economy is influenced by the amount of shade imposed by younger leaf cohorts and by daily ambient sun exposure.

Resumen

Las características fotosintéticas de las hojas de *C. micronesica* K.D. Hill de dos a 24 meses de edad fueron determinadas en plantas cultivadas. La fotosíntesis neta (ACO2) fue medida como intercambio neto de dióxido de carbono. La luz saturada ACO2 se declinó con la edad de las hojas al 75% de las hojas más jóvenes del estudio. El flujo fotosintético fotón del cual el ACO2 alcanzó su capacidad fue alrededor de 1550 mmol m-2 s-1 para hojas de dos meses de edad y 1350 mmol m-2 s-1 para hojas de 22 meses de edad. Estos resultados indican que la capacidad fotosintética de las hojas de *C. micronesica* queda alta por lo menos dos años después su emergencia. Sin embargo, estos resultados son relevantes solamente al nivel de campo con la ausencia de cualquier cambio en la intensidad del daño a las hojas por predadores de hojas de las cícadas presentes en Guam así como la ausencia de nuevos predadores exóticos estableciéndose en la isla. El número de horas durante el fotoperíodo en donde la capacidad fotosintética sostenida se alcanza también se declina con la edad de la hoja por varios razones. La contribución a la economía de carbono de la planta por estas hojas más viejas está influenciada por la intensidad de sombra impuesta por lo cohortes más jóvenes de hojas y por la exposición diaria de sol.

Introduction

Cycas micronesica K.D. Hill is the only native gymnosperm in the Mariana Islands. It is a prominent member of all forested habitats throughout the island of Guam underlain by the Ritidian soil series and in riparian habitats within intact forests in southern Guam. The leaves on this arborescent cycad are produced in pulses that are synchronized among many individuals within a population. Moreover, leaves of this and other arborescent cycad species are relatively large and long-lived (Hill, 1994; Norstog & Nicholls, 1997; Marler, 2004).

A leaf's rate of decline in initial photosynthetic capacity is highly correlated with leaf life span capacity. Species producing short-lived leaves lose photosynthetic capacity rapidly, and species producing long-lived leaves lose photosynthetic capacity more slowly (Chabot & Hicks, 1982; Reich et al., 1991; Kitajima et al., 2002).

To my knowledge, the literature is devoid of any data relating cycad leaf age to leaf function. The objectives of this study were to determine the timing and extent of decline in photosynthetic capacity of *C. micronesica* leaves over a two-year period in Guam.

Materials and Methods

Experimental Site and Plant Material

This study was conducted in northern Guam (13.5°N, 145.7°E) from 1997 to 1999. *Cycas micronesica* plants used in the study were under cultivation in soils classified as clayey, gibbsitic, nonacid, isohyperthermic Lithic Ustorthents (Young, 1988). A minimal measure of weed control was used, but no fertilizer, irrigation, or pesticide applications were part of the management practices. Thus, lack of neighbor competition for edaphic and light resources was the primary difference between this site and the typical habitat in northern Guam.

Two methods were used to determine the decline in photosynthetic capacity with leaf age:

1. The first scheme used repeated measurements on the same cohort of leaves as they aged throughout the study. These leaves emerged in a synchronized manner in January 1997, and measurements were made from March 1997 until January 1999 when the leaves reached 24 months of age. No particular pattern was used to select each of the dates for these repeated measurements. In general, one month after each measurement date I began waiting for a day with appropriate conditions as defined by the combination of adequate soil moisture resulting from recent rainfall and relatively clear skies resulting from a high-pressure weather system.
2. The second scheme used measurements from various leaf cohorts of known ages on the same plants. The ability to use this scheme was achieved by recording leaf

flush emergence dates beginning January 1997. Measurements on these individuals were made on several days in November 1998 when the youngest leaves were 3 months of age and oldest leaves were 22 months of age.

Physiology Measurements

Leaf gas exchange was measured using an open system (CIRAS-1, PP Systems, Stotfield, Hitchin, Herts, U.K.) as described in Marler (2004). The ambient conditions for each of the measurement days for scheme 1 were similar because of my selection of days with similar weather conditions throughout the study. Measurements commenced about 0900 and were terminated by 1100 on each of the measurement dates to avoid the possibility of any midday depression in A_{CO2}. Gas exchange was measured on leaflets near the base, midrange, and tip of each leaf. These three values were averaged to make up each replication. There were 10 replications for each measurement period. The range in relevant leaf cuvette conditions was 32–35°C, 328–337 $\mu L \cdot L^{-1}$ CO_2, 1600– 200 $\mu mol\ m^{-2}\ s^{-1}$ photosynthetic photon flux (PPF), and 0.8–1.5 kPa vapor pressure deficit.

Gas exchange measurements for photosynthetic capacity in scheme 2 were conducted as with scheme 1, and there were eight replications. I also measured leaf A_{CO2} of 3-month-old and 22-month-old leaves as a function of irradiance to determine the influence of leaf age on light saturation for A_{CO2}. This was accomplished by placing shadecloth of varied percentages above the cuvette to reduce PPF from full sun conditions to approximately 300 $\mu mol\ m^{-2}\ s^{-1}$. This process ensured that each measured leaf was light-limited in the lower PPF range. Leaf cuvette conditions were 31–34°C, 329–336 $\mu L \cdot L^{-1}$ CO_2, and 0.8–1.3 kPa vapor pressure deficit.

Statistics

The decline in photosynthetic capacity with age was linear for scheme 1 and scheme 2 data. Analysis of covariance was used to test for homogeneity of slope for the two data sets. The linear slope was consistent for the two sets. Thus, the data were combined to determine the influence of leaf age on the decline in photosynthetic capacity.

The relationship between A_{CO2} and incident PPF greater than 300 μmol m^{-2} s^{-1} was determined using nonlinear regression. The response curves were defined by the equation $A_{CO2} = a(1-e^{-b*PPF})$. The light saturation point for A_{CO2} was calculated as the PPF at 95% of maximum A_{CO2}.

Results

Maximum leaf age for scheme 1 measurements was 24 months. Scheme 2 measurements included leaves from flushes that occurred in January 1997, June 1997, February 1998, and August 1998. Thus, maximum leaf age for scheme 2 was 22 months. Photosynthetic capacity decreased linearly with leaf age for both measurement schemes, with a negative slope of 0.091 for scheme 1 and 0.095 for scheme 2. Analysis of covariance for homogeneity of slope was not significant ($p \leq .6052$), indicating that the two data sets could not be used to estimate different functions. Thus, the data were combined, and the A_{CO2}–age relationship revealed a y-intercept of 9.64 and a linear loss in photosynthetic capacity of 0.094 μmol m^{-2} s^{-1} CO_2 per month (Fig. 13-1).

The dependence of A_{CO2} on irradiance followed the same trend for the two leaf age categories and within the PPF range used in this study (Fig. 13-2). The PPF at which A_{CO2} reached capacity was about 1550 μmol m^{-2} s^{-1} for 2-month-old leaves and 1350 μmol m^{-2} s^{-1} for 22-month-old leaves.

Discussion

Relatively little is known about physiology of cycads (Nortstog & Nicholls, 1997), and very few reports include information about photosynthesis of cycads. Within a relatively homogeneous ex situ setting, Marler and Willis (1997) compared gas exchange characteristics of 16 diverse species, and A_{CO2} ranged from approximately 5 to 10 μmol m^{-2} s^{-1}. In situ net photosynthesis of *C. micronesica* leaves was 9 to 10 μmol m^{-2} s^{-1} (Marler, 2004). *Macrozamia communis* L.A.S. Johnson leaf photosynthesis was 7.2 μmol m^{-2} s^{-1} (Wright et al., 2001; I. J. Wright, pers. comm.). *Zamia skinneri* Warsz. ex Dietrich leaves exhibited net photosynthesis of 2.4 to 3.0 μmol m^{-2} s^{-1} (Fetcher et al., 1994).

Photosynthetic capacity of young *C. micronesica* leaves as earlier reported (Marler, 2004) was confirmed in the present study. This moderate photosynthetic capacity is substantially lower than deciduous or pioneer species with short life spans but higher than many species with leaves exhibiting extended life spans. For example, Reich et al. (1991) reported photosynthetic capacity of 16 $\mu mol\ m^{-2}\ s^{-1}$ for species with leaf life spans of one and one-half months and 5.3 $\mu mol\ m^{-2}\ s^{-1}$ for species with leaf life spans of four years.

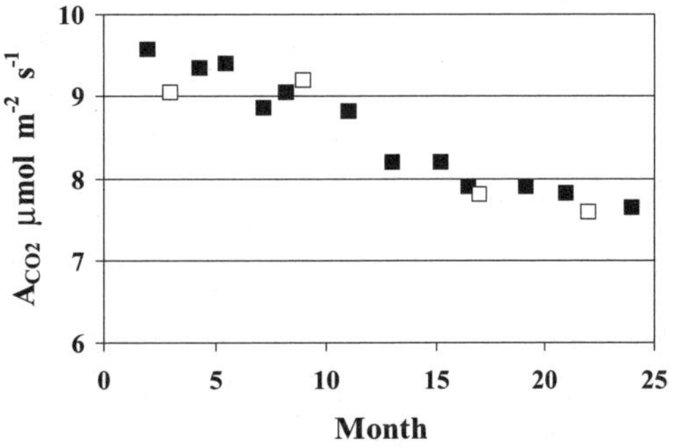

Figure 13-1. Net CO_2 assimilation (A_{CO2}) of *Cycas micronesica* leaves as influenced by leaf age in months. Closed squares are from repeated measurements from the same cohort of leaves (n=10). Open squares are from one measurement period using different leaves of known ages within the same plants (n=8). Response equation was A_{CO2}=9.64–0.09*month.

Median leaf life span of *Z. skinneri* leaves was reported to be 4.6 years (Clark et al., 1992) and that of *Zamia debilis* L.f. was reported to be 2.3 years (Negron-Ortiz & Breckon, 1989). Leaf loss rates were not recorded in the present study. Thus, neither the ultimate leaf life span for *C. micronesica* nor the major causes of mortality is known to date. Extrapolating the linear decline of the function in Figure 13-1 beyond the 24-month data set reveals an x-intercept of 107 months, or 8.9 years. However, this is not a defendable estimate of leaf life span. For example, pioneer species with leaves with short life spans exhibited greater x-intercepts for this function than directly measured mean leaf life span (Kitajima et al., 2002). Another incidental comparison for estimating life span is placing photosynthetic capacity of 9–10 $\mu mol\ m^{-2}\ s^{-1}$ for this cycad species within the published relationships between photosynthetic capacity and measured leaf life span for other species. Using the relationships of Reich et al. (1991), for example, *C. micronesica* would have a leaf life span of about 13.5 months. Obviously, this is also not a defendable estimate of life span.

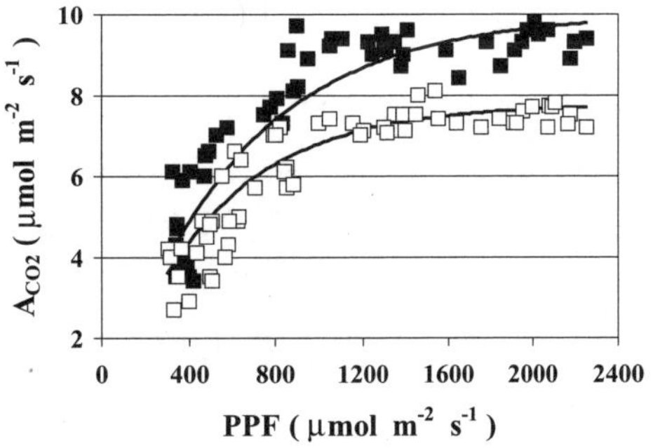

Figure 13-2. Net CO_2 assimilation (A_{CO2}) of *Cycas micronesica* leaves as influenced photosynthetic photon flux (PPF). Closed squares are 3-month-old leaves, open squares are 22-month-old leaves. Response equations were $A_{CO2} = 10.01(1-e^{-0.0016*PPF})$ for young leaves and $A_{CO2}=7.77(1-e^{-0.0021*PPF})$ for old leaves.

The decline in photosynthetic capacity of the Guam cycad was 0.094 μmol m^{-2} s^{-1} CO_2 per month (Fig. 13-1). Comparisons of this slope with that of other species is limited to the report of Kitajima et al. (2002), who reported slopes for two pioneer species of –0.2 to –0.25 μmol m^{-2} s^{-1} CO_2 per day. The rate of decline in photosynthetic capacity for these two species with leaf life spans of 74 to 93 days was 65 to 80 times greater than that for the Guam cycad. This stark difference reveals one advantage of lengthy leaf life span: a sustained photosynthetic capacity.

Prolonged leaf life span may be advantageous for several reasons other than sustained photosynthetic capacity. Chapin (1980) noted nutrient conservation and nutrient use efficiency as two particular advantages. Waring and Franklin (1979) considered that being evergreen, which involves maintaining a cohort of leaves for more than one year (Chabot & Hicks, 1982), to be an advantage for maintaining positive plant carbon balance. Plants possessing a long leaf life span are competitive in resource-limited habitats (Chabot & Hicks, 1982; Coley et al., 1985; Wright et al., 2001).

The decline in photosynthetic capacity following full leaf expansion is not considered a passive descent in function due to aging. Rather, an active redistribution of resources such as nitrogen to younger leaves aids in optimizing whole-plant level carbon gain (Field & Mooney, 1983; Hikosaka et al., 1994; Wright & Westoby, 2003). The association of *C. micronesica* with cyanobionts and the resulting access to fixed nitrogen may account for the sustained photosynthetic capacity of leaves from this cycad species.

Clark et al. (1992) attribute major forms of leaf death in *Z. skinneri* to deterioration of leaf carbon balance, physical damage, and attack of specialist herbivores. Determining causal factors for leaf death in *C. micronesica* deserves attention; this would put the data presented in the present paper into context. Anecdotal observations indicate that tropical cyclones and alien leaf consumers are two major potential causes of cycad leaf death on Guam.

Tropical cyclones are frequently responsible for leaf death or damage among Guam's *C. micronesica* population. Almost 200 tropical cyclones have passed near the island during the past 50 years (Marler, 2001; Hirsh & Marler, 2002). Partial to complete defoliation is a common form of damage. For example, during Typhoon Paka in 1997, an estimated 75% of the population suffered from this form of damage (Marler & Hirsh, 1998). Tropical cyclone damage is a naturally occurring phenomenon that has also helped shape the ability of *C. micronesica* to sustain life on Guam throughout evolutionary history. However, anthropogenic factors are also relevant. Sambar deer (*Cervus unicolor* Kerr) were introduced to Guam in the late 1700s (Rogers, 1995). In recent years, these feral deer have become a primary herbivore of cycad foliage in some locations. Moreover, predation of cycad leaves by alien arthropods has become one of the greatest threats to *C. micronesica*. Two species have been identified as major threats subsequent to the 1999 dates of data collection reported herein. The first is an *Erechthias* leaf miner not yet identified to the species level, and the second is the Asian cycad scale (*Aulacaspis yasumatsui* Takagi). Plant response to the scale is so acute that mature cycad individuals are typically dead within one year of initial infestation.

The photosynthetic capacity, slope of decline in photosynthetic capacity, and leaf life span for the Guam cycad were defined in the absence of these alien cycad leaf consumers. Thus, the extent to which deer, leaf miner, scale, and other as yet unidentified alien herbivores have had an impact on leaf longevity at the landscape level is a subject that deserves attention. The incredible ability of *C. micronesica* to sustain photosynthetic capacity for years provides little benefit in the face of these epidemic outbreaks of alien species.

The issue of sun versus shade leaves also deserves mentioning. *Cycas micronesica* grows well in exposed or shaded locations, and leaves exhibit phenotypic plasticity to developmental light level (Marler, 2004). Sun leaves generally have shorter life spans than shade leaves (Nilsen & Orcutt, 1996). In the present study, all leaves were sun leaves; thus, the results do not apply to shade leaves.

Acknowledgments

This research was supported by the U.S. Department of Agriculture under Cooperative State Research, Education, and Extension Services (CSREES) Tropical and Subtropical Agricultural Research program, which is administered by the Pacific Basin Administrative Group.

Literature Cited

Chabot, B. R. & D. J. Hicks. 1982. The ecology of leaf life spans. Annual Rev. Ecol. Syst. 13: 229–259.

Chapin, F. S. III. 1980. The mineral nutrition of wild plants. Annual Rev. Ecol. Syst. 11: 233–260.

Clark, D. B., D. A. Clark & M. H. Grayum. 1992. Leaf demography of a neotropical rain forest cycad, *Zamia skinneri* (Zamiaceae). Amer. J. Bot. 79: 28–33.

Coley, P. D., J. P. Bryant & F. S. Chapin III. 1985. Resource availability and plant anti-herbivore defense. Science 230: 895–899.

Fetcher, N., S. F. Oberbauer & R. L. Chazdon. 1994. Physiological ecology of plants at La Selva. Pp. 128–141 *in* L.A. McDade, K.S. Bawa, H.A. Hespenheide, and G. S. Hartshorn (eds.), La Selve: ecology and natural history of a neotropical rainforest. University of Chicago Press, Chicago, IL.

Field, C. B. & H. A. Mooney. 1983. Leaf age and seasonal effects on light, water, and nutrient use efficiency in a California shrub. Oecologia 56: 348–355.

Hikosaka, K., I. Terashima & S. Katoh. 1994. Effects of leaf age, nitrogen nutrition and photon flux density on the distribution of nitrogen among leaves of a vine (*Ipomea tricolor* Cav.) grown horizontally to avoid mutual shading of leaves. Oecologia 97: 451–457.

Hill, K. D. 1994. The *Cycas rumphii* complex (Cycadaceae) in New Guinea and the Western Pacific. Austral. Syst. Bot. 7: 543–567.

Hirsh, H. & T. Marler. 2002. Damage and recovery of *Cycas micronesica* after Typhoon Paka. Biotropica 34: 598–602.

Kitajima, K. S. S. Mulkey, M. Samaniego & S. J. Wright. 2002. Decline of photosynthetic capacity with leaf age and position in two tropical pioneer tree species. Amer. J. Bot. 89: 1925–1932.

Marler, T. E. 2001. Tropical cyclones and perennial species in the Mariana Islands. HortScience 36: 264–268.

———. 2004. Leaf physiology of shade-grown *Cycas micronesica* leaves following removal of shade. Bot. Rev. 70: 63–71.

——— & H. Hirsh. 1998. Guam's *Cycas micronesica* population ravaged by Supertyphoon Paka. HortScience 33: 116–118.

——— & L. E. Willis. 1997. Leaf gas-exchange characteristics of sixteen cycad species. J. Amer. Soc. Hort. Sci. 122:38–42.

Negron-Ortiz, V. & G. J. Breckon. 1989. Population structure in *Zamia debilis* (Zamiaceae). I. size classes, leaf phenology, and leaf turnover. Amer. J. Bot. 76:891–900.

Nilsen, E. T. & D. M. Orcutt. 1996. Physiology of plants under stress. Wiley, New York.

Norstog, K. J. & T. J. Nicholls. 1997. The biology of the cycads. Cornell University Press, Ithaca, NY.

Reich, P. B., C. Uhl, M. M. Walters & D. S. Ellsworth. 1991. Leaf lifespan as a determinant of leaf structure and function among 23 amazonian tree species. Oecologia 86: 16–24.

Rogers, R. F. 1995. Destiny's landfall—a history of Guam. University of Hawaii Press, Honolulu, Hawaii.

Waring, R. H. & J. F. Franklin. 1979. Evergreen coniferous forests of the Pacific Northwest. Science 204: 1380–1386.

Wright, I. J. & M. Westoby. 2003. Nutrient concentration, resorption and lifespan: leaf traits of Australian sclerophyll species. Funct. Ecol. 17: 10–19.

———, P. B. Reich & W. Westoby. 2001. Strategy shifts in leaf physiology, structure and nutrient content between species of high- and low-rainfall and high- and low-nutrient habitats. Funct. Ecol. 15: 423–434.

Young, F. J. 1988. Soil survey of the territory of Guam. U.S.D.A. Soil Conservation Service, National Cooperative Soil Survey.

GENOMICS, PHYSIOLOGY, AND STRUCTURE

CHAPTER 14

Water Economy of Macrozamia riedlei in the Jarrah (Eucalyptus marginata) Forest of Southwestern Western Australia

John K. Marshall

Abstract

Macrozamia riedlei is found mainly in the jarrah (*Eucalyptus marginata*) forest of southwestern Western Australia at latitudes ranging from 32° to 35°S. The climate is extreme mediterranean-type with cool, wet winters and hot, dry summers. Annual rainfall is 500 to 1200 mm, with more than 70% falling between April and November. Dawn leaf water potential (DLWP) of *M. riedlei*, an indicator of the potential of the water source, was measured twice a year over five years. Average rainfall for the field site was 720 mm. Measurements were made in October, when soil water was fully recharged from winter rains, and in April, when soil water was at its annual minimum. At these times, DLWP was also measured on plants representing other growth forms (herbs, shrubs, grass trees, midstory trees, and canopy trees). In October, DLWP of *M. riedlei* was similar to, although general higher than, that of other growth forms. However, by April, the herbs were no longer alive, and the DLWP of the shrubs was very low, indicating a shallow root system. The taller grass trees exhibited less water stress than the midstory trees. Of these other growth forms, the deepest-rooting forest trees forming the canopy exhibited the least water stress. *Macrozamia riedlei* was the exception; it maintained a comparatively high DLWP potential by April. This cycad seems to be managing its water economy by having a large internal subterranean water storage reservoir. This interpretation is discussed in relation to survival and distribution.

Resumen

Macrozamia riedlei se conoce principalmente en el bosque de jarrah (*Eucalyptus marginata*) del suroeste de Australia, latitudes 30 a 35°S. El clima es de tipo mediterráneo extremoso con inviernos frescos y húmedos y veranos calientes y secos. La precipitación anual es de 500 hasta más de 1,200 mm con el 70% de la precipitación entre abril y septiembre. El potencial hídrico foliar al amanecer es un indicador del contenido de la fuente de agua en *M. riedlei,* esto fue medido dos veces por año durante 5 años. La precipitación promedio fue 720 mm. Las medidas fueron tomadas en octubre cuando el agua del suelo es recargada plenamente por las lluvias invernales, y en abril cuando el nivel del agua fue mínima durante el año. En estos períodos, también se midió el potencial hídrico foliar de plantas con flor representando otras formas de vida: es decir, hierbas, arbustos, pastos arbóreos, árboles del subdocel y árboles del docel. El potencial hídrico foliar de *M. riedlei* fue similar, aunque generalmente más alto, que en las otras formas de vida en octubre. Sin embargo, en abril, cuando las hierbas están secas y que el potencial hídrico foliar al amanecer de los arbustos fue muy bajo, esto indica un sistema radicular somero de ambas formas de vida. Los pastos arbóreos más altos fueron menos estresados que los arbustos, siendo más similares a los árboles del subdocel. Los árboles del docel, con raíces más profundas fueron los menos estresados de todos. *M. riedlei* fue excepcional ya que mantuvo un nivel comparativo alto del potencial hídrico foliar al amanecer desde octubre hasta abril. Esto fue interpretado como que *M. riedlei* mantiene su economía hídrica por presentar una gran capacidad de almacenar agua internamente por debajo del nivel del suelo. Esta interpretación fue discutida en relación a su supervivencia y distribución.

Introduction

Macrozamia Miq. is unusual among cycad genera in that it is distributed almost entirely outside the tropics. In Western Australia, it is confined to the South West Botanical Province as first identified by Diels (1906) and refined by Beard (1990). *Macrozamia* is also unusual in Western Australia in having a wide, semicontinuous distribution over 600 km from north to south and 800 km from west to east. Three closely similar species are recognized (Hill & Osborne, 2001). *Macrozamia fraseri* Miq. occurs on and near the coastal plain mainly north of Perth to Eneabba. *Macrozamia riedlei* (Gaudich.) C. A. Gardner, the most widespread, occurs in the jarrah forest, the more restricted karri forest, and the *Banksia*-sheoak woodland of the coastal plain from north of Perth to the southwest corner of Western Australia and along the south coast to Albany. *Macrozamia dyeri* (F. Muell.) C. A. Gardner occurs in an isolated area that is 300 km to the east of Albany from west of

Esperance to Cape Arid in the east. In this account, all plants will be referred to as *M. riedlei* with details of their localities added. Further information on several aspects of *M. riedlei* is contained in Ladd and Connell (1995).

The climate of southwestern Western Australia is extreme mediterranean-type. In the region of distribution of *M. riedlei* described above, annual rainfall varies from 500 to more than 1200 mm. More than 70% of the rain falls from April to September. Temperatures vary from summer average daily minima and maxima of 12° to 33°C to winter average daily minima and maxima of 3° to 18°C. Subzero temperatures occur in winter away from the coastal plain but are relatively uncommon near the ocean.

Macrozamia riedlei, like many other cycads, survives an annual wet season and an annual dry season. Less commonly among cycads, however, does the wet season coincide with the cooler months of the year. Little rain falls in the six warmer months from October to March in areas where *M. riedlei* occurs. Despite this, plants retain a crown, sometimes of more than 50 leaves, each in excess of 2 m long, which remains dark green for three or more years. New leaves appear in spring and early summer (September to December) following an increase in temperature and a soil profile wet from winter rains.

This study aimed to examine how *M. riedlei* manages its water economy so as to survive in the summer-dry climate of southwestern Western Australia. The relatively large belowground part (stem and roots) of *M. riedlei* suggests that it may function both as a storage for water as well as for reserves for growth. To gain insight into the function of the belowground part as water storage, dawn leaf water potential (DLWP) of *M. riedlei* was compared with that of flowering plant species of other growth forms in the jarrah forest. DLWP is a measure of the potential of the water that serves as the water source for the plants. Measurements were made twice a year over a period of five years, in October when the soil water was fully recharged from winter rain and in April when the soil water was at its lowest following the dry summer. Supplementary studies were carried out on the relationships between the parts of *M. riedlei* plants and on soil water storage over the range of distribution of *Macrozamia* in southwestern Western Australia.

Material and Methods

Dawn Leaf Water Potential

The study site was in the jarrah forest (33°23'S, 116°42'E) between Collie and Williams, 180 km SSE of Perth, Western Australia. The site is in the 750-mm rainfall zone. Rainfall averaged 720 mm for the study period from 1977 to 1983.

The vegetation was dominated by jarrah (*Eucalyptus marginata*) and marri (*Eucalyptus calophylla*) on the ridges and slopes, with wandoo (*Eucalyptus wandoo*) appearing in the drainage lines. The canopy top was about 16 m above the ground, with the trees forming more of a woodland than a forest. A midstory composed of *Allocasuarina fraseriana, Banksia grandis*, and *Persoonia longifolia* was present. The understory was dominated by shrubs, mainly members of the Myrtaceae, Proteaceae, and Papilionaceae. *Macrozamia riedlei* was present in the understory at midslope in the landscape. On winter wet ground, perennial rushes (Restionaceae) and sedges (Cyperaceae) occurred along with ephemerals (Centrolepidaceae, Droseraceae, Stylidiaceae). Nomenclature follows Marchant et al. (1987).

The soils were typical of the Darling Range. The full laterite profile is found on the ridges comprising ironstone over mottled and pallid kaolinitic clays (Gilkes et al., 1973). This weathered saprolite can be some tens of meters deep. The laterite profile, truncated by erosion, forms the slopes where a superficial layer of progressively smaller ironstone gravels covers the clay layer with colluvium in the drainage lines, often comprising yellow sands.

Sites for measurement of DLWP were selected on the slopes where *M. riedlei* was present along with flowering plant species of contrasting growth forms. These included the ephemeral herb *Drosera gigantea* (Droseraceae); five shrub species, *Acacia pulchella* (Mimosaceae), *Allocasuarina humilis* (Casuarinaceae), *Bossiaea ornata* (Papilionaceae), *Daviesia decurrens* (Papilionaceae), and *Hakea lissocarpha* (Proteaceae); the grass tree *Xanthorrhoea preissii* (Xanthorrhoeaceae); three midstory tree species, *A. fraseriana* (Casuarinaceae), *B. grandis* (Proteaceae), and *P. longifolia* (Proteaceae); as well as two canopy tree species, jarrah (*E. marginata*) and marri (*E. calophylla*).

DLWP was selected as the measurement to indicate the water status of the different growth forms. This measurement, using the Scholander bomb, is well established in studies of plant water relations and is a robust technique for field use (Scholander et al., 1965).

Leaf water potential measured just about dawn gives an indication of the full extent to which a plant has recovered from water loss (transpiration) the previous day. By dawn, recovery is complete, and so leaf water potential at that time indicates the water potential of the source of the water. Adjustment for the height of the leaf sample above the source of water (–1 bar per 10 m; Slatyer, 1967) has not been made in the data presented. Generally, the potential of the water source is the average for the soil layer in which the roots are accessing water. An exception may be where there is a source of water for uptake by the plant stored elsewhere. The prospect that such a source would be indicated by a leaf water potential differing noticeably from the water potential of the soil layers was relevant to the present study. DLWP is the preferred measurement for this study because at other times during the day, leaf water potential is dependent both on the water potential of the source and the prevailing climatic conditions.

DLWP was measured on four or more samples from the selected species. Sampling took place about dawn over 14 days. Samples beyond reach from the ground, such as trees, were obtained by using a shotgun and flashlight to find them when they fell to ground. Otherwise, the samples were cut from the plants.

There were two sampling periods each year from 1977 to 1983. These coincided with a soil profile fully recharged from winter rains in spring (October) or when the soils were driest in autumn (April).

Relationships between Plant Parts

Macrozamia riedlei plants from the jarrah forest have predominantly belowground stems. That is, there is a crown of leaves above ground connected by leaf bases to a stem below ground, comprising persistent leaf bases, cortex, and pith. The other belowground parts are attached to the underside of the stem and comprise nontuberous, tuberous, and coralloid roots. To investigate the contributions of the plant parts to the entire plant, advantage was taken of sites at which plants of varying sizes were being excavated from cattle pastures or for road widening. Twenty-nine plants were examined from the sandy coastal plain near Gingin, 31°21'S, 115°54'E, north of Perth, and three plants were examined from alluvium near Denmark, 34°57'S, 117°11'E, near the south coast of Western Australia.

To assess the contribution of the different plant parts, leaves, stems, and roots were separated from plants representative of the size range available and weighed fresh. The persistent leaf bases were separated from the stem and weighed. Nontuberous, tuberous, and coralloid roots were separated and weighed.

To assess the demand for water and the capacity for water storage, estimates of leaf area and stem volume were made. To estimate leaf area, the leaves were counted, rachis and petiole length recorded, the leaflets counted, and their length and breadth recorded. To estimate water storage volume, stem diameter and depth were measured and tuberous roots counted, measured, and weighed.

The 32 plants available for study had an average fresh weight of 97 kg (range, 5 to 288 kg), thus increasing the range of fresh weights for which published data were available (Halliday & Pate [1976]; 0.2 to 99 kg; Grove et al. [1980]; 0.2 to 8.3 kg).

Water Capture By *M. riedlei*

Observations on the length, size, and distribution of roots of *M. riedlei* were made opportunistically at the time of excavation of the plants referred to in the previous section. More detailed measurements—fresh weights and dimensions—of the roots of excavated plants were made. The contribution of intact roots in supplying water to an additional 28 plants, which were excavated and transplanted, was indicated by the varying number of leaves that dried off in the weeks following severing of the roots and transplanting.

A secondary feature of *M. riedlei* that contributes to its water economy involves its aboveground parts. The funnel-shaped display of leaves and incubous leaflets concentrates water at the leaf bases. Measurements of leaf length and average angle were used to estimate this catchment effect. Also, leaf bases, both from the stem and from current leaves, were hydrated over 48 hours to demonstrate their capacity to take up water. The water-retaining capacity of the brown hairs associated with the leaf bases, which form a progressively prominent mat as plant size increases, was also determined.

Soil Water Storage

Maps recently published by the Bureau of Meteorology depict the distribution of monthly rainfall and monthly actual evapotranspiration (Bureau of Meteorology, 2000, 2001). From these, monthly soil water surplus (rainfall less evapotranspiration plus surplus rainfall from the previous month) was estimated for locations within and beyond the distribution of *Macrozamia* to add to the interpretation of that distribution.

Results and Comments

Dawn Leaf Water Potential

Rainfall for the six months prior to each sampling period is given in Figure 14-1. The cool, wet–warm, dry sequence of the climate is apparent.

DLWP of *M. riedlei* and of representatives of other growth forms in the jarrah forest—herbs, shrubs, grass trees, midstory trees, and canopy trees—is given in Figure 14-2.

In spring, when the soil profile was recharged from winter rains, DLWPs of *M. riedlei* and associated forest species were at their most similar, in the range of about 5 bars. Even so, *M. riedlei* always had the highest water potential, ranging from −1.5 to −2.6 bars over the five years of measurement, with the exception of spring 1980 when the ephemeral *D. gigantea* had a higher DLWP of −0.34 bar.

In autumn, when the soil profile was at its driest, DLWPs of *M. riedlei* and associated forest species were least similar, ranging from 8 to more than 25 bars depending on the year. There were no measurements for *Drosera* because the plants had no living aboveground parts in autumn. In the autumns of 1978 and 1979, the DLWPs of the shrubs, midcanopy trees, and canopy trees were progressively higher, reflecting the greater rooting depths of these growth forms and hence access to more water. With the exception of the autumn of 1981, the shrubs always had the lowest DLWP. The grass trees, represented by *X. preissii*, occupied the least consistent position of the other growth forms in autumn, sometimes with higher DLWPs than the midstory trees (1978, 1979, 1982) or with lower water potentials than the midstory trees (1980, 1981). The consistent and notable difference in DLWPs occurred between *M. riedlei* and all the other growth forms. The DLWP of the cycad in autumn was always highest, varying from −2.3 to −10.7 bars, depending on the year.

Relationships between Plant Parts

The percentage, by fresh weight, of the different plant parts is related to total plant weight in Figure 14-3. The underground stem is by far the main contributor to plant fresh

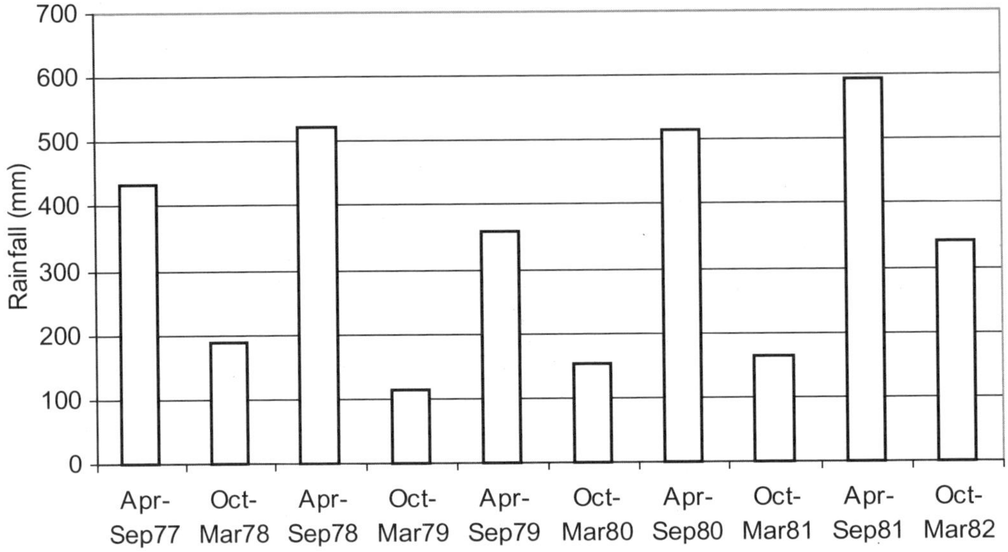

Figure 14-1. Rainfall for the six-month periods, April to September and October to March 1977 to 1982, prior to measurement of dawn leaf water potential of jarrah forest species between Collie and Williams, southwestern Western Australia.

weight, at about 80% for plants that weigh more than 100 kg. Tuberous roots contributed about 3% to total plant weight, nontuberous roots about 13%, and coralloid roots less than 1%.

Stem volume, although variable as a proportion of leaf area, was comparatively unchanged over the range of stem weight, averaging about 8 mm (Fig. 14-4). This would equate to up to 8 days of water use in summer, implying continuing replenishment of the store of water by uptake by the roots.

Water Capture

Plants of *M. riedlei* have a deep and extensive root system. Roots extend at least 4 m, predominantly downward but also laterally from the plant. At some sites, the deep roots take the plant within reach of groundwater. However, at other sites, the water table is not reached because of its great depth or absence.

The importance of the extensive root system was demonstrated by the reaction of

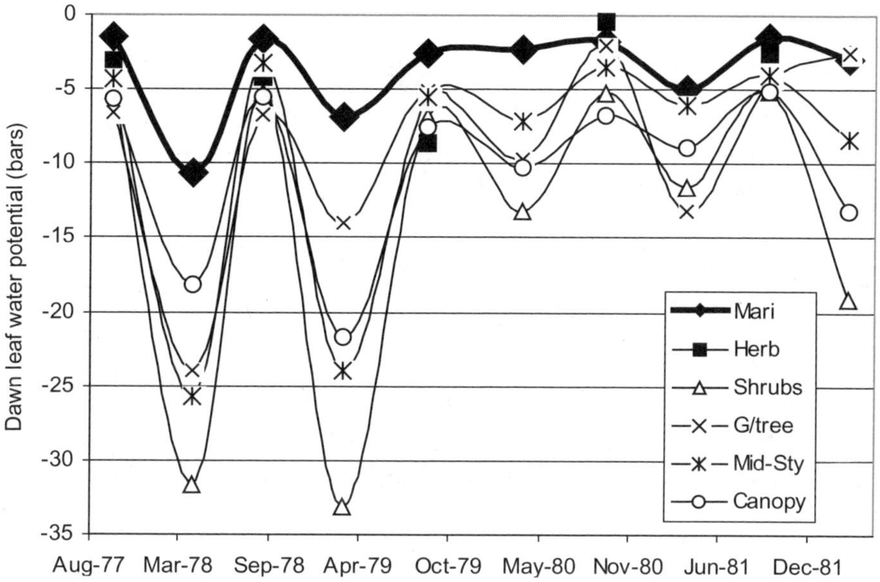

Figure 14-2. Dawn leaf water potential of *Macrozamia riedlei* compared with other growth forms of flowering plant in the jarrah forest measured in spring (October) and autumn (April) from 1977 to 1982. Mari=*M. riedlei*; herb=*Drosera gigantea*; shrubs=*Acacia pulchella, Allocasuarina humilis, Bossiaea ornata, Daviesia decurrens,* and *Hakea lissocarpha*; g/tree=*Xanthorrhoea preissii*; mid-sty=*Allocasuarina fraseriana, Banksia grandis,* and *Persoonia longifolia*; canopy=jarrah (*Eucalyptus marginata*) and marri (*Eucalyptus calophylla*).

28 transplanted plants that ranged in size from 150 to 700 mm in stem diameter to the severing of the root system during excavation. A varying proportion of leaves, typically about 70%, dried off, indicating a dependence on water supply by the deep root system to the stem. Notably, however, at least some leaves remained green while those in the process of flushing continued to grow, indicating a contribution of the stem to the water supply. The reduction in leaf area equates to a threefold increase (i.e., 24 days) in water storage duration relative to water use.

Tuberous roots were observed on practically all of the excavated plants (Fig. 14-5). These numbered up to 20 but were more commonly in the range of 7 to 15. They were parsnip- to sweet potato–shaped, tuberous, sessile organs attached to the underside of the stem. Fully developed tubers were typically about 250 cm^3 in volume, occasionally up to 600 cm^3. Remains of tubers were common, indicating they are periodically

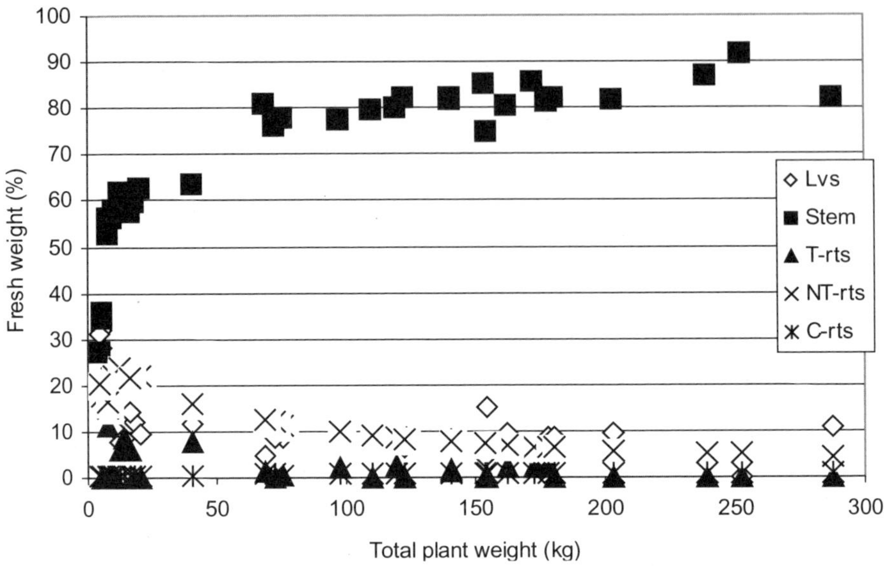

Figure 14-3. Fresh weight of parts of *Macrozamia riedlei* as percentage of whole plant plants of different weights. Lvs = leaves; stem = stem including persistent leaf bases; T-rts = tuberous roots; NT-rts = nontuberous roots; C-rts = coralloid roots. Twenty-nine plants from near Gingin; three plants from near Denmark, Western Australia.

replaced. Given their reaction to iodine, they contained some starch, but less than the leaf bases, cortex, and pith of the stem. The tubers continued at their distal end as a root, although generally of smaller diameter than the nontuberous roots. Although the capacity of the tuberous roots to store water was small compared to that of the stem, these roots probably contribute importantly to the water economy of the plant. A study of the seasonal dynamics of the tuberous roots would be instructive.

Less important and, at this stage, only suggestive, is the interception of rain by the aboveground parts. The funnel-shaped catchment presented by the angled leaves typically concentrates rainfall about a 100-fold on the apex of the stem at the point of entry of the petioles. The incubous arrangement of the leaflets on the rachis causes water to be ducted down the abaxial side of the leaves. The brown hairs on the abaxial sides of the leaf bases trap the water very effectively. The capacity of the mat of brown hairs to trap water becomes substantial in larger plants: about 1 L of water in stems of volume 20 L and more than 10 L in stems of volume 100 L. It is not known whether,

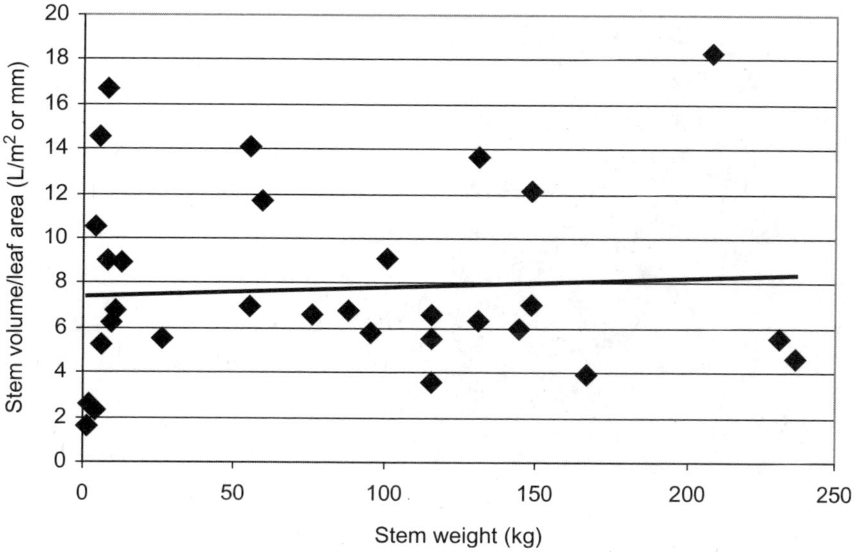

Figure 14-4. Ratio of capacity of belowground stem (L) to leaf area (m^2) (equivalent to mm) related to stem weight of plants of *Macrozamia riedlei*.

once trapped, the water is absorbed into the leaf bases. However, following immersion in water for 48 hours, the leaf bases were found to be some of the most absorbent organs of the plant, increasing 53±25% (range, 31–124%) in weight. Although these features of water capture by aboveground parts of *M. riedlei* can be demonstrated, their contribution to the water economy, even assuming the trapped water is absorbed, is likely to be much less than that of water uptake from the soil and storage in the tuberous roots and stem.

Soil Water Surplus

Monthly soil water surplus at localities within and beyond the range of distribution of *Macrozamia* in southwestern Western Australia is plotted in Figure 14-6. Within the range of distribution of *Macrozamia*, soil water surplus occurs in at least four and one-half months each year. Beyond the drier limit, soil water surplus is either negligible (southeastern extreme of distribution) or persists for fewer than about four months (northern and southwestern extremes of distribution).

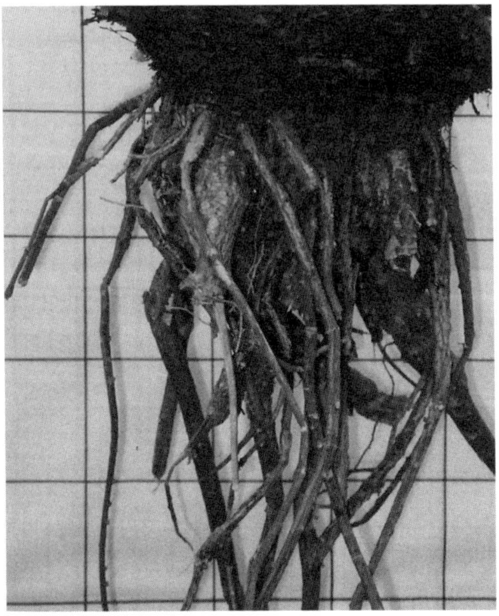

Figure 14-5. Roots of excavated *Macrozamia riedlei*. Grid size: 100mm×100mm. Note current (LHS) and spent (foreground, RHS) tuberous roots.

Discussion and Conclusions

Water Strategy Of *M. riedlei*

The DLWP of *M. riedlei* compared with that of other growth forms in the jarrah forest indicates that *M. riedlei* accesses a source of water other than soil water, especially in autumn. DLWP in autumn of grass trees and midstory, subdominant trees, the two growth forms most likely to exploit a similar depth of soil to *M. riedlei*, was 7 to 25 bars lower than that of *M. riedlei*. DLWP of the dominant, canopy trees, the deepest rooting plants of the forest, was 7 to 12 bars lower than that of *M. riedlei*. Water accessed by *M. riedlei* in autumn, when the uppermost soil layers were at their driest, was at least in part from a source of water of higher potential. It is concluded that this source was the belowground parts of *M. riedlei*.

These observations suggest a different strategy in the water economy of *M. riedlei* in the jarrah forest to that of its flowering plant associates. These associated species rely on the soil to store water for use in transpiration, the amount accessed depending primarily on rooting depth and reflected in stature aboveground (Crombie et al., 1988). Ephemeral herbs have particularly shallow root systems. Slightly deeper rooting is associated with shrubs, at least some of which have a partly deciduous habit or folding leaves to aid survival. The deepest rooting plants are evergreen, perennial trees. *Macrozamia riedlei*, although relatively small, maintains a crown of green leaves year round, and although the soil is the major source of water, it differs in having significant internal storage to supplement its water supply when soil water is largely depleted at the end of the dry summer.

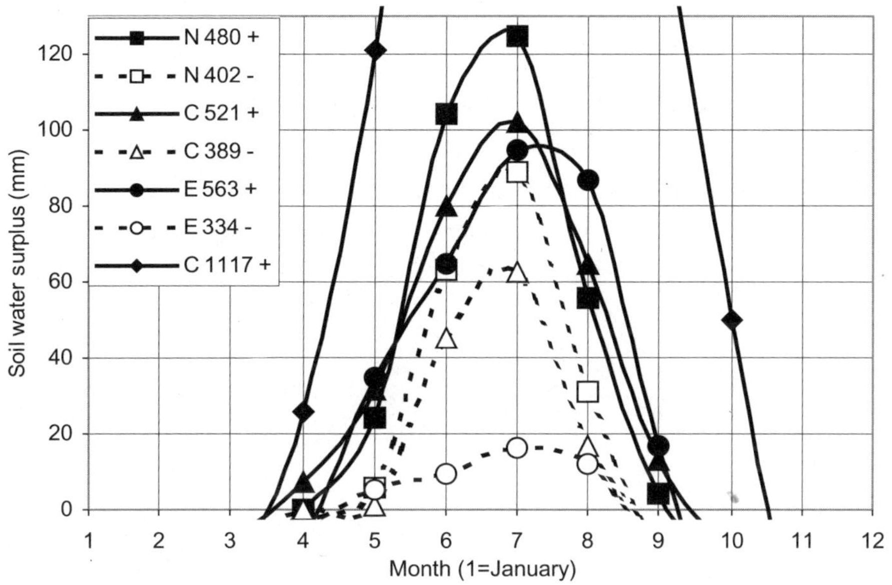

Figure 14-6. Monthly soil water surplus from locations within and beyond the range of distribution of Macrozamia in Western Australia. Part of range: N = northern (squares); C = central (triangles and diamonds); E = eastern (circles). Number = annual rainfall in mm. Macrozamia distribution = + within range (continuous line) or – beyond range (broken line).

Survival of *M. riedlei*

Plants of *M. riedlei* are deceptive, with a modest crown of leaves, giving little indication of the massive belowground parts. *Macrozamia riedlei* is deep rooting; most of its mass is in its subterranean stem, with temporary tuberous roots adding to its ability to store water. The deep roots, which grow to more than 4 m, give the plant access to soil water at depths bettered only by the dominant trees of the forest. Although water at these depths is decreasing by the end of summer, the DLWP of *M. riedlei* indicates that it is accessing water at potentials higher than any of the soil layers. The inferred internal storage of water is supported by the presence of the large underground stem and tuberous roots. The temporary nature of the tuberous roots and their small size compared with the much larger stem suggests that they act as a supplementary reservoir to storage in the stem. As soil water is depleted over the predominantly rainless summer and early autumn (January to April), it appears that water stored in the underground stem and tuberous roots is used to progressively supplement soil water for survival.

Unlike the woody flowering plant species with which *M. riedlei* has been compared, the leaves of *M. riedlei* are directly attached to the water storage organ. Survival is aided by this and by the low rate of water loss (transpiration). Stomatal conductance of leaves of *M. riedlei* was less than one fifth that of jarrah leaves (Crombie, 1992; Veneklaas & Poot, 2003).

Distribution of *Macrozamia* in Western Australia

Past Distribution

All species of *Macrozamia* are more widely distributed in Western Australia than the forest dominants, jarrah and marri, especially at the lower rainfall limits of distribution (Churchill, 1968; Ladd & Connell, 1995). This wider distribution is consistent with the superior water relations of *M. riedlei* compared to the forest dominants. Assuming that this relative distribution of *Macrozamia* and the forest dominants has persisted over time, changes in the distribution of *Macrozamia* may be inferred from the results of the palynological studies of Churchill (1968). The varying amounts of pollen of jarrah, marri, and karri (*Eucalyptus diversicolor*) in deposits of known age indicated a markedly wetter period from 6000 to 5000 BP, followed by a drier period from 4500 to 2500 BP. A return to a somewhat wetter period occurred from then until about 1300 BP, followed by a drier period from 800 to 500 BP. From then until recent times, the tendency has been for conditions to become wetter again. These results suggest that *Macrozamia* was more widely distributed in the wetter period 6000 to 5000 years ago, which may have been the most recent period that the now disjunct distribution of *M. dyeri* was continuous with the distribution of *M. riedlei* on the south coast. Since then, with intermittent drier periods and relatively short returns to slightly wetter conditions, the distribution of *Macrozamia*, while fluctuating, has probably tended to decrease overall.

Reduction in *Macrozamia* has accelerated over the past 100 years, particularly toward the drier limit of distribution, with clearing for agriculture and active extermination of *Macrozamia* because of its association with livestock poisoning. State government policy in the 1960s of clearing 1,000,000 acres of native vegetation per year for agriculture probably added markedly to the decrease of *Macrozamia*. Also, over the past 50 or so years, increased clearing for urban development, mainly in the wetter part of the range of distribution, has resulted in a further decrease in *Macrozamia*.

Present Distribution

The present boundary at the drier edge of distribution of *Macrozamia* occurs somewhere between the 500 and 400 mm rainfall isohyets (Ladd & Connell, 1995). When examined in terms of soil water surplus, regimes on either side of this boundary are distinctly different. Within the distribution of *Macrozamia*, sufficient rain falls in the cooler months for a surplus of soil water beyond the requirements of evapotranpiration to occur for four and one-half months or more each year. Beyond the drier limit of the range of *Macrozamia*, rainfall in the cooler months about equals evapotranspiration with little or no surplus, or the duration of surplus is fewer than four months per year. The suggested strategy of *Macrozamia* in storing water from the cooler months for use in the warmer months would consequently be ineffective in the absence of sufficient water surplus from the rains of the cooler months.

Future Distribution

The effect of climate change on the survival of biota is a current topic. Trends in rainfall along with the results of global climate modelling indicate southwestern Western Australia is, uniquely in Australia, becoming drier (Indian Ocean Climate Initiative, 2002). The effect is largely confined to the period from May to July inclusive, the early part of the wet season. Adjacent regions appear to be showing little change or slight increases in rainfall (Bureau of Meteorology, 2000). The neutral region of no change about coincides with that of the present boundary of distribution of *Macrozamia*. Therefore, *Macrozamia* in Western Australia may not be greatly affected by climate change. Much will depend on the reduction in rainfall in the early part of the wet season. Continuing clearing for agriculture and, increasingly, urban development present a more immediate threat. Transplanting of otherwise doomed plants would be desirable. However, larger plants are at a disadvantage because their substantial belowground parts require large earth-moving equipment to excavate. A change in attitude is required, particularly in urban development, to recognize *Macrozamia* plants for what they are and to encourage their retention or transplanting. Fortunately, the most extensive habitat for *Macrozamia*, the jarrah forest, is unlikely to be cleared, apart from state-controlled mining, or subjected to threatening climate change because it occurs in the higher rainfall region. Consequently, the forest form of *M. riedlei* is least likely to have a foreseeable decrease in its distribution. In the longer term, while climate change may not appear to be a threat, its effect should be kept in mind, particularly for *M. dyeri*, where the species is trapped in its disjunct distribution between the drier limit of its range and the ocean.

Summary

When compared with flowering plant species in the jarrah forest, *M. riedlei* manages its water economy differently. It has an internal store of water belowground to draw on at the end of the dry summer–autumn period with its leaves directly connected to that store and with low stomatal conductance. Comparison of the weights of parts of *M. riedlei* plants indicates that the underground stem is the main permanent organ for water storage with the tuberous roots acting as temporary storages.

The limit of distribution of *Macrozamia* in Western Australia corresponds with a change from surplus soil water storage for at least four and one-half months to negligible surplus or a surplus for fewer than four months each year. Thus, the internal water storage strategy may determine the drier limit to the range of distribution of *Macrozamia*.

Macrozamia was probably more widely distributed 6000 to 5000 years ago, when the climate was wetter. At that time, the now disjunct distribution of *M. dyeri* may have been continuous with *M. riedlei* on the south coast. Over the past 100 years, decrease in distribution of *Macrozamia* in Western Australia has been accelerated by clearing for agriculture and, increasingly, urban development.

Future distribution may be relatively unaffected by the current drying of the climate, except for *M. dyeri*. A decrease in distribution is likely to continue, however, increasingly because of urban development and to a lesser extent because of clearing for agriculture. The core of distribution of *M. riedlei* in the jarrah forest is least likely to be affected, in spite of mining there.

Acknowledgments

Thanks go to Bill Dewar for his skillful excavation of plants for examination and transplanting as well as to Marion Marshall for assistance with weighing plant parts.

Literature Cited

Beard, J. S. 1990. Plant life of western Australia. Kangaroo Press, Kenthurst, NSW, Australia.

Bureau of Meteorology. 2000. Climatic atlas of Australia. Rainfall. Bureau of Meteorology, Melbourne, Australia.

———. 2001. Climatic atlas of Australia. Evapotranspiration. Bureau of Meteorology, Melbourne.

Churchill, D. M. 1968. The distribution and prehistory of *Eucalyptus diversicolor* F. Muell., *E. marginata* Donn ex Sm., and *E. calophylla* R.Br. in relation to rainfall. Austral. J. Bot. 16: 125–151.

Crombie, D. S. 1992. Root depth, leaf area and daytime water relations of jarrah (*Eucalyptus marginata*) forest overstorey and understorey during summer drought. Austral. J. Bot. 40: 113–122.

———, J. T. Tippett & T. C. Hill. 1988. Dawn water potential and root depth of trees and understorey species in southwestern Australia. Austral. J. Bot. 36: 621–631.

Diels, L. 1906. Die Pflanzenwelt von West-Australien sudlich des Wendeskreises. Vegetation der Erde 7. Leipzig, Germany.

Gilkes, R. J., G. Scholz & G. M. Dimmock. 1973. Lateritic deep weathering of granite. J. Soil Sci. 24: 523–536.

Grove, T. S., A. M. O'Connell & N. Malajczuk. 1980. Effects of fire on the growth, nutrient content and rate of nitrogen fixation of the cycad *Macrozamia riedlei*. Austral. J. Bot. 28: 271–281.

Halliday, J. & J. S. Pate. 1976. Symbiotic nitrogen fixation by coralloid roots of the cycad *Macrozamia riedlei*: physiological characteristics and ecological significance. Austral. J. Pl. Physiol. 3: 349–358.

Hill, K. & R. Osborne. 2001. Cycads of Australia. Kangaroo Press, East Roseville, NSW, Australia.

Indian Ocean Climate Initiative. 2002. Climate variability and change in south west Western Australia. Department of Environment, Perth, Australia.

Ladd, P. G. & S. W. Connell. 1995. Vicariant *Macrozamia* species in southern Australia. Pp. 225–240 *in* P. Vorster (ed.), Proceedings of the third international conference on cycad biology. Cycad Society of South Africa, Stellenbosch, South Africa.

Marchant, N. G., J. R. Wheeler, B. L. Rye, E. M. Bennett, N. S. Lander & T. D. Macfarlane. 1987. Flora of the Perth region. Western Australian Herbarium, Perth, Australia.

Scholander, P. F., H. T. Hammel, E. A. Hemmingsen & D. Bradstreet. 1965. Hydrostatic pressure and osmotic potential in leaves of mangroves and some other plants. Proc. Nat. Acad. Sci. U.S.A. 52: 119–125.

Slatyer, R. O. 1967. Plant-water relationships. Academic Press, London, England.

Veneklaas, E. J. & P. Poot. 2003. Seasonal patterns in water use and leaf turnover of different plant functional types in a species-rich woodland, south-western Australia. Pl. & Soil 257: 295–304.

The Vascularization of the Seed of Ceratozamia mexicana (Zamiaceae)

María Ydelia Sánchez-Tinoco, E. Mark Engleman, and Stephen D. Koch

Abstract

The vascular supply of the seed of *Ceratozamia mexicana* Brongn. consists of internal and external vascular systems. Four to six vascular bundles from the megasporophyll enter the seed through the hilum and form a plexus. Within the plexus, the bundles divide, anastomose, and form a coroniform arrangement. This corona is observable prior to lignification of the sclerotesta. Vascular bundles emerge from the plexus and pass through the sclerotesta to form the internal and external vascular systems. The external system consists of

10 bundles and the inner system of up to 40 at the level where the nucellus becomes free from the integument. The term "pachychalaza" is introduced for the description of gymnosperm seeds. These vascular systems are illustrated in detail. Finally, the implications of this vascular structure are discussed with respect to the phylogenetic origin of the integument.

Resumen

El aprovisionamiento vascular de la semilla de *Ceratozamia mexicana* Brongn. consiste de los sistemas interno y externo. Cuatro a seis haces vasculares provenientes del megasporófilo penetran la semilla en la región del hilo y forman un plexo. Dentro del plexo, los haces se dividen, se anastomosan y se organizan en una disposición de corona. Esta corona se observa en las etapas previas a la lignificación de la esclerotesta. Haces vasculares salen del plexo y atraviesan la esclerotesta hacia el interior y exterior, conformando los sistemas vasculares interno y externo. El sistema externo consiste de diez haces vasculares, y en el sistema interno llega aproximadamente a cuarenta haces vasculares al nivel en que la nucela se libera del tegumento. Se introduce el término "paquicálaza" para la descripción de semillas de gimnospermas. Los sistemas de haces vasculares se ilustran con detalle. Las implicaciones de esta estructura vascular se discuten con respecto al origen filogenético del tegumento.

Introduction

The seed coat of all gymnosperms, with the exception of most conifer genera, is vascularized (Singh, 1978). The vascularization patterns differ among taxa and have been used to establish the typology of the gymnosperm seed (Meyen, 1984). In plants of the Cycadales, the seed coat is characterized by the existence of separate internal and external vascular systems (Stopes, 1905). This character ("double vascularization") has been included in cladistic studies of the major gymnosperm taxa (Crane, 1985a,b, 1988; Doyle & Donoghue, 1986, Loconte & Stevenson, 1990; Nixon et al., 1994; Rothwell & Serbet, 1994).

The vascular supply leading to the ovule in all cycad genera except *Microcycas* and *Ceratozamia* was studied by Worsdell (1898), who showed that two vascular bundles diverge from the axial vascular cylinder of the megastrobilus and divide, producing four vascular bundles in the pedicel of the megasporophyll. Of these, two internal bundles supply the central part of the megasporophyll and two external bundles supply the lateral parts, including the two ovules. Each of these external bundles, in turn, divides twice, so that four vascular bundles enter each ovule.

Stopes (1905) traced the vascular bundles within the cycad ovule. She described and illustrated the way in which the internal and external bundle systems are distributed with respect to the layers of the seed coat in *Cycas* and *Ceratozamia*. These systems come from a single central vascular bundle in the hilum in *Cycas*. This coincides with the observations of Matte (1904) for other genera but not with those of Worsdell (1898), who reported that the ovule is supplied by four bundles.

Singh (1978), in his review of the literature on the seed anatomy of living and fossil gymnosperms, emphasized three aspects: the distribution pattern of the vascular bundles in the seed coat, the vascularization of the nucellus, and the number and distribution of the vascular bundles that penetrate the chalaza. With respect to the distribution of vascular bundles in the seed coat in cycads, he focused his attention on the double vascular system. The external system is limited to the sarcotesta and extends from the chalaza to the micropyle. The internal system passes through the sclerotesta and extends to the point of separation between the integument and the nucellus. In the majority of cycads, the nucellus is not vascularized; the internal vascular bundles are found in an internal parenchymatous layer of the integument ("endotesta"), ending where the integument separates from the nucellus. However, in some genera, the nucellus is vascularized to various degrees and with varying frequency. With respect to the vascular bundles that penetrate the chalaza, Singh (1978) presented details for gymnosperm taxa but stated only that the mode of entry is different for cycads.

Matte (1904) studied the transverse differentiation of the vascular bundles in the ovule. According to this report, the vascular bundles are mesarch, both in the megasporophyll and in the external bundle system of the ovule. In the internal system, the protoxylem is difficult to distinguish, but Matte stated that the bundles are collateral and have little phloem. The single large vascular bundle that Stopes (1905) described in *Cycas* and that Matte found in other genera is reportedly concentric or nearly so.

Chamberlain (1935) stated that the number of external vascular bundles is constant for at least some species of cycads and generally does not exceed 12. On the other hand,

Crane (1988) stated that the external system of cycads consists of only 10 vascular bundles, but he did not specify in which species.

In *Ceratozamia mexicana* there is extensive fusion between the integument and the nucellus; this fused zone occupies more than four-fifths of the surface of the seed and is called the "pachychalaza" (Sánchez-Tinoco & Engleman, 2004). Thus, in a transverse section of the seed, the following layers can be distinguished: an external fleshy layer or sarcotesta; a stony layer or sclerotesta; and an internal parenchymatous layer of the testa that is fused with the nucellus, forming the pachychalaza. All of these surround the gametophyte (Sánchez-Tinoco & Engleman, 2004).

In this study we present new information about the vascular system of the seed of *C. mexicana*. The term "pachychalaza" is introduced for the description of gymnosperm seeds. In addition, the implications of this vascular structure are discussed with respect to the origin of the integument. Finally, the phylogenetic implications of double vascularization in the Cycadales and related groups of higher gymnosperm taxa are analyzed, and the potential usefulness of this character in cladistic studies is discussed.

Materials and Methods

The plant material was collected from a wild population of *C. mexicana* on Cerro Coacoatzintla, in the municipality of Coacoatzintla, 13 km due north of Xalapa, Veracruz, Mexico.

Whole megastrobili in their initial stages, sections of megastrobili in more advanced stages, as well as ovules and seeds from different plants were fixed in formalin-acetic acid-ethanol (FAA) and chromic acid–acetic acid–formalin (CRAF III) solutions (Sass, 1958). Sections were made until hardening of the stony layer began, which occurred in May 1995, nine months after the emergence of the megastrobili (Sánchez-Tinoco et al., 2000). The sections were prepared using common microtechnique (embedding in paraffin, making 15-μm sections, and staining with safranin–fast green). To observe the course of the vascular bundles in the seed coat, 50 μm sections were made using a hand microtome. Both longitudinal and transverse serial sections of the complete seed were obtained. In addition, radial sections were made of the megasporophyll and the seed with respect to the strobilus axis. These serial sections were cleared using 5% NaOH for about 18 hours at room temperature. After the sections were washed in water, they were mounted in glycerol or glycerol jelly. To observe cuticles, freehand sections and 60-μm sections with a hand microtome were made of the micropylar region. These were stained with oil red O. The preparations were studied and some photographed using light microscopy.

Results

Vascular Bundles as Seen in Cross Section of the Seed

The external vascular system consists of 10 longitudinal vascular bundles of the external system located within the sarcotesta but outside of the sclerotesta (Fig. 15-1). These bundles are endarch and collateral.

At approximately the middle of the seed, 32 longitudinal vascular bundles of the internal system and a few layers of collapsed parenchyma cells occur under the sclerotesta. Some vascular bundles are endarch, and others are mesarch (Fig. 15-2A, 15-2B); a few are amphicribral, whereas others are collateral. Large cells containing small particles of precipitated tannins commonly occur between the bundles (Fig. 15-2A, 15-2B).

Hilar Region

Four to six mesarch and amphicribral vascular bundles from the megasporophyll enter the seed through the hilum (Fig. 15-1). When the sarcotesta is removed by animals or decay, a lignified prominence of the sclerotesta can be observed in the center of the hilar region (Fig. 15-1). Here are four to six central orifices through which these vascular bundles enter the prominence (Fig. 15-3A). Surrounding this prominence are 10 orifices, which correspond to the exit of the 10 vascular bundles that make up the external system (Fig. 15-3A, 15-3B). These bundles are endarch and collateral.

Within the prominence of the sclerotesta, the vascular bundles form a plexus (Fig. 15-1; sclerified part of prominence shown in outline) where the bundles divide, anastomose, and are organized into a coroniform arrangement. This corona was observed prior to lignification of the sclerotesta, which occurred in March 1995 (Sánchez-Tinoco et al., 2000), at which time its structure is indicated by the distribution of vascular tissue accompanied by sclereids and tannin cells. Vascular bundles emerge from the plexus and pass through the sclerotesta to form the internal and external bundle systems (Fig. 15-1).

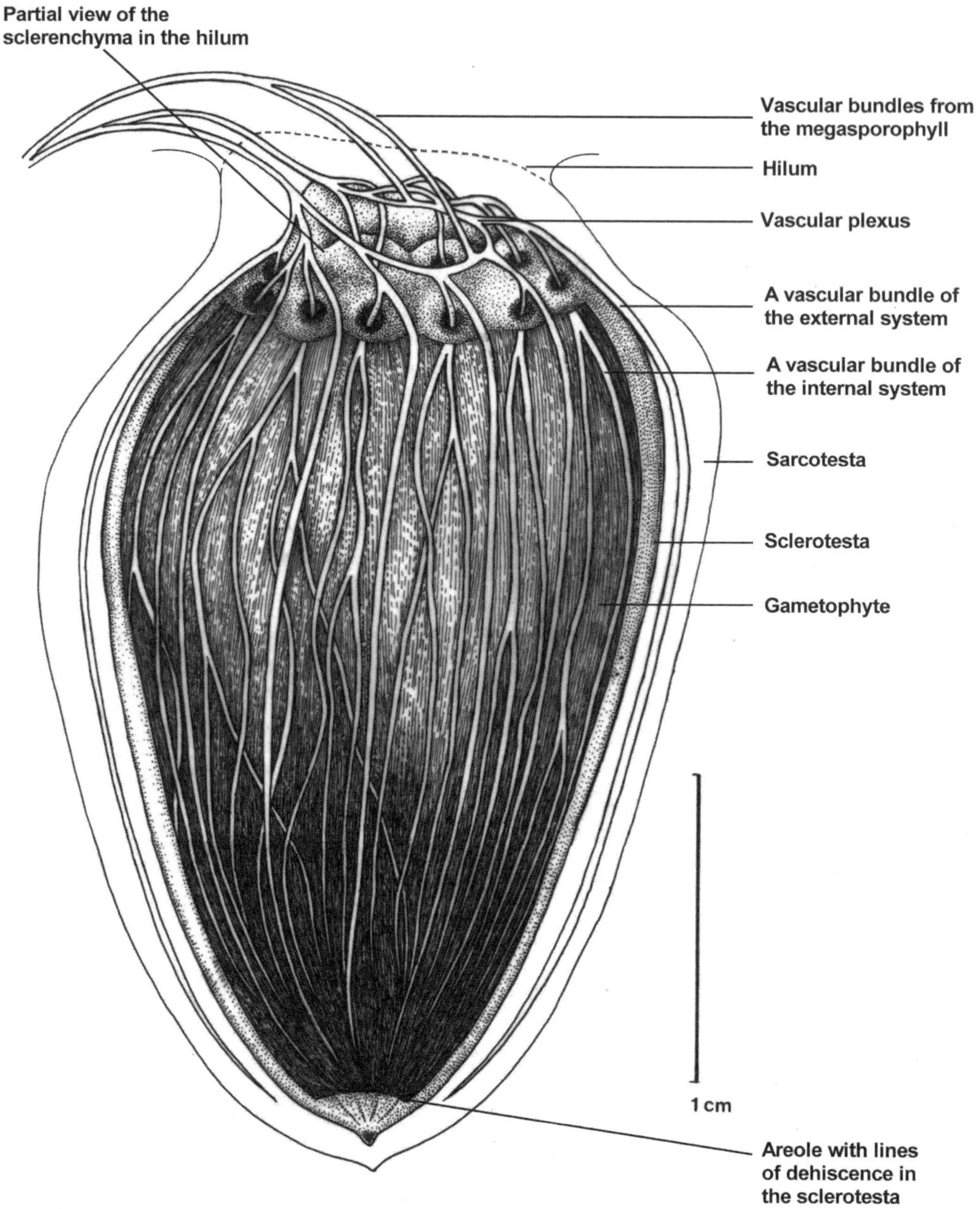

Figure 15-1. Vascularization of the mature seed of *Ceratozamia mexicana* Brogn. The sclerenchyma layer (stippled) is cut away on the side facing the viewer to expose the inner and outer longitudinal bundles and partly cut away to expose the vascular plexus at the upper end of the seed. The external bundles in an intact seed occupy the sarcotesta, which covers the sclerotesta. Neither of these layers is represented in the drawing. The tips of the external bundles reach almost to the micropylar areola. The inner bundles are seen in the drawing emerging underneath the orifices in the mass of sclerenchyma of the vascular plexus; they branch at various levels in their trajectory and end where the nucellus becomes free from the integument.

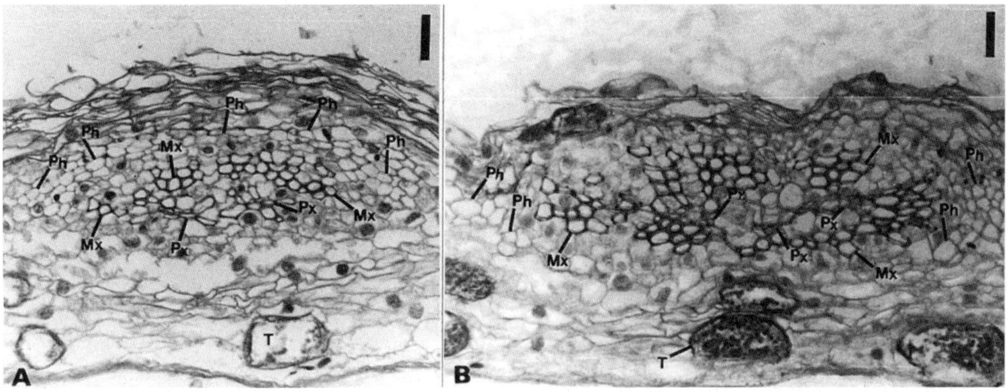

Figure15-2. Vascular bundles in cross section from the system inside the sclerotesta. Neither sclerotesta nor gametophyte appears in these preparations. The center of the seed is oriented downward in these views. T=tannin cell. Bar length=40 µm. **A.** Phloem (Ph) and metaxylem (Mx) form arcs on the upper side of the bundle. The left protoxylem (Px) pole has only parenchyma on the downward side; this is characteristic of an endarch xylem ontogeny. **B.** Metaxylem (Mx) is located on both the upward and downward sides of the protoxylem (Px) poles; this is characteristic of a mesarch xylem ontogeny.

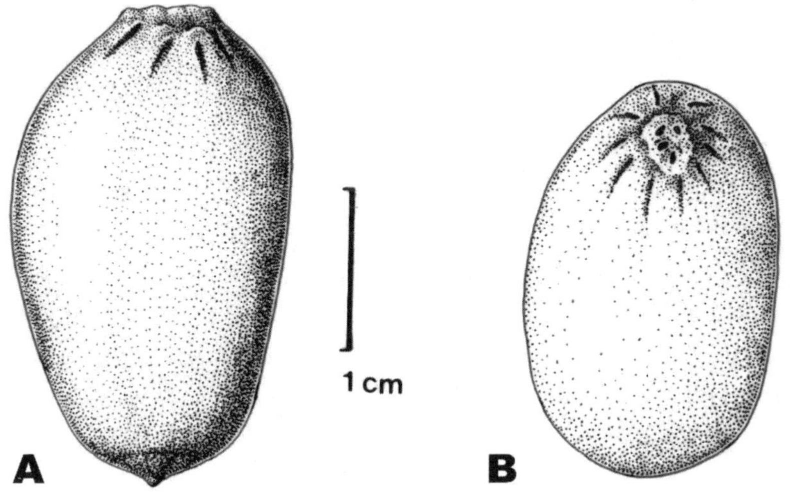

Figure 15-3. External views of the mature sclerotesta after the sarcotesta was removed. **A.** Oblique view showing the orifices for entrance of the vascular bundles from the cone scale. Within the sclerenchyma tissue, the vascular bundles merge and branch in the vascular plexus. **B.** Side view with the micropylar end down. The radiating lines at the top are the orifices for the emergence of the 10 vascular bundles of the external system.

Once the sclerotesta has become lignified, the plexus lies within the coroniform prominence located in the hilum (Fig. 15-3A, 15-3B).

The internal vascular bundles that emerge from the plexus divide several times, ultimately producing up to 40 bundles at the level where the nucellus becomes free from the integument. Some internal bundles are endarch (Fig. 15-2A), and others are mesarch (Fig. 15-2B); a few are amphicribral, and others are collateral. Details of the vascularization of the seed are illustrated in Figure 15-1.

Micropylar Region

The 10 vascular bundles of the external system run longitudinally the entire length of the seed without branching, ending at a circle with a radius of 2.4 mm around the micropyle with the remaining portion of the micropyle being free of vascualture (Fig. 15-1). The approximately 40 vascular bundles of the internal system end at the level where the nucellus becomes free from the integument. The free nucellus is not vascularized. The epidermis of the free portion of the nucellus is covered by a cuticle that is approximately 6 μm thick. The epidermis of the internal side of the free integument is covered by a cuticle 2 μm thick.

Discussion

In the present study, a vascular plexus (Fig. 15-1) where the vascular bundles divide and anastomose on entering the sclerotesta is described and illustrated in detail for *C. mexicana*. This vascular plexus of *C. mexicana* is comparable to the "chalazal pad of tracheids" mentioned by Singh (1978). Both the vascular plexus and the chalazal pad of tracheids refer to the course and organization of the vascular bundles in the hilar region.

For the fossil seed genera *Rhabadospermum, Mitrospermum,* and *Taxospermum,* Singh mentioned this plexus or chalazal pad of tracheids. For the cycads he stated only that "the mode of entry of vascular bundles into the chalazal, and their number at the base is quite different . . ." For other gymnosperm taxa, including fossils, the organization of the chalazal region of the seed is not described.

According to Singh (1978), the number of vascular bundles that enter the hilar region in gymnosperm seeds is variable. In the cycads, Worsdell (1898), Matte (1904), and

Stopes (1905) showed that the number varies from one to four. In *C. mexicana,* four to six bundles were found to enter the seed.

The presence of 10 vascular bundles in the external system of *C. mexicana* confirms the observations of Chamberlain (1935). These vascular bundles emerge from the 10 easily observed orifices surrounding the prominence on the sclerotesta in the hilar region (Fig. 15-3) and the bundles extend to the areola, a well-defined structure surrounding the micropyle (Fig. 15-1). In *C. mexicana,* the vascular bundles of the external system were found to be mesarch. This contrasts with the reports that they are endarch in *Cycas* (Stopes, 1905) and in other genera (Matte, 1904).

Crane (1985a), referring to the Cycadales in general, and the present study of *C. mexicana,* both suggest that the internal system is located in a precisely delimited zone. This zone corresponds to the area where the nucellus and integument are fused; there are no vascular bundles in the free portion of the nucellus. This agrees with Singh's (1978) statement that in cycads, the vascular bundles of the internal system traverse the parenchyma ("endotesta") just inside the sclerotesta and end just below the free nucellus. However, the existence of vascular bundles of the internal system extending into the free portion of the nucellus has been reported for *Cycas* (Stopes, 1905), *Dioon edule* (Chamberlain, 1935), *Microcycas* (Singh, 1978), and occasionally for *Zamia* and *Bowenia* (Singh, 1978). Clearly, further investigation of other genera and species of cycads is needed.

For dicotyledons, Corner (1976) introduced the term "pachychalaza" to refer to a vascularized zone at the chalazal end of the ovule where there is no distinction between the nucellus and the integument(s); that is, these two structures are congenitally fused. In *C. mexicana,* the pachychalaza is quite large, occupying four-fifths of the length of the seed. In the remaining fifth, the integument and nucellus are differentiated and free. In this free zone, the internal side of the integument and the free portion of the nucellus are each covered with an epidermis, which bears a cuticle.

Anatomically, the pachychalaza has been described in seeds for a century, but it has not received a name that distinguishes it from a more restricted chalaza. Clearly, the lack of a name had often prevented the recognition of this structure through the years. In particular, the vascularized zone around the gametophyte of the cycads has not been recognized as the result of a possible congenital fusion between the nucellus and the integument.

The significance of the fact that the vascular bundles of the two systems end at different levels has not been appreciated previously. As already pointed out, this strongly supports the argument for the presence of a pachychalaza in the cycad seed.

Sánchez-Tinoco and Engleman (2004) described 10 fissures in the areole of the sclerotesta radiating from the micropyle, where the radicle emerges. The 10 vascular bundles of the external system end at the edge of the areole, apparently between the adjacent radiating fissures (Fig. 15-1). It was not, however, possible to determine with precision the position of the vascular bundles with respect to the fissures.

If it is confirmed that the vascular bundles are located between the fissures, this would constitute strong support for the synangial hypothesis (Benson, 1904) for the origin of the integument in the Cycadales. Benson postulated that the ovule is the result of the transformation of a group of sporangia in which the peripheral sporangia become sterilized and form a ring of free lobes surrounding a single fertile central sporangium. This hypothesis received support from Takhtajan (1981) and more recently from Kenrick and Crane (1997), who proposed the neosynangial hypothesis of the origin of the ovule.

Among the additions made by the neosynangial hypothesis to the synangial hypothesis is the presence of a single, unbranched vascular bundle in each integumentary lobe. As shown by Rothwell and Scheckler (1988), this condition is found in the earliest Pteridospermophyta (e.g., *Genosperma kidstonii*). In *C. mexicana*, the 10 segments between the fissures, each apparently supplied by a single unbranched vascular bundle of the external system, may correspond to the tips of the integumentary segments in some fossil plants. Thus, the seed of *C. mexicana* has features in common with some of the most primitive seeds.

The phylogenetic implications of the double vascularization system in the Cycadales and related groups are analyzed as follows:

- Crane (1985a), in his character matrix for the seed plants, refers to the double vascularization system ("double vascular supply to ovules") and considers that character 7 is present only in the Medullosaceae and Cycadales, which he places as sister groups in his cladogram. This is in agreement with Meyen (1984).
- Doyle and Donoghue (1986) do not include character 7 in their data matrix, but their character 46 refers to "nucellar vasculature."

- Crane (1988) does not include double vascularization as a character in his data matrix. However, his character 1.9, radiospermic seeds, is based on the arrangement of vascular bundles in the seed. As part of the definition of this character, he describes double vascularization in the cycads, contrasting it with the vascular system in the Medullosaceae; he concludes by stating that homology between these two taxa is improbable (Crane, 1985a).
- Loconte and Stevenson (1990; character 26) include vascularization of the ovule in their data matrix. Based on Stevenson's proposed classification (1985) of the Cycadales and its later modification (1990), these researchers state that double vascularization in the integument is an autapomorphy in the Cycadales and one of the synapomorphies of the Stangerineae and Cycadineae.
- Rothwell and Serbet (1994) consider that the integument can lack vascularization or can have from two to several vascular bundles. For the Medullosaceae and Cycadales, they assign a variable number of vascular bundles and state that character 40 is a modification of character 6 of Crane (1985b).
- Nixon et al. (1994) propose double vascularization in character 79 of their data matrix, referring to a "dual vascular system" in the seed of the Medullosaceae and the Cycadales. These authors base their description on Stopes (1905).

From the above, it is clear that following the initial description by Stopes (1905), cladists have considered double vascularization to be of major importance in establishing phylogenetic relationships among the major groups of seed plants, notably the relationship between the Medullosaceae and Cycadales (Crane, 1985a).

Acknowledgments

The authors thank Andrew P. Vovides and an anonymous reviewer for critically reviewing this manuscript.

Literature Cited

Benson, M. 1904. *Telangium scotti*, a new species of *Telangium* (Calymmatotheca) showing structure. Ann. Bot. 18: 161–177.

Corner, E.J.H. 1976. The seeds of the dicotyledons. Vol. 2. Cambridge University Press, London.

Chamberlain, C. J. 1935. Gymnosperms. Structure and evolution. University of Chicago Press, Chicago, IL.

Crane, P. R. 1985a. Phylogenetic analysis of seed plants and the origin of angiosperms. Ann. Missouri Bot. Gard. 72: 716–793.

———. 1985b. Phylogenetic relationships in seed plants. Cladistics 1: 329–348.

———. 1988. Major clades and relationships in the "higher" gymnosperms. Pp. 218–271 *in* C. B. Beck (ed.), Origin and evolution of gymnosperms. Columbia University Press, New York.

Doyle, J. A. & M. J. Donoghue. 1986. Seed plant phylogeny and the origin of angiosperms: an experimental cladistic approach. Bot. Rev. 52: 321–431.

Kenrick, P. & P. R. Crane. 1997. The origin and early diversification of land plants: a cladistic study. Smithsonian Institution Press, Washington DC.

Loconte, H. & D. W. Stevenson. 1990. Cladistics of the spermatophyta. Brittonia 42: 197–211.

Matte, H. 1904. Recherches sur l'appareil libéro-ligneux des cycadées. Mem. Soc. Linn. Normandie 22: 1–233.

Meyen, S.V. 1984. Basic features of gymnosperm systematics and phylogeny as evidenced by the fossil record. Bot. Rev. 50: 1–111.

Nixon, K. C., W. L. Crepet, D. W. Stevenson & E. M. Friis. 1994. A reevaluation of seed plant phylogeny. Ann. Missouri Bot. Gard. 81: 484–533.

Rothwell G. W. & S. E. Scheckler. 1988. Biology of ancestral gymnosperms. Pp. 85–134 *in* Beck CB (ed.), Origin and evolution of gymnosperms. Columbia University Press, New York.

Rothwell, G. W. & R. Serbet. 1994. Lignophyta phylogeny and the evolution of spermatophytes: a numerical cladistic analysis. Syst. Bot. 19: 443–482.

Sánchez-Tinoco, M.Y. & E. M. Engleman. 2004. Seed coat anatomy of *Ceratozamia mexicana* (Cycadales). Bot. Rev. 70(1): 24–38.

———, ——— & A. P. Vovides. 2000. Cronología reproductora de *Ceratozamia mexicana* (Cycadales). Bol. Soc. Bot. México. 66: 15–23.

Sass, J. E. 1958. Botanical microtechnique. Ed. 3. Iowa State University Press, Ames.

Singh, H. 1978. Embryology of gymnosperms. Handbuch der Pflanzenanatomie. Band X. Teil 2. Borntraeger, Berlin.

Stevenson, D. W. 1985. A proposed classification of the Cycadales. Am. J. Bot. 72: 971–972.

———. 1990. Morphology and systematics of the Cycadales. Mem. New York Bot. Gard. 57: 8–55.

Stopes, M. C. 1905. On the double nature of the cycadean integument. Ann. Bot. 19: 561–566.

Takhtajan, A. 1981. Flowering plants: origin and dispersal. Bishen Singh Mahendra Pal Singh, Dehra Dun, India. Otto Koeltz Science Publishers, Koenigstein, Germany. [Translated from the Russian].

Worsdell, W. C. 1898. The vascular structure of the sporophylls of the Cycadaceae. Ann. Bot. 12: 203–241.

NEUROBIOLOGY
CHAPTER 16

Defining the Role of BMAA and Its Potential Glutamate Receptor Targets in Arabidopsis thaliana and Cycas rumphii

Eric D. Brenner, Dennis Wm. Stevenson,

Joanna Chiu, Suzan J. Runko, Peter M. Palenchar,

and Gloria M. Coruzzi

Abstract

The presence of compounds in plants with neuroactive properties has always raised speculation regarding their endogenous function. One such compound is S(1)-β-methyl-α, β-diaminopropionic acid (BMAA), which is found in all cycads. BMAA has been shown to activate glutamate receptors in animal neuronal cells, and it has been subsequently argued that it is the causative agent of neurological disorders in people who consume cycads or cycad-eating animals. Although it is still not clear the role what BMAA might play with regard to neurological disorders in humans, recent evidence has shown that BMAA has dramatic effects on plant development when applied to the model organism, *Arabidopsis*. Here we utilize BMAA in a screen to study the role of *Arabidopsis* glutamate receptors (*AtGLRs*) by isolating plants with BMAA-insensitive mutant (*bim*) involved in a BMAA-activated pathway. Then we present analysis on the presumptive target of BMAA—the *AtGLRs*. We further examine the expression of three different AtGLRs, which represent three separate clades of these genes. Finally, we use a genomics approach successfully to isolate a glutamate receptor gene from *Cycas rumphii*.

Resumen

La presencia de compuestos con propiedades neuroactivas en plantas siempre ha generado especulación en relación a su función endógena. El compuesto BMAA, que se encuentra en todas las cícadas, es uno de ellos. Se ha demostrado que BMAA activa exógenamente receptores de glutamato en células neuronales animales y se ha propuesto como el agente causante de desórdenes neurológicos en humanos que consumen cícadas o consumen animales que comen estas plantas. Aunque el papel que tiene el BMAA en los desórdenes neurológicos en humanos todavía no está claro, la evidencia que se presenta abajo muestra que el BMAA tiene efectos dramáticos en el desarrollo de las plantas cuando se aplica al organismo modelo *Arabidopsis*. En este estudio utilizamos BMAA en una búsqueda genética para estudiar el papel de los receptores de glutamato de *Arabidopsis* (*AtGLRs*) por medio de la identificación de *bim* (mutantes insensibles a BMAA) involucrados en una vía activada por BMAA. Además, en este estudio presentamos un análisis del supuesto blanco de BMAA, los AtGLRs. Examinamos la expresión de tres AtGLRs diferentes que representan tres clades distintos de estos genes. Finalmente, utilizamos enfoques genómicos para aislar un gen del receptor de glutamato de *Cycas rumphii*.

Introduction

In the human brain, the amino acid glutamate serves as the predominant neurotransmitter where it acts on glutamate receptors (GLRs). The GLR genes are ligand-gated ion channels that mediate neurotransmissions in animal nervous systems (Oswald, 2004). Two important neurological pathologies associated with iGLR malfunction are Alzheimer and Parkinson dementia (Ikonomidou & Turski, 1996; Forsythe & Barnes-Davies, 1997). Proper iGLR response is necessary for operations in animals varying from vision to memory. The iGluRs are separated into two defined subgroups based on their pharmacological responsiveness to different agonists. The two main iGluR subfamilies are *N*-methyl-d-Asp (NMDA) and α-amino-3-hydroxy-5-methylisoxazole-4-propionic acid (AMPA) /kainite (KA) (non-NMDA) activated iGluRs (Ikonomidou & Turski, 1996; Forsythe & Barnes-Davies, 1997).

Because of their fundamental role in the nervous systems of animals, it came with surprise that genes with homology to iGluRs were discovered in the plant *Arabidopsis thaliana* (Lam et al., 1998; Lacombe et al., 2001; Brenner, 2002). Based on computer predictions, plant glutamate receptor (*GLR*) genes have a similar protein motif organization to those of animal glutamate receptors, including ligand-binding and membrane-spanning domains (Lam et al., 1998; Chiu et al., 1999; Chiu et al., 2002). In the completely sequenced genome of *Arabidopsis,* 20 GLRs have been detected and annotated (Lacombe et al., 2001; Chiu et al., 2002). Based on gene tree analysis and common sequence similarity, these 20 genes have been subdivided into three clades (Lacombe et al., 2001; Chiu et al., 2002). The origin of GLRs in plants is not clear, but it is clearly ancient. GLRs have even been found in the prokaryotic organism *Cyanobacterium* (Chen et al., 1999).

One path toward understanding the role of GLRs in plants requires examination of their expression patterns. A combination of reverse transcriptase–polymerase chain reaction (RT-PCR) and reporter expression studies have looked at GLR expression in *A. thaliana* and have found some correlation between the expression of certain genes and clades in which they belong (Chiu et al., 2002). Understanding where the GLRs act may further help understand what their specific localized roles might be.

Beyond examining their expression patterns in *Arabidopsis*, the function of *GLR* genes has not yet been clearly defined. Pharmacological studies have shown that treatment of *Arabidopsis* plants with the iGLR antagonist 6,7-dinotropuinoxaline-2,3(1H, 4H)-dione (DNQX) causes an increase in hypocotyl length and a reduction in chlorophyll. This response is reminiscent of *Arabidopsis* plants that have impaired photoperception, suggesting that DNQX may be blocking the action of native GLR proteins involved in transmitting light signals (Lam, 1990). A similar experiment found that plants treated with the cycad-derived agonist BMAA also exhibited hypocotyl elongation and impaired cotyledon opening (Brenner et al., 2000). BMAA-mediated hypocotyl lengthening and decreased cotyledon opening could be reversed with the addition of increasing concentrations of glutamate, further indicating the mode of BMAA action at a glutamate-responsive site (Brenner et al., 2000). The similar response between an agonist, BMAA, and an antagonist, DNQX, at first might seem contradictory, but it can be most simply explained with the supposition that BMAA causes GLR impairment via desensitization. Desensitization is an important and common response to excessive exposure to agonist (Geoffroy et al., 1991; Yelshansky et al., 2004). Because of its dramatic effect on *Arabidopsis* growth, presumably on endogenous GLRs, BMAA is an attractive pharmacological tool to study the role of GLRs in plants.

To study further the mechanism by which BMAA mediates changes in *Arabidopsis* and to determine if this mechanism operates through endogenous GLRs in *Arabidopsis*, a forward genetic screen was designed to isolate mutants that were impaired in BMAA-mediated activity. Ethylmethanylsulfate (EMS) mutagenized *Arabidopsis* seeds that displayed a BMAA-insensitive morphology (*bim*) were isolated. These *bim* mutants have a short hypocotyl and open cotyledons in the presence of BMAA when compared to wild type. Such *bim* mutants may be caused by mutations either in glutamate receptor genes or genes involved in a GLR-mediated pathway.

BMAA is ubiquitous to all species and families of cycads tested (Pan et al., 1997). However, the role that BMAA plays in the physiology of cycads is unknown. In cycads, BMAA is sequestered in specialized idioplast known as "gold cells," where it has been hypothesized to play a role in defense (Norstog & Nicholls, 1997). As described above, BMAA shows a dramatic ability to alter the growth of *Arabidopsis*. This discovery raises two questions. May BMAA also play a role in the physiology and development of cycads? Or is BMAA sequestered in gold cells to protect certain plant tissues from BMAA toxicity

(Brenner et al., 2003)? Although *Arabidopsis* was first found to express GLRs, a variety of other plant species have since been found to express *GLR* genes (J. Chiu, unpublished data), suggesting that cycads may also have these receptors. If GLRs are present in cycads, do they have an interactive role with BMAA? Finally, what are the genes that are responsible for the biosynthesis of BMAA?

As a first step to define the role of BMAA in cycads, a genomic approach was taken to screen for a likely endogenous target of BMAA: a cycad GLR. Toward that end, an expressed sequence tag (EST) library was constructed from the leaves of *Cycas rumphii* Miq. This library was subsequently characterized with BLAST (Altschul et al., 1990) comparative analytical algorithm to screen for the presence of genes with homology to plant *GLR* genes. The next step was to screen this library for the presence of genes that would likely play a role in the biosynthesis of BMAA.

This report has three parts. First, we describe the isolation and characterization of *Arabidopsis* mutants resistant to BMAA. Second, we describe the expression patterns of three *Arabidopsis GLR* genes. Third, we report the purification of a *GLR* gene from cycads as well as candidate genes that might be involved in the biosynthesis of BMAA. Although cycads and *Arabidopsis* have been separated by more than 200 million years of evolution, both species can be used together in a unified approach to understand the role of neuroactive compounds in plants.

Methods

A Screen for Arabidopsis Mutants with bim Phenotype

EMS mutagenized *A. thaliana* seeds were plated on MS media with or without 50 μM BMAA, according to Brenner et al. (2000). Briefly, plants that exhibited a *bim* phenotype after 5 to 14 days were transferred onto MS media lacking BMAA and were allowed to recover (generate healthy roots and new leaves). The *bim* mutants were self fertilized and retested in the M3 generation to confirm resistance to BMAA. Tests on these M3 seedlings were simultaneously conducted on media containing or lacking BMAA and grown in either light or dark conditions. Plants that repeatedly scored resistant to the effects of BMAA in the M3 generation were considered to be stable *bim* mutants and subjected to further analysis. All mutants were backcrossed into the *Arabidopsis* ecotype Columbia to produce F1 plants. F1 plants were tested as dominant or recessive for the *bim* phenotype as described in Brenner et al. (2000). Hybrid *bim* F1 plants were self fertilized to produce F2 seeds and subsequently grown on BMAA to reselect the *bim* phenotype in cases where the mutation could only be observed as a recessive trait. Mutants of interest were subsequently backcrossed for four generation cycles to remove deleterious background mutations. A mapping population was created by outcrossing the *bim* mutants into the Landsberg ecotypic background to create an F1 hybrid, which was subsequently grown to maturity to produce F2 seeds for mapping and positional closing, according to Jander et al. (2002).

Promoter Expression Analysis of Three *AtGLR* Genes in *Arabidopsis*

Transcriptional fusions were made between the promoters of *GLR1.1, GLR2.1,* and *GLR3.1,* with the reporter gene β-glucuronidase, and transformed into *A. thaliana,* accession Columbia (Col-0). A 1713–base pair (bp) region upstream of the start codon of *GLR1.1,* a 1553-bp region upstream of the start codon of *GLR2.1,* and a 1780-bp upstream region from *GLR3.1* was fused upstream of the reporter gene and transformed into *Arabidopsis* according to Chiu et al. (2002).

A Screen for a *GLR* Gene in *C. rumphii*

A cDNA library from young, developing leaves of *C. rumphii* was prepared and analyzed as described in this volume and according to Brenner et al. (2003).

Results

A Screen for Arabidopsis Mutants Resistant to BMAA

Arabidopsis seedlings grown in the presence of BMAA in the light exhibited approximately a 2.5- fold increase in hypocotyl length when compared to untreated plants (Fig.

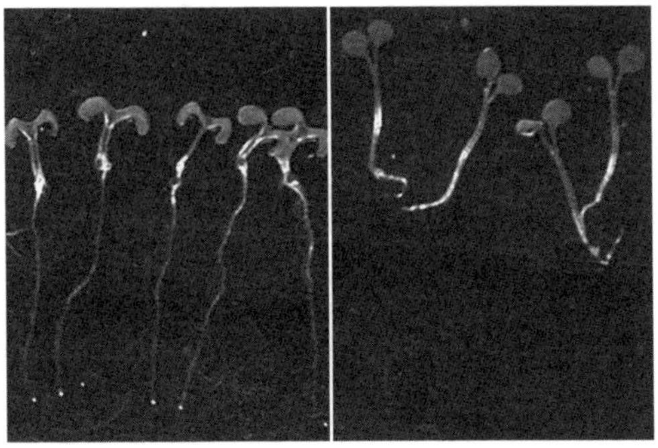

Figure 16-1. ***Arabidopsis*** **seedlings were grown in the light on MS containing agar media (left) or MS media containing 50 µM BMAA. BMAA treated seedlings exhibit a two-to-threefold increase in hypocotyl length, inhibition of cotyledon development and separation, and virtually no primary root growth when compared with untreated plants.**

16-1). In contrast, BMAA did not induce hypocotyl elongation in dark-grown plants when compared to wild type. Higher concentrations than 50 µM BMAA inhibited hypocotyl growth and even lower concentrations (2–5 µM) inhibited root growth (Brenner et al., 2000). BMAA also inhibited cotyledon development in the light (Fig. 16-1; Brenner et al., 2000). The combination of these three morphological effects caused by BMAA (elongated hypocotyl growth, inhibited cotyledon development, and reduced root growth) "phenocopied" the morphology of *Arabidopsis* phytochrome mutants, which are impaired in light perception (Chory et al., 1996; Fankhauser & Chory, 1997).

As a means to examine the genes involved in BMAA-induced developmental changes, a pharmacological approach was taken to screen for mutants in *Arabidopsis* that are completely or partially immune from the effects of this drug. Such BMAA-resistant plants are likely to have mutations in genes that are in GLRs, or in genes downstream of a GLR-mediated pathway. To screen for these types of mutants, *Arabidopsis* seeds, collected as progeny from EMS-mutated parents, were sown on MS agar media containing 50 µM BMAA. Plants that displayed a *bim* were isolated, grown to produce seed, and used in further tests. In general, *bim* mutants displayed a short hypocotyl and/or expanded cotyledons when cultivated in the presence of BMAA in comparison to the wild type.

The progeny of these mutants were tested for resistance to BMAA to confirm that the *bim* phenotype was maintained in the next generation. Plants that exhibited a *bim* phenotype were analyzed further. It was noted that among these *bim* mutants, three separate phenotypes were distinguishable among seedlings cultivated in the dark (Fig. 16-2). Class I mutants have completely normal hypocotyls when compared to wild type seedlings. Class II mutants have short hypocotyls and closed cotyledons. Class III mutants have short hypocotyls but exhibit open cotyledons when cultivated in the dark.

EXPRESSION ANALYSIS OF *AtGLR* GENES

To better understand the role of *AtGLR* genes in *Arabidopsis,* reporter constructs were transformed into *Arabidopsis* plants to evaluate the transcriptional expression of one *GLR* gene from each of the three phylogenetic classes as determined by (Chiu, 2002). In Figure 16-3 one can see distinct expression patterns for each of these genes. *AtGLR1.1* and *AtGLR2.1* appear to have minimal expression in the shoot, except in the stipules (Fig. 16-3A, 16-3E), and both genes show expression in the root, which is weak in the case of *AtGLR1.1* (Fig. 16-3B) and pronounced in the case of *AtGLR2.1* (Fig. 16-3F). *AtGLR1.1* also appears to show strong expression at the base of the hypocotyl at the root shoot junction. A very different expression pattern is seen for *AtGLR3.1,* which is present in both the shoot (Fig. 16-3I, 16-3K, 16-3L) and in the root (not shown); however, *AtGLR3.1* expression is delayed during early development in that it is not present newly germinated seedlings (Fig. 16-3J). When *AtGLR3.1* expression becomes detectable, it appears most

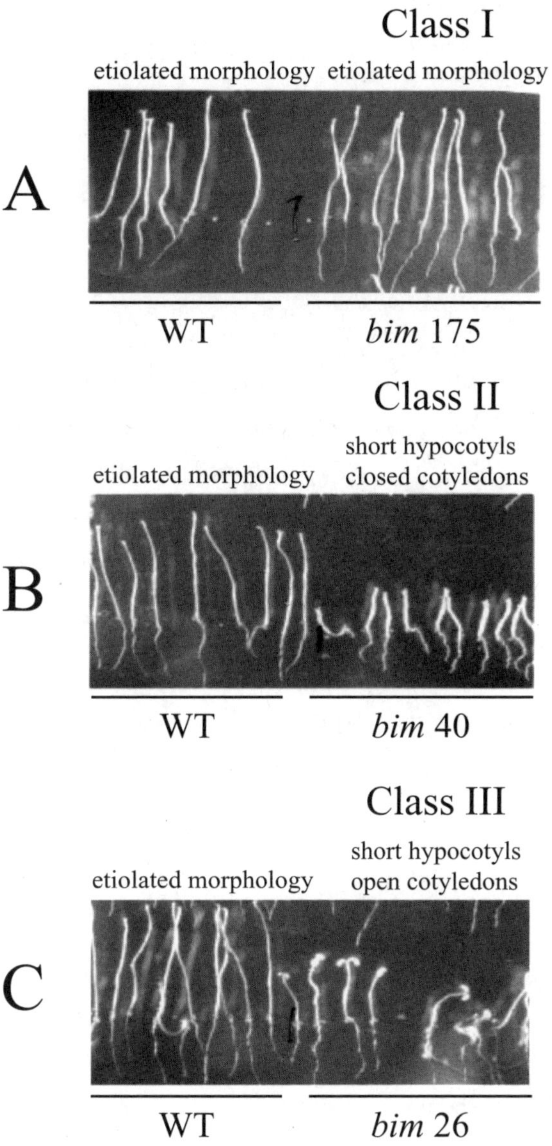

Figure 16-2. *Arabidopsis* mutants with *bim* mutants grown in the dark. *Arabidopsis* seedlings cultivated on MS media in the dark can be separated into three classes of mutant. **A.** Class I has a normal etiolated phenotype. **B.** Class II mutants have short hypocotyls 0and undeveloped cotyledons. **C.** Class III mutants have short hypocotyls and developed cotyledons.

prominent in the vasculature and in the tissues associated with the vasculature in the leaves (Fig. 16-3I), the developing vascular network of the petals and ovules of the young flower (Fig. 16-3K), and in the peduncle of the developing seeds (Fig. 16-3L).

Creation of a *C. rumphii* EST Library and Detection of a Cycad *GLR* Gene

To begin an analysis on the role of BMAA in cycads, a cDNA library was constructed from the developing organs of a mature specimen of *Cycas rumphii* Miq. growing in the Enid A. Haupt Conservatory at the New York Botanical Garden. To date, more than 9000 ESTs have been collected from cycad tissues, including young leaves (Brenner et al., 2000), sporophylls (data not given), and ovules (data not given). Of these ESTs, 4210 isolated from developing leaves have been fully analyzed and sorted into a unigene set of 2458 contigs (or unigenes), which can downloaded at http://www.nypgenomics.org. Each EST was subjected to BLAST analysis (Altschul et al., 1990) to assess their homology to genes in the *Arabidopsis* database. Among the 4210 leaf ESTs, one of them showed a strong similarity score to a ligand-gated ion channel gene. This clone, GenBank number CB091652, was found to have strongest BLAST homology to a plant GLR.

EST Analysis and Support for a Predicted Pathway for BMAA Synthesis in *C. rumphii*

Although the presence of BMAA has been known in cycads for nearly four decades (Vega & Bell, 1967), its mode of biosynthesis has not been explored at either the biochemical or molecular level. Because BMAA is structurally similar to beta-substituted alanine variant molecules (Warrilow & Hawkesford, 2000, 2002), we hypothesize that BMAA biosynthesis requires either phosphoserine, cysteine, o-acetylserine, or cyanoalanine as an initial substrate. Therefore, a BMAA biosynthetic pathway (Fig. 16-4) would occur as a two-step reaction initiated with the transfer of NH_3 to the β-carbon of the substituted alanine as a first step (Fig. 16-4A), followed by addition of a methyl (CH_3) group (Fig. 16-4B) to form BMAA as a final product (Fig. 16-4C). The transfer of NH_3 would require a nucleophilic reaction catalyzed by a cysteine synthase–like protein. Our

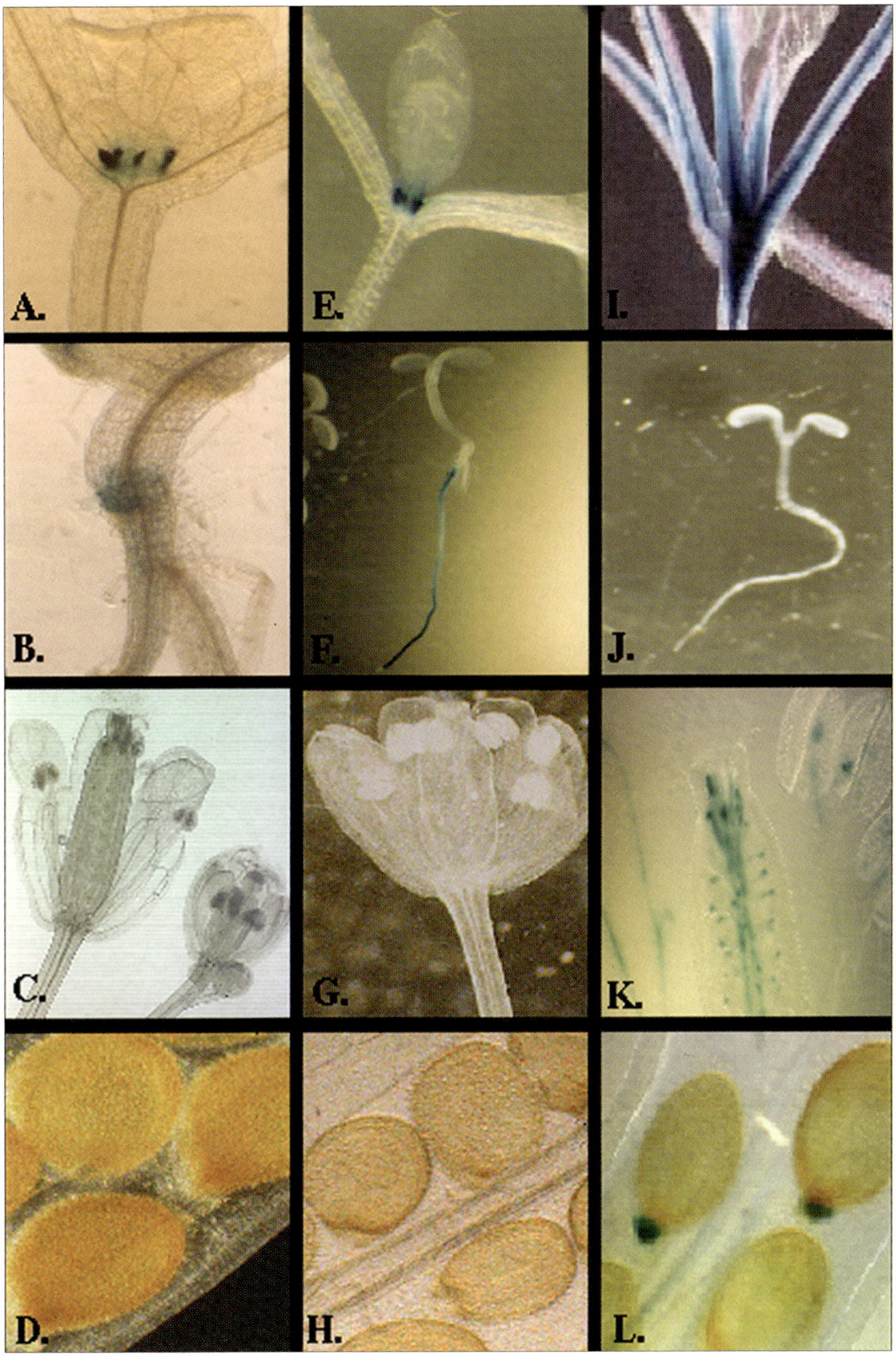

Figure 16-3. GUS (β-glucuronidase) reporter expression of three *AtGLR* genes in *Arabidopsis*. The promoter regions of *AtGLR1.1, AtGLR 2.1,* and *AtGLR3.2* were cloned upstream of GUS and transformed into *Arabidopsis* via *Agrobacterium tumafaciens*. Each of these three genes represent the separate clades determined by gene tree analysis. A-D. *AtGLR1.1.* E-H. *AtGLR 2.1.* I-L. *AtGLR3.2.*

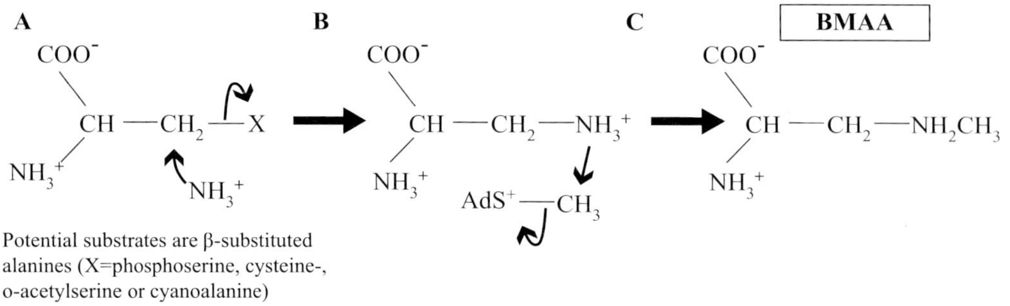

Figure 16-4. Likely pathway in *Arabidopsis* that might lead to the formation of BMAA. **A.** Transfer of alanine. **B.** Addition of a methyl group. **C.** Formation of BMAA. Candidate genes for the first step from compound A to B for cysteine synthase and from B to C as a methyl transferase were cloned and are shown in Table 16-1.

survey of genes (Table 16-1) from the cycad EST library identified candidates for each of these enzymatic steps. The library has two ESTs for genes encoding cysteine synthase. The EST library contains two potential methyltransferases (caffeic acid O-methyltransferase II and caffeol-CoA 3-O-methyltransferase), which would catalyze the second step of BMAA synthesis. A methyl donor would be required in the second step, with the most likely molecule being S-adenosylmethionine (SAM). Consumption of SAM would require the presence of enzymes for SAM regeneration. Candidate genes were found in the cycad leaf EST library that could be involved in SAM recycling: adenosylhomocysteinase, SAM synthetase and homocysteine methyltransferase. Taken together, the cycad EST library contains candidate genes for all of the enzymes predicted to be present during the biosynthesis of BMAA.

Conclusion

Although it has been more than 100 years since action (electrical) potentials were discovered (interestingly enough, first in plants), their role in their overall growth and physiology of plants is unclear (Brenner, 2002). Similarly, the mystery behind the presence of neuroactive compounds in plants has been the subject of numerous studies and considerable speculation: postulated roles include protection against predation, production of

Table 16-1 Biosynthetic enzymes of cycad-specific phytochemicals (BMAA)[a].

GenBank #	Biosynthetic enzymes of cycad specific phytochemicals (BMAA)	e-value	% identified	Score
Cysteine synthase				
CB089577	Cysteine synthase (O-acetylserine sulfhydrylase)	3.00e-50	0.75	128
CB092214	Plastidic cysteine synthase 2 *Solanum tuberosum*	5.00e-27	0.64	83
Methyl transferases				
CB091906	Caffeic acid O-methyltransferase II *Nicotiana tabacum*	3.00e-35	0.56	122
CB090738	Caffeoyl-CoA 3-O-methyltransferase. *Oryza sativa*	1.00e-37	0.47	188
SAdM metabolism; adenosylhomocystei-nase (S-adenosyl-L-homocysteine hydrolase)				
CB091477	Adenosylhomocysteinase *Phalaenopsis*	1.00e-87	0.84	185
CB091821	Adenosylhomocysteinase *Triticum aestivum*	3.00e-78	0.90	156
CB090818	Adenosylhomocysteinase *Medicago sativa*	2.00e-18	0.68	66
S-adenosylmethionine synthase				
CB091682	S-adenosylmethionine synthetase *Brassica juncea*	4.00e-90	0.94	167
CB090997	S-adenosylmethionine synthetase (methionine adenosyltransferase) *Petunia*	1.00e-69	0.94	133
CB090407	S-adenosyl-L-methionine synthetase *Elaeagnus umbellata*	1.00e-93	0.88	191
Homocysteine methyltransferase				
CB092344	Methionine synthase protein *Sorghum bicolor*	4.00e-94	0.90	190
CB091647	5-methyltetrahydropteroyltriglutamate—homocysteine S-methyltransferase	3.00e-79	0.76	205

[a]Expressed sequence tags (ESTs) with homology to expected enzymes that would be found in a BMAA biosynthetic pathway in cycads as shown in Figure 16-4.

metabolic intermediates, or as signaling molecules, which possibly are involved in regulating action potential (for review, see Roschina, 2001). In contrast, analysis of action potentials and their regulation by endogenous receptors as well as agonists in the nervous system in animals developed into the field of neurobiology. The development of molecular biology and genomics is a useful way to unravel the lingering mystery behind the presence of neuroactive compounds in plants.

Besides its well-known role as a metabolite, glutamate is hypothesized to function as a signaling molecule (Lam et al., 1994; Stitt et al., 2002). The discovery of genes similar to glutamate receptors in the nervous system of animals in *Arabidopsis* has provided an important new angle to understand the role of neuroactive compounds in plants (Lam et al., 1998; Lacombe et al., 2001). To study these *Arabidopsis GLR* genes, reverse genetics has provided a number of functional clues. Overexpression of *Arabidopsis GLR* 3.2 genes in *Arabidopsis* causes symptoms of calcium deficiency, which can be reversed by exogenous calcium supplements. Furthermore, these overexpression lines are also hypersensitive to salt stress, which can be ameliorated with additional calcium (Kim et al., 2001). As of yet, this work does not prove conclusively that *Arabidopsis* GLRs operate as calcium channels as NMDA receptors have been shown to operate in animals. Further experimentation is necessary to define which ions are transferred through the GLR protein pore. A reverse genetic approach on a different *GLR* gene, *GLR* 1.1, shows involvement in a separate category of physiological processes. In antisense *GLR*1.1 *Arabidopsis* lines, there was a significant increase in endogenous abscisic acid (ABA) levels. Hence, antisense *GLR*1.1 *Arabidopsis* plants show ABA-sensitive phenotypes, including enhanced sensitivity to the inhibitory effects of DNQX on germination (Kang & Turano, 2003; Kang et al., 2004). DNQX inhibition of germination can be reversed by supplementation of the media with exogenous glutamate or BMAA. Evidence from analysis of these antisense plants suggests that repression of GLR1.1 impairs physiological glutamate detection, as manifested by diminished ability to detect and respond to the carbon/nitrogen ratio in the environment (Kang & Turano, 2003; Kang et al., 2004). Considered together, *Arabidopsis* GLRs are implicated in a variety of physiological and developmental process, including germination; nitrogen sensing; salt sensitivity; and through our work presented here and previously (Brenner et al., 2000), in seedling morphogenesis.

Consequently, we have taken a both a forward genetic as well as a genomics approach to study the role of *GLR* genes in plants. Our efforts have yielded a variety of *bim* mutants at a number of loci that are impaired in BMAA-mediated responses. Our forward genetic screen has succeeded in isolating a variety of mutants that have three different phenotypes when grown in the dark (Fig. 16-2). The first class of *bim* mutants has a normal etiolated morphology in the dark. The second class of *bim* mutants has short hypocotyls in the dark and closed cotyledons; this phenotype is similar to mutants affecting cell wall formation (Nicol et al., 1998). The third class of *bim* mutants has short hypocotyls and open cotyledons in the dark characteristic of constitutively photomorphogenic "cop, det, fus" mutants (Kwok et al., 1996). Class I mutant morphology is evident only when the plants are grown in the light in the presence of BMAA. Therefore, class I mutants are excellent candidates in which to look for genes operating specifically in a BMAA-mediated response, presumably through the action of GLRs. Future studies on *bim* mutants can be used to reveal the genetic regulators that act within a BMMA–GLR pathway.

Our data reveal that other plant-derived agonists can substantially affect plant development, even at low (micromolar) concentrations. BMAA, a GLR agonist found in all cycads, has been suggested as the cause of Guam dementia—a neurological disorder prevalent among the Chamorros of Guam who consume cycad products (Spencer et al., 1987; Duncan et al., 1989; Banack & Cox, 2003; Murch et al., 2004). The role of BMAA in cycads has not been deduced, although it has been suggested as a possible defensive compound (Norstog et al., 1993). With our discovery of a GLR in cycads, this opens up the possibility that BMAA may act on these receptors. Alternatively, BMAA in cycads could be kept sequestered to protect GLRs, or alternatively, cycad GLRs may be structurally resistant to the effects of BMAA. Only future testing will define the role of GLRs as well as BMAA in cycads.

Acknowledgments

Funding for this work comes from the Plant Genomics Consortium. The Plant Genomics Consortium is funded from the generosity of the Altria Group, Inc.; The Mary Flagler Cary Charitable Trust; The Eppley Foundation for Research, Inc.; The Leon Lowenstein Foundation, Inc.; The Ambrose Monell Foundation; and The Wallace Genetic Foundation, Inc. Funding also came from the National Science Foundation (U.S.A).

Literature Cited

Altschul, S. F., W. Gish, W. Miller, E. W. Myers & D. J. Lipman. 1990. Basic local alignment search tool. J. Mol. Biol. 215: 403–410.

Banack, S. A. & P. A. Cox. 2003. Biomagnification of cycad neurotoxins in flying foxes: implications for ALS-PDC in Guam. Neurology 61: 387–389.

Brenner, E. D. 2002. Drugs in the plant. Cell 109: 680–681.

———, N. Martinez-Barboza, A. Clark, Q. Liang, D. Stevenson & G. Coruzzi. 2000. *Arabidopsis* mutants resistant to S(+)-β-methyl-α-β-diaminopropionic acid, a cycad-derived glutamate receptor agonist. Pl. Physiol. 124: 1615–1624.

———, D. W. Stevenson, R. W. McCombie, M. S. Katari, S. A. Rudd, K. F. Mayer, P. M. Palenchar, S. J. Runko, R. W. Twigg, G. Dai, R. A. Martienssen, P. N. Benfey & G. M. Coruzzi. 2003. Expressed sequence tag analysis in *Cycas rumphii*, the most primitive living seed plant. Genome Biol. 4: R78.

Chen, G.-Q., C. Cui, M. L. Mayer & E. Gouaux. 1999. Functional characterization of a potassium-selective prokaryotic glutamate receptor. Nature : 409: 817–821.

Chiu, J., R. Desalle, H. Lam, L. Meisel & G. Coruzzi. 1999. Molecular evolution of putative plant glutamate receptors and their relationship to animal ionotropic glutamate receptors. Molec. Biol. Evol. 16: 826–838.

Chiu, J. C., E. D. Brenner, R. DeSalle, M. N. Nitabach, T. C. Holmes & G. M. Coruzzi. 2002. Phylogenetic and expression analysis of the glutamate-receptor-like gene family in *Arabidopsis thaliana*. Molec. Biol. Evol. 19: 1066–1082.

Chory, J., M. Chatterjee, R. K. Cook, T. Elich, C. Fankhauser, J. *Li, P. Nagpal, M. Neff, A. Pepper, D. Poole, J. Reed* & V. *Vitart.* 1996. From seed germination to flowering, light controls plant development via the pigment phytochrome. Proc. Natl. Acad. Sci. U.S.A. 93: 12066–12077.

Duncan, M., I. Kopin, J. Crowley, S. Jones & S. Markey. 1989. Quantification of the putative neurotoxin 2-amino-3-(methylamino) propanoic acid (BMAA) in Cycadales: analysis of the seeds of some members of the family Cycadaceae. J. Anal. Toxicol. 13: 189–212.

Fankhauser, C. & J. Chory. 1997. Light control of plant development. Ann. Rev. Cell Dev. Biol. 13: 203–229.

Forsythe, D. I. & M. Barnes-Davies. 1997. Synaptic transmission: well-placed modulators. Curr. Biol. 7: 362–365.

Geoffroy, M., B. Lambolez, E. Audinat, B. Hamon, F. Crepel, J. Rossier & R. T. Kado. 1991. Reduction of desensitization of a glutamate ionotropic receptor by antagonists. Molec. Pharmacol. 39: 587–591.

Ikonomidou, C. & L. Turski. 1996. Neurodegenerative disorders: clues from glutamate and energy metabolism. Crit. Rev. Neurobiol. 10: 239–263.

Jander, G., S. R. Norris, S. D. Rounsley, D. F. Bush, I. M. Levin & R. L. Last. 2002. *Arabidopsis* map-based cloning in the post-genome era. Pl. Physiol. 129: 440–450.

Kang, J. & F. J. Turano. 2003. The putative glutamate receptor 1.1 (AtGLR1.1) functions as a regulator of carbon and nitrogen metabolism in *Arabidopsis thaliana*. Proc. Natl. Acad. Sci. U.S.A. 100: 6872–6877.

Kang, J., S. Mehta & F. J. Turano. 2004. The putative glutamate receptor 1.1 (AtGLR1.1) in *Arabidopsis thaliana* regulates abscisic acid biosynthesis and signaling to control development and water loss. Pl. Cell Physiol. 45: 1380–1389.

Kim, S. A., J. M. Kwak, S. K. Jae, M. H. Wang & H. G. Nam. 2001. Overexpression of the AtGluR2 gene encoding an *Arabidopsis* homolog of mammalian glutamate receptors impairs calcium utilization and sensitivity to ionic stress in transgenic plants. Pl. Cell Physiol. 42: 74–84.

Kwok, S. F., B. Piekos, S. Miséra & X.-W. Deng. 1996. A complement of ten essential and pleiotropic *Arabidopsis COP/DET/FUS* genes is necessary for repression of photomorphogenesis in darkness. Pl. Physiol. 110: 731–742.

Lacombe, B., D. Becker, R. Hedrich, R. DeSalle, M. Hollmann, J. M. Kwak, J. I. Schroeder, N. Le Novere, H. G. Nam, E. P. Spalding, M. Tester, F. J. Turano, J. Chiu & G. Coruzzi. 2001. The identity of plant glutamate receptors. Science 292: 1486–1487.

Lam, H., S. Peng & G. Coruzzi. 1994. Metabolic regulation of the gene encoding glutamine-dependent asparagine synthetase in *Arabidopsis thaliana*. Pl. Physiol. 106: 1347–1357.

Lam, H. M., J. Chiu, M. H. Hsieh, L. Meisel, I. C. Oliveira, M. Shin & G. Coruzzi. 1998. Glutamate-receptor genes in plants. Nature 396: 125–126.

Murch, S. J., P. A. Cox, S. A. Banack, J. C. Steele & O. W. Sacks. 2004. Occurrence of beta-methylamino-l-alanine (BMAA) in ALS/PDC patients from Guam. Acta Neurol. Scand. 110: 267–269.

Nicol, F., I. His, A. Jauneau, S. Vernhettes, H. Canut & H. Höfte. 1998. A plasma membrane-bound putative endo-1,4-β-D-glucanase is required for normal wall assembly and cell elongation in *Arabidopsis*. EMBO 17: 5563–5576.

Norstog, K. J. & T. J. Nicholls. 1997. The biology of the cycads. Cornell University Press, Ithaca, NY.

———, P. K. S. Fawcett, A. P. Vovides & M. W. Dunn. 1993. Toxicity and insect-cycad symbiosis in *Zamia furfuracea*. Pp. 42–50 *in* K.J. Norstog(ed.), Proceedings of CYCAD 90, the second international conference on cycad biology. Palm and Cycad Societies of Australia Ltd., Milton, Queensland, Australia.

Oswald, R. E. 2004. Ionotropic glutamate receptor recognition and activation. Adv. Protein Chem. 68: 313–349.

Pan, M., T. J. Mabry, P. Cao & M. Moini. 1997. Identification of nonprotein amino acids from cycad seeds as N-ethoxycarbonyl ethyl ester derivatives by positive chemical-ionization gas chromatography-mass spectrometry. J. Chromatogr. A. 787: 288–294.

Roschina, V.V. 2001. Neurotransmitters in plants. Science Publishers, Enfield, New Hampshire.

Spencer, P. S., P. B. Nunn, J. Hugon, A. C. Ludolph, S. M. Ross, D. N. Roy & R. C. Robertson. 1987. Guam amyotrophic lateral sclerosis-parkinsonism-dementia linked to a plant excitant neurotoxin. Science 237: 517–522.

Stitt, M., C. Muller, P. Matt, Y. Gibon, P. Carillo, R. Morcuende, W. R. Scheible & A. Krapp. 2002. Steps towards an integrated view of nitrogen metabolism. J. Exp. Bot. 53: 959–970.

Vega, A. & E. A. Bell. 1967. Alpha-amino-beta-methylaminopropionic acid, a new amino acid from seeds of *C. circinalis*. Phytochemistry 6: 759–762.

Warrilow, A. G. & M. J. Hawkesford. 2000. Cysteine synthase (O-acetylserine (thiol) lyase) substrate specificities classify the mitochondrial isoform as a cyanoalanine synthase. J Exp. Bot. 51: 985–993.

——— & ———. 2002. Modulation of cyanoalanine synthase and O-acetylserine (thiol) lyases A and B activity by beta-substituted alanyl and anion inhibitors. J. Exp. Bot. 53: 439–445.

Yelshansky, M. V., A. I. Sobolevsky, C. Jatzke & L. P. Wollmuth. 2004. Block of AMPA receptor desensitization by a point mutation outside the ligand-binding domain. J. Neurosci. 24: 4728–4736.

NEUROBIOLOGY
CHAPTER 17

Cyanobacteria, Cycads, and Neurodegenerative Disease among the Chamorro People of Guam

Paul Alan Cox, Sandra Anne Banack,

and Susan J. Murch

Abstract

Since the 1960s several candidate cycad neurotoxins have been discovered, including a neurotoxic non-protein amino acid, BMAA, which was proposed by Spencer and coworkers as the cause of ALS/PDC in Guam. Interest in BMAA waned when Duncan and coworkers reported that Chamorro washing of the flour removes most of the free BMAA amino acid. However, it was recently discovered that significant amounts of protein-bound BMAA are released when the protein fraction of cycad flour is hydrolyzed. Quantification of BMAA in samples should be based on the most sensitive technique, fluorescent reverse-phase HPLC detection of ACQ-tagged derivatives, with LC-Mass Spec confirmation of peaks; inappropriate techniques can result in loss of the BMAA signal. The discovery of protein-bound BMAA, together with the finding that BMAA accumulates in flying foxes and other animals that forage on cycad seeds, suggests multiple inputs of BMAA within the traditional Chamorro diet. BMAA is produced by cyanobacterial symbionts of cycad coralloid roots and then biomagnifies within the Guam ecosystem. Diverse taxa of other free-living cyanobacteria also produce BMAA, which raises the possibility of human exposure to BMAA in other areas of the world. The discovery of BMAA in brain tissues of ALS/PDC and other neurodegenerative patients, but not in patients who died of causes unrelated to neurodegeneration, is interesting, but the sample sizes remain small. Until an animal model of BMAA-induced progressive neurodegeneration is published, the BMAA hypothesis should be considered tentative.

Resumen

Desde los 1960 varios candidatos de cicada neurotoxico han sido descubiertos, incluyendo un neurotoxico aminoácido non-proteína, BMAA, el cual fue propuesto por Spencer y socios ser la causa de ALS/PDC en Guam. Interés en BMAA disminuyo cuando Duncan y socios reportaron que el lavado de harina Chamorro remueve la mayoría del BMAA amino acido. Sin embargo, recientemente a sido descubierto que cantidades significantes de BMAA con proteína son separados cuando la fracción de proteína de harina cicada es hidrolada. La calificación de ejemplos de BMAA debe de ser basada en la técnica más sensitiva, la detención fluorescente fase reversa HPLC de derivativos marcados de ACQ, con LC-Mass Spec confirmación de alturas; técnicas inapropiadas pueden resultar en la pérdida de señal de BMAA. El descubrimiento de BMAA conectado con proteína, junto con el descubrimiento de BMAA acumulado en murciélagos grandes y otros animales que forrajean en semillas de cicada, sugiere entradas múltiples de BMAA en las comidas tradicionales de la gente Chamorro. BMAA es uno producto de los cianobacteria que viven simbionte dentro los raíces coralina de cicada y es bio-amplificada en el ecosistema de Guam. Tipos diversos de otra cianobacteria también produce BMAA, el cual invita la posibilidad de exposura de humanos a BMAA en otras partes del mundo. El descubrimiento de BMAA en la materia del cerebro de ALS-PDC y otros pacientes neurodegerativos, pero no en pacientes que ha muerto por causas sin relación al la neuro-degeneración, es interesante, pero el tamaño de las muestras sigue siendo pequeño. Hasta que un modelo animal inducido de degeneración neura progresiva de BMAA sea publicado, la hipótesis de BMAA debe de ser considerada tentativa.

Introduction

The possible relationship between cycads and neurological illness among the Chamorro people of Guam has been of interest since the 1960s, when Marjorie Whiting identified flour manufactured from cycad seeds as a candidate source of neurotoxins in the traditional Chamorro diet (Whiting, 1963). The Chamorro call the disease lytico-bodig, whereas physicians refer to it as the amyotrophic lateral sclerosis/parkinsonism-dementia complex (ALS/PDC). In the 1950s, at the height of the epidemic, ALS/PDC in Guam was 50 to 100 times more prevalent than amyotrophic lateral sclerosis (ALS) elsewhere in the world (Kurland & Mulder, 1954). ALS/PDC is clinically and pathologically like classical ALS, with additional similarities to Alzheimer disease (AD) and Parkinson disease (PD). One key feature of the disease is the manifestation of a significant latency subsequent to any possible environmental exposure: Chamorros raised on Guam but who left the island after adolescence subsequently were afflicted with the disease (Garruto et al., 1980). Thus, Peter Spencer proposed that any causative environmental insult must be a "slow toxin" (Spencer et al., 1991). The exact cause of the disease remains uncertain, but epidemiological research on Guam has implicated environmental factors rather than genetic predisposition, infectious agents, or mineral imbalances. Because other cultures resident in Guam do not have an elevated risk of neurological disease, and because epidemiological evidence indicates that consumption of a traditional Chamorro diet is significantly associated with an increased risk of ALS/PDC in Guam, the "cycad hypothesis" initially generated a tremendous amount of interest (Torres et al., 1957; Brody et al.,1978; Anderson et al., 1979; Garruto et al., 1981; Reed et al., 1987). Five different international conferences were held on the possible role of cycads in ALS/PDC. Chemical analysis of cycad flour and cycad seeds led to the discovery of several potentially neurotoxic compounds, including the unusual nonprotein amino acid β-N-methylamino-L-alanine (BMAA) (Vega & Bell, 1967).

The cycad hypothesis was controversial from the start. Pioneers in the study of ALS/PDC describe intense rivalries and hostilities, far transcending the occasional antagonisms that arise in scientific debates common to other academic circles. Criticisms sometimes were *ad hominem* in nature, attacking the individual investigators and their character rather than their scientific arguments, leaving a residue of extraordinary bitterness among different groups. Unfortunately, such antagonisms led to sequestering of specimens, particularly brain tissues obtained from the Chamorro people, as well as heated contests over funding and publication.

Serious scientific questions arose concerning the BMAA hypothesis. The relatively low concentrations of free-BMAA in washed cycad flour led to questions of whether consumption of cycad products could deliver a toxic dose of the amino acid (Rosenthal & Bell, 1979). The Chamorros knew that cycad seeds could be acutely toxic to livestock and so detoxified the flour by multiple washings in water (Whiting, 1963). In the late 1980s, quantification of BMAA in Chamorro flour showed that the washing procedures used by the Chamorro people to prepare the cycad flour reduced the free BMAA content so that one would have to eat hundreds of kilograms to absorb anything approaching a toxic dose (Duncan et al., 1988). After three decades of National Institutes of Health–funded research into cycads, and with the retirement of cycad pioneer Leonard Kurland, the medical study of cycads languished.

By the early 1990s, with the ultimate cause of ALS/PDC still undetermined, and the incidence of the disease reportedly receding, the study of cycads and their possible role in the etiology of human neurodegenerative disease became academically peripheral; it was pursued only by a small group of diligent investigators. Our medical hypothesis was first published in 2002 (Cox & Sacks, 2002a) and immediately rekindled old controversies (Chen et al., 2002). That hypothesis, and the evidence for it, have been developed in a series of nine papers (Cox & Sacks, 2002a; Monson et al., 2003; Banack & Cox, 2003a; Banack &

Cox, 2003b; Cox et al., 2003; Murch et al., 2004a; Murch et al., 2004b; Cox et al., 2005; Banack & Cox, 2006). In this chapter, we present students of Cycadaceae with an overview of our hypothesis, which links cyanobacteria, cycads, flying foxes (and other animals that forage on cycad seeds) to the high incidence of neurodegenerative disease in Guam. Although the role of cycad neurotoxins as a cause of human neurodegeneration is far from proven, we here review our methods and the data that support our hypothesis, as well as candidly point out the places where the hypothesis is weak.

Cyanobacteria

Cyanobacteria, prokaryotes of ancient lineage, are believed to have played a significant role in establishing the early atmosphere of the earth. Geological evidence suggests that cyanobacteria date back at least 3.5 billion years (Schopf et al., 2002). Cyanobacteria are among the most abundant organisms in the earth's oceans and continue to play a major role in maintaining the global oxygen balance. Many (but not all) species of cyanobacteria fix atmospheric nitrogen, thus facilitating colonization of otherwise nutrient-deprived habitats. Ecologically broad, some species of cyanobacteria have specialized in harsh habitats, including deserts, glaciers, thermal pools, and high-salinity waters. It has been hypothesized that modern plant chloroplasts arose from endosymbiosis with ancient cyanobacteria (Douglas & Turner, 1991). The ecological diversity of cyanobacteria is mirrored by the diversity of molecules they produce, including cyanotoxins, which have injured livestock and human beings. Many cyanotoxins, including microcystins and nodularins, are cyclic peptides, whereas others are heterocyclic structures (e.g., anatoxin-α), more complex carbohydrates (e.g., saxitoxins), or unusual nonprotein amino acids and peptides (e.g., anabaenopeptins and cyanopeptolins) (Chorus, 2001).

In cycads, cyanobacteria infect specialized root tissues known as coralloid roots and provide organic nitrogen to the plant (Pate et al., 1988). Although production of cyanotoxins occurs sporadically within the cyanobacterial radiation, we recently found that taxa from all five major cyanobacterial sections, including nonnitrogen fixing taxa, produce the nonprotein amino acid, BMAA (Cox et al., 2005). Comparison of the 30 cyanobacterial strains we assayed with published putative molecular phylogenies of the cyanobacteria suggests that the production of BMAA may represent a synapomorphy for the entire group (discussed further below, in "BMAA and Other Geographical Foci of

ALS/PDC"). Cyanobacterial specialists at the Stockholm University, the University of Dundee, and the University of Hawaii have independently replicated in their laboratories our finding of BMAA production by cyanobacteria.

BMAA

Discovery and Chemical Characterization

Historically, Vega and Bell (1967) first isolated BMAA from seeds of Guam cycads, collected by Marjorie Whiting, by high-voltage ionophoresis. Their initial extractions were performed on gametophytes, separated from the sarcotesta and extracted with ethanol for 17 hours. They found gametophytes to be a rich source of BMAA (Vega & Bell, 1967; Vega et al., 1968).

Chemically, BMAA is a polar amino acid with a molecular weight of 118.13 and two functional nitrogen groups. Also known by the synonyms α-amino-β-methylaminopropionic acid (MeDAP) and S(+)-m-methyl-α, β-diaminopropionic acid, the exact mechanisms of BMAA biosynthesis or catabolism have not been determined. BMAA can be produced either through synthesis from an α-acetamidoacrylic acid precursor or by extraction from the seeds of cycads (Vega & Bell, 1967). There is no evidence to suggest that BMAA is synthesized as a result of tissue fixation, and the fact that tissues fixed in this manner routinely do not show BMAA strongly suggests otherwise. In addition, findings of BMAA in some but not all tissues prepared in other manners, such as frozen, fresh, and dried forms, also suggests that finding BMAA is not a random occurrence somehow related to artifacts of tissue preparation. BMAA is a colorless, crystalline compound with a melting point of 168°C and relatively high chemical stability. No detectable change in composition or structure can be observed after heating at 100°C for 64 hours (Vega & Bell, 1967). Quantification of BMAA, like that of all amino acids, can be accomplished with a variety of methods, including cation-exchange chromatography with postcolumn derivatization and spectrophotometric detection, reverse-phase high-performance liquid chromatography with precolumn derivatization and spectrophotometric detection, high-performance anion-exchange chromatography with integrated pulsed amperometric detection, micellar electrokinetic chromatography with laser-induced fluorescence, thin-layer chromatography, or liquid chromatography–mass spectrometry. The range of different potential techniques reflects the complexity of amino acid analyses, but methods are increasingly highly developed for both clinical and research applications.

Techniques for Detection

BMAA is a relatively stable molecule (particularly if inserted into the polypeptide sequence—see below) and can be detected in fresh, frozen, dried, or fluid-preserved specimens. In general terms, amino acids are derivatized with a fluorescent absorbent chemical tag that both allows for detection and improves the efficiency of chromatographic separation. As such, each method is dependent on the efficiency of the derivatization and the stability of the tagged compound (Micallef et al., 1989). In early studies, BMAA was quantified by detection of the fluorescent tag 9-fluorenylmethoxycarbonyl chloride (FMOC) (Kisby et al., 1988). Recently, more sensitive techniques have been developed for analysis of amino acids using the fluorescent tag 6-aminoquinolyl-N-hydroxysuccinimidyl carbamate (AQC). AQC universally tags amino acids at primary and secondary nitrogens producing complex molecules that do not degrade during high pressure separation and are stable for a period of up to six months of normal storage (Cohen & Michaud, 1993). Further advantages of the AQC reaction include the increased sensitivity over previous techniques, including phenylisothiocyanate (PITC), opthalaldehyde (OPA), and FMOC derivatizations, and increased efficiency of separation by reverse-phase chromatography. The resulting compound eluting from the C18 column is a stable derivative with consistent chemical characteristics (Fig. 17-1). Interestingly, BMAA can derivatize at both of the reactive nitrogen moieties, resulting in both single and double derivatized compounds depending on the pH of the reaction and the relative ratios of the fluorescent tag to the total amino acid content of the samples. Failure to optimize the reaction can result in either doubling of peaks or loss of signal. As a result, naive investigators without significant previous experience in the analysis of amino acids may miss the BMAA signal, particularly if a fluorescent detector is not utilized, or if the procedures are not guided by an expert in amino acid analysis. For example, Montine et al. (2005) failed to detect to free BMAA in Chamorro brain tissues, but the researchers did not search for protein-bound BMAA, did not repeat the validated AQC derivatization procedure used in the original studies, and failed to verify their observation by any second analytical technique such as TLC (Thin Layer Chromatography) or LC-MS (Liquid Chromatography-Mass Spectroscopy) (Cox et al., 2005). In contrast, a team led by

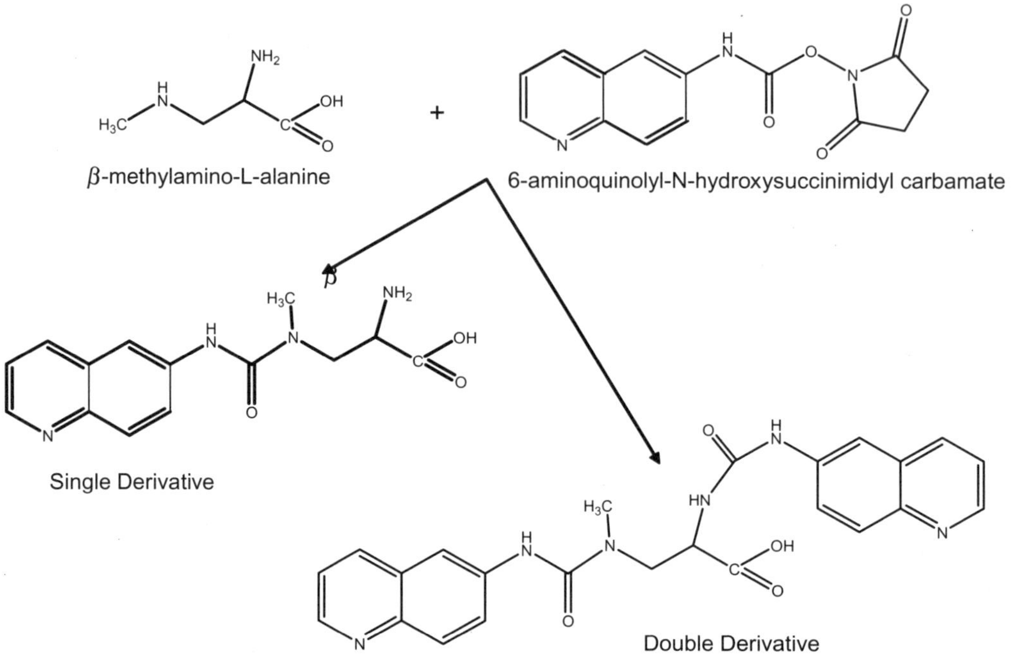

Figure 17-1. 6-Aminoquinolyl-N-hydroxysuccinimidyl carbamate (ACQ) derivatization of the BMAA molecule.

Deborah Mash at the Miami Brain Bank (unpublished data) detected protein-bound BMAA in brain tissues of neurodegenerative patients from outside of Guam by independently duplicating the validated AQC derivatization procedure.

Neurotoxicity

In the case of BMAA and neurotoxicity, a large number of studies of acute and chronic toxicity in animal model systems and cell cultures have produced highly characteristic results (Ross & Spencer, 1987; Ross et al., 1987; Zeevalk & Nicklas, 1989; Seawright et al., 1990; Rakonczay et al., 1991; Chang, et al., 1993; Matsuoka et al., 1993; Brownson et al., 2002). In initial studies, rhesus monkeys showed neurological impairment when fed washed cycad flour, but no attempt was made to quantify BMAA, particularly protein-bound BMAA, which we will later discuss (Dastur, 1964). However, in more recent whole animal studies, BMAA was shown to be acutely neurotoxic at relatively high concentrations when administered to chicks, rats, and mice, but chronic toxicity was not observed when low doses were administered to growing rats (Polsky et al., 1972). Further studies with primates confirmed the toxicity of the compound at high doses (Spencer et al., 1987; Nunn et al., 1987) and characterized the damage to primate neurological tissues (Ludolph et al., 1988). These in vivo reports of toxicity are supported by a myriad of studies done in cell cultures. In spinal cord cultures, BMAA induced a pattern of acute postsynaptic neuronal vacuolation, consistent with the patterns observed with other excitotoxins and glutamate analogues (Nunn et al., 1987). Protein degradation has been demonstrated in isolated sections of mice brains exposed to BMAA (Romlong et al., 1989), whereas in cultured cortical neurons, BMAA exposure resulted in rapid cell death (Weiss et al., 1989). Biochemically, BMAA has been shown to alter significantly the physical state of cytoskeletal proteins in a dose-dependent manner (Butterfield et al., 1993). The gradually emerging scenario of BMAA neurotoxicity is quite complex, potentially involving a requirement for bicarbonate as a cofactor (Weiss & Choi, 1988), the inactivation of free BMAA through conjugation (Polsky et al., 1972), as well as biochemical alternation of the BMAA molecule to produce a Schiff base (Mahmood et al., 1998) or possibly other related neurotoxic carbamates. More recently, BMAA has shown to be selectively neurotoxic to motor neurons at extraordinarily low concentrations ranging from 10 to 30 μM (Rao et al., 2006, Lobner et al., 2007).

Interesting work reported at this symposium includes that by Brenner at the New York Botanical Garden, who, with his colleagues has mapped the glutamate receptor genes in cycads (Brenner et al., 2003). Most recently, he reported that BMAA strongly binds to glutamate receptor genes in *Arabidopsis,* producing profound morphological effects in the phenotypes.

Protein Binding

The incorporation of BMAA into protein was first suggested by Polsky et al. (1972), who detected BMAA in the fraction remaining after free amino acids were removed in *Cycas* samples. The researchers further found that acid hydrolysis releases BMAA in quantities greater than were found in the free fraction. More recently, Seebach et al. (1994) demonstrated the insertion of BMAA into small artificial peptides. Further, the nonribosomal incorporation of other nonprotein amino acids has been demonstrated in cyanobacterial proteins (Neilan et al., 1999). These earlier studies led to the hypothesis that BMAA may be present in the protein fraction of cyanobacteria, plant, and human samples. Analysis of the amino acid content of proteins is a well-established technique and the foundation of all protein sequencing and proteome analyses. Determination of the amino acid profile of any protein requires two essential steps: hydrolysis of the protein and derivatization of the amino acids for HPLC (High Pressure Liquid Chromatography) separation and quantification. A comparison of methods of hydrolysis of proteins for amino acid analysis was recently reviewed (Fountoulakis & Lahm, 1998). Hydrochloric acid (HCl) hydrolysis is by far the most commonly used method of releasing amino acids from proteins, and there have been more than 6000 manuscripts describing results generated by this technique (Moore et al., 1958; Biflingmeyer et al., 1984; Sarwar et al., 1988; Brooks et al., 1995; Fountoulakis & Lahm, 1998; and references therein). The conventional acidic hydrolysis uses 6 normal HCl for 18 to 24 hours in vacuo and provides high recovery of all protein-bound amino acids except tryptophan. Using these classical, well-proven techniques to the study of bound BMAA, we found that BMAA is released on acid hydrolysis from the protein fractions of cyanobacteria, cycads, and components of the traditional Chamorro diet, including cycad flour, flying foxes, and feral deer, as well as from brain tissues of Chamorro patients with ALS/PDC. The detection of BMAA in these analyses is significant, because the methods are consistent with those used to determine the complete amino acid content of foodstuffs (Sarwar et al., 1988).

The incorporation and storage of BMAA in protein explains not only how biomagnification of a water-soluble molecule can occur but also may account for the disease latency due to the intermittent release of the endogenous neurotoxic reservoir stored in the proteins. The effect of the release of low levels of BMAA through time cannot be easily studied in neuronal cultures due to the short-lived nature of the cultures. Furthermore, neurodegenerative diseases are well known to be associated with protein misfolding and protein truncation—and thus are called "tangle diseases." These two processes, protein misfolding and truncation, are likely occurrences in the case of insertion of a foreign amino acid during protein synthesis.

Symbiosis Between Cyanobacteria and Cycads

The symbiotic relationship between cyanobacteria and cycads may be ancient, because all extant taxa of cycads harbor cyanobacterial symbionts (Costa et al., 1999; Zheng et al., 2002). These are housed in specialized, negatively geotropic roots. Because of extensive ramification by the apical meristems of these specialized secondary roots and subsequent medial enlargement of the resultant distal axes, these roots superficially resemble small pieces of coral. The highly ramified distal ends of such coralloid roots often are present near or above the soil surface or leaf litter, thus placing their cyanobacterial symbionts in a photosynthetically favorable environment. Cycads are unique among seed plants in their production of negatively geotropic, coralloid roots. The symbiotic relationship between cyanobacteria and cycads has long been known and was well documented in older monographs of the Cycadaceae (Schuster, 1932). However, the discovery that symbiotic cyanobacteria produce BMAA is very recent (Cox et al., 2003, 2005).

Variability in Cycad Symbionts

Detailed studies of *Cycas micronesica* physiology have been conducted in Guam to investigate physiological parameters of cycad growth and survival. Measurements of tree diameters, light environments, reproductive activity, and photosynthetic capacity have been described (Marler & Hirsh, 1998; Hirsh & Marler, 2002; Marler, 2004), but these studies do not describe the presence, diversity, or importance of cyanobacterial symbionts in the Guam cycads. These coralloid roots are highly apparent in the field, with some specimens collected from above the ground exceeding 0.5 dm in diameter and 4 kg in weight (Fig. 17-2). Such direct studies of the cyanobacterial symbionts are necessary to understand patterns of BMAA production in the cycads.

To begin the investigation of the physiological role of cyanobacteria in the growth, survival, and chemical composition of *C. micronesica*, we studied BMAA levels in cycads collected using samples from 38 individuals spanning five populations in Guam as well as a single population on Rota Island. Seeds were collected from all reproductive cycad trees on a 100-m transect through native forest, with the first reproductive cycad tree representing the initial point on the transect. Cyanobacteria were isolated from selected cycad root samples and cultured onto selective media for growth of axenic

Figure 17-2. Coralloid root of *C. micronesica* in Guam.

cultures. Tissues were excised from coralloid root clusters (Fig. 17-3), and all subsamples were individually categorized to account for differences in the culture isolate as a result of location of the population on the island or the morphological location within the root. Root tissues were surface sterilized and cultured into a sterile liquid BG11 medium for transport. Root tissues were split into replicates for establishment of axenic cultures and genetic evaluation of the symbioses. Replicates for establishment of cyanobacterial cultures were surface sterilized a second time and cultured on solidified BG11 medium (Cox et al., 2005). Replicate roots for genetic analysis were sliced in thin sections with a sterile razor blade, and the cyanobacteria were collected by needle dissection. The genetic analysis, performed in collaboration with Stockholm University, used short tandem repeated repetitive sequence primers (5GCGCCCCAATCC3') previously shown to effectively distinguish different cyanobacterial symbionts from cycads (Zheng et al., 2002).

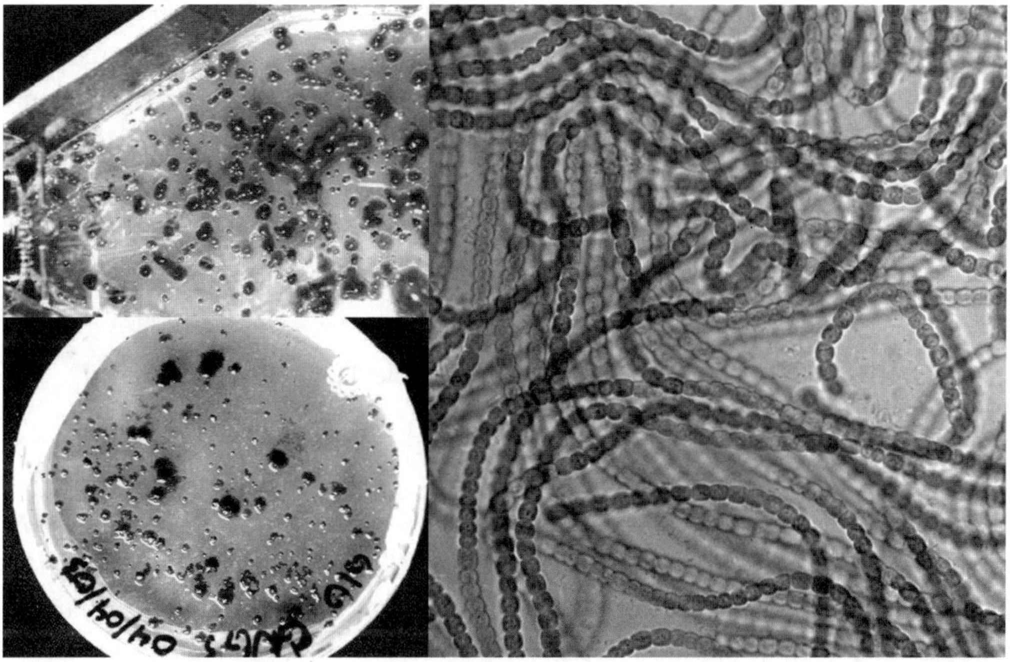

Figure 17-3. ***Nostoc*** **sp. cultured from** ***C. micronesica.***

Previous studies have shown that different coralloid roots of a single cycad plant can harbor different cyanobacterial symbionts but that only one strain of cyanobacteria is commonly found in individual coralloid roots (Costa et al., 1999). Consistent with these earlier findings, there were no significant differences among the cultures or isolates from different regions within individual roots. However, significant differences were observed between culture isolates from different geographic locations on Guam and Rota. The evidence that significant genotypic variation of the cyanobacterial symbionts occurs between cycad trees, between populations, and across different geographical locations provides an interesting perspective for ecological adaptations within the species. Interestingly, the variability in the genetic complement of the cyanobacterial symbionts was reflected in additional variability in the BMAA content of axenic cyanobacteria cultures isolated from the roots collected on Guam and Rota as well as in the root symbioses, where the BMAA content ranged from levels below detection to 112 μg/g across the collections. Before a full understanding of the role of cyanobacteria in the physiology of Guam cycads can be elucidated , much more work needs to be done and many interesting research questions need to be asked. Do different cyanobacterial symbionts affect the growth of the plants? Do differences in the cyanobacterial strains contribute to cycad survival after hurricanes or change the biochemistry of the plants? How do conditions within the plant affect the production of BMAA? Has there been (as suggested by D. Stevenson and E. Brenner (pers com.) transfer of BMAA-producing genes from cyanobacterial symbionts to their cycad hosts?

Variability of BMAA Concentrations in Cycads

Given the underlying genotypic variability in cyanobacterial symbionts, it is of interest to compare variability in BMAA levels in the host cycads. For *C. micronesica*, we are currently examining variability of total (free plus protein-bound) BMAA production within individuals, within populations, and between populations. We used seed samples from 37 individuals from six populations in Guam and Rota. Seeds were collected from all reproductive cycad trees on 100-m transects through native forest, with the first reproductive cycad tree representing the initial point on the transect. We found a mean density of 43 trees per transect in the native Guam forests, and an average of 11% of the cycads are reproductive females. Ongoing research from within a single seed, within a single tree, between trees in a single population, and between different populations will appear elsewhere. However, our initial results suggests that the importance of variability should be examined with respect to the population because an individual (flying fox or Chamorro) sampling the population of available seeds will be exposed to the entire range of BMAA toxin variation if they consume some minimum number of seeds. The exact number can be effectively estimated using a running mean of the variation because each new seed consumed alters the mean value only until it reaches the true population mean. After this point, continued sampling will not alter the mean exposure. Our initial analysis suggests that this point is reached in gametophyte tissue at a sample size of eight to ten seeds. Similarly, if four to five trees are sampled within a population, the variability becomes dampened, suggesting that the population mean has been reached at this point. We suggest that it is not necessary to know the BMAA concentration of each seed consumed; rather, knowing the population mean of BMAA will provide a reasonable estimate of the BMAA concentration consumed if the individual eats the seeds on several occasions.

In other words, both the Chamorros (who are likely to use many gametophytes from seeds within a population in the manufacture of cycad seed flour) and flying foxes (which forage on sarcotesta from trees within and between populations) are likely to have a sufficient sample size that their ingestion of BMAA will reflect the running mean of BMAA contents of gametophytes and sarcotesta, respectively. Thus variation in ingested BMAA likely reflects individual preferences far more than underlying cycad diversity in BMAA. Evidence for this assertion comes from published analyses of free BMAA content in cycad flour samples (Murch et al., 2004a, Table 17-1). Totally different groups of investigators (Duncan et al., 1990; Kisby et al., 1992; Banack & Cox, 2003a) found remarkably consistent levels of free-BMAA in cycad flour samples, ranging up to 146 μg/g (with the exception of Khabazian and his colleagues who detected a maximum of 0.003 μg/g) (Khabazian et al., 2002).

Flying Foxes and Cycads

The seeds of *Cycas* species have long been known to be dispersed by bats (Van der Pijl, 1957, 1969; Jones, 1993). Van der Pijl recorded *Cycas rumphii* as a bat-dispersed plant. Accounts of bat dispersal of cycads seeds range from South Africa (Giddy, 1984, referring to *Encephalartos* sp.) to Sri Lanka (Vorster, 1995, referring to *Cycas circinalis* L.). Individual observations of bat dispersal of cycad seeds can also be found in scattered web sites (e.g. *Encephalartos ferox* is "relished by . . . fruit eating bats" [www.plantzafrica.com/plnteg/encephferox], as well as an observation by one of the editors of this volume, Roy Osborne, who has noted, "*Encephalartos laevifolius* is eaten and the seeds are dispersed by fruit-eating bats" [www.cycadsociety.org/laevifolius/laevifolius]). The most authoritative web site on cycads (www.plantnet.rbgsyd.gov.au/PlantNet/cycad/ecol) notes, "The fleshy sarcotesta attracts animals, mainly birds, rodents, small marsupials, and fruit-eating bats, which serve as dispersal agents. In most cases, the fleshy coat is eaten off the seed, and the entire seed is not consumed." An authoritative review of bat–cycad relationships, similar to Dobat's review of the relationships between bats and flowers, would be useful (Dobat, 1985).

On Guam, Wiles listed cycads as one of the top 10 food items of the flying fox *Pteropus mariannus*:

> Information on the foods of *P. mariannus* was gathered by direct observations, finding evidence (discarded fruit, chewed pellets of fruit pulp) of bats feeding, examining faeces [*sic*] collected under roosts and interviewing island residents. Thirty species of plants are known to be used by fruit bats in the Marianas. These include 21 different fruits, 11 flowers and one leaf. Favoured [*sic*] foods include the fruits of *Artocarpus mariannensis*, *A. altilis*, *Pandanus tectorius*, *Cycas circinalis*, *Mammea odorata*, *Ficus prolixa*, and *Terminalia catappa* as well as the flowers of *Ceiba pentandra*, *Erythrina variegata* and *Cocos nucifera*. The stems of leaves and tips of small twigs on *A. mariannensis* are also frequently eaten. (Wiles 1987a, p. 94.)

We should note that prior to Hill's monograph, the taxonomy of Guam cycads was "unstable," with the Guam species being referred to as *C. rumphii* by authorities such as Jones (1993) and Norstog and Nichols (1997), whereas Safford (1905) and Stone (1970) termed it *C. circinalis*. Wiles' observations of flying foxes feeding on cycad seeds extends not only to Guam (Wiles & Fujita, 1992) but also to Palau (Wiles & Conry, 1990). Through our own field observations in Guam, we confirmed Wiles' assertion that flying foxes eat cycad seeds (Monson et al., 2003). Evidence of cycad feeding can be determined from chewed fruits or seeds because of (1) the unique dentition patterns of flying foxes, (2) longitudinal scratch marks from the nails of the digits, and (3) ejecta pellets formed in the flying fox oral cavity when juice and fluid are extracted and the dried pulp is spit out (Banack, 1998).

To see if flying foxes accumulate BMAA, we did not seek samples from the small remaining Guam colony at Anderson Air Force base because *P. mariannus* is a federally listed endangered species. Unwilling to disturb this colony, we initially chose to test accessioned museum specimens collected in 1951 and deposited at the Museum of Vertebrate Zoology (MVZ), University of California, Berkeley. Samples of dried skin were tested for the presence of BMAA. We found that all three specimens contained BMAA, with values ranging from 1287 to 7502 μg/g (Banack & Cox, 2003a). This represents a comparable dose of eating somewhere between 174 and 1014 kg of cycad flour for each flying fox ingested. This is within the range, suggested by Duncan et al. (1990), necessary to consider BMAA a possible causative factor of ALS/PDC. We have analyzed an additional 21 flying fox specimens from the MVZ and the U.S. National Museum, including dried, fluid-preserved, and frozen tissues. We detected free- and protein-associated BMAA in these samples ranging up to 272 μg/g (free) and 727 μg/g (protein-bound). Free- and protein-associated BMAA at significant levels were also found in flying fox stew prepared as part of a Chamorro meal; consumption of 1 L of the broth alone would result in ingestion of 12.4 mg of BMAA (Banack, et al., 2006).

In the original publication of our hypothesis, we suggested that the high incidence followed by a subsequent decline of ALS/PDC was concomitant with the decline of the Guam flying fox population. Flying foxes were once abundant in Guam. This normally nocturnal species "was not an uncommon sight during the daytime" (Crampton, 1921, p. 135) and was so abundant, that they were fed to hogs in the early twentieth century (Lujan, 1992). Flying foxes were sold in the 1930s in Guam markets, where demand often exceeded supply (Thompson, 1969). However, subsequent to the acquisition of firearms by the Chamorros, the flying fox population was hunted "continually and persistently" (Baker, 1948, p. 54) and had declined to what D.H. Woodside estimated as 3000 animals by 1958 (Wiles, 1987b). Hunting and habitat destruction continued to have an impact on the population in the early 1960s (Perez, 1973), and by 1971 the population had fallen to fewer than 1000 animals (Wheeler, 1979; Wiles, 1987b). By the mid-1970s, the population had declined to fewer than 100 animals (Wiles, 1987b). By 1978, the population hovered on the edge of extinction, with fewer than 50 animals (Wiles, 1987b).

With no reliable surveys of flying fox densities being conducted in Guam prior to 1958, we estimated the population of flying foxes in the early twentieth century by calculating two values (Monson, et al., 2003). First, based on published accounts of flying fox densities within forested areas of nearby islands in the Marianas that were not heavily affected by hunting (Wiles et al., 1989), we estimated the typical flying fox population density in undisturbed forests of Guam to be 186 animals/sq. km. This seems to be a conservative value, given that Stinson et al. (1992) estimated the flying fox density on Rota to be 195 animals/sq. km. Second, using Fosberg's determination that the majority of Guam was still forested in 1960 (Fosberg, 1960), we suggest that 60% of Guam's 541 sq. km. were forested at the beginning of the century. Therefore, a population density of 186 flying foxes/sq. km. for 60% of 541 sq. km. yields 60,375 flying foxes in the early 1900s (Monson et al., 2003). Given space constraints, these details were not published in the original medical hypothesis in *Neurology* (Cox & Sacks, 2002a).. However, they were detailed in the *Conservation Biology* article (Monson, et al., 2003), which was submitted prior to publication of the first *Neurology* article.

After the decimation of Guam's flying fox populations, *Pteropus tokudae* became extinct, and *Pteropus mariannus* was placed on the U.S. endangered species list. However, the continuing commercial demand for flying foxes led to significant importation of dead flying foxes from other islands. The magnitude of Guam's commercial traffic in flying foxes, including the importation of an estimated 230,000 dead flying foxes from 1975–1990 from other Pacific islands, is a matter of public record (Wiles, 1992; Wiles, 1994). It was a major factor in the Appendix I and II listing of Pacific island flying fox species by the parties to the Convention on International Trade in Endangered Species (CITES) (Bräutigam & Elmqvist, 1990; Mickleburg et al., 1992). There is no question that a significant decline in the size of Guam's flying fox populations occurred prior to the reported decline in ALS/PDC incidence rates among the Chamorros. Although, given the nature of the population estimates, no attempt was made to calculate a correlation coefficient, as noted in our original *Neurology* hypothesis, "the decline of ALS-PDC among the Chamorro mirrors the decline of flying foxes in Guam (Cox & Sacks 2002a, p. 958)."

Consumption of Flying Foxes by the Chamorro People of Guam

Extensive ethnobotanical interviews were conducted with Chamorros in Umatac, Merizo, Agat, and Rota concerning their dietary practices, with data recorded both on audiotape and on a standardized form. Chamorros were asked to list their favorite foods. With the exception of small children, Chamorros in these villages overwhelming identified flying foxes (*fanihi* in the Chamorro language) as the most salient and desired traditional food. When flying foxes were abundant, informants indicated eating up to seven to nine flying foxes per meal. The results from our interviews are consistent with Sheeline's (1991) survey of more than 200 Chamorros,in which 53% of those interviewed like to eat flying fox. In addition, 55% of respondents indicated that other people within their households ate flying foxes, 49% indicated that eating flying fox was an important part of being a Chamorro, and 19% of those surveyed could distinguish different species of flying fox by taste. In our interviews, some Chamorros told us that nothing in the world "tastes as good as a fanihi," a statement that might reflect the possible stimulation of glutamate receptors in the tongue by BMAA.

Figure 17-4. Advertisement for flying foxes in a Guam newspaper. (Photograph by Merlin Tuttle, Bat Conservation International.)

The importance of flying fox meals in traditional diets was noted by early European explorers of the Pacific. In 1819, Arago noted that flying fox consumption was limited to the Chamorros in Rota, rather than the Caroline islanders: "We find here also monstrous bats, similar to those of Guam, perhaps even larger. The Carolinians would not eat them." (Arago, 1823, p. 276). The concept of flying fox consumption, which might appear unappealing to ethnocentric investigators, is not a cryptic phenomenon in Guam, where advertisements for imported flying foxes have prominently appeared in the local paper (Fig. 17-4). In Washington D.C., in the 1980s, the U.S. Congress appropriated a line-item budget increase to fund U.S. Fish and Wildlife inspection in Guam of incoming international shipments for flying foxes. The Guam flying fox trade raised alarm in the international community, which in 1988 used the Guam commercial traffic in flying foxes as one of the primary reasons to extend Appendix I and Appendix II CITES protection to flying foxes in the Pacific Islands (Bräutigam & Elmqvist, 1990).

Prior to the militarization of Guam, Chamorros used traditional techniques and tools to harvest flying foxes, including the use of handheld nets at night, to capture flying foxes foraging on breadfruit trees. The efficacy of these indigenous techniques was eclipsed by the advent of pump shotguns and semiautomatic weapons. With such firearms, a single evening hunt at a flying fox roost could result in harvests of dozens of flying foxes (Graham, 1992). As a result, flying fox consumption accelerated, particularly in the aftermath of World War II. Overharvesting and habitat loss led to extinction of one flying fox species and near extinction of the other. In addition to cooked flying foxes, there are other significant inputs of BMAA in the traditional Chamorro diet. A large amount of protein-bound BMAA occurs in the cycad seed flour, a finding that tends to corroborate the suggestions of Spencer et al. (1987) and Dastur (1964) that there is a link between Chamorro consumption of cycad flour and ALS/PDC. BMAA also occurs in feral deer in Guam and possibly other animals that occasionally forage on cycad seeds, including pigs and land crabs.

What is the fate of BMAA in traditional food items consumed by the Chamorros? We obtained small (<1g) samples from the Kinsmen Neurological Laboratory, British Columbia, of brain tissue (superior frontal gyrus) of six Chamorro patients with ALS/PDC. Five of the six samples had free BMAA present at a mean level of 6 μg/g (standard deviation, 3.8) (Cox et al., 2003). Subsequent analysis showed that all six patients also had protein-bound BMAA in their neuroproteins at a mean level of 627 μg/g (standard deviation, 345) (Murch et al., 2004a, 2004b). By comparison, none of the 13 Canadians who had died of causes unrelated to neurodegeneration had detectable BMAA in their brains (Murch et al., 2004b). These small samples we studied were largely consumed in our analysis and we were able to exhaust these samples because we did not receive any requests from other investigators for these or any other samples from our laboratory.

BMAA and Other Geographical Foci of ALS/PDC

The idea that BMAA can accumulate in successive levels of a food chain from cyanobacteria to cycads to herbivores and finally into carnivores is now well documented in Guam. However, two other geographical loci of ALS/PDC have been described: a few villages in the western part of the island of New Guinea (Gajdusek & Salazar, 1982) and the Kii peninsula of Japan (Yase, 1979). Peter Spencer reported that the indigenous people in that part of New Guinea use cycad poultices (Spencer et al., 1987a), and Garruto is currently investigating the possibility that the indigenous people there eat flying foxes. However, the West Papua geographical focus still requires further investigation. In the Kii peninsula, Spencer (Spencer et al., 1987b) also recorded the historical use of cycad seeds in the ethnopharmacopoeia, but it is unclear if this use has survived in recent times.

We have now made four expeditions to the remote villages in the mountainous Kii peninsula of Japan, where the original descriptions of ALS/PDC were made. Although it once appeared that ALS/PDC may have been disappearing from the area, more recent and intensive epidemiological studies by Kokubu and collaborators (Kuzuhara et al., 2001) have produced an incidence rate of 800/100,000, about twice the incidence rate of the Chamorro people in Guam. In our ethnobotanical studies in Kii, we have not found any current use of cycads by the indigenous people, and we have performed exhaustive studies of the water, rice, miso, and other traditional food items consumed in the area. We have been unable to detect BMAA in any common terrestrial dietary inputs, and all the unusual terrestrial food items we found in our ethnobotanical surveys proved to be devoid of BMAA. We therefore turned to the marine ecosystem and immediately detected BMAA in significant levels in certain unusual fish consumed by the villagers. We are still attempting to piece together the marine Kii food web that results in BMAA biomagnification, which may be based on a marine cyanobacterium.

Our findings of BMAA in brain tissues from eight of the nine Canadian patients with AD we tested (Cox et al., 2003; Murch et al., 2004a) have suggested the possibility that BMAA may sometimes be ingested by North Americans. Because cyanobacteria are present worldwide in diverse ecological environments, we wondered if cyanobacteria other than *Nostoc* species (which are the symbionts of *C. micronesica)* could produce BMAA. Through collaboration with our colleagues at the University of Hawaii, the University of Dundee, and Stockholm University, we analyzed 30 laboratory cultures of cyanobacteria originally derived from freshwater, marine systems, brackish water, soil, and geothermal geysers throughout the world, representing all five morphological sections of cyanobacteria. In our recently published analysis (Cox et al., 2005) we found BMAA produced by representative taxa from all five morphological sections, in 95% (20/21) of the genera, and in 97% (29/30) of the strains we analyzed. Because some cyanobacterial species form blooms spanning hundreds of square kilometers, BMAA input through cyanobacterially contaminated water supplies and biomagnification in marine and freshwater ecosystems should be studied. Given the broad taxonomic and ecological breadth of these cyanobacterial species, we believe it possible that human exposure to BMAA could occur in areas far distant from Guam. This may account for our finding of BMAA in a few neurodegenerative patients from North America.

Weaknesses in the BMAA Hypothesis

Almost immediately after its publication, criticisms of the reformulated BMAA hypothesis began to appear in newspapers, popular literature, and in the peer-reviewed scientific literature of what was clearly labeled as a "medical hypothesis" (Cox & Sacks, 2002a). Chen et al. (2002, p. 1664) suggested that (1) "*C. circinalis* grows in the Marianas but not *C. rumphii*; (2) bats "eat only cycad skin when forests are devoid of fruit," (3) hunting of *P. marianas* [*sic*] was banned in 1973, and afterwards flying foxes "that do not eat cycads" were brought in from the Carolinian islands and Samoa, (4) bats were uncommon in Guam in 1930, and (5) "one can plot the epidemic of brown tree snakes from 1978 to 1993 against the decline in ALS incidence" as well as against "the rapid increase in numbers of civil automobiles." The researchers also reported the case of a Japanese soldier who spent 28 years hiding in the Guam jungle, who was believed to eat "cycads, grasses, taro, yam, wild boar, rodents, including fruit bats [*sic*]," but who on examination was "mentally sound" even though on autopsy he "had classical PD with some Lewy bodies" but "no neurofibrillary tangles or senile plaques." In general, the authors believed the hypothesis to be far-fetched, without sufficient pertinent data to satisfy Koch's postulates.

We responded in the same issue of *Neurology* (Cox & Sacks 2002b, pp. 1664–1665) that (1) Ken Hill's (1994) monograph excluded *C. circinalis* from Guam and that "the *C. rumphii* species complex has been placed into subsection Rumphiae of the genus *Cycas*" and that Hill "proposes the new binomial *C. micronesica* as part of the *C. rumphii* species complex," (2) the fleshy sarcotesta of cycad seeds is attractive to animal dispersers and "is easily ingested by flying foxes and is rich in BMAA," (3) hunting; we did not comment because this observation supports the hypothesis, (4) flying foxes were numerous in Guam in the early twentieth century, and our estimate of population size is consistent with densities reported from nearby islands, and (5) "snake densities or number of automobiles correlate [*sic*] with ALS/PDC incidence, but such factors do not have a rise and subsequent decline as does the human consumption of cycad-eating flying foxes." The issue raised by Chen et al. (2002) concerning the failure of the BMAA hypothesis to satisfy Koch's postulates is the most serious weakness of the hypothesis, and one that has plagued it from its early inception by such distinguished and accomplished chemists, physiologists, and neurotoxicologists as Arthur Bell, Peter Nunn, and Peter Spencer. Briefly stated, until an animal model of progressive neurodegeneration is published and verified, Koch's postulates (Cox & Knox, 1988) remain unfulfilled, and the link between BMAA and human disease remains unproven. Although Nunn produced an acute toxic response in chicks and rats using BMAA, and although Spencer produced acute toxicity in primates through administration of BMAA, to our knowledge no investigator has succeeded in inducing progressive neurodegeneration in a mammal through low, chronic doses of BMAA. Perry et al. (1989) gave mice an 11-week course of BMAA (15.5 g/kg) but noted no neurochemical or neuropathological changes consistent with ALS or PD; however, this result contrasts with the earlier work of Polsky et al. (1972). It is indeed possible that BMAA is merely a biomarker for neurological disease and that some other toxic molecule is biomagnified along a similar route.

A second weakness with the BMAA hypothesis is the low sample size, particularly of the human tissues we have examined in blinded experiments. We have found BMAA in only six of six of the Chamorro patients with ALS/PDC, s one of two asymptomatic Chamorro individuals, and eight of nine North American patients with AD. We have not detected BMAA in 14 of 14 North American patients who died of causes unrelated to neurodegeneration. Although these results are interesting, more study is needed. Deborah Mash at Miami Brain Bank, precisely reproducing our analytical techniques, has reported that her team has succeeded in replicating our findings of BMAA in tissues of patients who died of neurological illness (unpublished data). Further criticisms of the BMAA hypothesis (Duncan & Marini, 2006) based on claims of low toxicity of BMAA, specificity of the BMAA assay, possible effects of different preservation methods, and claims that flying foxes were not a major dietary component on Guam were refuted (Cox & Banack, 2006). However, their objection that flying foxes (and by implication other cycad-based products and food chains) are not eaten in the Kii (where flying foxes do not occur) and the West New Guinea foci of ALS/PDC is cogent and remains unanswered until some other route of BMAA biomagnification from cyanobacteria is identified in those areas (see "BMAA and Other Geographical Foci of ALS/PDC" above).

Other thoughtful criticisms of the BMAA hypothesis have been summarized by

Ince and Codd (2005). As we noted in our response "the BMAA hypothesis is far from proven but remains a viable possibility. We think it worthwhile to continue multidisciplinary investigations on this neurotoxin and its possible link to neurodegenerative illness in Guam and other locales" (Cox et al., 2006, p. 680).

Conclusions: BMAA and the Scientific Process

As botanists, we participate in a polite and learned discourse community that dates back centuries. Botanists, including members of societies such as that sponsoring this symposium on the biology of cycads, are no less rigorous than other scientists in debating new ideas and hypotheses. One of the intellectually compelling attractions of science is that data rather than personalities are the ultimate arbiters of truth. Experiments can be replicated, data points can be verified, and conceptual linkages between different intellectual realms can be carefully examined. Therefore, scientific controversies are best resolved in the peer-reviewed literature.

Here we have proposed that BMAA may be linked to certain patterns of neurological disease in Guam through biomagnification, an elaboration of earlier ideas generated by Bell, Nunn, and Spencer (Vega & Bell, 1967; Nunn et al., 1987; Spencer et al., 1987). We have discovered that BMAA accumulates in the Guam ecosystem and that it is bound to protein. In addition, we have found that earlier suggestions that cycad flour washed by the Chamorros is largely devoid of BMAA are, in fact, incorrect; cycad flour contains a high amount of BMAA in the protein fraction, and represents a significant input of BMAA into the Chamorro diet. Finally, we have discovered that cyanobacteria of diverse taxa and disparate ecological communities produce BMAA. This indicates a possibility for human exposure to BMAA in areas far from Guam, as well as in cultures that do not consume cycads. All of these observations have been published in the peer-reviewed, scientific literature and are subject to rigorous testing; precise reproduction of the detailed experimental methods will facilitate replication of the data points.

In the process of our investigations, we have discovered some fascinating connections between cyanobacteria, cycads, and people. It is our hope that this work, and the broader ethnobotanical perspective that links plants and people, may prove of interest to the cycad biologists at this symposium.

Literature Cited

Anderson, F. H., E. P. Richardson Jr., H. Okazaki & J. A. Brody. 1979. Neurofibrillary degeneration on Guam frequency in Chamorros and non Chamorros with no known neurological disease. Brain 102: 65–77.

Arago, J. 1823. Narrative of a voyage round the world in the Uranie and Physicienne corvettes, commanded by Captain Freycinet during the years 1817, 1818, 1819, and 1820. Treuttel and Wurz, Treuteel, Jun, and Richter, London.

Baker, R. H. 1948. Comments on conservation probelms in Micronesia. Pp. 53–55 *in* J.J. Coolidge (ed.), Conservation in Micronesia. A report on two conferences held under the auspices of the Pacific Science Board in Honolulu, T.H. and Washington, DC, April–May 1948, National Research Council, Washington, DC.

Banack, S. A. 1998. Diet selection and resource use by flying foxes (Genus Pteropus). Ecology 79: 1949–1967.

——— & P. A. Cox. 2003a. Biomagnification of cycad neurotoxins in flying foxes: implications for ALS-PDC. Neurology 61: 387–389.

——— & ———. 2003b. Distribution of the neurotoxic non-protein amino acid BMAA in *Cycas micronesica* Hill. Bot. J. Linn. So. 143: 165–168.

———, S. J. Murch & P. A. Cox. 2006. Neurotoxic flying foxes as dietary items for the Chamorro people, Marianas Islands. J. Ethnopharmacol. 106: 97–104.

Biflingmeyer, B. A., S. A. Cohen & T. L. Tarvin. 1984. Rapid analysis of amino acids using pre-column derivatization. J. Chromatogr. 336: 93–104.

Bräutigam, A. & T. Elmqvist. 1990. Conserving Pacific island flying foxes. Oryx 24: 81–89.

Brenner, E. D., D. W. Stevenson, R. W. McCombie, M. S. Katari, S. A. Rudd, K. F. Mayer, P. M. Palenchar, S .J. Runko, R. W. Twigg, G. Dai, R. A. Martienssen, P. N. Benfey & G. M. Coruzzi.. 2003. Expressed sequence tag analysis in *Cycas,* the most primitive living seed plant. Genome Biol. 4(12): R78.

Brody, J. A., A. H. Edgar & M. M. Gillespie. 1978. Amyotrophic lateral sclerosis: no increase among US construction workers in Guam. JAMA. 240: 551–552.

Brooks, S.P.J., B. J. Lampi, G. Sarwar & H. G. Botting. 1995. A comparison of methods for determining whole body protein. Anal. Biochem. 226: 26–30.

Brownson, D. M., T. J. Mabry & S. W. Leslie. 2002. The cycad neurotoxic amino acid β-N-methylamino-L-alanine (BMAA), elates intracellular calcium levels in dissociated rat brain cells. J. Ethnopharmacol. 82:159–167.

Butterfield, D. A., N. C. Hall & S. J. Cross. 1993. Effects of β-(N-methylamino)-L-alanine on cytoskeletal proteins of erythrocyte membranes. Chem Res. Toxicol. 6: 417–420.

Chang, Y. C., S. J. Chiu & K. P. Kao. 1993. Beta-N-methylamino-L-alanine (L-BMAA) decreases brain glutamate receptor number and induces behavioral changes in rats. Chin J Physiol. 36(2): 79–84.

Chen, K. M., U. K. Craig, C. T. Lee & R. Haddock. 2002. Cycad neurotoxin, consumption of flying foxes, and ALS/PDC disease in Guam: To the Editor. Neurology 59: 1664–1665.

Chorus, I. 2001. Cyanotoxins: occurrence, causes, consequences. Springer, Berlin.

Cohen, S. A. & D. P. Michaud. 1993. Synthesis of a fluorescent derivatizing reagent, 6-aminoquinolyl-N-hydroxysate amino acids via high-performance liquid chromatography. Anal. Biochem. 211: 279–287.

Costa, J-L., P. Paulsrud & P. Lindblad. 1999. Cyanobiont diversity within coralloid roots of selected cycad species. FEMS Microbiol. Ecol. 28: 85–91.

Cox, P. A. & S. A. Banack. 2006. A nonprotein amino acid and neurodegeneration. Science 314: 1242.

——— & R. B. Knox. 1988. Pollination postulates and two-dimensional pollination in hydrophilous monocotyledons. Ann. Missouri Bot. Gard. 75: 811–818.

——— & O. W. Sacks. 2002a. Cycad neurotoxins, consumption of flying foxes, and ALS-PDC disease in Guam. Neurology 58: 956–959.

——— & ———. 2002b. Cycad neurotoxin, consumption of flying foxes, and ALS/PDC disease in Guam: Reply from the authors. Neurology 59: 1664–1665.

———, S. A. Banack & S. J. Murch. 2003. Biomagnification of cyanobacterial neurotoxins and neurodegenerative disease among the Chamorro people of Guam. Proc. Natl. Acad. Sci. U.S.A. 100(23): 13380–13383.

———, ———, ——— & O. Sacks. 2006. Commentary on: return of the cycad hypothesis—does the amyotrophic lateral sclerosis/Parksinsonism dementia complex (ALS/PDC) of Guam have new implications for global health? Neuropathol. Appl. Neurobiol. 32: 679–682.

———, ———, ———, U. Rasmussen, G. Tien, R. R. Bidigare, J. S. Metcalfe, L. F. Morrison, G. A. Codd & B. Bergman. 2005. Diverse taxa of cyanobacteria produce β-

N-methylamino-L-alanine, a neurotoxic amino acid. Proc.. Natl. Acad. Sci. U.S.A. 102(14): 5074–5078.

———, ———, ———, A. Bell, P. Nunn, W. Bradley, D. Mash, S. Papapetropoulos, J. Pablo, L-O Ronnevi, S. Conradi, B. Bergman, U Rasmussen, R. Bidigare, G. A. Codd, J. S. Metcalf, H. Johnson, R. Speth & J. Weiss. 2005. Correspondence to: Lack of β-methylamino-L-alanine in brain from controls, AD, or Chamorros with PDC. Neurology 65: 768–769.

Crampton, H. E. 1921. A journey to the Mariana Islands, Guam and Saipan. J. Amer. Mus. Nat. Hist. 21: 126–145.

Dastur, D. K. 1964. Cycad toxicity in monkeys: clinical, pathological and biochemical aspects. Federation Proc. 23: 1368–1369.

Dobat, K. 1985. Blüten und Fledermäuse. Verlag von Waldemar Kramer, Frankfurt.

Douglas, S. E. & S. Turner. 1991. Molecular evidence for the origin of plastids from a cyanobacterium-like ancestor. J. Mol. Evol. 33(3): 267–273.

Duncan, M. W. & A. M. Marini. 2006. Debating the cause of a neurological disorder. Science 313: 1737.

———, I. J. Kopin, L. Lavine, R. Garruto & S. P. Markey. 1988. The putative neurotoxin BMAA in cycad-derived foods is an unlikely cause of ALS-PD. Lancet 2:631–632.

———, J. C. Steele, I. J. Kopin & S. P. Markey. 1990. 2-Amino-3-(methylyamino)-propanoic acid (BMAA) in cycad flour: an unlikely cause of amyotrophic lateral sclerosis and parkinonism-dementia of Guam. Neurology 40: 767–772.

Fosberg, F. R. 1960. The vegetation of Micronesia. Bull. Am. Mus. Nat. Hist. 119: 1–75.

Fountoulakis, M. & H-W. Lahm. 1998. Hydrolysis and amino acid composition analysis of proteins. J. Chromatogr. A. 826: 109–134.

Gajdusek, D. C. & A. M. Salazar. 1982. Amyotrophic lateral sclerosis and parkinsonian syndromes in high incidence among the Auyu and Jakai people of West New Guinea. Neurology 32: 107–126.

Garruto, R. M., D. C. Gajdusek & K. M. Chen. 1980. Amyotrophic lateral sclerosis among Chamorro migrants from Guam. Ann. Neurol. 8: 612–619.

———, ——— & ———. 1981. Amyotrophic lateral sclerosis and parkinsonism-dementia among Filipino migrants to Guam. Ann. Neurol. 10: 341–350.

Giddy, C. 1984. Cycads of South Africa. Ed. 2. C. Struik, Cape Town, South Africa.

Graham, G. L. 1992. Conservation and subsistence harvesting of Pacific Island flying foxes. Pp. 46–49 *in* D.E. Wilson and G.L. Graham (eds.), Pacific island flying foxes: proceedings of an international conference. Biological report 90(23). United States Fish and Wildlife Service, Washington, DC.

Hill, K. D. 1994. The *Cycas rumphii* complex (Cycadaceae) in New Guinea and the Western Pacific. Austral. Syst. Bot. 7: 543–567.

Hirsh, H. & T. Marler. 2002. Damage and recovery of *Cycas micronesica* after Typhoon Paka. Biotropica 34: 598–602.

Ince, P. G. & G. A. Codd. 2005. Annotation: return of the cycad hypothesis—does the amyotrophic lateral sclerosis/parkinsonism dementia complex (ALS/PDC) of Guam have new implications for global health? Neuropathol. Appl. Neurobiol. 31: 345–353.

Jones, D. L. 1993. Cycads of the world. Smithsonian Institution Press, Washington, DC.

Khabazian, I., J. S. Bains, D. E. Williams, J. Cheung, J.M.B. Wilson, B. A. Pasqualotto, S. L Pelech, R. J. Andersen, Y.-T. Wang, L. Liu, A. Nagai, S. U. Kim, U-K. Craig & C. A. Shaw. 2002. J. Neurochem. 82: 516–528.

Kisby, G. E., M. Ellison & P. S. Spencer. 1992. Content of the neurotoxins in cycasin (methylazoxymethanol beta-D-glucoside) and BMAA (beta-N-methylamino-L-alanine) in cycad flour prepared by Guam Chamorros. Neurology 42: 1336–1340.

———, D. N. Roy & P. S. Spencer.. 1988. Determination of β-N-methylamino-L alanine (BMAA) in plant (*Cycas circinalis* L.) and animal tissue by precolumn derivatization of with 9-fluorencylmethyl chloroformate (FMOC) and reversed-phase high-performance liquid chromatography. J. Neurosci. Methods 26: 45–54.

Kurland, L. & D. Mulder. 1954. Epidemiological investigations of amyotrophic lateral sclerosis: 1. Preliminary report on geographic distribution with special reference to the Mariana Islands, including clinical and pathological observations. Neurology 4: 355–378, 438–448.

Kuzuhara, S. R. Sasaki, Y. Kokubo & Y. Narita. 2001. Amyotrophic lateral sclerosis/parkinsonism-dementia complex of the Kii peninsula of Japan (Kii ALS/PDC) may be a familial tauopathy. Epidemiological trends, clinical features, neuropathology and molecular genetics. Pp. 85–93 *in* C. Tanaka, P L. McGeer, and Y. Ihara (eds.), Neuroscientific basis of dementia. Birkhaeuser, Basel, Switzerland.

Lobner, D., P.M.T. Piana, A. K. Salous & R. W. Peoples. 2007. β-N-methylamino-L-alanine enhances neurotoxicity through multiple mechanisms. Neurobiol. Diseases 25: 360–366.

Ludolph, A., J. Hugon, P. B. Nunn, D. N. Roy & P. S. Spencer. 1988. Clinical and electrophysiological features of the primate motor system disease induced by BMAA—a neurotoxin in the Guam cycad plant. Electroencephalogr. Clin. Neurophysiol. 69: P18–P19.

Lujan, R. J. 1992. Local policies and protection by the government of Guam. P. 131–134 *in* D.E. Wilson and G.L. Graham (eds.), Pacific island flying foxes. Proceedings of an International Conference. Biological report 90(23). United States Fish and Wildlife Service, Washington, DC.

Mahmood, S., M. A. Malik, M. Motevalli, P. B. Nunn & P. O'Brien. 1998. Synthesis and x-ray crystal structures of Schiff bases prepared from salicyladehyde and the diamino acids L-2-amino-3-methylaminopropionic acid, DL-2,4-diamino-butanoic acid and SL-2,3-diaminopropanoic acid. Tetrahed. 54: 5721–5730.

Marler, T. & H. Hirsh. 1998. Guam's *Cycas micronesica* populations ravaged by Supertyphoon Paka. HortScience 33: 1116–1118.

Marler, T. E. 2004. Leaf physiology of shade-grown *Cycas micronesica* leaves following removal of shade. Bot. Rev. 70: 63–71.

Matsuoka, Y., Z. Rakonczay, E. Giacobini & D. Naritoku. 1993. L-β-methylamino-alanine-induced behavioral changes in rats. Pharmacol. Biochem. Behavior 44: 727–734.

Micallef, B. J., B. J. Shelp & R. O. Ball. 1989. Quantification of C-14-labelled amino acids by reverse phase high performance liquid chromatography. J. Liq. Chrom. 12(7): 1281–1300.

Mickleburg, S. P., A. M. Hutson & P. A. Racey. 1992. Old World fruit bats: an action plan for their conservation. World Conservation Union, Chiroptera specialist group, Gland, Switzerland.

Monson, C. S., S. A. Banack & P. A. Cox. 2003. Conservation implications of Chamorro consumption of flying foxes as a possible cause of ALS-PDC in Guam. Conservation Biol. 17: 678–686.

Montine T. J., K. Li, D. P. Perl & D. Galasko. 2005. Lack of β-methylamino-L-alanine in brain from controls, AD, or Chamorros with PDC. Neurology 65: 768–769.

Moore, S., D. H. Spackman & W. H. Stein. 1958. Chromatography of amino acids on sulfonated polystyrene resins. An improved system. Anal. Chem. 30:1185–1190.

Murch, S. J., P. A. Cox & S. A. Banack. 2004a. A mechanism for slow release of biomagnified cyanobacterial neurotoxins and neurodegenerative disease in Guam. Proc. Natl. Acad. Sci. U.S.A. 101: 12228–12231.

———, ———, ———, J. C. Steele & O. W. Sacks. 2004b. Occurrence of β-methylamino-L-alanine (BMAA) in ALS/PDC patients from Guam. Acta Neurol. Scand. 110: 267–269.

Neilan, B. A., E. Dittmann, L. Rouhiainen, R. A. Bass, V. Schaub, K. Sivonen & T. Börner. 1999. Nonribosomal peptide synthesis and toxigenicity of cyanobacteria. J. Bacteriol. 181: 4089–4097.

Norstog, K. J. & T. J. Nicholls. 1997. *The Biology of the Cycads*. Cornell University Press, Ithaca, New York.

Nunn, P. B., M. Seelig, J. C. Zagoren & P. S. Spencer. 1987. Stereospecific acute neuronotoxicity of uncommon plant amino acids linked to human motor-system diseases. Brain Res. 410(2): 375–379.

Pate, J. S., P. Lindblad & C. A. Atkins. 1988. Pathways of assimilation and transfer of fixed nitrogen in coralloid roots of cycad-nostoc symbioses. Planta 176:461–471.

Perez, G.S.A. 1973. Notes on the ecology and life history of Pteropidae on Guam. Periodicum Biologorum 75: 163–168.

Perry, T. L., C. Bergeron, A. J. Biro & S. Hansen. 1989. β-*N*-Methylamino-L-alanine chronic oral administration is not neurotoxic to mice. J. Neurolog. Sci, 94: 173–180.

Polsky, F. I., P. B. Nunn & E. A. Bell. 1972. Distribution and toxicity of α-amino-β-methylaminopropionic acid. Federation Proc. 31: 1473–1475.

Rakonczay, Z., Y. Matsuoka & E. Giacobini. 1991. Effects of L-beta-N-methylamino-L-alanine (L-BMAA) on the cortical cholinergic and glutamatergic systems of the rat. J. Neurosci. Res. 29(1): 121–126.

Rao, S. D., S. A. Banack, P. A. Cox, P.A. & J. H. Weiss. 2006. BMAA selectively injures motor neurons via AMPA/kainate receptor activation. Exp. Neurol. 201: 244–252.

Reed, D., D. Labarthe, K. M. Chen & R. Stallones. 1987. A cohort study of amyotrophic lateral sclerosis and parkinsonism-dementia on Guam and Rota. Amer. J Epidemiol. 125: 92–100.

Romlong, U., D. M. Reece, P. B. Nunn & A. R. Hipkiss. Protein degradation in mouse brain slices. Biochem. Soc. Trans. 17: 214.

Rosenthal, G. A. & E. A. Bell. 1979. Naturally occurring toxic nonprotein amino acids. Pp. 353–385 *in* G.A. Rosenthal and D.H. Janzen (eds.), Herbivores: their interaction with secondary plant metabolites. Academic Press, New York.

Ross, S. M. & P. S. Spencer. 1987. Specific antagonism of behavioral action of "uncommon" amino acids linked to motor-system diseases. Synapse 1(3): 248–253.

———, M. Seelig & P. S. Spencer. 1987. Specific antagonism of excitotoxic action of 'uncommon' amino acids assayed in organotypic mouse cortical cultures. Brain Res. Safford, W. E. 1905. The useful plants of the island of Guam. Contr. U. S. Natl. Herb. 9: 1–416.

Sarwar, G., H. G. Botting & R. W. Peace. 1988. Complete amino acid analysis in hydrolysates of foods and feces by liquid chromatography of precolumn phenylisothiocyanate derivatives. J. Assoc. Off. Anal. Chem. 71: 1172–1175.

Schopf, J. W., A. B. Kudryavtsev, D. G. Agresti, T. J. Wdowiak & A. D. Czaja. 2002. Laser-Raman spectroscopy (Communication arising): images of the Earth's earliest fossils? Nature 420: 477–477.

Schuster, J. 1932. Cycadaceae. IV. 1 *in* A. Engler (ed.), Das Pflanzenreich. Verlag von Wilhelm Engelmann, Leipzig.

Seawright, A. A., A. W. Brown, C. C. Nolan & J. B. Cavanagh. 1990. Selective degeneration of cerebellar cortical neurons caused by cycad neurotoxin, L-beta-methylaminoalanine (L-BMAA), in rats. Neuropathol. Appl. Neurobiol. 16(2): 153–169.

Seebach, D., A. Studer, E. Pfammatter & H. Widmer. 1994. Synthesis of tri-, penta-, and heptapeptides containing an (R)-2-alkyl-2-amino-3-(methylamino)-propionic acid residue in the central position. Helv. Chim. Acta 77: 2035–2050.

Sheeline, L. 1991. Cultural significance of Pacific fruit bats (*Pteropus*) to the Chamorro people of Guam. Conservation implications. Report. World Wildlife Fund/Traffic U.S.A., Washington, DC.

Spencer, P. S., G. E. Kisby & A. C. Ludolph. 1991. Slow toxins, biological markers, and long-latency neurodegenerative disease in the western Pacific region. *Neurology* 41(supplement 2): 62–66.

———, P. B. Nunn, J. Hugon, A. C. Ludolph, S. M. Ross, D. N. Roy & R. C. Robertson. 1987. Guam amyotrophic lateral sclerosis Parkinsonism dementia linked to a plant excitant neurotoxin. Science 237: 517–522.

———, M. Ohta & V. S. Palmer. 1987a. Cycad use and motor neurone disease in Kii Peninsula of Japan. Lancet 2: 1462–1463.

———, V. S. Palmer, A. Herman & A. Asmedi. 1987b. Cycad use and motor neurone disease in irian Jaya. Lancet 2: 1273–1274.

Stinson, D. W., P. O. Glass & E. M. Taisacan. 1992. Declines and trade in fruit bats on Saipan, Tinian, Anguijan, and Rota. Pp. 61–67 *in* D.E. Wilson and G.L. Graham (eds.), Pacific island flying foxes: proceedings of an international conference. Biological report 90(23) United States Fish and Wildlife Service, Washington, DC.

Stone, B. C. 1970–1971. The flora of Guam. Micronesica 6: 1–659.

Thompson, L. 1969. Guam and its people. Greenwood Press Publishers, New York.

Torres, J., L. L. G. Iriate & L. T. Kurland. 1957. Amyotrophic lateral sclerosis among guamanians in California. Calif. Med. 86: 385–388.

Van Der Pijl, L. 1957. The dispersal of plants by bats (chiropterochory). Acta Bot. Neerl. 6: 291–315.

Van Der Pijl, L. 1969. Principles of dispersal in higher plants. Springer-Verlag, Berlin. Vega, A. & E. A. Bell. 1967. α-Amino-β-methyl-aminopropionic acid, a new amino acid from seeds of *Cycas circinalis*. Phytochemistry 6: 759–762.

Vega, A., E. Bell, & P. B. Nunn. 1968. The preparation of L- and D-α-amino-β-methylaminopropionic acids and the identification of the compound isolated from *Cycas circinalis* as the L-isomer. Phytochemistry 7: 1885–1887.

Vorster, P. 1995. Aspects of the reproduction of cycads. 2. An annotated review of known information. Proceedings of the third international conference on cycad biology. Stellenbosch, South Africa: The Cycad Society of South Africa. pp.379–389.

Weiss, J. H. & D. W. Choi. 1988. Beta-N-methylamino-L-alanine neurotoxicity: requirement for bicarbonate as a cofactor. Science 241: 973–976.

———, J. Koh & D. W. Choi. 1989. Neurotoxicity of β-N-methylamino-L-alanine (BMAA) and β-N-oxalylamino-L-alanine (BOAA) on cultured cortical neurons. Brain Res. 497: 64–71.

Wheeler, M. E. 1979. The Marianas fruit bat: management history, current status, and future plans. Cal-Neva Wildlife Trans. 979: 149–165.

Whiting, M. G. 1963. Toxicity of cycads. Econ. Bot. 17: 270–302.

Wiles, G. J. 1987a. Current research and future management of Mariana fruit bats on Guam. Australian Mammalogy 10: 93–95.

———. 1987b. The status of fruit bats on Guam. Pacific Science 41: 148–157.

———. 1992. Recent trends in the fruit bat trade on Guam. Pp. 53–60 *in* D.E. Wilson and G.L. Graham (eds.), Pacific island flying foxes: proceedings of an international conference. Biological report 90(23). United States Fish and Wildlife Service, Washington, DC.

———. 1994. The Pacific flying fox trade: a new dilemma. Bats 12: 15–18.

——— & P. J. Conry. 1990. Terrestrial vertebrates of Ngerukewid Islands Wildlife Preserve, Palau Islands. Micronesica 23: 41–66.

——— & M. Fujita. 1992. Food plants and economic importance of flying foxes on Pacific Islands. Pp. 24–35 *in* D.E. Wilson and G.L. Graham (eds.), Pacific island flying foxes: proceedings of an international conference. Biological report 90(23). United States Fish and Wildlife Service, Washington, DC.

———, T. O. Lemke & N. H. Payne. 1989. Population estimates of fruit bats, *Pteropus mariannus* in the Mariana island. Conservation Biol. 3: 66–75.

Yase, Y. 1979. ALS in the Kii Peninsula: one possible etiological hypothesis. Pp. 307–318 *in* T. Tsubaki, Y. Toyokura, (eds.), Amyotrophic lateral sclerosis. University Tokyo Press, Tokyo.

Zeevalk, G. D. & W. J. Nicklas. 1989. Acute excitotoxicity in chick retina caused by the unusual amino acids BOAA and BMAA: effects of MK-801 and kynurenate. Neurosci. Lett. 102: 284–290.

Zheng, W. W., T. Y. Song, X. Bao, B. Bergman & U. Rasmussen. 2002. High cyanobacterial diversity in coralloid roots of cycads revealed by PCR fingerprinting. FEMS Microbiol. Ecol. 40: 215–222.

NEUROBIOLOGY
CHAPTER 18

Cycad-Induced Neurodegeneration in a Mouse Model of ALS-PDC: Is the Culprit Really BMAA, or is a Novel Toxin to Blame?

Christopher A. Shaw, Jason M. B. Wilson, Reyneil Cruz-Aguado, Swaraj Singh, Erin L. Hawkes, Vivian Lee, and Thomas Marler

Abstract

The strongest epidemiological correlate for amyotrophic lateral sclerosis -Parkinsonism dementia complex (ALS-PDC) is consumption of the seed of the cycad (*Cycas micronesica*) known to contain various toxins. Much previous work focused on cycasin and its aglycone methylazoxymethanol (MAM) as well as on various excitatory amino acids, notably β-N-oxalylamino-L-alanine (BOAA) and β-methylamino-alanine (BMAA). None of these compounds is contained in significant concentrations in cycad seed "chips" prepared by traditional Chamorro washing procedures. Cycasin/MAM and BOAA do not induce neurological symptoms resembling ALS-PDC. BMAA, a weak N-methyl-D-aspartate (NMDA) receptor agonist, has also routinely failed to induce ALS-PDC–like outcomes. Recent work by Cox and colleagues suggests that BMAA contained in cycads reaches levels that are able to induce neurodegeneration in ALS-PDC following biomagnification up the food chain.

We have reexamined the "cycad hypothesis" for ALS-PDC in in vivo experiments on adult male mice fed washed cycad pellets (up to 25% of total diet). The cycad pellets contained nil to trace concentrations of cycasin/MAM, BOAA, or BMAA. Cycad-fed mice rapidly developed typical motor and cognitive behavioral phenotypes strongly resembling both amyotrophic lateral sclerosis and parkinsonian dementia. Histological analysis of the central nervous systems of these mice showed neuronal loss with apoptosis, as well as the presence of activated astrocytes in the spinal cord, substantia nigra, hippocampus, cortex, and olfactory bulb. In contrast, pure BMAA at high dosages did not give rise to behavioral deficits or pathological outcomes. Other neurotoxins such as methionine sulfoximine or 3-nitropropionic acid showed spinal cord– or basal ganglia–associated deficits, respectively. BMAA thus appears unlikely to play a significant role in ALS-PDC. The bound BMAA reported in restricted regions of the brains of neurological disease victims may be an artefact of fixation techniques. We have isolated other potential cycad neurotoxins and find that the most toxic are steryl glucosides that induce Small Heat-Shock Protein (HSP), tau expression, and caspase-3 in neural cell cultures. These data support the original hypothesis that cycad consumption is a cause of ALS-PDC on Guam but not a primary role for BMAA in disease onset or progression.

Resumen

El complejo demencia-ELA-Parkinson (ELA-PDC) tiene como mejor correlato epidemiologico al consumo de semillas de cícadas (*Cycas micronesica*), que es conocida por contener varias toxinas. Las investigaciones iniciales estuvieron enfocadas en la cicasina y su aglicona (NAM), así como en varios aminoácidos excitatorios, particularmente, BOAA y BMAA. Ninguno de estos compuestos están presentes en concentraciones suficientemente altas en las galletas de semillas de cícadas preparadas por el procedimiento de lavado tradicional de los Chamorros. Ni la cicasina/BMAA ni la BOAA inducen síntomas neurológicos semejantes a la ELA-PDC. La BMAA, un agonista débil del receptor NMDA, tampoco induce cambios parecidos a los que ocurren en la ELA-PDC. Los trabajos recientes de Cox y sus colegas sugieren que la BMAA en las Cycas, mediante la biomagnificación a lo largo de la cadena alimenticia, alcanza niveles suficientemente altos como para inducir la neurodegeneración asociada a la ELA-PDC. Nosotros hemos reexaminado la "hipótesis de la Cycas" mediante experimentos *in vivo* con ratones machos adultos alimentados con harina lavada de cícadas en una dosis de hasta el 25% de la dieta total. La harina de cícadas contuvo niveles nulos o trazas de cicasina/MAM, BOAA o BMAA. Los ratones alimentados con cícadas desarrollaron rápidamente un robusto fenotipo conductual, tanto motor como cognitivo, que semejó a la ELA y a la demencia parkinsoniana. El análisis histológico del SNC de estos ratones mostró perdida neuronal con apóptosis, así como la presencia de astrocitos activados en la médula espinal, la substantia nigra, el hipocampo, la corteza cerebral y el bulbo olfatorio. Sin embargo, la BMAA pura en dosis altas no produjo ni déficits conductales ni cambios neuropatológicos. Otras neurotoxinas como la metionín-sulfoximina o el ácido 3-nitropropiónico indujeron deterioros asociados a la médula espinal y a los ganglios basales, respectivamente. Así, la BMAA no parece desempeñar un papel significativo en la ELA-PDC. El resultado de que la BMAA se encuentra localizada en regiones discretas del cerebro de humanos con enfermedades neurológicas puede haber sido originado por un artefacto en las técnicas de fijación. Nosotros hemos aislado otras potenciales neurotoxinas de las cícadas, de las cuales los esteril-glucósidos resultaron los mas tóxicos e indujeron HSP, expresión de tau, y caspasa-3 en cultivos de células neuronales. Estos datos apoyan la hipótesis original de que el consumo de cícadas tiene un rol causal en la ELA-PDC de Guam, pero no apoya el papel primario de la BMAA en el inicio de la enfermedad ni en su progreso.

Introduction

The scientific community that studies neurological disease has been traditionally divided into those favoring either genetic or epigenetic explanations. Those who hold to a genetic explanation tend to take the view that gene mutations or deletions create the key circumstances for disease initiation. The classical example usually given is for Huntington disease, which arises due to a gene mutation that produces abnormal repeat copies of a CAG nucleotide sequence (Gardian & Vecsei, 2004). Those studying amyotrophic lateral sclerosis (ALS), Parkinson disease (PD), or Alzheimer disease (AD) can point to familiar inherited forms of each in which gene mutations seem to be causal. Consider, for example, mutant SOD (superoxide dismutase-1) for ALS (Morrison & Morrison, 1999; Andersen, 2001); mutations in genes for alpha-synuclein, parkin, and ubiquitin carboxyhydrolase L1 associated with familial PD (Huang et al., 2004; Betarbet et al., 2005); or genes responsible for amyloid beta metabolism for familial AD (Kowalska, 2004). However, considerably fewer than 10% of any of these diseases appears to have a genetic etiology. The proponents of a genetic etiology for these age-dependent diseases acknowledge this fact, but they also claim that regardless of overall percentages affected, the gene abnormality evokes a biochemical pathway culminating in neuronal cell death and that this pathway is likely to be the same for as for nongenetic factors. There is, as yet, no evidence that this assertion is correct. The second school of thought posits an epigenetic factor or factors leading to cell death, with the likely candidates arising in the human environment via toxins in food, air, or water. Although it seems likely that this notion is basically correct, to date there have been remarkably few clear examples of toxins able to induce the appropriate forms of neurodegeneration in anything approaching a natural manner. In recent years, there has been increasing cross talk between the two schools of thought, increasingly, scientists on both sides of the debate are acknowledging that a combination of genes and environment is likely key to solving the puzzle. The impact of aging is now widely considered to be the third crucial ingredient. This notion is illustrated in Figure 18-1.

In brief, the notion is that various toxins, perhaps present ubiquitously in subacute concentrations, act against a backdrop of genetic susceptibility factors. These gene factors alone are not sufficient to induce disease outcomes in the nervous system but somehow facilitate or inhibit the action of potential toxins. To complete the general hypothesis, the notion is that as people age, biochemical processes of detoxification protection change, thus allowing toxins that might have had no impact when people were younger years to induce damage when they become older.

As plausible as this combination of environmental toxins, susceptibility genes, and age seems, the outstanding difficulty is to translate this general notion into specific toxins and genes. The dimensions of the problem can be appreciated when we consider that the human genome contains close to 30,000 genes, most of unknown function. In addition, thousands of toxins, natural and synthetic, have been described, with more entering the

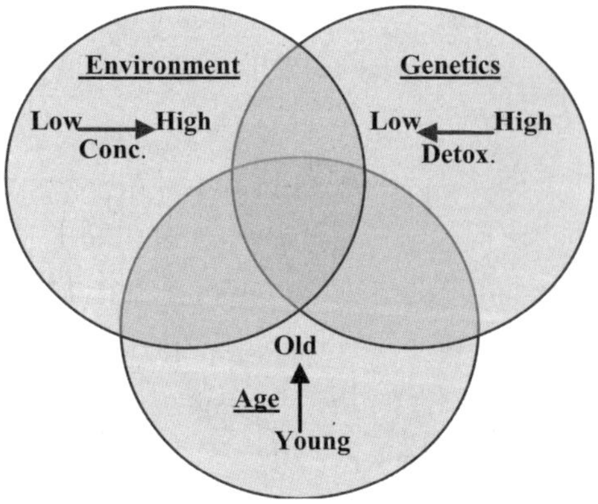

Figure 18-1. Potential synergies of causal and risk factors in sporadic neurological disease. Causal factors involved in such diseases may reflect exposure to some toxin(s) that can arise from the environment or as a result of individual biochemical processes. In the former case, the toxins may be synthetic or naturally occurring. The toxic factor is represented by the set on the left side of the Venn diagram. The range of toxin's effects run from left to right as "low to high." Intersecting this is a set consisting of a genetic susceptibility that could arise due to genetic polymorphism in efficiency of detoxification mechanisms (from right to left, expressed as "high to low"). Genes with other features such as transport could also be involved (e.g., ApoE alleles). The intersection of these two sets describes the individuals who may be at risk of developing the neurological disorder. Note that the intersecting region can increase or decrease depending on strength of either variable. Intersecting these two sets is the variable of age with the risk factor increasing from young to old (bottom to top).

environment each year. Taken together, the number of possible gene–environmental interactions is vast, making the search for the causal events in neurological disease a daunting task. In fact, without a reasonable starting point, the task may be impossible.

There is, however, one solution to the above problem: to find a "cluster" of neurological disease of sufficient numbers, in a relatively constrained locality and over a relative constrained time period, and to use this cluster to provide clues about the genes and environmental factors that are most likely to operate in this cluster. Those from outside the neurological sciences might be forgiven for thinking that clusters that could provide these clues would be abundant. Unfortunately, this is not true, in fact the opposite. For ALS, PD, and AD, there is only one widely acknowledged cluster, the unusual disease of the Western Pacific, ALS-PD complex (ALS-PDC). ALS-PDC thus stands out as the neurological equivalent of the "Rosetta Stone," the key to deciphering the causal factors of neurological disease worldwide.

ALS-PDC represents a spectrum of often overlapping neurological disease symptoms recognized as early as the 1950s (Whiting, 1963; see Kurland, 1988 for summary).Although the disease has been documented for hundreds of years from relatively early in the period of the Spanish occupation of Guam, it became the subject of intense neurological study only after World War II. Kurland and others noted the unusual combination of features in some individuals, the extremely high incidence rates, and the early onset of symptoms compared to North American neurological diseases. The early epidemiological data supported a cultural/dietary link between the incidence of the disease and the consumption of flour made from the seed of a local variety of cycad (*Cycas micronesica* K.D. Hill, formerly termed *Cycas circinalis* L.), long used as traditional and/or famine food and as a medicinal (Kurland, 1988), both on Guam and elsewhere in the western Pacific (for a more comprehensive review of cycad toxicity and epidemiology, see Schulz et al., 2005a). Cycad seeds were found to contain various toxins, and early studies focused on potential links between these compounds and the disease. The nitrogenous glucoside cycasin, was found to exert cytotoxic actions, acutely affecting the liver via its aglycone, methlyazoxymethanol (MAM). (Another principal MAM glycoside, macrozamin, is ubiquitous in the Cycadales at generally higher concentrations than cycasin but is innocuous [De Luca et al., 1980; Moretti et al., 1981; Moretti et al., 1983].) Although both cycasin and MAM induced developmental abnormalities of the nervous system in rodents and could produce death in various animals, they did not otherwise mimic ALS-PDC in either behavioral or neuropathological outcomes (Wilson et al., 2002). Other toxins in cycads, for example, β-N-oxalylamino-L-alanine (BOAA) and β-methylamino-alanine (BMAA), evoked differing forms of neurodegeneration through activation of either alpha-amino-3-hydroxy-5-methyl-4-isoxazole propionic acid (AMPA) or N-methyl d-aspartate (NMDA) glutamate receptor subtypes (Weiss et al., 1989). BMAA has also been shown to be neurotoxic when injected intraperitoneally in rodents; however, only cerebellar neurons were affected, not areas typically associated with ALS-PDC (Seawright et al., 1990). BOAA, an AMPA receptor agonist, gives rise to a transient loss of motor function in the chronic disorder lathyrism (Spencer & Schaumburg, 1983; Ravindranath, 2002). BMAA, a weak NMDA agonist, reportedly induced neuronal abnormalities in spinal motor neurons and motor behavior in monkeys (Spencer et al., 1987). These latter data could not be reproduced in mice (Perry et al., 1989; Schulz et al., 2005b). Spencer et al. did not publish additional evidence for in vivo BMAA neurotoxicity. Further attempts to duplicate BMAA neurotoxicity by scientists from the National Institutes of Health in various in vivo experiments were unsuccessful.

The failure of first cycasin/MAM and latter BMAA to induce ALS-PDC–like outcomes in in vivo behavioral and histological studies led to overall scepticism that cycad seed contained the neurotoxins responsible for the disease. Ultimately, the "cycad hypothesis" lost favor, finally due to studies showing that the extensive washing that was part of traditional Chamorro flour preparation removed the water-soluble toxins from cycad flour (Duncan et al., 1990; see also Wilson et al., 2002). In addition, the incidence of ALS-PDC had been in decline for years. (The incidence of ALS is now near North American levels, and that of PDC remains three to four times higher, but nowhere near its previous level [Plato et al., 2003].)

Although such outcomes proved nearly fatal for research into cycad toxicity generally and ALS-PDC specifically, epidemiological studies both then, and now, continued to show strong links between the two. Lessell et al. (1962) noted that cycad consumption on Guam was not uniform and that regions of Guam with the highest levels of consumption were ALS-PDC "hotspots." In addition, Kurland and others noted that while the Chamorro peoples of Guam and Saipan are of similar ethnic/genetic stock, the rates of ALS-PDC were dissimilar, which correlated with the fact that the Chamorros on Guam consumed cycads, especially during World War II, whereas their fellow Chamorros on Saipan did not. Another reason it may have been premature to abandon the cycad hypothesis was that, until recently, few studies had provided detailed behavioral and histological analyses of neural function and morphology following prolonged exposure to washed cycad as part of diet (i.e., in experiments designed to mimic Chamorro cycad consumption completely). The exceptions were the preliminary studies performed by Dastur (Dastur, 1964; Dastur et al., 1990) that demonstrated motor deficits and motor neuron loss in rhesus monkeys fed washed cycad flour obtained from Guam. Our laboratory has now performed such experiments on three distinct strains of mice, and these data provide the basis for our claim that a toxin contained in cycad flour can induce neurological outcomes essentially analogous to those seen in ALS-PDC. Because the cycad flour used in our experiments contains at most trace amounts of BMAA, BOAA, and MAM (Khabazian et al., 2002; Wilson et al., 2002), the ALS-PDC–like outcomes that we observe occur independently of these other toxins. These data have led us to seek other potential neurotoxins in cycad, as described below.

ALS-PDC is endemic to Guam and Rota and linked to consumption of *C. micronesica*. Similarly, consumption of other *Cycas* species has been linked to similar neurological outcomes (e.g. in Irian Jaya, the Kii Peninsula of Japan). Neurological outcomes following cycad consumption have not been identified elsewhere in the world. For example, Australian aboriginal people consume cycads but are not afflicted with neurological disease. It is notable that the cycad species in Australia are different from those in

Guam and that the processing of seed flour also differs significantly. Thus, variations in toxicity may relate to any or all of the following: cycad species, processing methods, unknown genetic susceptibility factors of the local population, or all of these factors.

The above description of ALS-PDC thus links it to the cycads and the potential toxins contained within the plants. Numerous, very different approaches have been taken to identifying the toxins causal to ALS-PDC. As cited above, various investigators followed several of the water-soluble toxins first isolated from the plants. In spite of enormous effort, neither cycasin/MAM nor BMAA could be shown to induce ALS-PDC–like neurological outcomes in experimental animals, and this entire field of research lay dormant for the better part of 20 years. Early in the twenty-first century, two groups of researchers began reexamining the "cycad hypothesis," taking quite different approaches. Our group started from the simple notion that the washed cycad seed flour could alone induce neurological deficits and noted that a previous, and largely forgotten study (Dastur, 1964) had already demonstrated such outcomes. The data provided by our laboratory will be described in detail below (Wilson et al., 2002, 2003, 2004; Khabazian et al., 2002). First, however, we will turn to the second hypothesis, that of Cox and colleagues, which sought to resurrect the notion that the causal toxin in ALS-PDC was BMAA.

The BMAA–Fruit Bat Hypothesis of Cox and Sacks

A key critique of the original BMAA hypothesis of Spencer et al. (1987) was that BMAA as a weak NMDA receptor agonist was found in too low a concentration in processed cycad flour to cause damage to the nervous system. Cox and Sacks (Cox & Sacks, 2002; Monson et al., 2003) proposed that this problem could be solved by a form of animal biomagnification of BMAA concentration. The basic notion has three parts. First, BMAA in cycad seeds is concentrated in the sarcotesta that is consumed by fruit bats once endemic to Guam. Second, the Chamorro people of Guam hunted and ate fruit bats in abundance, in effect hunting them to near extinction. Third, the BMAA consumed somehow became incorporated into proteins, in time leaching out of the proteins to induce neural damage. Cox and colleagues went on to propose that BMAA in cycad originated in cyanobacteria that are common symbionts in cycad coralloid roots (Banack & Cox, 2003; Cox et al., 2003). Most importantly, these investigators claimed that BMAA could

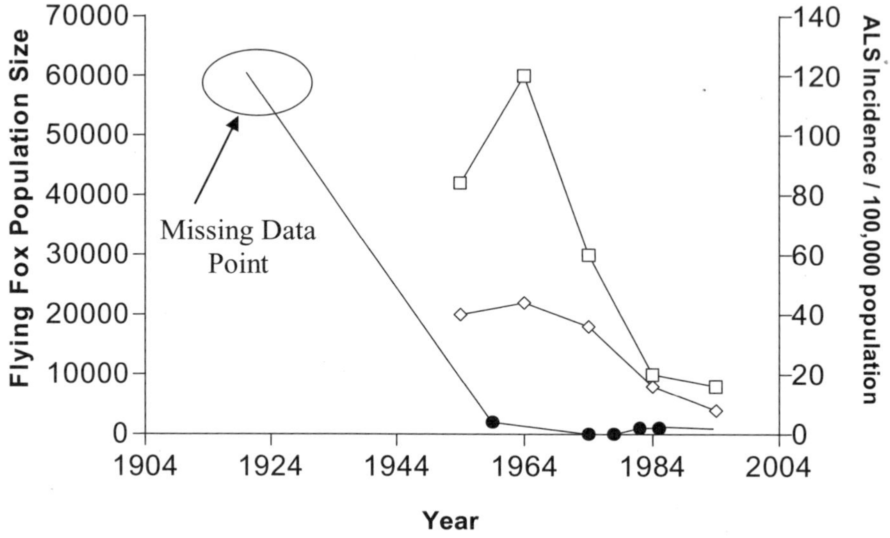

Figure 18-2. Reproduction of figure from Cox and Sacks (2002). Close examination shows that it actually contains nothing that supports the general notion of a linkage between bat population and ALS-PDC levels. Red circle: highlighted area showing the lack of data point. (Recreated based on data from Cox and Sacks, 2002).

be found in high concentrations in "bound" form within the brains of Chamorro victims of ALS-PDC (and further, that similar levels could be found in AD samples from Canada (Cox et al., 2003; Murch et al., 2004).

To evaluate this hypothesis, it is essential to examine the component parts, starting with the initial claim in the Cox and Sacks (2002) paper, which claimed that the rise and fall of ALS-PDC were related to the decline in bat populations on Guam. Scrutiny of this article shows that there is actually only one data figure in that original article (Fig. 18-2).

Cox and Sacks use this figure as the basis for all that follows, but a close examination of the figure shows that it really contains nothing that supports the general notion of a link between bat populations and ALS-PDC levels. The latter data on the graph (plotted after approximately 1950) are from the work of various epidemiologists of ALS-PDC and can be accepted as being generally accurate. The corresponding graph for the bat population plots data points derived from measurements of the bat populations on Guam from the 1950s. This portion of the curve is essentially flat, showing little variation over this time period. However, the authors link this portion of the curve to assumed bat populations in the 1920s, giving the curve a dramatic upward deflection that would appear to mirror the incidence levels of ALS-PDC during the twentieth century. The reader will notice that neither does the leftmost point on the curve contain a valid data point nor is the curve drawn to indicate that this point is an assumed point. Nowhere in the article (Cox & Sacks, 2002) do the authors describe how this curve was drawn. In fact, this was not justified until one year later as well as in a publication by Monson et al. (2003) in which the authors describe their calculations. In brief, the authors take the measured number of bats in the 1960s, calculating a population density for the type of forest that the bats inhabit. They then extrapolate to an assumed level of forest cover in the 1920s to come up with a bat population for this time point. The time point on which the Cox hypothesis rests is thus based on an unsupported assumption, one that at least one wildlife biologist on Guam rejects (A. Brooke, pers. comm.). Without an actual measurement of bat numbers, the correlation of bat population to ALS-PDC cannot be justified, because the curve for the former could have taken any shape. Cox and Sacks have been unable to justify this omission, one that we sought to address in a recent letter directed to Dr. Sacks. As is well known, Dr. Sacks, is a trained neurologist who would be aware of the standards for scientific reporting of data. Dr. Sack's reply was that he was interested in "theory," not the quantitative measures that he left to Cox (O. Sacks, pers. comm.).

The above would alone normally be sufficient to render the bat hypothesis untenable, but Cox and colleagues have so magnified these initial results into numerous manuscripts that some of the other assumptions of their hypothesis must now also be addressed. These are:

- Bats consume cycads as a major part of their diet.
- Cycads concentrate BMAA in the sarcotesta.
- People hunted and ate BMAA-containing bats in high numbers.
- BMAA entered the human brain via consumption of bats and assumed a protein-bound form.
- BMAA leached out of the protein over time to induce neuronal damage, ultimately culminating in ALS-PDC.
- High levels of "bound" BMAA can be found in the brains of victims of ALS-PDC on Guam and victims of AD from Canada.

As noted above, the misleading curve showing bat population levels on Guam in Cox and Sacks (2002) puts the initial hypothesis on shaky ground. We will therefore note only some of the more egregious problems associated with the remaining notions as already cited above:

1. Bats consume cycad as a major part of diet. Evidence to support this assertion is based on a misrepresentation of a wildlife study by Wiles (Wiles, 1987, as cited in Monson et al., 2003). This is the same study that made the calculation of the bat population in 1920 based on numbers measured decades later.
2. Cycads concentrate BMAA in the sarcotesta. These data are based on a study of extremely small sample, much of it not from Guam. The conclusions reached on the basis of this limited sample do not achieve the minimum standards of the field and are thus not supportable.
3. People hunted and ate BMAA-containing bats in high numbers. The notion that bat consumption was a frequent component of the Chamorro diet is not universally accepted. Ulla Craig (Bodig-Lytico Research Project, Agana, Guam) and Douglas Galasko (ALS-PDC Consortium, University of California, San Diego), both neurologists with widespread and long-term experience with ALS-PDC and Chamorro culture, cite first-hand reports from Chamorros refuting the assumptions of Cox and colleagues (U. Craig, D. Galasko, pers. comm.).
4. BMAA enters the human brain in a protein-bound form. The notion that bats contain high levels of "bound" BMAA is derived from measurements taken from a very limited number of museum specimens (Cox et al., 2003). Evidence for BMAA incorporation into protein is taken solely from a study of artificial proteins (Seebach et al., 1994). A previous assertion that BMAA might be bound into animal or human protein was raised by Bell (see, response to the question by Dr. Matsumoto, p. 1475, Polsky et al., 1972) who could not rebut the suggestion that such data reflected an artifact of the preparation. Binding of BMAA to mammalian protein not only requires a currently unknown biochemical process but further raises the question of how such incorporation of an alien protein would fail to induce widespread protein dysfunction. Normal protein function crucially depends on the precise types and sequences of amino acids; hence, it

seems likely that BMAA would have a profound and deleterious impact on all proteins. Cox and colleagues have not addressed this matter (Murch et al., 2004).

5. BMAA leaches out of the protein over time to induce neuronal damage, ultimately culminating in ALS-PDC. Cox and colleagues have focused on a mechanism by which BMAA is biomagnified up the food chain to reach levels able to induce neurological damage in humans. This requirement arises from the fact that BMAA is, at best, a weak NMDA receptor agonist with a LD50 in neuronal culture of approximately 1 mM. Given this, high levels of NMDA would be required to induce neuronal death. The BMAA hypothesis requires a mechanism of action in which a slow leaching process occurs. The outstanding problem with this notion is that BMAA is as weak a neurotoxin as it is. Cox and colleagues require that BMAA be both biomagnified and bound into protein to achieve the levels that might cause neurological damage. They then require that BMAA be released slowly in concentrations that do not induce neurological damage. Clearly, they want to have it both ways, but the illogic of this proposal is obvious to most readers. In addition, a slow-release process requires yet another unknown biochemical process.
6. High levels of "bound" BMAA are found in the brains of victims of ALS-PDC and AD. Beyond the problems cited above, the data cited by Cox and colleagues suffer from many of the same sampling errors that occur in their cycad and bat studies. The BMAA levels cited by these authors have not been replicated by other laboratories (Galasko et al., pers. comm.).

In conclusion, for the reasons cited above, the BMAA–bat hypothesis seems untenable. We will compare this hypothesis to our own after first describing the latter in some detail.

ALS-PDC Outcomes Induced by Cycad Flour Consumption

According to Dastur (1964), rhesus monkeys that consumed of washed cycad flour from Guam developed the behavioral and histopathological features of ALS (see also Dastur et al., 1990). These data remained largely unknown and were not repeated until we began a similar experiment feeding washed cycad flour to various strains of male colony mice.

In our studies, adult male outbred colony mice (CD-1) were fed washed cycad seed pellets as approximately 25% of diet daily for up to 4 months (for full details, see Wilson et al., 2002, 2004). This flour contained at best trace quantities of BMAA, or the other known cycad toxins cycasin or MAM (Khabazian et al., 2002). Control mice received equivalent pellets made of commercial white flour. Both groups received the bulk of their diets from Purina mouse chow. All animals were pretrained on a number of motor, cognitive, and sensory tests, and they were then tested on these measures at regular intervals during and after cycad feeding. At certain points during the proceedings, cycad-fed and control mice were sacrificed, and their brains and spinal cords removed for morphological and/or histological analysis (Wilson et al., 2002). Brains and cords from some mice of each group were analyzed by magnetic resonance microscopy for volume and area of selected central nervous system (CNS) structures (Wilson et al., 2004). Other brains and spinal cords were used for histological and/or biochemical analyses. This "four dimensional" analysis allowed for multiple levels to be examined over time (see Shaw & Wilson, 2003). Some of these data are shown in Figure 18-3. The data reveal profound and significant behavioral changes of motor, cognitive, and olfactory function in cycad-fed mice.

In addition to these data, we have demonstrated both astrocytes proliferation in affected regions of the CNS, along with evidence for microglial proliferation (Fig. 18-4), both key features of ALS-PDC, as well as other neurological diseases.

Cycad-fed mice also showed the presence of tau protein in regions of the CNS, a feature consistent with ALS-PDC (Fig. 18-5A). We have also demonstrated tau expression in neurons exposed to cycad toxins in culture (Fig. 18-5B) (Shaw et al., in preparation). The identification of the putative toxin, a sterol glucoside, is documented elsewhere (Khabazian et al., 2002).

Overall, the data from cycad-fed mice duplicate the full spectrum of features of ALS-PDC and further replicate the earlier work of Dastur (1964). These data allow reasonable confidence that ALS-PDC arises due to the presence of water-insoluble neurotoxins in cycad seed flour. The possibility of synergistic interaction of such toxic molecules with trace amounts of BMAA and/or cycasin/MAM cannot be ruled out. Neither do the effects that we report depend on biomagnification via other vectors to achieve the levels of CNS dysfunction of ALS-PDC nor do they require the addition of the multiple

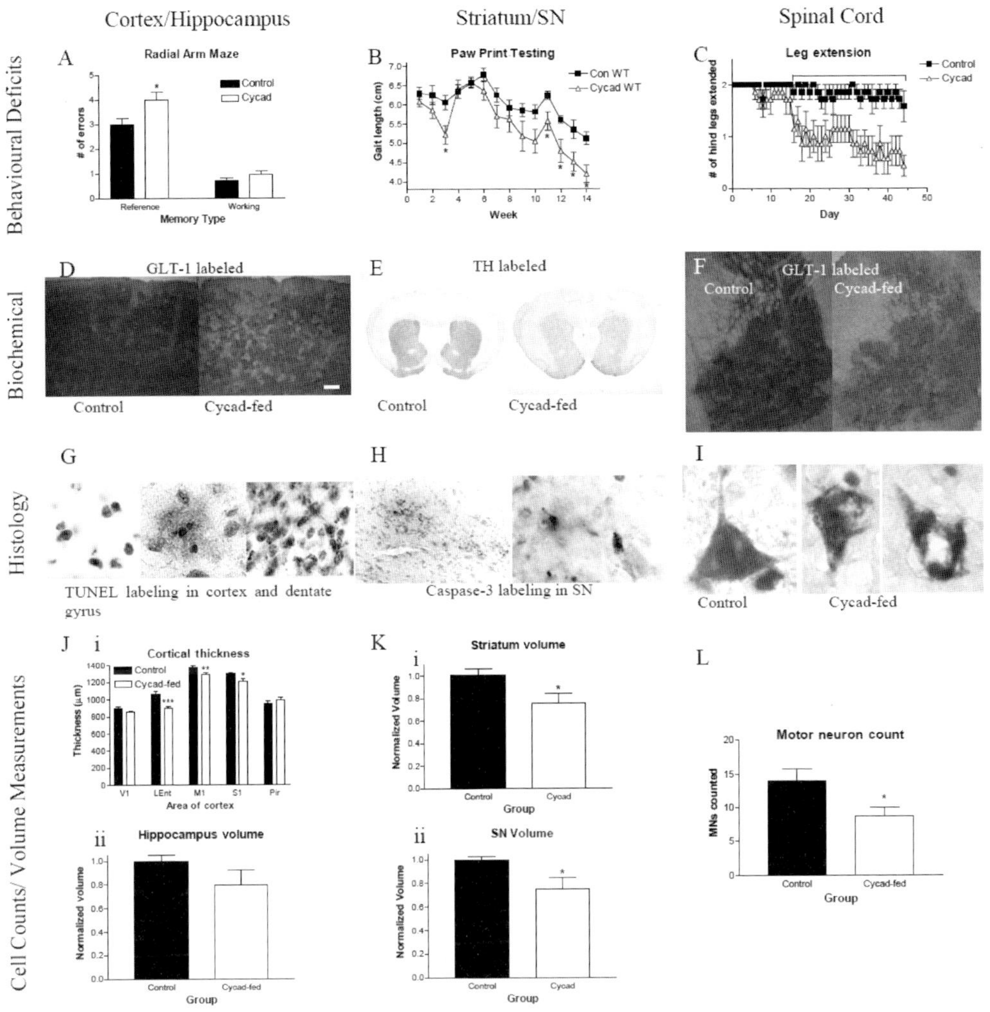

Figure 18-3. ALS-PDC mouse model summary. **A–C.** Behavioral deficits displayed by cycad-fed mice (□) compared to controls (■). **A.** Radial arm maze test errors by cycad-fed and control mice. **B.** Paw print testing for gait length. **C.** Leg extension reflex test. **D–F.** Samples of biochemical changes in CNS tissue from cycad-fed mice. **D.** Immunolabeling of GLT-1 in primary motor cortex, scale bar = 80 μm. **E.** Tyrosine Hydroxalase labeling of brain section. **F.** Immunolabeling of GLT-1 of spinal cord. **G–I.** Histological display of cell pathology in CNS of cycad-fed mice. **J–L.** Cell counts or MRI volume measurements in specific CNS areas of cycad-fed mice compared to controls. Graphs show means ± S.E.M., * $p < .05$.

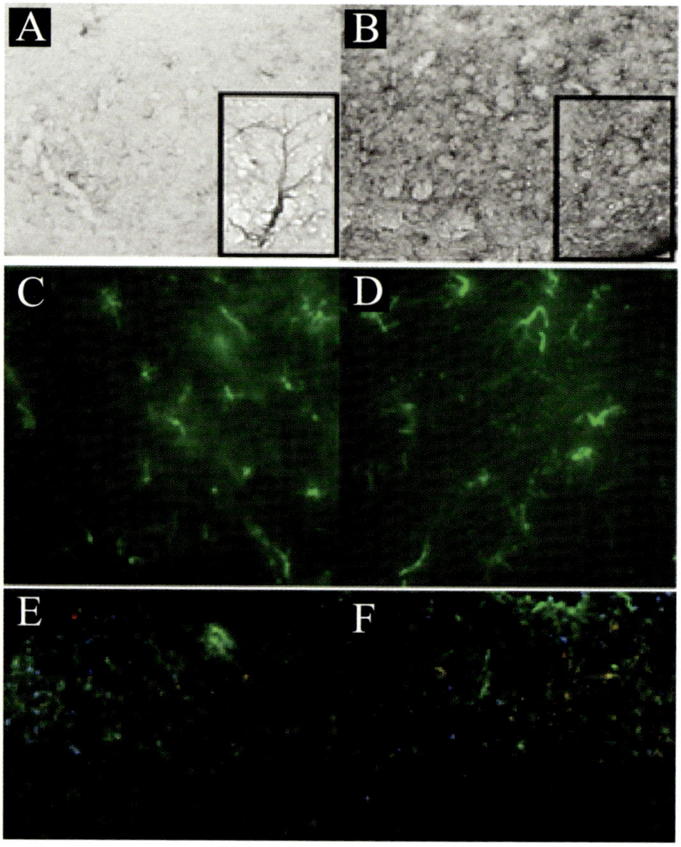

Figure 18-4. GFAP and microglial immunoreactivity in the cycad-fed mice. **A, B.** Brightfield images of reactive astrocytes (GFAP) in substantia nigra. **A.** Control. **B.** Cycad-fed mouse. Inset: high magnification images of individual GFAP-labeled astrocytes. Magnification left panels: 10×; right panels: 20× (inset 100×). All cycad-fed mice showed this effect. **C, D.** Fluorescent labeling of spinal cord. **C.** GFAP-stained ventral horn tissue from lumbar spinal cord of control mouse. GFAP is diffuse, with no clear labeling of astrocytic cell processes. **D.** GFAP staining shows increased labeling of astrocytic cells in ventral horn tissue from lumbar spinal cord of cycad-fed mouse. **E, F.** GFAP staining and MAPK labeling colocalized in lumbar spinal cord of control mouse (20× objective). **E.** Control. **F.** Cycad-fed mouse. Slides show an increased GFAP and MAPK staining in cycad animals. Colocalization of GFAP and MAPK suggests increased inflammation and presence of microglial cells in the spinal cord sections of cycad-fed mice.

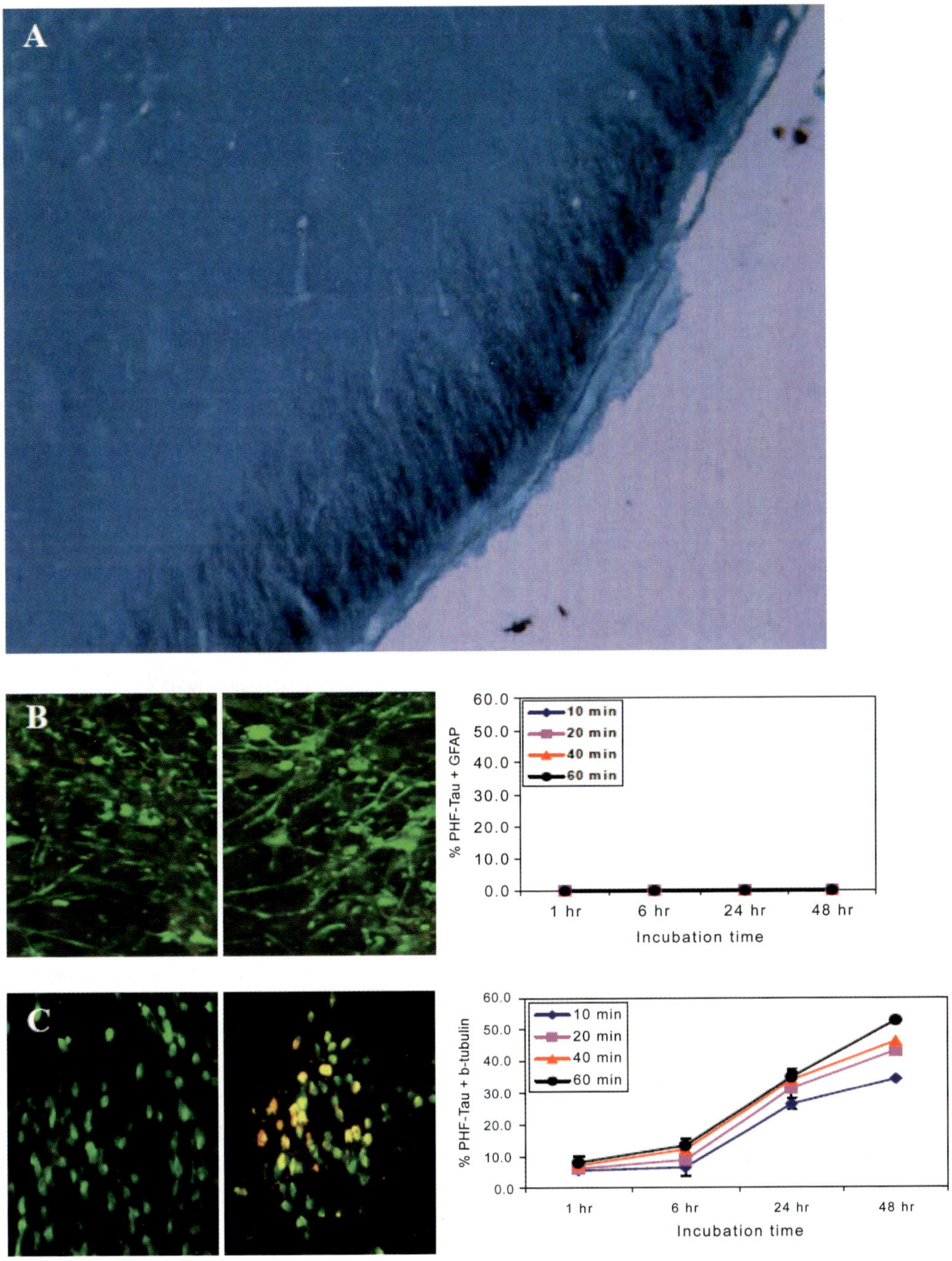

Figure 18-5. Tau labeling of neural cells. **A.** Piriform cortex from cycad-fed mouse showing early stage tangle formation (Gallyas stain). **B.** Astrocytes treated with 50 μM cholesterol β-D-glucoside for 10, 20, 40, or 60 minutes. GFAP (green) labeling. There is no labeling with anti-PHF-Tau (red) antibody at any exposure period measured after 24 hours. The graph to the right shows a complete absence of tau labeling at all time points, as expected. **C.** Neurons labeled with β-tubulin (green); anti-PHF-Tau (red) shows tau labeling in many neurons at 24 hours following a 60-minute exposure to CG. CG-induced tau production in neurons is both time and duration dependent. Each point represents five separate determinations.

factors required by the Cox hypothesis. A consideration of the key components of our model versus that of the Cox group is provided below.

Subjecting the Two Hypotheses to Occam's Razor

Occam's Razor (also called the Law of Parsimony) has been used by generations of scientists to evaluate competing hypotheses. Basically, this law states that if two hypotheses account for a body of data equally well, we are obliged to choose the simplest as the one most likely to be correct. As we have demonstrated above, the Cox and Sacks hypothesis does not account for the data very well but adds multiple levels of complexity in addition. Table 18-1 provides a comparison of the two hypotheses, focusing on some possible points of comparison.

For example, our model is consistent with a simple route of administration of a water-insoluble toxin contained in the seed of the cycad and does not require the addition of either biomagnification or multiple transition steps from BMAA synthesis by *Nostoc* bacteria through cycad to fruit bats to people. Next, our identified sterol glucoside toxins are directly neurotoxic in cell culture at concentrations in the low micromolar range; in contrast, BMAA requires millimolar concentrations to achieve neurotoxic impacts. In addition, most in vivo animal studies have utterly failed to find any impact of BMAA on behavior or neuronal heath (e.g., Spencer et al., 1987, versus Perry et al., 1989, and Schulz et al., 2005b). Simple cycad seed consumption is consistent with the other loci for ALS-PDC, notably Irian Jaya and the Kii Peninsula of Japan. The Cox hypothesis fails in both regards, in the latter being forced to suggest that the *Nostoc* are directly responsible rather than cycad. Thus, to make the BMAA hypothesis work in both locales, bats must be eaten on Guam (after having biomagnified BMAA), but BMAA from bacteria without biomagnification provide the disease in Kii.

Summary and Conclusions

The hypothesis put forward initially by Cox and Sacks (2002) is based on misrepresented information. The further elaborations of the hypothesis have been largely promoted in the popular press without concern for scientific rigor. The primary technique used to promote the hypothesis is that of press release (J. Morton, pers. comm.). That process

Table 18-1 BMAA versus cycad/SG? Comparison of two hypotheses for the cause of ALS-PDC[a].

Question	BMAA	Cycad/SG
1. Does the model account for effect without biomagnification?	No	Yes
2. Is the molecule present in all Chamorros in highest levels in those who have disease?	No: ALS-PDC victims/some AD victims reported to have highest "bound" BMAA levels; controls nil	Yes: CG is present in all samples in a preliminary study of ALS and control patients
3. Can the molecule be stored in tissue?	Maybe, but requires incorporation into protein by unknown mechanism	Yes
4. Can the molecule be found elsewhere in environment?	Yes: reported in *Nostoc*, biomagnification up food chain?	Yes: SG and CG present in other plants in high concentrations, in various bacteria, and CG may be made endogenously in human cells
5. Can the molecule generate other features of ALS-PDC?		
a. Male > female?	a. No?	a. Yes
b. ApoE interaction?	b. No	b. Yes
c. Tau production?	c. No	c. Yes
6. Can BMAA give ALS-PDC–like outcomes in a model system?	No/maybe: Perry et al., 1989 (16 mg/g); Shaw et al., 2003 (0.75 mg/g);. Spencer et al., 1987 (15 mg/g)	Washed cycad; yes
7. Are neural changes progressive?	No	Yes
8. Is neurodegeneration observed at "typical" neurotoxin levels?	No	Yes: low μM
9. Can the molecule produce oxidative stress?	? Not directly	Yes: loss of dopamine, D2r up, DAT down
10. Can the molecule account for delayed onset?	Maybe?	Yes: late-onset ALS phenotype observed

[a]Considers which is simpler and best accounts for the factors of the disease: BMAA or washed cycad with no BMAA.

SG = steryl glucoside; CG = cholesterol glucoside.

continues largely with only cursory scrutiny of the claims or whether such claims make any sense in context to existing scientific knowledge. Each claim becomes more complex without any apparent concern for distinguishing facts from speculation. Although the BMAA hypothesis has begun to attract scientific scrutiny, the continual changes in what is actually being claimed by Cox and colleagues make serious attempts at analysis difficult to achieve. Those who attempted to determine if BMAA could alone induce neurological deficits resembling ALS-PDC have found that the Cox hypothesis no longer required such data because it had moved to encompass a claim that BMAA achieved an unknown form of protein chemistry. Those who attempt to test this new realm will discover that "bound" BMAA in bats or elsewhere no longer matters because the neurotoxin is found universally in *Nostoc* bacteria in drinking and sea water. Given such a moving target, such a hypothesis can never be disproved. This is not to say, however, that it cannot continue to be promoted by this group through the media. It is notably easy to make extravagant claims, but more difficult toback these with rigorous scientific observation. The pattern of the Cox group has been just such a process, and whatever else this process may be it is not scientific. The unsupported claims by this group are also a discredit to many of conscientious scientists who have devoted themselves to finding the actual causes of neurological disease and a further egregious insult to the victims of such diseases.

In spite of the hyperbole demonstrated by this group, we believe that cycads continue to hold the key to ALS-PDC and neurological diseases worldwide. The toxins contained in cycad seeds, steryl glucosides or other toxic compounds, alone or in synergy with other toxins and genetic susceptibility factors, might indeed be the neurological "Rosetta Stone." We have only to decipher the types and actions of these toxins to open a new chapter in our understanding and treatment of these diseases.

Acknowledgments

The authors thank Dr. Ulla Craig for her continued support. Funding for this project is acknowledged from the following sources: U.S. Department of Defense (#DAMD17-02-1-0678), Scottish Rite Charitable Foundation of Canada, and the Natural Science and Engineering Research Council of Canada (CAS). We dedicate this article to the memory of two great scientists who first recognized cycad as the key to the mystery of neurological disease: Dr. Leonard T. Kurland and Dr. Darab K. Dastur.

Literature Cited

Andersen, P. M. 2001. Genetics of sporadic ALS. Amyotroph. Lateral Scler. Other Motor Neuron Disord. 2 Suppl 1: S37–41.

Banack, S. A. & P. A. Cox. 2003. Biomagnification of cycad neurotoxins in flying foxes: implications for ALS-PDC in Guam. Neurology 61(3): 387–379.

Betarbet, R., T. B. Sherer & J. T. Greenamyre. 2005. Ubiquitin-proteasome system and Parkinson's diseases. Exp, Neurol. 191 Suppl 1: S17–27.

Cox, P. A. & O. W. Sacks. 2002. Cycad neurotoxins, consumption of flying foxes, and ALS-PDC disease in Guam. Neurology 58(6): 956–959.

Cox, P. A., S. A. Banack & S. J. Murch. 2003. Biomagnification of cyanobacterial neurotoxins and neurodegenerative disease among the Chamorro people of Guam. Proc. Natl. Acad. Sci. U.S.A. 100(23): 13380–13383.

Dastur, D. K. 1964. Cycad toxicity in monkeys: clinical, pathological, and biochemical aspects. Fed. Proc. 23: 1368–1369.

Dastur, D. K., Palekar, R. S. & Manghani, D. K. 1990. Toxicity of various forms of cycas circinalis in rhesus monkeys: Pathology of brain, spinal cord and liver. Pp. 129–141 *in* F.C. Rose and F.H. Norris. Amyotrophic lateral sclerosis: new advances in toxicology and epidemiology. Smith-Gordon/Nishimura Press, London.

De Luca, P., A. Moretti, S. Sabato & G. Siniscalco Gigliano. 1980. The ubiquity of cycasin in cycads. Phytochemistry 19: 2230–2231.

Duncan, M. W., J. C. Steele, I. J. Kopin & S. P. Markey. 1990. 2-Amino-3-(methylyamino)-propanoic acid (BMAA) in cycad flour: an unlikely cause of amyotrophic lateral sclerosis and parkinonism-dementia of Guam. Neurology 40: 767–772.

Gardian, G. & L. Vecsei. 2004. Huntington's disease: pathomechanism and therapeutic perspectives. J. Neural Transm. 111(10–11): 1485–1494.

Huang, Y., L. Cheung, D. Rowe & G. Halliday. 2004. Genetic contributions to Parkinson's disease. Brain Res. Brain Res. Rev. 46(1): 44–70.

Khabazian, I., J. S. Bains, D. E. Williams, J. Cheung, J. M. Wilson, B. A. Pasqualotto, S. L. Pelech, R. J. Andersen, Y. T. Wang, L. Liu, A. Nagai, S. U. Kim, U. K. Craig & C. A. Shaw. 2002. Isolation of various forms of sterol beta-d-glucoside from the seed of *Cycas circinalis*: neurotoxicity and implications for ALS-parkinsonism dementia complex. J. Neurochem. 82(3): 516–528.

Kowalska, A. 2004. Genetic aspects of amyloid beta-protein fibrillogenesis in Alzheimer's disease. Folia Neuropathol. 42(4): 235–237.

Kurland, L. T. 1988. Amyotrophic lateral sclerosis and Parkinson's disease complex on Guam linked to an environmental neurotoxin. Trends Neurosci. 11(2): 51–54.

Lessell, S., A. Hirano, J. Torres & L. T. Kurland. 1962. Parkinsonism-dementia complex. Epidemiological considerations in the Chamorros of the Mariana Islands and California. Arch. Neurol. 7: 377–385.

Monson, C. S., S. A. Banack & P. A. Cox. 2003. Conservation implications of Chamorro consumption of flying foxes as as possible cause of amyotrophic lateral sclerosis–parkinsonism dementia complex in Guam. Conservation Biol. 17(3): 678–686.

Moretti, A., S. Sabato & G. Siniscalco Gigliano. 1981. Distribution of macrozamin in Australian cycads. Phytochemistry 20: 1415–1416.

———. ——— & ———. 1983. Taxonomic significance of methylazoxymethanol glycosides in the cycads. Phytochemistry 22: 115–117.

Morrison, B. M. & J. H. Morrison. 1999. Amyotrophic lateral sclerosis associated with mutations in superoxide dismutase: a putative mechanism of degeneration. Brain Res. Brain Res. Rev. 29(1): 121–135.

Murch, S. J., P. A. Cox & S. A. Banack. 2004. A mechanism for slow release of biomagnified cyanobacterial neurotoxins and neurodegenerative disease in Guam. Proc. Natl. Acad. Sci. U.S.A. 101(33): 12228–122231.

Perry, T. L., C. Bergeron, A. J. Biro & S. Hansen. 1989. Beta-N-methylamino-L-alanine. Chronic oral administration is not neurotoxic to mice. J. Neurol. Sci. 94(1–3): 173–180.

Plato, C. C., R. M. Garruto, D. Galasko, U. K. Craig, M. Plato, A. Gamst, J. M. Torres & W. Wiederholt. 2003. Amyotrophic lateral sclerosis and parkinsonism-dementia complex of Guam: changing incidence rates during the past 60 years. Am. J. Epidemiol. 157(2): 149–157.

Polsky, F. I., P. B. Nunn & E. A. Bell. 1972. Distribution and toxicity of alpha-amino-beta-methylaminopropionic acid. Fed. Proc. 31 (5): 1473–1475.

Ravindranath, V. 2002. Neurolathyrism: mitochondrial dysfunction in excitotoxicity mediated by L-beta-oxalyl aminoalanine. Neurochem. Int. 40(6): 505–509.

Seawright, A. A., A. W. Brown, C. C. Nolan & J. B. Cavanagh. 1990. Selective degeneration of cerebellar cortical neurons caused by cycad neurotoxin, L-beta-methylaminoalanine (L-BMAA), in rats. Neuropathol. Appl. Neurobiol. 16(2): 153–169.

Shaw, C. A. & J. M. Wilson. 2003. Analysis of neurological disease in four dimensions: insight from ALS-PDC epidemiology and animal models. Neurosci. Biobehav. Rev. 27(6): 493–505.

Schulz, J. D., R. Cruz-Aguado, E. L. Hawkes, S. Singh & C. A. Shaw. 2005a. Comparative study of the roles of novel cycad toxins versus BMAA in the etiology of ALS-PDC. Soc. Neurosci. Abstract.: 34.

———, E. L. Hawkes & C. A. Shaw, 2005b Cycad consumption and neurological disease. Pp. 233–270 *in* V.R. Preedy and R.R. Watson (eds.), Reviews in food and nutrition toxicity. Taylor & Francis, New York.

Seebach, D., A. Studer, E. Pfammatter & H. Widmer, 1994 Synthesis of tri-, penta-, and heptapeptides containing and (R)-2-alkyl-2-amino-3-(methylamino)-propionic acid residue in the central position Helv. Chim. Acta 77: 2035–2050.

Spencer, P. B. Nunn, J. Hugon, A. C. Ludolph, S. M. Ross, D. N. Roy & R. C. Robertson. 1987. Guam amyotrophic lateral sclerosis-parkinsonism-dementia linked to a plant excitant neurotoxin. Science 237 (4814): 517–522.

——— & H. H. Schaumburg. 1983. Lathyrism: a neurotoxic disease. Neurobehav Toxicol Teratol 5(6): 625–629.

Weiss, J. H., J.Y. Koh & D. W. Choi. 1989. Neurotoxicity of beta-N-methylamino-L-alanine (BMAA) and beta-N- oxalylamino-L-alanine (BOAA) on cultured cortical neurons. Brain. Res. 497(1): 64–71.

Whiting, M. G. 1963. Toxicity of cycads. Econ Bot. 17: 271–302.

Wiles, G. J. 1987. Current research and future management of Mariana fruit bats on Guam. Australian Mammalogy 10: 93–95.

Wilson, J. M., I. Khabazian, M. C. Wong, A. Seyedalikhani, J. S. Bains, B. A. Pasqualotto, D. E. Williams, R. J. Andersen, R. J. Simpson, R. Smith, U. K. Craig, L. T. Kurland & C. A. Shaw. 2002. Behavioral and neurological correlates of ALS-parkinsonism dementia complex in adult mice fed washed cycad flour. Neuromolecular Med. 1(3): 207–221.

———, ———, D.V. Pow, U.-K. Craig & C. A. Shaw. 2003. Neuromolecular Med. 3: 105–117.

———, M. S. Petrik, S. C. Grant, S. J. Blackband, J. Lai & C. A. Shaw. 2004. Quantitative measurement of neurodegeneration in an ALS-PDC model using MR microscopy. Neuroimage 23(1): 336–343.

REPRODUCTIVE BIOLOGY
CHAPTER 19

Hot and Smelly Sex! Theoretical Considerations of the Adaptive Significance of Cone Thermogenesis and Volatile Odors in Cycad Reproduction

John S. Donaldson

Abstract

Cone thermogenesis and the production of volatile odors during the pollination period are almost ubiquitous in cycads. The incidence of both events at a time when insect pollinators are active has led to speculation about their functional role in cycad pollination. Most often, heating and odors are assumed to attract pollinators in the same way that they do in some flowering plants. However, all the known cycad pollinators are phytophagous insects that mate, breed, and feed on cycads and have varying levels of host specificity within cycad genera. The function of cone thermogenesis and volatile odors therefore needs to be explained within the context of pollination systems that have evolved from cycad–herbivore interactions in which insects may be preadapted to finding their host plants and spending most of their life cycles associated with the host plant. In this paper, several possible models for the role of cone thermogenesis and volatile compounds in cycad pollination are given, with the expectations associated with each model. The limited available data indicate that these models are credible explanations of cycad pollination systems but are insufficient to refute or support any of the models.

Resumen

La termogénesis del cono y/o la producción de olores volátiles durante el período de la polinización parece estar ampliamente distribuida en las cícadas. La incidencia de ambos eventos en un período cuando los insectos polinizadores son más activos, ha llevado a la especulación sobre su papel funcional en la polinización de las cícadas. Muy a menudo, se asume que el calentamiento y los olores son para atraer los polinizadores y/o mantener la fidelidad del polinizador, de la misma forma como se presenta en la plantas con flores. Sin embargo, actualmente se conoce que todos los insectos polinizadores de la cícadas son fitofagos, con niveles variables de especificad de anfitrión al interior de cada género de cícada. Esto levanta la posibilidad, que la polinización de las cícadas haya evolucionado en el contexto de interacciones especializadas, y que la termogénesis de los conos y olores volátiles tienen una función alternativa. En este trabajo, examino algunas hipótesis alternas de termogénesis de los conos y olores volátiles, particularmente 'la hipótesis de polinizador moroso.'

Introduction

The release of volatile odors and periods of thermogenesis (except in *Stangeria* T. Moore) typically occur in cycad cones during pollen shed (males) and pollen receptivity (females) (Tang, 1987a, 1993; Tang et al., 1987; Pellmyr et al., 1991; Skubatz et al., 1993; Seymour et al., 2004; Terry et al., 2004a,b). Insect pollinators are active on the cones at this time, pointing to the possibility that volatile compounds attract pollinators (Pellmyr et al., 1991; Stevenson et al., 1998) and that thermogenesis either helps volatilize attractant compounds (Tang, 1993) or acts independently on pollinators. Volatile odors in cycads could therefore have a similar role to floral odors in some flowering plants (Pellmyr et al., 1991). Some of the compounds emitted by cycads are the same as those that occur in flowering plants (e.g., linalool), indicating a possible convergence in the function of volatile odors in these lineages (Pellmyr et al., 1991).

Despite the inherent appeal of this theory, the adaptive significance of cone volatiles and thermogenesis in cycad pollination systems has not been rigorously tested. A preliminary look at the available data raises some important questions regarding the relationship between cycads and their insect pollinators.

First, all the experimental tests of cycad pollination systems have identified nursery pollination interactions in which the pollinators are herbivores that feed exclusively on cycad cone tissues (species of Curculionoidea and Cucujoidea) or pollen (*Cycadothrips*) (Norstog et al., 1986; Tang, 1987b; Donaldson et al., 1995; Donaldson, 1997; Hall et al., 2004; Terry et al., 2004a,b, 2005;). This excludes untested observations of other insects on cones, such as *Trigona* bees on *Macrozamia* Miq. and *Cycas* L. cones (P. I. Forster, A. Lindstrom, pers. comm.) and flies on *Stangeria* cones (author, personal observation). Nursery pollination systems appear to have evolved from cycad-herbivore interactions that are host specific, at least within cycad genera (Donaldson, 1992, 1993; Oberprieler, 1995). Insect herbivores often respond to specific cues associated with their host plants, so that cycad pollinators may have been preadapted to finding cycad reproductive structures at the time of pollen shed prior to the evolution of pollinator mutualisms. The specific function of cone thermogenesis and volatile odors needs to be explained within this context.

Second, initial observations of beetle activity on the Australian cycad *Macrozamia machinii* P.I. Forst. and D.L. Jones and thrips on *Macrozamia lucida* L.A.S. Johnson showed that pollinators were consistently leaving the male cone during the peak period of heating and odor production (Mound & Terry, 2001; Terry et al., 2004b; J. Donaldson, pers. observation). This is contrary to the expectation that insects would be attracted to male cones at this time. Other studies of insect responses to volatile compounds show that these chemicals can act as attractants or repellents. Some of the compounds emitted by cycad cones during pollen dehiscence, such as terpenoid vapors and short chain esters (Pellmyr et al., 1991; Terry et al., 2004a,b), are known to be toxic or repellent to insects (Renwick, 1988; Pellmyr et al., 1991), and exposure to monoterpenes can even be fatal to many insects (Renwick, 1988). In contrast, some bark beetles detoxify monoterpenes through oxidization, and the resultant terpene alcohols are key chemicals in aggregation pheromones (Renwick, 1988). Terpene alcohols such as linalool, which is also emitted by cycad cones (Pellmyr et al., 1991; Terry et al., 2004a,b), increase host attraction responses in a range of insects (Pellmyr et al., 1991; Bruce & Cork, 2001). Similarly, cone heating could attract pollinators by volatilizing attractant compounds (Tang, 1993; Terry et al., 2004b) or repel pollinators by increasing metabolic activity and/or volatilizing deterrent compounds. These observations of insect behavior suggest that a more rigorous analysis of cycad pollination systems is required to test the function of volatile compounds and cone thermogenesis.

The aim of this paper is to provide a theoretical framework for assessing the function of cone thermogenesis and volatile compounds in cycad pollination. I describe several possible models for the role of thermogenesis and the release of volatile compounds in cycad pollination, suggest tests of these models, and then examine the limited data that are available to determine what support exists for the different models.

Models of Cycad Pollination

Cycad Pollination Systems

The few detailed studies of cycad pollination systems show that they comprise complex interactions between cycad-specific insects and their host plants and that it is not possible to generalize from a few studies to all cycads. The roles of cone volatiles and thermogenesis may contribute to this complexity and already the generalization that these attributes act as attractants for pollinators is being challenged (Oberprieler et al., 2007; Terry et. al., 2007).

Based on the premise that volatile odors and thermogenesis are either adaptations that actively influence pollinator behavior or that they are the effects of other processes and have no functional role in influencing pollinator behavior, then cycad pollination systems can be classified according to whether their role is active or neutral (e.g., if cycads simply rely on insects that already visit cones as herbivores). Because cycads are dioecious, pollinators have to visit both male and female cones to be effective, so that the classification of active or neutral pollinator interactions must be applied to both male and female cones. In addition, successful pollination requires that pollinators move from male to female cones and, in at least some systems, pollinators may move back and forth between male and female cones as part of a daily cycle. Dispersal from the cone may also be either actively facilitated by changes in the plant (e.g., by release of repellent volatiles) or result from events that are not under the control of the host plant (e.g., changes in ambient light or changes in nutrient availability due to herbivory), meaning that cone volatiles and thermogenesis may also be neutral in relation to pollinator dispersal. The combination of attraction or repellent functions as well as neutral effects of cone attributes means that there are at least 16 possible interactions with pollinators for each set of cone attributes (Table 19-1). The combinations in Table 19-1 make no assumptions about which attributes of the plant influence pollination, so that the combinations apply whether volatile odors and thermogenesis act together or independently of one another.

The 16 possible interactions between cone volatiles, and/or thermogenesis, and pollinator behavior can be divided into four broad groups. The first group includes those interactions in which volatiles and/or thermogenesis act only to attract pollinators to male or female cones (attraction model). The second group includes those interactions in which cone attributes act only to repel or disperse pollinators (repellent model). The third group is a mixed model in which cone volatiles and/or cone heating may attract pollinators in some instances and disperse or repel them in others. Terry et al. (in press) have referred to these as push-pull interactions, and this terminology is used here to be consistent with other studies. The fourth and final group involves a single type of interaction in which cone volatiles and cone heating play no role in either attracting or dispersing pollinators (neutral model). For the purposes of further analysis, pollination systems are grouped simply according to whether volatile odors or thermogenesis actively attract pollinators (attraction models), actively disperse pollinators (repellent models), or do not affect pollinator behavior (neutral models).

Table 19-1 Combinations of possible influences of cone volatiles odors and/or thermogenesis on cycad pollinator behavior[a].

Male cone		Female cone		Interaction type
Attraction	Dispersal	Attraction	Dispersal	
Active	Active	Active	Active	Mixed
Active	Active	Active	Neutral	Mixed
Active	Active	Neutral	Neutral	Mixed
Active	Neutral	Active	Active	Mixed
Active	Active	Neutral	Active	Mixed
Active	Neutral	Neutral	Active	Mixed
Active	Neutral	Active	Neutral	Attraction
Active	Neutral	Neutral	Neutral	Attraction
Neutral	Active	Active	Active	Mixed
Neutral	Active	Active	Neutral	Mixed
Neutral	Active	Neutral	Neutral	Repellent
Neutral	Active	Neutral	Active	Repellent
Neutral	Neutral	Active	Active	Mixed
Neutral	Neutral	Active	Neutral	Attraction
Neutral	Neutral	Neutral	Active	Repellent
Neutral	Neutral	Neutral	Neutral	Neutral

[a]"Active" implies that the attribute has a specific influence on pollinator behavior, whereas "neutral" implies that the attribute does not influence that behavior. Pollinator behavior is divided into attraction to the male or female cone and dispersal from the male or female cone. The interactions are grouped according to whether cone attributes only attract pollinators, only repel pollinators, have a mixed effect, or have no effect (neutral).

Attraction Models

Attraction models assume that cycads have evolved specific mechanisms to attract pollinators at the time of pollen shed and female receptivity. Even if cycad pollinators are preadapted to finding their host plants for the purposes of mating, feeding, and breeding, there are still many reasons why cycads may have evolved mechanisms to attract pollinators.

- In situations in which pollinators are regularly a limiting factor in a population, selection would be expected to favor plants that can attract a greater number of pollinators or increase the effectiveness of pollination such as by attracting specific pollinators. In flowering plants, pollinator limitation has been proposed as one of the selective forces for the evolution of floral traits (Johnson, 1996).

- In cases in which cones in a population could be at different stages of development, attraction may be linked to specific stages of cone development. For example, heating and volatile odors could serve to attract pollinators to those cones in the population that are shedding pollen (males) or receptive for pollination (females).
- Many cycad species cone infrequently, and coning plants may be widely separated in the landscape. Cone heating and volatile odors could provide a "lighthouse effect" to attract pollinators (Tang, 1993) to isolated cones. According to Stevenson et al. (1998), volatiles may act as a cue for *Rhopalotria* pollinators of *Zamia furfuracea* L.*f.* to emerge from diapause and seek out male cones.

Expectations of Attraction Models

Volatile compounds released by cycad cones at the time of pollination and/or the thermogenic activity of cones either attracts insect pollinators that would not be attracted in the absence of these attributes or increases levels of attraction. However, volatiles and cone heating would not be expected to influence dispersal.

Repellent Models—The "Jonah Effect"

Repellent models (Table 19-1) assume that there is an adaptive advantage to plants that actively disperse pollinators. Male cones offer shelter, food, and a larval host, so the expected behavior of the insect herbivore/pollinator would be to locate a suitable cone, mate, and lay eggs. Until the cone deteriorates or is saturated with eggs or larvae, it represents a secure host, and insects may tend to remain mostly within a single cone. As such, they would be ineffective pollinators. Cone heating and volatiles could function to make the cone less habitable for insect herbivores, with cone heating mobilizing insects in the male cone (although according to Terry et al. [2004b], this is unnecessary) and repellent volatile odors (e.g., monoterpenes) could disperse otherwise sedentary insects from the cone. The term used here for the mobilizing of otherwise "reluctant" pollinators is the Jonah effect, after the biblical character who was prodded into action by a series of divine interventions.

In addition to providing an alternative explanation for cone odors and thermogenesis, this theory offers a plausible explanation for one of the enigmas of cycad pollination, that is, why pollinators visit female cones when there is apparently little or no reward for doing so (Stevenson et al., 1998; Terry et al., 2004a). If male cones heat up and emit odors that repel pollinators for a brief period each day, the dispersed insects would be looking for suitable hosts. If female cones are not repellent at the same time, they offer temporary shelter and may therefore attract insect pollinators at a time when male cones are repellent.

Expectations of Repellent Models

Male and/or female cones should be repellent to insect pollinators at some stage during development or some period during the day because of cone heating and/or the release of cone volatiles. However, volatiles and cone heating should not act to attract pollinators.

Mixed Models—"Push-Pull" Effects

The majority of possible interactions in Table 19-1 involves a combination of attraction and dispersal. This implies either that cone volatiles act as attractants in some instances and repellents in others because of changes in chemical composition or concentration, or that cone volatiles and cone heating have separate functions to attract and repel pollinators. Following Terry et al. (in press), these are push-pull interactions based on the terminology used in integrated pest management systems in which a combination of repellent plants and trap crops are used to manage insect pests. These are the most complex interactions in Table 19-1 and will be the most difficult to test. Nevertheless, the diversity of compounds isolated from cone volatiles (Pellmyr, et al. 1991; Terry et al., 2007; Oberprieler et al., 2007), as well as the presence of compounds that have been shown to both attract and repel insect herbivores in other studies, indicates that push-pull models are possible. There is considerable variation in the number of compounds that make up cone volatiles, from only two dominant compounds in *Z. furfuracea* to more than 15 in *Zamia pumila* and *M. lucida* (Pellmyr et. al., 1991; Tang, 1993; Terry et. al., 2005, and it will be important to determine whether cycads with push-pull interactions have more volatile compounds than species that fit attraction or repellent models.

Expectations of Mixed Models

Cone volatiles and/or thermogenesis should attract pollinators to male or female cones at some stage during development or some period during the day, and these features should act to repel pollinators at other times.

Neutral Models

Neutral models assume that there is no attractant or repellent function associated with cone volatiles or thermogenesis and that pollinating insects visit and move between cones for other reasons. Most cycad pollinators are herbivores that feed on male cone tissues (Curculionoidea and Cucujoidea) or pollen (some Curculionoidea and *Cycadothrips*) and may therefore locate cones using the same cues as other nonpollinator herbivores (i.e., the plant has not evolved specific attributes to attract pollinators). Several nonpollinator herbivores routinely attack cycad cones at the time of pollination, including taxa that have no relationship with known pollinators. Examples include species of *Antliarhinus* (Coleoptera: Curculionoidea) as well as sister taxa to known pollinator groups such as species of *Amorphocerus* (Coleoptera: Curculionoidea) (Donaldson, 1993; Oberprieler, 1995). This behavior indicates that a wide range of insects respond to cues associated with cycad cones at the time of pollination and that these cues need not have evolved as attractants for pollinators.

The dispersal of pollinators, particularly from male cones, may also be influenced by factors other than cone volatiles or periods of increased cone temperature. Thermogenesis and herbivory steadily reduce the starch and nutrient content of developing male cones (Tang, 1993) so that the availability of brood sites diminishes as the cone matures. Pollinators would be expected to move away from older cones (as they deteriorate) in search of other more suitable brood sites and mating opportunities (for males). In this case, movement away from the male cone would not be associated directly with daily peaks of heating and/or odor emissions. Nonpollinator insects, such as species of *Antliarhinus,* spend less time on cones that have been previously oviposited into than on cones of the same age that have no history of oviposition (J. Donaldson, unpublished data).

Nonpollinator herbivores cannot necessarily be used as a neutral group to test whether cone attributes are adaptations to attract pollinators until the evolutionary history of these interactions is better known. At present, it is not known whether similarities in behavior between pollinators and nonpollinators arose from relationships that preceded the evolution of pollinator mutualisms or if nonpollinators have adapted to cues that evolved to attract specific groups of pollinating insects. Although cycad–pollinator mutualisms were initially thought to have originated from ancient interactions (e.g., Crowson, 1981; Norstog & Nicholls, 1997), recent analyses suggest that many cycad–insect interactions may be more recent (Leschen, 2003; Oberprieler, 2004); some authors have argued that cycad pollinators are derived from insect herbivores feeding on angiosperms (Oberprieler, 2004). Cycad pollinators may therefore be responding to general cues associated with potential host plants.

Expectations of Neutral Models

Pollinators should be unaffected, in terms of either attraction or dispersal, by peaks in thermogenesis or odor release. As a corollary, movement between cones must be explained by other factors such as levels of herbivory.

Tests of Cycad Pollination Models

The main test for the above models depends on whether cones are attractive or repellent during periods of odor release, thermogenesis, or both. Until recently, most cycad pollination studies focused on establishing whether wind or insects were the principal pollen vectors (Norstog et al., 1986; Tang, 1987b; Forster et al., 1994; Wilson, 1994; Donaldson et al., 1995; Donaldson, 1997; Hall et al., 2004; Terry et al., 2004a, 2005). A few studies have examined the volatile compounds emitted by cycad cones (Pellmyr et al., 1991; Tang, 1993; Terry et al., 2004a, b) as well as the responses of pollinators to cone odors (Tang, 1993; Terry et al., 2004b), but rigorous tests of attraction or dispersal models have only recently begun (Terry et al., 2007; Oberprieler et al., 2007; J. Donaldson, unpublished data). Therefore, it is appropriate to look at which tests are likely to provide the most meaningful insights into the roles of cone volatiles and thermogenesis in cycad pollination.

Test 1. Detailed observations of insect behavior are required to show whether pollinator activity (attraction or dispersal) is correlated with patterns of cone heating or an increase in particular volatile compounds. To date, observations have mostly taken place during peaks in thermogenesis, but activity at this time needs to be compared to activity at other times. This means observing cones over 24 hours from before the male cones dehisce pollen until all the male cones have shed their pollen (i.e., the effective period of pollination is over). Time-lapse photography or continuous video recording are likely to give the best results. However, if this is not possible, live observations need to be timed to include at least two periods during cone development (i.e., before and during pollen dehiscence in males, as well as before and during pollen receptivity in female cones) and over at least three periods during the daily cycle (i.e., several hours before the peak in odor release or thermogenesis, during the peak, and several hours after the peak).

Test 2. A combination of gas chromatography and electroantennal responses is required to identify the range of volatiles emitted by cycad cones as well as the specific compounds to which the pollinators are responding. Such a method has been used successfully by Oberprieler et al. (2007). This test does not determine whether the insect is responding to an attractant or repellent compound.

Test 3. Olfactometer experiments provide one of the best tests for the role of volatile compounds in attracting pollinators. Tests with both male and female cones at different stages of development are required to determine whether the cones are attractive to known pollinators and at what stage they are attractive. An olfactometer can also be used to test the responses of pollinators to specific compounds.

Test 4. Putatively repellent compounds need to be tested by exposing cones with insects present to a source of the specific compound. This could be done by running an airstream with or without the compound over a cone containing pollinators. The behavior of the insects would indicate whether the compound acts as a repellent. In other systems, repellent compounds have also been tested by giving insect herbivores a choice between plants treated with repellent compound and control plants that are left untreated or that are treated only with the relevant solvent. This could only be replicated in cycads with relatively small cones (where the entire cone could be treated) or in cases in which insect pollinators feed on individual sporophylls.

Test 5. If pollinators respond to specific odors in the laboratory, field tests need to be carried out to determine whether the responses are consistent under less artificial conditions. Simple traps using volatile compounds have not been successful (Tang, 1993), but this may be because insect responses are dose-dependent. More sophisticated traps that control the release of compound are likely to be more effective. The literature on such traps related to insect pest management is considerable, and traps are also commercially available. Nevertheless, these tests should take place only before or in conjunction with tests 1 and 2; otherwise, it becomes impossible to determine whether a negative result means that the compound has no effect or that it acts as a repellent.

Test 6. The role of cone thermogenesis in attracting pollinators, either as an independent attribute or in conjunction with cone volatiles, needs to be tested. One way to do this is to compare pollinator attraction between taxa that are thermogenic (most species tested to date) and those that are not (e.g., *Stangeria*). However, more direct tests, in which cone temperature follows its daily cycle, but in which pollinators cannot detect cone odors, would be more meaningful. This would mean having to enclose the cone in a sealed bag.

Existing Data and Tests of Pollination Models

There is limited information on the activity patterns of cycad pollinators, and known behavior is summarized in Table 19-2. The relationships between species of *Rhopalotria* (Curculionoidea, Belidae) and cycads in the genus *Zamia* are the best known cycad–pollinator interactions (Norstog et al., 1986; Norstog & Fawcett, 1989). Adult *Rhopalotria mollis* congregate on the male cones of *Z. furfuracea* before the cone extends and dehisces pollen. The weevils spend most of their life inside the male cycad cones, mating and laying eggs in the starch-rich microsporophylls, and they may only leave the male cone after a few days. The eggs hatch into larvae that vigorously eat the microsporophyll tissue and then pupate in the sporophyll. The entire life cycle is repeated approximately every 10 days during the coning season, and new adults may emerge before the cone has finished dehiscing pollen. Toward the end of the coning season, a small proportion of the larvae develop waxy cuticles and estivate in thick-walled cases within the male cycad cones until the following season, or even some years later.

Although the biology of other pollinators is less well known, there appears to be a remarkable convergence in the biology of different weevil groups associated with cycads. Unrelated species of *Porthetes* weevils (pollinating *Encephalartos villosus* Lem.) and *Tranes* weevils (pollinating *Lepidozamia peroffskyana* Regel and *Macrozamia communis* L.A.S. Johnson) exhibit similar behavior to *R. mollis* and also congregate in large numbers on male cones prior to dehiscence (Donaldson, 1997; Hall et al., 2004; R. Oberprieler, pers.

Table 19-2 Behavior of insect pollinators on male cycad cones in relation to daily patterns of heat and odor production.

Species	Pollinator	Pollinator Behavior	Reference(s)
Encephalartos villosus	*Porthetes pearsonii*	Beetles were attracted to male cones two days before dehiscence.	Donaldson, 1997
		In olfactometer experiments, beetles were attracted to male cones (before and during pollen shed) throughout the day.	Donaldson, unpublished data
Lepidozamia peroffskyana	*Tranes* sp.	Beetles were active on male cones at least 24 h before sporophylls separated and shed pollen	Hall et al., 2004
		Beetles were active on male cones throughout daylight hours	
		Beetles left male cones 24–48 h after the start of pollen shed	
Macrozamia communis	*Tranes* and *Cycadothrips*	Thrips moved from male cones around midday, to coincide with a first peak in thermogenesis	Terry, 2001
		Beetles left the male cones in a massed group after sunset when cones experienced a second peak in thermogenesis	
Macrozamia lucida	*Cycadothrips*	Thrips left male cones in swarms between 1100 h and 1400 h when emissions of volatile compounds were highest	Terry et al., 2004b
Macrozamia macleayii	*Cycadothrips*	Thrips left male cones between 1000 and 1300 h, which coincides with peaks in volatile emissions	Terry et al., 2004b
M. machinii	*Tranes* sp.	Beetles were present in male cones during the day	Terry et al., 2004b
		Beetles dispersed from male cone during peak heating and odor release (1900 h to 2000 h) and were observed on female cones over the same period.	Personal observation
Zamia furfuracea	*Rhopalotria mollis*	Beetles were active on male cones before dehiscence	Norstog & Fawcett, 1989
		Beetles dispersed after a few days or once the next generation emerges from cone	
Zamia pumila	*Rhopalotria slossoni*	Beetles attracted to male cone just before dehiscence	Tang, 1995
		In experiments, beetles were not attracted to cones during volatile release or to cone extracts	

observation). At first this behavior seems to be consistent with the neutral model of cycad pollination in that pollinators congregate on male cones before pollen shed and remain in the male cone for several days, at which stage the sporophylls are being rapidly consumed by developing larvae. However, male cones of several cycad taxa become thermogenic before pollen shed (Tang, 1993; Terry et al., 2004b; J. Donaldson, unpublished data) and may also produce volatiles at this time, suggesting that attraction to male cones is actively influenced by these cone attributes (i.e. supports one of the attraction models). Circumstantial evidence for strong attraction to female cones also comes from attempts to exclude pollinators from female cones during pollination experiments. Even though beetle pollinators do not typically develop in female cones, in almost all cases where exclusion bags have been used, beetles have burrowed through gaps in the bags to reach the cones (Norstog et al., 1986; Tang, 1987a; Donaldson et al., 1995; Donaldson, 1997; Hall et al., 2004).

Data from experimental tests of cone attraction are equivocal. Pollinators of *Z. furfuracea* were not attracted to odor traps in the field or laboratory or to cones in the laboratory, but the lack of response may have been an artifact of experimental conditions (Tang, 1993). Olfactometer experiments with weevil pollinators of *E. villosus* (*Porthetes pearsonii*) showed that weevils were consistently attracted to male cones, indicating that cone volatiles were attractive at least during some stages of cone development (J. Donaldson, unpublished data); similar attraction was noted by Terry et al. (in press) with *Macrozamia* pollinators. Electroantennograms also show that *Tranes* weevils responded to volatiles from *M. communis* male cones, although these tests cannot determine whether the response is positive (attraction) or negative (repellent) (Oberprieler et al., 2007).

The biology of *R. mollis* on *Z. furfuracea* (Norstog & Fawcett, 1989), as well as other weevils (Hall et al., 2004), in which beetles leave the male cone as it deteriorates, also appear to support the neutral model of pollination. However, there is strong circumstantial support for the Jonah effect from studies of *Macrozamia* species in Australia (Terry, 2001; Terry et al., 2004a, b) (Table 19-2). *Tranes* beetles were observed dispersing from male cones of *M. communis* (Terry, 2001) and *M. machinii* (Terry et al., 2004b) during evening peaks in cone temperature and odor production. Similarly, swarms of thrips were observed leaving male cones during midday peaks in temperature and volatile emissions in *M. lucida* L.A.S. Johnson and *Macrozamia macleayi* Miq. (Terry et al., 2004b). A variety of terpenoid compounds have been isolated from male cones in which pollinators disperse during peaks in odor emissions, including high concentrations of monoterpenes (e.g., β-myrcene, cineole, (Z)-β-ocimene), and this is consistent with the expectations of the Jonah effect.

These results do not exclude the possibility that attraction to cones and dispersal form part of push-pull models, but there are currently no data to test these models.

Conclusions

Since 1986, studies of cycad pollination systems have shown the following: that insects are often essential pollen vectors; that complex pollination mutualisms have evolved from cycad–herbivore interactions; and, more recently, that these interactions have evolved separately on different continents and may involve dynamic shifts in pollinator interactions (e.g., from weevils to thrips) and even shifts of pollinators from angiosperms to cycads (Oberprieler, 2004). The role of cone volatiles and thermogenesis in these interactions has yet to be fully explored, and the models presented in this paper are meant to provide a framework for assessing their possible roles. Rigorous experiments and more focused field observations are now required to test these different models.

Acknowledgments

I would like to thank Irene Terry and Rolf Oberprieler for fruitful discussions on pollination systems and an anonymous reviewer for helpful comments on the manuscript.

Literature Cited

Bruce, T. J. & A. Cork. 2001. Electrophysiological and behavioural responses of female *Helicoverpa armigera* to compounds identified in flowers of African marigold *Tagetes erecta*. J. Chem. Ecol. 27: 1119–1131.

Crowson, R. A. 1981. *The Biology of the Coleoptera*. Academic Press, London.

Donaldson, J. S. 1992. Adaptation for oviposition into concealed cycad ovules in the cycad weevils *Antliarhinus zamiae* and *A. signatus* (Coleoptera: Curculioniodea). Biol. J. Linn. Soc. 47: 23–35.

———. 1993. Insect predation on ovules in the South African species of *Encephalartos* (Cycadales, Zamiaceae). Pp. 103–108 *in* D. W. Stevenson and K. J. Norstog (eds.), Proceedings of the second international conference on cycad biology. Palm and Cycad Societies of Australia. Milton, Queensland, Australia.

———. 1997. Is there a floral parasite mutalism in cycad pollination: the pollination biology of *Encephalartos villosus* (Zamiaceae). Amer. J. Bot: 1398–1406.

———, I. Nanni & J. de Wet Bösenberg. 1995. The role of insects in the pollination of Encephalartos cycadifolius. Pp. 423–434 *in* P. Vorster (ed.), Proceedings of the third international conference on cycad biology. Cycad Society of South Africa, Stellenbosch, South Africa.

Forster, P. I., P. J. Machin, L. Mound & G. W. Wilson. 1994. Insects associated with reproductive structures of cycads in Queensland and northeast New South Wales, Australia. Biotropica 26: 217–222.

Hall, J., G. H. Walter, D. M. Bergstrom & P. Machin. 2004. Pollination ecology of the Australian cycad *Lepidozamia peroffskyana*. Austral. J. Bot. 52: 333–343.

Leschen, R. A. 2003. Erotylidae (Insecta: Coleoptera: Cucujoidea): Phylogeny and Review. Manaaki Whenua Press, Canterbury, New Zealand.

Mound, L. A. & I. Terry. 2001. Thrips pollination of the central Australian cycad, *Macrozamia macdonnellii* (Cycadales). Int. J. Pl. Sci. 162: 147–154.

Johnson, S. D. 1996. Pollination, adaption and speciation models in the Cape flora of South Africa. Taxon 45: 59–66.

Norstog, K. J. & P. K. S. Fawcett. 1989. Insect-cycad symbiosis and its relation to the pollination of *Zamia furfuracea* (Zamiaceae) by *Rhopalotria mollis* (Curculionidae). Amer. J. Bot. 76: 1380–1394.

——— & T. J. Nicholls. 1997. *The Biology of the Cycads*. Cornell University Press, Ithaca, NY.

———, D. W. Stevenson & K. J. Niklas. 1986. The role of beetles in the pollination of *Zamia furfuracea* L.*f.* Biotropica 18: 300–306.

Oberprieler, R. G. 1995. The weevils (Coleoptera: Curculionoidea) associated with cycads. 1. Classification, relationships and biology. Pp. 295–334 *in* P. Vorster (ed.), Proceedings of the third international conference on cycad biology. Cycad Society of South Africa, Stellenbosch, South Africa.

———. 2004. "Evil weevils"—the key to cycad survival and diversification? Pp. 170–194 *in* A. Lindstrom (ed.), Proceedings of the sixth international cycad conference on cycad biology. Nong Nooch Tropical Botanical Garden, Thailand.

Pellmyr, O., W. Tang, I. Groth, G. Bergstrom & L. B. Thien. 1991. Cycad cone and angiosperm floral volatiles: inferences for the evolution of insect pollination. Biochem. Syst. Ecol. 19: 623–627.

Renwick, J.A.A. 1988. Comparative mechanisms of host selection by insects attacking pine trees and crucifers. *In* K.C. Spencer (ed.), *Chemical Mediation of Co-evolution*. Academic Press, San Diego, CA.

Seymour, R. S., I. Terry & R. B. Roemer. 2004. Respiration and thermogenesis by cones of the Australian cycad *Macrozamia machinii*. Funct. Ecol. 18(6): 925–930.

Skubatz, H., W. Tang & B.J.D. Meeuse. 1993. Oscillatory heat production in the male cones of cycads. J. Exp. Bot. 44: 489–492.

Stevenson, D. W., K. J. Norstog & K. S. Fawcett. 1998. Pollination biology of cycads. Pp. 277–294 *in* S. J. Owens and P. J. Rudall (eds.), Reproductive biology. Royal Botanic Gardens, Kew, England.

Tang, W. 1987a. Heat production in cycad cones. Bot. Gaz. 148: 165–174.

———. 1987b. Insect pollination in the cycad *Zamia pumila* (Zamiaceae). Amer. J. Bot. 74: 90–99.

———. 1993. Heat and odor production in cycad cones and their role in insect pollination. Pp. 140–147 *in* D. W. Stevenson and K. J. Norstog (eds.), Proceedings of the second international conference on cycad biology. Palm and Cycad Societies of Australia. Milton, Queensland, Australia.

———, L. Sternberg & D. Price. 1987. Thermogenesis in male cones of five cycad species. Amer. J. Bot. 74: 1555–1559.

Terry, I. 2001. Thrips and weevils as dual, specialist pollinators of the Australian cycad *Macrozamia communis* (Zamiaceae). Int. J. Pl. Sci. 162: 1293–1305.

———, C. J. Moore, P. I. Forster, G. H. Walter, P. J. Machin & J. S. Donaldson. 2004a. Pollination ecology of the genus *Macrozamia*: cone volatiles and pollinator specificity. Pp. 155–169 *in* A. Lindström (ed.), Proceedings of the sixth international conference on cycad biology. Nong Nooch Tropical Garden, Thailand.

———, ———, G. H. Walter, P. I. Forster, R. B. Roemer, J. S. Donaldson & P. J. Machin. 2004b. Association of cone thermogenesis and volatiles with pollinator specificity in *Macrozamia* cycads. Pl. Syst. Evol. 243: 233–247.

———, G. Walter, J. S. Donaldson, E. Snow, P. I. Forster & P. J. Machin. 2005. Pollination of Australian *Macrozamia* cycads – effectiveness and behavior of specialist vectors in a dependent mutualism. Amer. J. Bot. 92: 931–940.

———, ———, C. Hull & C. Moore. 2007. Response of pollinating thrips and weevils to specific *Macrozamia* cycad cone volatiles. Mem. New York Bot. Gard. 97: 346–371.

Wilson, G. W. 1994. Initial observations of the reproductive behaviour and the insect pollination agent of Bowenia serrulata (W. Bull) Chamberlain. Encephalartos 36: 13–18.

REPRODUCTIVE BIOLOGY
CHAPTER 20

Evolution of Dioecy and Sex Chromosomes in Cycads

Root Gorelick and Roy Osborne

Abstract

Sex determination in cycads—and most other plants and animals—is a function of cytosine methylation and/or chromatin formation downregulating genes that control female or male function. Sex change probably occasionally occurs in cycads because there is no genetic assimilation of these epigenetic signals. In plants, allopolyploidy is the only escape from such canalized dioecy, but allopolyploidy never occurs in cycads. Cycads never developed a pair of unequal length sex chromosomes because their haploid stages are large, complex, and long-lived.

Resumen

La determinación del sexo en las cícadas—y la mayoría de otras plantas y animales—es una función de la metilación de la citosina y/o la formación de cromatina que inhibe a los genes que controlan la función masculina o femenina. El cambio de sexo en cícadas probablemente ocurre ocasionalmente por ausencia de la asimilación genética de estas señales epigenéticas. En las plantas, la alopoliploidía es el único escape de tal dioicismo canalizado, pero la alopoliploidía no ocurre en las cícadas. Las cícadas nunca han desarrollado un par de cromosomas de sexo in iguales de largo dado que sus etapas haploides son muy largas, complejas y longevas.

Introduction

Living cycads are considered strictly dioecious (i.e., have separate female and male reproductive parts on separate individuals) but contain no discernible sex chromosomes (Norstog & Nicholls, 1997). In this paper, we propose an evolutionary theory that describes how we believe this situation arose in cycads.

Charles Darwin hypothesized that ancestral lineages were hermaphroditic (each individual produces both functional female and male gametes) and that dioecy evolved as a derived condition (Darwin, 1873; Stauffer, 1975). In plants and animals, evolution from hermaphroditism to strict dioecy almost certainly occurred via an intermediate stage that involved individuals who were both hermaphroditic and either functional females or males (Darwin, 1873; Charlesworth & Charlesworth, 1978; Charlesworth & Guttman, 1999; Gorelick, 2003a). These intermediate stages are referred to as gynodioecy (functional females) or androdioecy (functional males).

Cycads are an ancient lineage. They have existed in roughly modern form for 275 to 300 million years (Zhifeng & Thomas, 1989). Cycads were never diverse (Harris, 1961, 1964). Dioecy appears to be quite ancient among cycads; no bisexual cycad cones or individuals have ever been found in the fossil record. The only instances of anything other than strict dioecy that we have ever seen in cycads are those examples of sex change discussed in the accompanying paper (Osborne & Gorelick, 2007). Thus, we must explain how cycads evolved strict dioecy, maintained it for hundreds of millions of years, yet managed to maintain enough phenotypic plasticity to permit sex change under extreme environmental conditions.

Gene Regulation and Sexual Function

In all living organisms, genes are regulated by epigenetic signals that bind to regulatory portions of the gene (Riggs et al., 1996). The most ancestral such epigenetic signal is cytosine methylation (Poole et al., 2001). If sufficient methylation occurs on the regulatory portion of a gene, then that gene is turned off (Finnegan et al., 1993; Tate & Bird, 1993). This is an essential process, because most genes are not expressed in any given cell. Thus, even though all cells in an individual plant contain identical DNA, only certain genes are turned on in leaf cells, whereas other genes may be turned on in root cells, and so on. Furthermore, which genes are turned on and off depends on the age of the tissue.

In almost all living organisms (there are a few highly derived exceptions among animals and fungi, but not plants), cytosine methylation controls gene regulation. Cytosine methylation and other chemicals (e.g., heterochromatic proteins and histones)—that attach to DNA nucleotides are types of epigenetic signals. Epigenetic simply means "on top of" genetic. These epigenetic signals act largely by blocking those sites on the chromosomes to which the enzymes that mediate transcription of messenger RNA would ordinarily bind (see Gorelick 2003a, 2003b, for review of these mechanisms).

Cytosine methylation regulates production of female and male sex hormones, and these sex hormones are believed to be the primary determiner of sex in plants (Jost & Saluz, 1993; Grant, 1999). Furthermore, sex hormones probably regulate the production of male and female gametes in plants (Charlesworth & Charlesworth, 1978; Gorelick & Osborne, 2002; Gorelick, 2003a). Female reproductive structures are those on which the genes for female function have been upregulated (unmethylated), whereas the genes for male function have been downregulated (methylated) (Janousek et al., 1996; Iannello et al., 2000). This is true in both hermaphroditic and dioecious lineages.

Evolution of Dioecy

Consider a hermaphroditic lineage with no distinct sex chromosomes. It is then possible to obtain the first female or male in that lineage by simply downregulating its gene for maleness or femaleness, respectively. This downregulation was almost certainly triggered by extra methylation near the sex-determining gene.

Often only one gene has to be downregulated to impede either female or male function. That is, sex of an individual is controlled by a single locus for each sex. This certainly appears to be true in most mammals, with their *SRY* gene for maleness (Graves, 2002). Likewise, turtles all have their sex determined by egg temperature, which turns their *Dmrt1* gene on and off (Murdock & Wibbels, 2003). It is, however, not known whether sex is determined by one or multiple genes in cycads.

The above hypothesis begs the question of how or why extra methylation originally arose on a gene for female or male function. We propose that this was somewhat serendipitous. Excess cytosine methylation is invariably incorporated onto and nearby those portions of the genome that are new (Yoder et al., 1997; Gorelick, 2003b). This is a mechanism by which cells protect themselves against genomic parasites, such as insertions of viral or bacterial DNA. If, say, a transposon (a piece of parasitic DNA) is inserted near or within the gene for maleness in a hermaphroditic individual, then that individual would likely become the first female of the lineage (Gorelick, 2003a).

Methylation patterns, including those that determine the sex of an individual, are highly heritable. Thus, once dioecy, androdioecy, or gynodioecy is established in a lineage, that sexual system is retained until other evolutionary forces intercede.

At this juncture, the lineage has evolved to a condition of gynodioecy or androdioecy. Other selective pressures then take over in driving the lineage to strict dioecy, especially in those lineages in which sex is determined by two or more genes (Charlesworth & Charlesworth, 1978; Dorken & Barrett, 2004). Strict dioecy could arise via either methylation of genes for femaleness in male individuals or methylation of genes for maleness in female individuals, and with, in either case, suppression of recombination (crossing over) between the genes for femaleness and maleness. Suppressed recombination arises as a by-product of increased cytosine methylation (Griffin et al., 2002; Gorelick, 2003b; Liu et al., 2004), making this process much more likely.

Once plants evolve strict dioecy via the epigenetic mechanism of cytosine methylation suppressing male function in females and suppressing female function in males, this condition could become more permanent over evolutionary time through genetic assimilation. Genetic assimilation means that epigenetic control of phenotypes is transferred to genetic control; that is, there is a switch from coding by cytosine methylation to coding by the nucleotides themselves (Waddington, 1957; Pál & Miklós, 1999). This may have occurred in some cycads but probably not in those lineages for which environmental shocks can cause sex change (Osborne & Gorelick, 2007). Cytosine methylation provides a convincing mechanism for generating and maintaining phenotypic plasticity (Gorelick, 2004), which has apparently not been fully lost in cycad sex determination. Because sex change has been documented in each of the extant families of cycads and in six of the ten extant genera, genetic assimilation probably did not occur in cycads (unless, in the very unlikely situation, the genetic assimilation occurred along with substantial nascent genotype-by-environment interactions).

In general, there is one ready escape from dioecy that evolved via cytosine methylation, namely allopolyploidy. Allopolyploids are formed via the hybridization of individuals from two distinct lineages (e.g., two different species). The allopolyploid offspring contain all of the chromosomes from both of their parents; hence, the offspring could simultaneously contain active (upregulated) genes for both femaleness and maleness. However, polyploidy has never been documented or even suspected in any cycads, extant or fossil. This may help explain why cycads have retained their strictly dioecious condition for so many millions of years, even without genetic assimilation arising in cycads.

Lack of Unequal Length Sex Chromosomes

In many animals (but only in a handful of plants), sex chromosomes of different lengths can be identified. For example, in humans, the Y chromosome is much shorter than the X chromosome. Evolution of sex chromosomes is usually explained by a population genetic model known as Muller's ratchet (Nei, 1970; Griffin et al., 2002; Gorelick, 2003a). Although the mathematics is complicated, the principle underlying Muller's ratchet is simple. Consider a population in which there are some individuals with no mutations. Follow that population over evolutionary time. If the mutation rate is large compared with the recombination rate, then eventually the population will reach a point at which every individual has at least one mutation. In fact, the minimal number of mutations among all individuals ratchets up over time. As the Y chromosome accumulates more mutations, there is strong selective pressure to expunge those portions of the Y chromosome, which thereby becomes shorter. This result depends crucially on there being very little recombination, such as occurs on Y chromosomes, especially if the Y chromosome is highly methylated. Instead, if recombination rates are high, then number of mutations in the population can decrease over time, because some gametes contain fewer mutations than expected due to fortuitous crossing over (some gametes also contain more mutations).

One of the prerequisites for Muller's ratchet for sexual organisms is that the haploid stage of the life cycle be largely immune from selection. This is the case in most animals, which generally have small and short-lived gametes. Plants, on the other hand, have large multicelled and long-lived haploid stages (gametophytes), with cycads being extreme in this regard. Cycad gametophytes are enormous, occupying much of the volume of what becomes a seed following fertilization (Norstog, 1987). Female cycad gametophytes are also long-lived; sometimes more than a year passes between pollination and fertilization (Norstog & Nicholls, 1997). Male gametophytes are also large and complex compared with other plants or animals (Norstog, 1993). Although there are no data concerning whether cycad haploid stages express most of the genes expressed by their diploid stages (as there are for some flowering plants; Willing & Mascarenhas, 1984; Willing et al., 1988; Drews & Yadegari, 2002), the structural complexity, size, and age of cycad gametophytes indicate that they should be largely immune from the effects of Muller's ratchet. Therefore, the chance of cycads (and most other plants) developing unequal length sex chromosomes is remote (Gorelick, 2003a, 2005).

Possession of different length sex chromosomes means that it should not be possible to change the sex of an individual, except possibly via large chromosomal rearrangements (Gorelick, 2003a). Because we know that sex change does occur in cycads, this implies that cycads must have sex chromosomes of the same length.

We expect cycad sex chromosomes to be only distinguishable by differential methylation and associated chromatin formation at sex-determining loci. Heterochromatic proteins typically bind to cytosine methylation, and this can be seen under a light microscope as a small chromosomal band that is different on female versus male plants (Gorelick, 2003a).

Summary

Occasional sex change in cycads provides us with a window into the evolution of their sex-determining mechanism and sex chromosomes. Cycads appear to have retained an ancestral form of dioecy, with the sex of an individual being determined by cytosine methylation downregulating genes responsible for production of gametes or sex hormones. Cycads have thereby retained the phenotypic plasticity to change sex via removing methylation in the face of large environmental perturbations. It is not obvious how many other plants have retained this plasticity in sex determination. It does not appear that any genetic assimilation of the epigenetic mechanism of sex determination has occurred in cycads. However, such canalization of dioecy may have been unnecessary because cycads cannot revert to a hermaphroditic condition via allopolyploidy. Finally, it appears that cycads (and most other plants) are immune from Muller's ratchet because they have haploid stages that express most of the genes expressed in their diploid stages. Microarray studies could be used to test this assertion at the molecular level. But until then, we base this inference of immunity from Muller's ratchet on the large size, complexity, and longevity of cycad gametophytes.

Literature Cited

Charlesworth, B. & D. Charlesworth. 1978. Model for evolution of dioecy and gynodioecy. Amer. Naturalist 112: 975–997.

Charlesworth, D. & D. S. Guttman. 1999. The evolution of dioecy and plant sex chromosome systems. Pp. 25–49 *in* C.C. Ainsworth (ed.), Sex determination in plants. Bios Scientific Publishers, Oxford, UK.

Darwin, C. R. 1873. On the males and complemental males of certain cirripedes, and rudimentary structures. Nature 8: 431–432.

Dorken, M. E. & S. C. H. Barrett. 2004. Sex determination and the evolution of dioecy from monoecy in *Sagittaria latifolia* (Alismataceae). Proc. Roy. Soc. London Ser. B 271: 213–219.

Drews, G. N. & R. Yadegari. 2002. Development and function of the angiosperm female gametophyte. Ann. Rev. Genet. 36: 99–124.

Finnegan, E. J., R. I. S. Bretell & E. S. Dennis. 1993. The role of DNA methylation in the regulation of plant gene expression. Pp. 218–261 *in* J.P. Jost and H.P. Saluz (eds.), DNA methylation: molecular biology and biological significance. Birkhäuser Verlag, Basel, Switzerland.

Gorelick, R. 2003a. Evolution of dioecy and sex chromosomes via methylation driving Muller's ratchet. Biol. J. Linn. Soc. 80: 353–368.

———. 2003b. Transposable elements suppress recombination in all meiotic eukaryotes, including automictic ancient asexuals: a reply to Schön and Martens. J. Nat. Hist. 37: 903–909.

———. 2004. Neo-Lamarckian medicine. Med. Hypotheses 62: 299–303.

———. 2005. Theory for why dioecious plants have equal length sex chromosomes. Amer. J. Bot. 92: 979–984.

——— & R. Osborne. 2002. Inducing sex change and organogenesis from tissue culture in the endangered African cycad *Encephalartos woodii* (Cycadales, Zamiaceae). S. African J. Sci. 98: 114–117.

Grant, S. R. 1999. Genetics of gender dimorphism in higher plants. Pp. 246–274 *in* M.A. Geber, T.E. Dawson, and L.F. Delph (eds.), Gender and sexual dimorphism in flowering plants. Springer-Verlag, Berlin, Germany.

Graves, J. A. M. 2002. The rise and fall of *SRY*. Trends Genet. 18: 259–264.

Griffin, D. K., S. C. Harvey, R. Campos-Ramos, L. J. Ayling, N. R. Bromage, J. S. Masabanda & D. J. Penman. 2002. Early origins of the X and Y chromosomes: lessons from tilapia. Cytogenet. Genome Res. 99: 157–163.

Harris, T. M. 1961. The fossil cycads. Palaeontology 4: 313–323.

———. 1964. The Yorkshire Jurassic Flora. II. Caytoniales, Cycadales and Pteridosperms. British Museum (Natural History), London, UK.

Iannello, R. C., J. A. Gould, J. C. Young, A. Giudice, R. Medcalf & I. Kola. 2000. Methylation-dependent silencing of the testis-specific *Pdha-2* basal promoter occurs through selective targeting of an activating transcription factor/cAMP-responsive element-binding site. J. Biol. Chem. 275: 19603–19608.

Janousek, B., J. Siroky & B. Vyskot. 1996. Epigenetic control of sexual phenotype in a dioecious plant, *Melandium album*. Mol. Gen. Genet. 250: 483–490.

Jost, J. P. & H. P. Saluz. 1993. Steroid hormone dependent changes in DNA methylation and its significance for the activation or silencing of specific genes. Pp. 425–451 *in* J.P. Jost and H.P. Saluz (eds.), DNA methylation: molecular biology and biological significance. Birkhäuser Verlag, Basel, Switzerland.

Liu, Z. Y., P. H. Moore, H. Ma., C. M. Ackerman, M. Ragiba, Q. Y. Yu, H. M. Pearl, M. S. Kim, J. W. Charlton, J. I. Stiles., F. T. Zee, A. H. Paterson & R. Ming. 2004. A primitive Y chromosome in papaya marks incipient sex chromosome evolution. Nature 427: 348–352.

Murdock, C. & T. Wibbels. 2003. Expression of Dmrt1 in a turtle with temperature-dependent sex determination. Cytogenet. Genome Res. 101: 302–308.

Nei, M. 1970. Accumulation of nonfunctional genes on sheltered chromosomes. Amer. Naturalist 104: 311–322.

Norstog, K. J. 1987. Cycads and the origin of insect pollination. Am. Scientist 75: 270–279.

———. 1993. Spermatogenesis in *Microcycas*: evolutionary significance of male gametes of seed plants. Pp. 270–278 *in* D.W. Stevenson and K.J. Norstog (eds.), The biology, structure, and systematics of the Cycadales. Palm and Cycad Societies of Australia Ltd, Milton, Queensland, Australia.

——— & T. J. Nicholls. 1997. The biology of the cycads. Cornell University Press, Ithaca, NY.

Osborne, R. & R. Gorelick. 2007. Sex change in cycads—Cases, causes, and chemistry. Mem. New York Bot. Gard. 97: 335–345.

Pál, C. & I. Miklós. 1999. Epigenetic inheritance, genetic assimilation and speciation. J. Theor. Biol. 200: 19–37.

Poole, A., D. Penny & B.-M. Sjöberg. (2001). Confounded cytosine! Tinkering and the evolution of DNA. Nature Rev. Mol. Cell Biol. 2: 147–151.

Riggs, A. D., R. Martienssen & V. E. A. Russo. 1996. Introduction. Pp. 1–4 *in*: V.E.A. Russo, R. Martienssen, and A.D. Riggs (eds.), Epigenetic mechanisms in gene regulation. Cold Spring Harbor Laboratory Press, Cold Spring Harbor, NY.

Stauffer, R. C. (ed.). 1975. Charles Darwin's natural selection: being the second part of his big book written from 1856 to 1858. Cambridge University Press, Cambridge, UK.

Tate, P. H. & A. P. Bird. 1993. Effects of DNA on DNA-binding proteins and gene expression. Curr. Opin. Genet. Dev. 3: 226–231.

Waddington, C. H. 1957. The strategy of the genes: a discussion of some aspects of theoretical biology. George Allen and Unwin, London.

Willing, R. P., D. Bashe & J. P. Mascarenhas. 1988. An analysis of the quantity and diversity of messenger RNAs from pollen and shoots of *Zea mays*. Theor. Appl. Genet. 75: 751–753.

——— & J. P. Mascarenhas. 1984. Analysis of the complexity and diversity of mRNAs from pollen and shoots of *Tradescantia*. Pl. Physiol. 75: 865–868.

Yoder, J. A., C. P. Walsh & T. H. Bestor. 1997. Cytosine methylation and the ecology of intragenomic parasites. Trends Genet. 13: 335–340.

Zhifeng, G. & B. A. Thomas. 1989. A review of fossil cycad megasporophylls, with new evidence of *Crossozamia* Pomel and its associated leaves from the Lower Permian of Taiyuan, China. Rev. Palaeobot. Palynol. 60: 205–223.

CHAPTER 21

REPRODUCTIVE BIOLOGY

Sex Change in Cycads—Cases, Causes, and Chemistry

Roy Osborne and Root Gorelick

Abstract

Thirty-three cases of sex reversals in six cycad genera and seventeen species are summarized. Unidirectional male-to-female and female-to-male changes have been observed and are typically associated with gross physical damage or severe environmental changes. We propose that sex expression in dioecious plants is governed by the presence or absence of methylation of DNA cytosine residues on genes that regulate production of gametes and/or sex hormones.

Resumen

Se resumen treinta y tres casos de reversión sexual en seis géneros de cícadas y 17 especies. Se han observado cambios unidireccionales masculino-femenino y *vice versa* que son típicamente asociados con daños físicos fuertes o cambios severos de ambiente. Proponemos que la expresión del sexo en plantas dioicas está gobernado por la presencia o ausencia de la metilación de residuos de ADN citosina sobre los genes que regulan la producción de gametos y/o hormonas sexuales.

Introduction

Sexual behavior in plants and animals is probably the most discussed, researched, and misunderstood of all biological processes. The eighteenth century concept of sex being genetically determined in organisms at the time of fertilization, and immutable thereafter, still finds support among some scientists. But sex changes are now known to be so widespread throughout the plant and animal kingdom that biologists accept that many organisms inherently have both male and female genetic propensity and that environmental influences can result in sex changes. In this paper, we discuss all known cases of sex reversals in cycads and attempt to find underlying causes for these changes. Furthermore, we propose a mechanism by which the sex expression in dioecious plants is controlled.

Known Cases of Sex Change in Cycads

Chronological Record of Documented Cases of Sex Change in Cycads

The earliest reference to sex change in cycads is that mentioned by Schuster (1932), who tells of a *Cycas revoluta* plant, which was longitudinally bisected (one wonders why), with both halves surviving but one being male and the other female. Charles Chamberlain (1935), although adamant on the absolute dioeciousness of cycads, records two incidents of sex reversal: (1) a cultivated plant of *C. revoluta* in Australia that changed from female to male and (2) a bud from a female specimen of *Cycas circinalis* (more probably *Cycas rumphii*) from a Chicago conservatory that later produced a male cone. Menniger (1967), in his book *Fantastic Trees*, also mentions two incidents of cycad sex reversal: a female specimen of *C. circinalis* (again, more probably *C. rumphii*) that changed to a male after being physically damaged and a male of the same species that produced a female cone after severe frost exposure.

Van Wyk and Claassen (1981) give a detailed account of a particular sex reversal that relates to one of several specimens of *Encephalartos umbeluziensis* growing in Dr. Claassen's garden in Pretoria, South Africa. The specimen in question produced a male cone in 1970 but a female cone in 1979 and thereafter. This plant was in a more exposed area, and the authors speculated that a freak cold weather spell in 1972 may have initiated the sex change.

Encephalartos, the journal of the Cycad Society of South Africa, has publicized numerous incidents of sex change in garden-grown cycads. A *C. revoluta* changed from male to female after being transplanted (Kemp, 1985). A female *Zamia* plant reportedly produced a side branch bearing a male cone (Osborne, 1985). A specimen of *Encephalartos latifrons* bore two successive crops of female cones with viable seeds after being transplanted in 1970 but then produced male cones for two years after a severe drought in 1983 in the eastern Cape (Osborne, 1988). A seed-grown *Stangeria eriopus* plant that had produced female cones for several years produced a side crown with a male cone (Anonymous, 1989). An *Encephalartos villosus* plant subjected to particularly dry conditions changed from female to male (Osborne, 1990). A female plant of the same species, which produced a cone and viable seeds in 1994, yielded an aborted cone in 1997 after being transplanted but then produced only male cones after 1998.

Osborne (1990) gave an account of a female specimen of *Encephalartos senticosus* that regularly produced cones and viable seeds. This plant produced nine basal suckers that were removed and rooted, and three of these suckers were subsequently monitored; one produced female cones, one produced male cones for three successive years, and the third did not produce cones at all. A female specimen of *Encephalartos middelburgensis*, which gave regular seeds crops in a garden in Durban, South Africa, produced six male cones one year (Fig. 21-1) (Osborne, 1991). A female plant of *C. revoluta* in a garden in Genoa, Italy, produced a male cone one year (Osborne, 1994). Four of 19 male *Zamia vazquezii* specimens at the Lowveld National Botanic Gardens in South Africa turned to females after being transplanted to a shadier site (Osborne, 1994). Two suckers removed from a known male plant of *E. senticosus* developed into female cone-producing plants (Steenkamp, 1995). A 40-year-old

Figure 21-1. Clear evidence of a sex change: this garden specimen of *Encephalartos middelburgensis* produced a female cone and seed crop in 1990—as demonstrated by the seed remnants trapped in the stem apex—but then produced six male cones in 1991.

female *C. revoluta* in Japan produced a lateral branch bearing a male cone (Tomiyama, 1995). Norstog and Nicolls (1997) tell of a female *Zamia fisheri* that began producing male cones after being moved and repotted.

A particularly significant case of sex change occurred in a planting of *Cycas taitungensis* at the Brisbane International Airport, Australia, and was reported in *The Cycad Newsletter* (Osborne, 1998). Twelve mature male plants were purchased from a local nursery in July 1997; the nursery was selectively selling known male plants and keeping known females for seed production. The plants were placed in a site adjacent to the international arrivals hall in low-light (200–400 lux) environments, resulting in etiolated foliage and a general decline in appearance. Within six months, 11 of the 12 plants had produced cones—3 had produced female cones. In 1998, all plants were retransplanted to nearby site in bright sunlight (>1000 lux). The plants soon recovered, and there has been no evidence of any "double" sex change.

Recent reports of cycad sex changes in *Palms and Cycads* (Osborne & Gorelick, 2002) included those of a male *Ceratozamia matudae* moved from a glasshouse to a garden site in Los Angeles, U.S.A, in 1998, which coned as a male in 1999 but as a female in 2000 and 2001. A known male specimen of *E. senticosus*, shipped from South Africa to Los Angeles, produced leaf flushes for seven years and then a female cone. A pot-bound male *Zamia fairchildiana*, rescued from a particularly dry glasshouse in which it had not produced cones in 10 to 15 years, produced a male cone after being repotted and a female cone the following year. A large specimen of *Cycas thouarsii* in a garden in the Comores Islands produced cones as a male for eight to nine years but then changed to a female after it was physical damaged.

Previously Unrecorded Cases of Sex Change in Cycads

Three new reports of cycad sex change have been brought to our attention. The first relates to a male specimen of *Lepidozamia peroffskyana*, planted in a low-light environment at the Mirage Casino in Las Vegas, U.S.A, which produced a male cone initially but then a female cone closely thereafter (D. Retzlaff, pers. comm.). The second is the history of a sucker removed from a well-established male specimen of *Encephalartos munchii* in a Brisbane, Australia, garden. This was removed from the parent in 1996, rooted, transplanted, but nearly destroyed by a weevil infestation in 2002. After the top section of the plant was rerooted, the apex produced three somewhat atrophied female cones in 2004 (S. Walkley, pers. comm.). The third and final case is that of a male *Zamia pygmaea* that produced a female cone after being moved and repotted (D. Stevenson, pers. comm.).

False Cases of Sex Change in Cycads

There are several caveats to alleged cases of sex changes in cycads, and we have been careful to be cognizant of these. In our analysis, we have not included any purely anecdotal accounts that are unsupported by written or photographic evidence. Some cases of cycads "bearing both male and female cones" have proved to be two plants growing in very close proximity (Osborne & Gorelick, 2002). We are also are mindful of the phenomenon of two embryos developing from a single seed and giving what appears to be one plant with both male and female characteristics. All such dubious cases have been excluded from our analysis.

Discussion

The cases of sex reversal detailed in the preceding paragraphs are summarized in Table 21-1. Several significant points emerge from these data.

1. The recorded cases span 6 of the 11 cycad genera and 16 species.
2. All recorded cases are from plants growing in the artificial (garden) environment; there are no known cases of sex change from plants in the wild.

Table 21-1 Known cases of sex change in cycads.

Cycad species	Change	Circumstances	Reference
Ceratozamia matudae	M to F	Moved from glasshouse to garden	Osborne & Gorelick, 2002
Cycas circinalis (*C. rumphii* ?)	F to M	Bud from female	Chamberlain, 1935
C. circinalis (*C. rumphii* ?)	F to M	Mechanical damage	Menninger, 1967
C. circinalis (*C. rumphii* ?)	M to F	Severe cold	Menninger, 1967
Cycas revoluta	Unknown	Mechanical damage	Schuster, 1932
C. revoluta	F to M	Unknown	Chamberlain, 1935
C. revoluta	M to F	Transplanted	Kemp, 1985
C. revoluta	F to M	Unknown	Osborne, 1994
C. revoluta	F to M	Lateral branch	Tomiyama, 1995
Cycas taitungensis (3 plants)	M to F	Transplanted to low light environment	Osborne, 1998
Cycas thouarsii	M to F	Mechanical damage	Osborne & Gorelick, 2002
Encephalartos latifrons	F to M	Severe drought	Osborne, 1988
Encephalartos middelburgensis	F to M	Unknown	Osborne, 1991
Encephalartos munchii	M to F	Sucker from male	S.J. Walkley, pers. comm.
Encephalartos senticosus	F to M	Sucker from female	Osborne, 1990
E. senticosus (2 plants)	M to F	Suckers from male	Steenkamp, 1995
E. senticosus	M to F	Transplanted	Osborne & Gorelick, 2002
Encephalartos umbeluziensis	M to F	Severe cold	Van Wyk & Claassen, 1981
Encephalartos villosus	F to M	Severe drought	Osborne, 1990
E. villosus	F to M	Transplanting	Osborne, 1990
Lepidozamia peroffskyana	M to F	Transplanted to low light environment	D. Retzlaff, pers. comm.
Stangeria eriopus	F to M	Lateral branch	Anonymous, 1989
Zamia fairchildiana	M to F	Repotted and watering increased	Osborne & Gorelick, 2002
Zamia fischeri	F to M	Repotted and environment	Norstog & Nicholls, 1997
Zamia pygmaea	M to F	Repotted and environment changed	D. W. Stevenson, pers.comm.
Zamia vazquezii (4 plants)	M to F	Transplanted to a shadier site	Osborne, 1994
Zamia sp.	F to M	Side crown	Osborne, 1985

3. All recorded cases are unidirectional (i.e., there are no known incidents in which any cycad has changed sex and then undergone a reversal of the primary change).
4. Of the cases recorded, 19 are male-to-female and 13 are female-to-male; changes can therefore occur in either direction.
5. Male-to-female changes are associated with exposure to low-light environments in eight cases and exposure to severe cold in two cases.
6. Female-to-male changes are associated with severe drought in two cases.

A Mechanism for Sex Change in Cycads

We have presented the theoretical aspects of the biochemical process (i.e., the cellular chemistry of sex changes) in the *South African Journal of Science* (Gorelick & Osborne 2002). A simplified summary of our theory is given in the following two paragraphs.

Sexual development in plants and animals is governed by a suite of genes that act both synergistically and antagonistically to dictate the "maleness" or "femaleness" of an organism. There are separate (albeit related) genes that code for primary and secondary sexual characteristics. The DNA code is contained in a sequence of the four nucleotide bases (cytosine, guanine, adenine, thymine) that comprise the axis of each DNA strand. In the classic Watson–Crick model, two strands of DNA are spirally interwoven in a right-handed helix. This can be best envisioned as a rope ladder that is twisted to form a double helix. For a gene to encode a product (typically proteins), the double helix must have a steady, gentle right-handed twist.

Of the four nucleotides, cytosine is susceptible to a process known as methylation, in which the hydrogen atom (H) at the C-5 position is replaced by a methyl group (CH_3) (adenine can also be methylated, but not in nuclear genomes). Cytosine methylation permanently alters the nature of the DNA double helix and all "new" copies of the molecule will carry the same change (Holliday, 1988; Vyskot et al., 1995). Methyl groups are bulky and water repellant. This changes the steady, gentle right-handed twist of the DNA in two ways. First, imagine a rope ladder, with some securely attached some large beach balls, representing the methylation. The bulkiness of the beach balls grossly alters any twisting of the double helix. Second, keep in mind that cells are largely made of water, so that water-repellant molecules may be forced into different conformations from other molecules. There is even evidence that methylation processes can convert a right-handed DNA helix to a left-handed one (Behe & Felsenfeld, 1981; McKay & Steitz, 1981). Heterochromatic proteins also attach to cytosine methylation, which thereby occupy binding sites that otherwise could be occupied by those enzymes that mediate transcription of messenger RNA, which ultimately codes for production of gene products. Methylation of cytosine, especially in promoter regions of genes, effectively shuts down the gene (Fig. 21-2). Thus, cytosine methylation can cause quantitative and qualitative changes to the synthesis of various gene products, including sex hormones. More simply put, cytosine methylation can cause sex changes in dioecious plants.

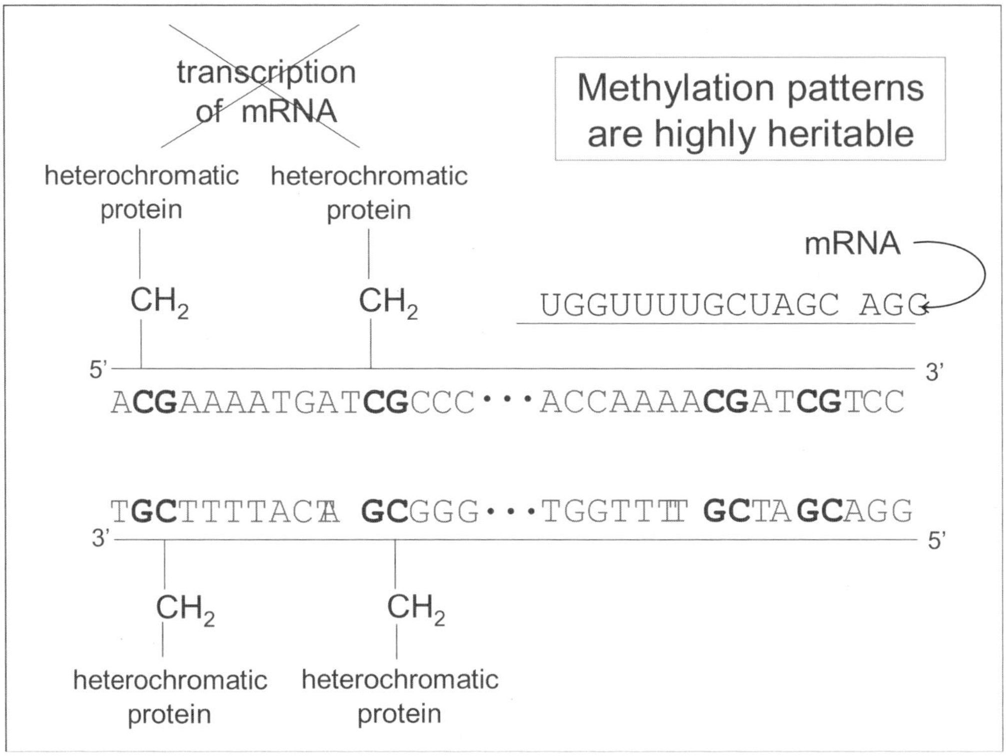

Figure 21-2. Methylation of cytosine shuts down genes. Methylated cytosine changes the confirmation of the double helix (not shown above), shutting down genes. Note that only cytosines in CG dinucleotides (or in CXG trinucleotides, where X can be A, C, G, or T) are usually methylated so as to maintain the methylation patterns following mitosis. Heterochromatic proteins bind to methylated cytosine, which thereby occupy binding sites that would otherwise be occupied by enzymes that mediate transcription. Gene activity is mediated by transcription of mRNA, which is then transported to the ribosomes where the mRNA signal gets converted (translated) into proteins. If, for example, the methylated gene was originally responsible for production of female gametes or sex hormones in a hermaphroditic lineage, then this methylated individual becomes a male.

Cytosine methylation can be removed by factors such as temperature (Dorazi et al., 1995; Demeulemeester et al., 1999), light (Tatra et al., 2000), and osmotic stress (Kovarich et al., 1997). The fact that variation in these environmental factors is associated with sex changes in cycads (as demonstrated in Table 21-1) is strongly supportive of the theory we have developed. Unidirectional sex change is consistent with this theory insofar because it is much simpler to remove methylation than to add it in a site-specific manner. We believe that most examples of sex change in cycads are probably due to removal of cytosine methylation as a consequence of environmental shock (Gorelick, 2003). Work is slowly progressing, although not with cycads, directly showing that methylation and other associated chromosomal changes turn genes for femaleness and maleness in plants on and off (Siljak-Yakovlev et al., 1996; Ma et al., 2004).

We believe that tissue culture techniques will allow researchers to manipulate methylation and demethylation artificially. We have already suggested that this avenue will ultimately permit induced sex changes in tissue from cycad species in which only one sex survives. In short, we have presented a potential recipe for the creation of a female *Encephalartos woodii*. By applying chemicals that effectively remove methylation, such as 5-azacytidine and ethionine, we believe it will be possible to change tissues from males of *E. woodii* and similarly endangered species into females or hermaphrodites.

Acknowledgments

We are indebted to the many members of the international cycad community who have taken an interest in this project and who have advised us of potential sex change cases to investigate.

Literature Cited

Anonymous. 1989. *Stangeria* sex change. Encephalartos 19: 27.

Behe, M. & G. Felsenfeld. 1981. Effects of methylation on a synthetic polynucleotide: the B-Z transition in poly(dG-m3dC)-poly(dG-m3dC). Proc. Natl. Acad. Sci. U.S.A. 78: 1619–1623.

Chamberlain, C. J. 1935. Gymnosperms: structure and evolution. (1966 Dover edition, facsimile of the 1935 edition.) University of Chicago Press, Chicago, IL.

Demeulemeester, M.A.C., N. Van Stallen & M. P. De Proft. 1999. Degree of DNA methylation in chicory (*Chicorium intybus* L.): influence of plant age and vernalization. Pl. Sci. 142: 101–108.

Dorazi, R., A. Chesnel & C. Dournan. 1995. Opposite sex determination of gonads in two *Pleurodeles* species may be due to temperature-dependent inactivation of sex chromosomes. J. Heredity 86: 28–31.

Gorelick, R. 2003. Evolution of dioecy and sex chromosomes via methylation driving Muller's ratchet. Biol. J. Linn. Soc. 80: 353–368.

——— & R. Osborne. 2002. Inducing sex change and organogenesis from tissue culture in the endangered cycad *Encephalartos woodii* (Cycadales, Zamiaceae). S. African J. Sci. 98: 114–117.

Holliday, R. 1988. A possible role for meiotic recombination in germ line reprogramming and maintenance. Pp. 45–55 *in* R.E. Michod and B.R. Levin, B.R. (eds.), The evolution of sex: an examination of current ideas. Sinauer Associates, Sunderland, MA.

Kemp, H. J. 1985. Sex change in *Cycas revoluta*. Encephalartos 2: 24.

Kovarich, A., B. Koukalová, M. Bezdek & Z. Opatrny. 1997. Hypermethylation of tobacco heterochromatic loci in response to osmotic stress. Theoret. Applied Genetics 95: 301–306.

Ma, H., P. H. Moore, Z. Liu, M. S. Kim, Q. Yu, M. M. M. Fitch, T. Sekioka, A. H. Paterson & R. Ming. 2004. High-density linkage mapping revealed suppression of recombination at the sex determination locus in papaya. Genetics 166: 419–436.

McKay, D. B. & T.A. Steitz. 1981. Structure of catabolite gene activator at 2.9Å resolution suggests binding to left-handed B-DNA. Nature 290: 744–749.

Menninger, E. A. 1967. Fantastic trees. Viking Press; 1975 reprint. Horticultural Books Inc., Stuart, FL.

Norstog, K. J. & T. J. Nicholls. 1997. The biology of the cycads. Cornell University Press, Ithaca, NY.

Osborne, R. 1985. Sex change in cycads – hope for *woodii*? Encephalartos 2: 20–22.

———. 1988. Another sex change. Encephalartos 13: 42.

———. 1990. Two new reports on cycad sex changes. Encephalartos 23: 18–20.

———. 1991. Evidence of a sex change in *Encephalartos middelburgensis*. Encephalartos 27: 33.

———. 1994. Further reports of cycad sex changes. Encephalartos 40: 20–21.

———. 1998. An update on cycad sex changes. The Cycad Newsletter 21(2): 3–5.

——— & R. Gorelick. 2002. Sex change in cycads. Palms and Cycads 76: 10–15.

Schuster, J. 1932. Cycadaceae. Pp. 1–168 *in* A. Engler & L. Diels (eds.), Das Pflanzenreich, Vol. IV, Part 1. Engelmann, Leipzig, Germany.

Siljak-Yakovlev, S., S. Benmalek, M. Cerbah, T. Coba de la Peña, N. Bounaga, S. C. Brown & A. Sarr. 1996. Chromosomal sex determination and heterochromatin structure in date palm. Sexual Pl. Reproduction 9: 127–132.

Steenkamp, K. 1995. Another sex change. Encephalartos 42: 25.

Tatra, G. S., J. Miranda, C. C. Chinnappa & D. M. Reid. 2000. Effect of light quality and 5-azacytidine on genomic methylation and stem elongation in two ecotypes of *Stellaria longipes*. Physiol. Pl. 109: 313–321.

Tomiyama, H. 1995. "Double sex" in *Cycas revoluta*. Encephalartos 44: 40–41.

Van Wyk, A. E. & M. I. Claassen. 1981. Sex reversal in *Encephalartos umbeluziensis*. Veld & Flora, December 1981: 120–122.

Vyskot, B., B. Koukalová, A. Kovarik, L. Sachambul, D. Reynolds & M. Bezdek. 1995. Meiotic transmission of a hypomethylated repetitive DNA family in tobacco. Theoret. Applied Genetics 91: 659–664.

REPRODUCTIVE BIOLOGY
CHAPTER 22

Responses of Pollinating Thrips and Weevils to Specific Macrozamia Cycad Cone Volatiles

Irene Terry, Gimme H. Walter, Craig Hull, and Chris Moore

Abstract

The primary cone volatiles of the cycads, *Macrozamia machinii*, pollinated solely by the weevils *Tranes* sp., and *Macrozamia lucida*, pollinated solely by the thrips *Cycadothrips chadwicki*, were studied for their effects on their species-specific pollinators. Two monoterpenes, linalool and β-myrcene, are the major components in the cone fragrance of *M. machinii*. Cones of *M. lucida* release numerous monoterpenes, including β-myrcene, the dominant component, and lower quantities of the ocimene isomers, (Z)-β-, (E)-β-, and (allo-)ocimene. No linalool has been detected in *M. lucida,* and no (E)-β-or (allo-) ocimene has been detected in *M. machinii*. Results of electroantennographic and gas chromatography–electroantennographic detection techniques demonstrated that *Tranes* sp. and *Cycadothrips* responded physiologically to components within the suite of volatiles generated by their host cones. In particular, *Cycadothrips* from *M. lucida* responded to β-myrcene as well as to (E)-β- and (allo-)ocimene but not to (Z)-β-ocimene. *Tranes* sp. from *M. machinii* responded to both linalool and β-myrcene. In Y-tube olfactometer behavioral tests, both insects responded positively to the volatiles of their own host (male cone sporophylls), but their response levels to nonhost sporophylls were not significantly different from their response levels to the control arm, which contained only filtered room air. The response of *Cycadothrips* to β-myrcene was dose dependent. Higher levels appeared to repel male and female thrips, whereas lower levels of β-myrcene slightly attracted male thrips but had no significant effect on female thrips (no repellency or attraction). *Cycadothrips* were attracted to a blend comprising three ocimene isomers, even in the presence of low levels of β-myrcene. The only level of β-myrcene tested on *Tranes* indicated a slightly repellent effect. These results are discussed in the context of cone thermogenesis studies and the nature of the *Macrozamia* pollination syndrome.

Resumen

Los principales volátiles de las cícadas *Macrozamia machinii*, polinizadas únicamente por Curculiónidos de *Tranes* sp. y *M. lucida* polinizada únicamente por el Thisanóptero *Cycadothrips chadwicki* fueron estudiados por sus efectos sobre los polinizadores específicos. Dos monoterpenos linalool y β-mycrene son los más importantes en la fragrancia de *M. machinii*. Los conos de *M. lucida* sueltan numerosos monoterpenos incluyendo β-myrcene, el componente dominante, y cantidades más bajas de los isómeros de ocimene, (Z)-β-, (E)-β-, y (allo-)ocimene. No se detectó linalool en *M. lucida* ní tampoco (E)-β-or (allo-) ocimene para *M. machinii*. Los resultados electroantenográficos (EAG) y técnicas de detección electroantenográfico por cromatografía de gases (GC-EAD) han mostrado que Curculiónidos de *Tranes* sp. y *Cycadothrips* responden fisiológicamente a los componentes dentro del rango de los volátiles de su cono hospedero. Particularmente, *Cycadothrips* de *M. lucida* respondió a β-myrcene y (E)-β- y (allo-)ocimene pero no a (Z)-β-ocimene. *Tranes* sp. de *M. machinii* respondió a ambos linalool y β-myrcene. Durante pruebas de olfactómetro de tubo-Y para comportamiento, ambos insectos respondieron positivamente a los volátiles de su propio hospedero (esporófilos del cono masculino) pero su respuestas a esporófilos de no-hospederos no mostraron diferencias significativas en los niveles de respuestas al brazo testigo, lo cual solamente contenía aire filtrado. La respuesta de *Cycadothrips* a β-myrcene fue dependiente al nivel de dosis. Niveles altos aparentemente repelaron a los machos y hembras de los Thisanópteros, mientras niveles más bajos de β-myrcene atrajo a los machos pero sin efecto significativo sobre las hembras (ni rechazo ni atracción). *Cycadothrips* estuvieron atraídos a una mezcla de tres isómeros de ocimene, aun en la presencia de niveles bajos de β-myrcene. El único nivel de β-myrcene probado con *Tranes* indicó un efecto ligero de rechazo. Los resultados fueron discutidos en el contexto de estudios de termogénesis de conos y la naturaleza del síndrome de polinización de *Macrozamia*.

Introduction

The pollination systems of some cycad species involve obligate mutualisms. Some cycads depend on only one species as a specialist insect pollinator (Norstog et al., 1986; Tang, 1987a, 1993; Oberprieler, 1995, 2004; Stevenson et al., 1998; Mound & Terry, 2001; Wilson, 2002; Hall et al., 2004; Terry et al., 2005), which provides the only visitors to cones of both sexes. In contrast, others may have two or more insect species that visit both male and female cones, such as some species in the cycad genera *Macrozamia* and *Encephalartos* (Connell & Ladd, 1993; Donaldson, 1997; Mound et al., 1998; Terry, 2001). Little is known about the specific interactive mechanisms that sustain these specialized obligate mutualisms. In particular, the host traits that attract the insects remain poorly known, despite their involvement in dictating which plants are pollinated and thus in setting the limits to cycad species. Such information about sympatric cycad species that have different pollinators is likely to be particularly informative about the diversification of cycads and their obligate pollinators. In addition, the observation that closely related species differ with respect to the number of pollinators raises questions about differences in the array of specific traits that attract only one insect pollinator and those that attract more than one.

The cycad genus *Macrozamia* Johnson is endemic to Australia and has about 40 species (Jones, 2002; Forster, 2004). More than 30 species are found in the eastern part of the country, several species in the far southwestern region, and one in the Macdonnell and Hart Ranges of central Australia. The insects that have been demonstrated to pollinate species of *Macrozamia* include members of the thrips genus *Cycadothrips* (family Aeolothripidae) and the weevil genus *Tranes* (family Curculionidae) (Terry, 2001; Mound & Terry, 2001; Terry et al., 2005). Some *Macrozamia* have only *Cycadothrips*, others only *Tranes*, and at least four other species have both insect types as pollinators (Mound, 1991; Forster et al., 1994; Mound et al., 1998; Terry, 2001; Mound & Terry, 2001; Terry et al., 2005). These pollinators live in an obligate mutualism with their host species; they feed, mate, lay eggs, and complete their larval development on their cycad host cones.

Some *Macrozamia* species occur sympatrically, with one species hosting only *Cycadothrips* and the other only *Tranes*. That no hybrids occur between them indicates that specific cone traits associated with each cycad species may be involved in the attraction of pollinators of one type and not the other. Cones of the *Cycadothrips*-associated species are thermogenic during the daytime along with increased volatile emissions and pollinator movement between cones. By contrast, *Tranes*-pollinated species are thermogenic during dusk and early evening, and at this time volatile output increases and insect movement takes place between cones (Terry et al., 2004a, 2004b). Conspecific male and female cone odors are similar in both the timing of emissions and in their major components, but female odor emissions and thermogenic metabolisms are much lower by comparison (Seymour et al., 2004; Terry et al., 2004b; Roemer et al., 2005).

Movement of thrips (*Cycadothrips chadwicki)* out of the cones of *Macrozamia lucida* L.A.S. Johnson seems to correlate more closely with cone metabolism than with cone temperatures, and peak cone temperatures lag behind estimated peak cone metabolic activity by an average of 46 minutes (Roemer et al., 2005). The major chemical component of the cone fragrance of several species pollinated only by *Cycadothrips* is β-myrcene, exceeding 90% of the mixture in many samples. The absolute amount of this component from some cones increases over many orders of magnitude between 30-minute samples taken during nonthermogenic periods (morning or early evening) and those taken mid-day during thermogenesis (Terry et al., 2004b). Other major identified components include (Z)- and (E)-β-ocimenes; ρ-mentha-3,6,-diene; β-myrcene; α-terpinene; and limonene (Terry et al., 2004b).

The major odor component (>80% of emissions) associated with the cones of *Macrozamia machinii* P. I. Forst. and D. L. Jones, a species pollinated only by *Tranes,* is linalool, with β-myrcene in much lower concentrations (<9% of the total volatile mix), as well as minor to trace amounts of cineole, limonene, and (Z)-β-ocimene (Terry et al., 2004b). Linalool concentrations increased approximately fourfold and β-myrcene levels threefold between (nonthermogenic) early afternoon samples and (thermogenic) evening ones (30 minutes duration each). The maximum β-myrcene levels measured in *M. machinii* cone emissions were several orders of magnitude smaller than the maxima of *M. lucida* samples (comparison of peak areas relative to that of an internal standard). The average peak cone metabolism occurs around dusk, but the average time difference between the peak metabolism and its peak temperature is only 18 minutes (Roemer et al., 2005).

To date, no linalool has been detected in any of the *Macrozamia* species that are pollinated only by *Cycadothrips*, and no (E)-β-ocimene has been detected in those pollinated only by *Tranes* sp. (Terry et al., 2004a, 2004b). However, all of these studies demonstrate only an association between these cone traits and their specific pollinator. No studies have shown any functional relationship between them, and many more *Macrozamia* species remain to be examined to determine their cone volatile component profiles and temporal dynamics. Some components of cone fragrances may simply be the by-products of secondary metabolism of other key components that are under selection, or they may result from exaptations of previous plant—animal interactions and thus may not be functioning in the current pollination mutualism (Armbruster et al., 1997; Levin et al., 2003).

The specific objective of this study was to determine whether *Tranes* sp. from *M. machinii* and *C. chadwicki* from *M. lucida* responded to specific cone fragrances from their usual host species. To accomplish this goal, complementary physiological and behavioral assays were performed on both *Cycadothrips* and *Tranes* to determine their responses to several cone fragrance components and to establish what cues, if any, encourage these insects to leave cones.

Materials and Methods

Study Species and Sites

Two southern Queensland *Macrozamia* species, *M. machinii* and *M. lucida*, were studied during their pollination periods (late October through early December) of 2001–2004. No species name is yet available for the *Tranes* on *M. machinii* (R. Oberprieler, pers. comm.). Insect voucher specimens have been deposited in the Australian National Insect Collection (ANIC; CSIRO) in Canberra, and plant vouchers from each of the *Macrozamia* species study sites have been deposited at the Queensland Herbarium in Brisbane. Samples of cones and insects were taken from coning populations of *M. machinii* in the Wondul Range north of Inglewood and of *M. lucida* in Brisbane Forest Park northwest of Brisbane, Queensland. The sites and species are described in more detail by Terry et al. (2004b).

Cone Volatile Collection and Analysis

Cones selected for study were in their active pollination phase (males dehiscing pollen and females with gaps between the sporophylls). Headspace adsorption techniques (Schiestl & Marion-Poll, 2002) were used for collecting cone volatiles in situ. This involved placing a polyacetate bag over the entire cone just prior to sampling. Air and volatiles from inside the bags were withdrawn by a portable field pump (Alpha Series 2, Ametek), calibrated at 60 to 100 mL min^{-1}, through glass Pasteur pipettes containing about 200 mg of Porapak Q adsorbent, plugged with glass wool. Volatiles of each species were collected for at least 30-minute periods during times when emissions were strongest as determined from previous studies: during midday for *M. lucida* and early evening (about 1900) for *M. machinii* (Terry et al., 2004b). Also, leaf volatile samples were taken. Further details are presented by Terry et al. (2004a, 2004b).

Adsorbed compounds were desorbed with 1 ml high-performance liquid chromatography (HPLC)-grade dichloromethane. Samples were concentrated to 0.5 mL, and 10 μL of a standard solution (2 mg/mL of nonane) were added to facilitate relative quantification of components between samples. Gas chromatographic (GC) analyses were performed by split injection of 1 μl of the sample solution into an HP Ultra-2 column

(30 m × 0.25 mm × 0.25 μm) housed in an HP 5890 gas chromatograph. Column temperature was programmed from 40–260°C at 10°C/minute, with a two-minute initial delay. Selected samples from 2001–2004 studies were used in the physiological experiments. A GC reanalysis was completed on older samples to confirm their consistency relative to the original sample analysis before these samples were used in any of the physiological tests. No attempt was made to determine the enantiomeric composition of linalool from the cones.

Standards of β-myrcene (Sigma, 90% purity) and racemic linalool (Aldrich, 97%) were purchased for use in physiological and/or behavioral tests. A sample of ocimene was derived from a natural source (C. Moore, pers. comm.), and identities of (Z)-β, (E)-β-, and (allo-) isomers were confirmed by GC retention times and GC-MS methods. The ocimene mix was first diluted 1:100 in dichloromethane for use as a standard. The standards were then added to castor oil before they were used in behavioral tests.

Maintaining Insects for Laboratory Experiments

Male cones of *M. lucida* and *M. machinii* were collected from field sites that had *Cycadothrips* (>1000 per cone) and *Tranes* (about 20 per cone), respectively. Insects were held on these cones in rearing cages at ambient laboratory conditions (22–24°C) until cone resources (pollen or sporophylls) were depleted. Additional cones and insects were collected from field sites as needed.

Physiological Experiments

Insects were tested for their physiological responses to cone volatile components using electroantennographic (EAG) and/ or GC-electroantennographic detection (GC-EAD) techniques. An adult *Cycadothrips* was captured and secured in Blu Tack (a putty-like adhesive; Bostik Findley Australia, PTY, Melbourne) with the head exposed. The reference electrode (glass/saline) was placed over one antenna, and the ground electrode was inserted into the thorax. For *Tranes*, the head was separated from the rest of the body, the reference electrode was placed over one antenna, and the ground electrode was inserted into the head.

To test the feasibility of preparing and performing electroantennography on

Cycadothrips and *Tranes* and to determine if they respond to any major cone volatiles, we performed preliminary EAG experiments. The responses of each insect type to linalool and β-myrcene, independently of one another, were monitored over an approximate 100-fold range from 5 μL to 400 μL of headspace (25°C). The headspace volatiles of each standard were extracted with a gas-tight syringe and were then manually injected into an air stream 10 cm from the delivery point to the antenna. In these experiments, hexanol, a green leaf volatile that many herbivorous insects can detect, was used as a standard to compare other odors and to correct for antennal decay over time. The ocimene sample containing a mixture of (Z)-β-, (E)-β-, and (allo-) isomers (see Fig. 22-5 for relative quantities) was tested at 400 μL headspace for a response by *Cycadothrips.* Because CO_2 concentrations in cones increase during cone thermogenesis as a result of the upsurge in metabolic activity (Seymour et al., 2004), EAGs were performed to determine whether *Cycadothrips* antennae can detect CO_2, because insects might use this as a signal to leave or locate cones. A parafilm-sealed flask was filled with gas from a CO_2 gas cylinder, and 400 μL of headspace was extracted and applied as described above.

Several samples were used to test for responses by both male and female adult *Cycadothrips* using GC-EAD techniques: cone volatiles from *M. lucida* and *M. machinii*, the sample containing the ocimene isomers, and leaf volatile sample from *M. lucida.* For each insect experiment, 1 μl of sample was injected into the gas chromatograph in splitless mode using a similar GC column and similar temperature–time program as for the original GC sample analysis. The retention time of volatile components from samples was 6 to 17 minutes using this protocol. The insects that successfully completed GC-EAD analysis lasted at least 30 minutes. To test whether the antennal response was adequate before and after each GC-EAD, a test EAG using 400 μL of the hexanol standard was applied to the antenna. If the antenna did not respond, the preparation and data were discarded.

Behavioral Assays—Olfactometer Experiments

A Y-tube glass olfactometer (Fig. 22-1; inside diameter 10 mm; each arm 11.5 cm long; stem of Y 10 cm long) was used to test various odor sources, which included cone sporophylls and chemical components of different concentrations (see below). Filters built into the Y-tube (see Fig. 22-1 for locations), consisting of finely molded ground glass, allowed

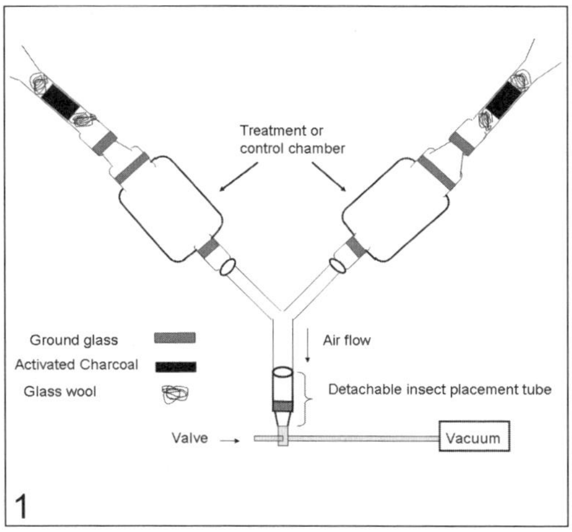

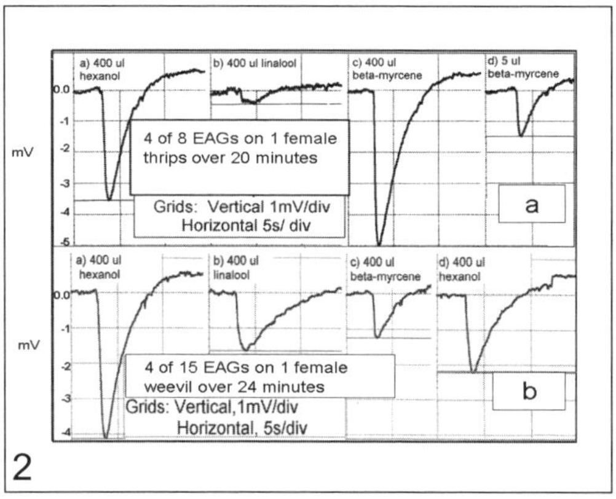

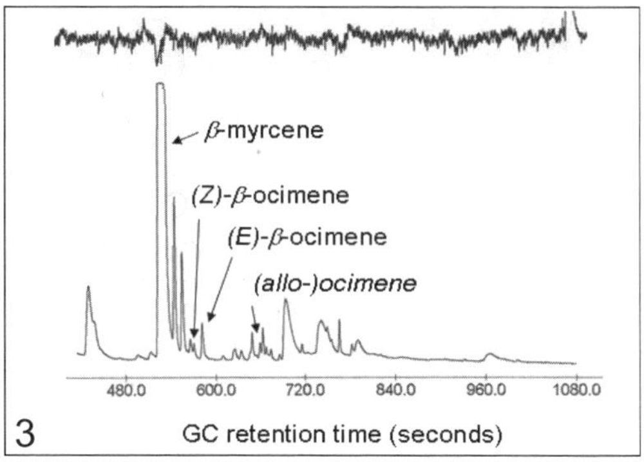

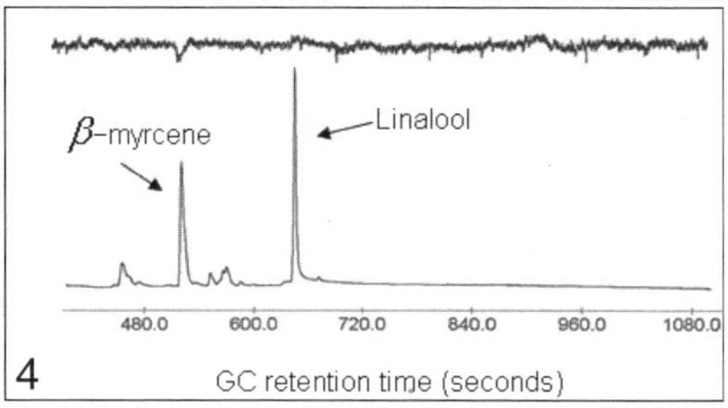

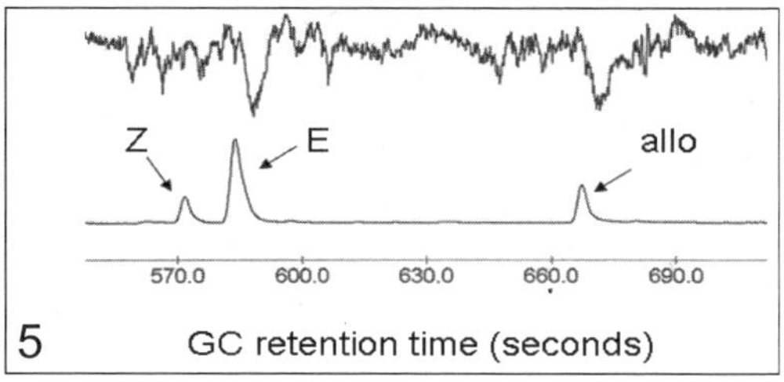

Figure 22-1. The Y-tube olfactometer system used to test the responses of thrips, *Cycadothrips chadwicki*, and weevils, *Tranes* sp., to various odors sources.

Figure 22-2. Examples of results of electroantennographic tests (EAGs) performed on **a** *Cycadothrips chadwicki* female and **b** *Tranes* sp. female. The figure shows a physiological response to β-myrcene by both insects and to linalool by *Tranes.* A final 400 μL hexanol response is included for *Tranes* (d, in Fig. 2B) to show the decay of the antenna. The decay rate for *Cycadothrips* was similar, but the final response to hexanol has not been included on the graph.

Figure 22-3. Gas chromatographic-electroantennograph detection technique performed on a male *Cycadothrips chadwicki,* showing a typical response to β-myrcene in a volatile sample from a *Macrozamia lucida* female cone. Upper trace = electroantennogram; lower trace = gas chromatogram associated with the gas chromatographic (GC) retention time.

Figure 22-4. Gas chromatographic-electroantennograph detection technique performed on a male *Cycadothrips chadwicki* showing a typical response to β-myrcene but not to linalool from a *Macrozamia machinii* male cone. Upper trace = electroantennogram; lower trace = gas chromatogram associated with the gas chromatographic (GC) retention time.

Figure 22-5. Gas chromatographic-electroantennograph detection technique performed on a female *Cycadothrips chadwicki,* showing a typical response to the ocimene isomers, (E)-β- and (allo-), but not to (Z)-β-ocimene. Upper trace = electroantennogram; lower trace = gas chromatogram associated with the gas chromatographic (GC) retention time.

air to flow through the system and prevented insect movement into control–treatment chambers (where odor sources were placed) or into the vacuum line. Air flow, created from a central laboratory vacuum line, was regulated (and calibrated on virtually every test day) at a rate of 45 cm^3 min^{-1}. All tests with cone sporophylls were conducted at times of the day when cones were not thermogenic (midday for *M. lucida* and in the early evening for *M. machinii* [Terry et al., 2004b]).

Several control and preliminary tests were conducted and established the following: (1) both *Cycadothrips* and *Tranes* could easily walk upwind at the above velocity setting and at room temperatures of 22–24°C; (2) there was no left-to-right bias in the glassware system; and (3) the light environment necessary to minimize bias in experiments with *Tranes* and *Cycadothrips* was standardized based on the insects' specific phototropic behavior (shaded windows and no artificial lighting for *Tranes* because they are photonegative, and shaded windows and the Y-tube placed directly underneath and perpendicular to overhead fluorescent lighting for uniformity in lighting for *Cycadothrips* because they are photopositive). A dark blue-gray background was placed underneath the olfactometer to minimize any background light reflectance into the tube. These standardized conditions resulted in no detectable bias in the system (see Results. Nevertheless, the left-to-right positions of the treatment arms were switched regularly between replicates of experiments. In addition, periodic control experiments were used to determine if the samples of insects were still active and moving upwind into the Y-tubes after holding them away from food for several hours. The olfactometer was washed thoroughly with soap and water, rinsed with acetone, and oven-dried at 100°C between treatment changes (different host cone sporophylls and chemical components).

Treatments (odor sources) included male cone sporophylls (from the host and non-host species,10 sporophylls), the β-myrcene standard, and the ocimene standard with isomers, (Z)-β-, (E)-β-, and (allo-)ocimenes. Standards were diluted in castor oil. The level of β-myrcene tested was determined by calculating the mass of this chemical in the β-myrcene standard (at 90% purity) added per mL castor oil multiplied by the amount of that mixture applied to Whatman filter paper. The filter paper was placed into one of the treatment-holding chambers connected to an arm of the Y-tube. The control arm had only filtered air. The insects were prepared for testing by first isolating and holding them without food before testing, for 1–4 hours (*Cycadothrips*) or 1–24 hours (*Tranes*). (Insects in the field spend time away from cones searching for new cones.) The thrips were readily captured for testing after they had left their host cones during the midday period of cone thermogenesis by aspirating them from the walls of the cage holding the cones. The weevils would leave their host cones when cones were placed in the dark, and then they were easily captured from the cage walls. Sexes were separated for testing, but insects from the field cannot be aged.

Thrips were tested in groups (8–15 individuals) of each sex to determine responses to an odor. At least three replicates of such groups were tested for each treatment. In addition, at least five follow-up replicates were done with individual thrips to ensure that the response was independent of the presence of other thrips or their responses (there was never an indication of a lack of independence). Weevils were tested individually. Insects of both types were observed for up to five minutes, and their positions in the Y-tube (one of four responses: treatment arm; control arm; "center"—stem of Y tube; and "start" chamber), were recorded. Representative odor samples were collected through the Y-tube for GC analysis (using the dynamic headspace techniques described in Materials and Methods, Cone Volatile Collection and Analysis) to determine if levels of volatile emissions were in an ecologically relevant range compared to those obtained in the field. A Pasteur pipette containing Porapak Q was inserted into the vacuum line between the Y-tube "insect placement chamber" and the vacuum line regulator. Volatiles were collected for 30 minutes at 45 mL min^{-1}. Previously described methods for eluting and preparing the volatile samples for GC analysis were used. All levels of the standard β-myrcene sample tested in olfactometer experiments and odors from cones were within ecologically relevant ranges found in field samples of headspace techniques measured on *M. lucida* cones. In situ 30-minute field volatile samples of *M. lucida* cones contained β-myrcene peak areas ranging from less than 1% to more than 12,400% of the internal standard, and the highest levels extracted from laboratory samples through the Y-tube (one arm with cone sporophylls, the other filtered air) were up to 13,000% (corrected for collecting air from both arms) of the internal standard.

Insect responses to odors were statistically compared to controls of the Y-tube using a two-sided binomial test with pooled numbers of individual responses (*Tranes*) or group plus individual responses (*Cycadothrips*) for each treatment (odor) source. The null hypothesis tested was that the distribution of the number choosing the control arm was equal to that of the treatment arm. To examine the null hypothesis that the distribution of the results for a particular treatment was the same for males and females, a χ^2 2×2 (to compare just treatment versus control arm) or 2×4 (to compare the distribution of all responses including those that do not move) contingency table test was used.

Results

Physiological Experiments

During GC and GC-MS analysis of each volatile sample to be used in physiological tests, a previously reported "unknown" component in *M. lucida* volatiles (Table 1, retention time of 11.35, [Terry et al., 2004b]) was identified as (allo-)ocimene. This component also has been detected in *Macrozamia macdonnellii* and *Macrozamia longispina*, both pollinated solely by *Cycadothrips*. This component was not detected in *M. machinii*.

In the EAG experiments, both *Cycadothrips* and *Tranes* responded to hexanol (Fig. 22-2A, 22-2B), which was used as a control in these tests to correct for antennal decay over time. The response of a single thrips female to the β-myrcene was strong across the 5 μL to 400 μl volumes (Fig. 22-2A). The response to 5 μL β-myrcene, corrected value for antennal decay, was quite strong, about one-third of the response to 400 μL β-myrcene. The experiment with 400 μL linalool headspace sample suggested that female *Cycadothrips* do not detect linalool or, at most, weakly respond to this compound (Fig. 22-2A). Alternatively, the very small response may have been stimulated by a puff of air over the antenna. On the other hand, *Tranes* antennae responded strongly to both β-myrcene and linalool at levels of 400 μL (Fig. 22-2B) and 100 μL (data not shown) headspace linearly corrected for decay using the relative response to hexanol (within Fig. 22-2B, labels a and d). However, the *Tranes* response to lower levels of linalool (25 μL) was much weaker than to the 25 and 5 μL levels of β-myrcene (data not shown). These exploratory results indicated that both insects detect specific components found in their own host cone volatiles and that GC-EAD tests were worth pursuing.

Two additional treatments were tested by EAGs for responses by male and female *Cycadothrips*. First, for the ocimene standard sample, the results showed that both male and female *Cycadothrips* respond to 400 μL headspace of the sample. Second, *Cycadothrips* (both males and females) did not respond to the 400 μL headspace sample of CO_2, suggesting that the antennae do not detect this gas.

GC-EAD experiments were successfully performed on *Cycadothrips*. Three insects of each sex were tested on one female and two male cone volatile samples of *M. lucida*, as well as one of each sex on a volatile sample of a *M. lucida* leaf. Two individuals of each sex were tested for any response to a volatile sample from a male cone of *M. machinii*. In all samples, both male and female *Cycadothrips* responded only to the β-myrcene peak in the *M. lucida* and *M. machinii* cone samples (Figs. 22-3, 22-4). Because all of the ocimene isomers in volatile samples had relatively low peak areas (e.g., GC chromatogram, Fig. 22-4), and previous EAG experiments showed a response to the ocimene standard, additional GC-EAD experiments were performed on a male and two female *Cycadothrips* for their response to the ocimene isomers in this standard mixture. The results revealed that both male and female *Cycadothrips* can detect (E)-β- and (allo-) ocimene, but not (Z)-β-ocimene (e.g., female response, Fig. 22-5).

Olfactometer Experiments

No bias was detected in the responses of weevils (Fig. 22-6) and thrips when tested in the olfactometer with both arms blank. All of these control data were combined; these data showed that (1) 76% (n=29) of the weevils (Fig. 22-6A, control tests) and 60% of the thrips (n=367) moved into one or the other arm, and (2) that their choice of arm was essentially equal (38% for each arm by *Tranes* and 29.3% and 30.5% across the arms for *Cycadothrips*). Both *Tranes* and *Cycadothrips* overwhelmingly selected the arm with volatiles drawn from sporophylls of their own host cones (Figs. 22-6B, 22-7A). Both males and females responded significantly in this way (Fig. 22-7A for *Cycadothrips*); in *Tranes* tests where sexes were separated, the ratio of those choosing the host arm over the control arm was 18 to 10 for males and 24 to 8 for females ($p=.09$ and $p<.012$, respectively, binomial test). In contrast, when one arm contained sporophylls from the nonhost cone, there was no statistical difference, at $p>.05$, in the number responding to the control or the treatment (nonhost) arm (Figs. 22-6C, 22-7B), although male *Cycadothrips* responses were slightly ($p=.054$) higher toward the *M. machinii* arm and the response by female thrips was slightly higher toward the control arm ($p=.08$).

The responses of *Cycadothrips* to β-myrcene depended on the concentration used. β-Myrcene at high concentrations (≥5 μg of standard mixed in oil applied to filter paper; Fig. 22-7C) appeared to be repellent. Most the insects moved into one of the Y-tube arms

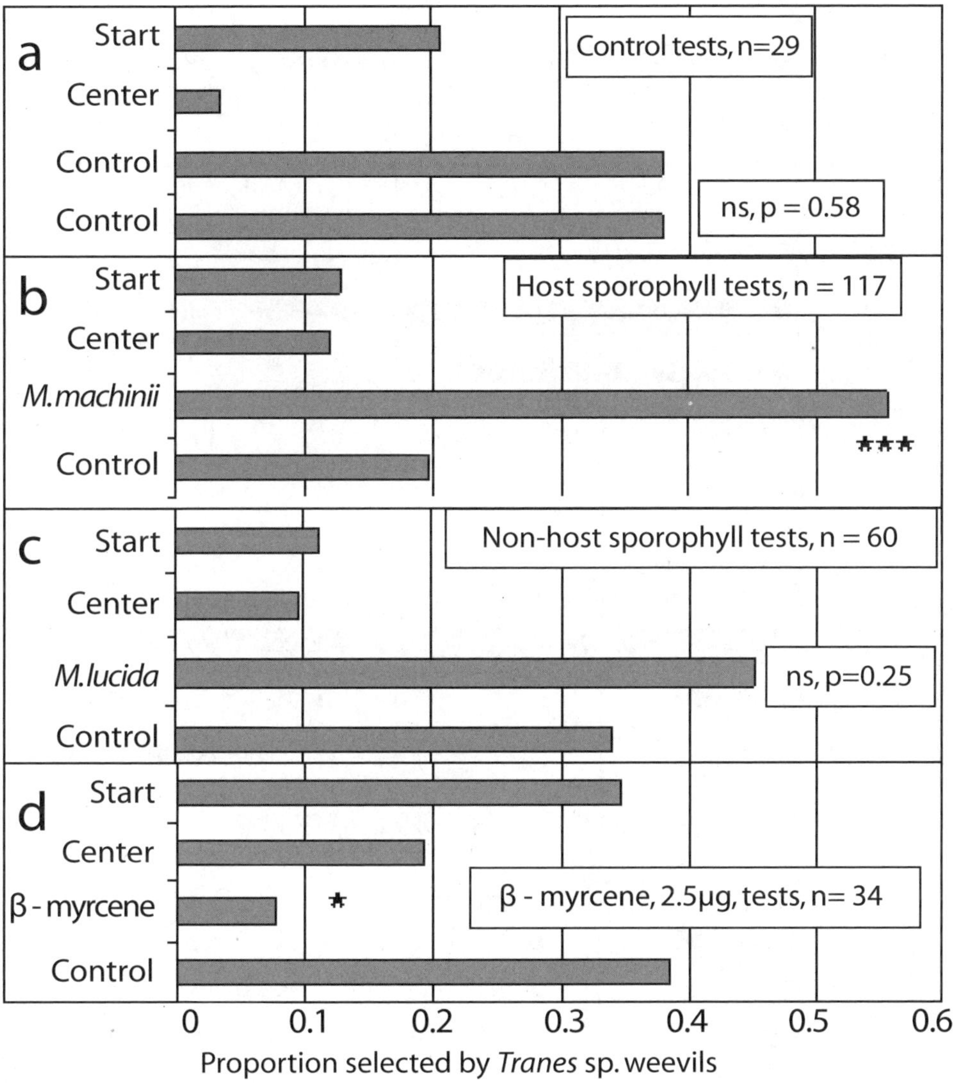

Figure 22-6. Proportion of *Tranes* sp. weevils (from *Macrozamia machinii)* in Y-tube olfactometer tests counted as not moving (start arena); in the base of the Y-tube (center); responding to treatment (odor) arm; or control arm. Odors tested are: **a.** none, control tests; **b.** host (*M. machinii*) or **c.** non-host (*M. lucida*) male cone sporophylls; and **d.** β-myrcene standard at 2.5 µg. n=number of individuals tested. The significance of the statistical comparison, a two-sided binomial, between the treatment and control (ns=not significant) is $p>.05$; *= $p\leq.05$; **= $p\leq.01$; ***,=$p\leq.001$.

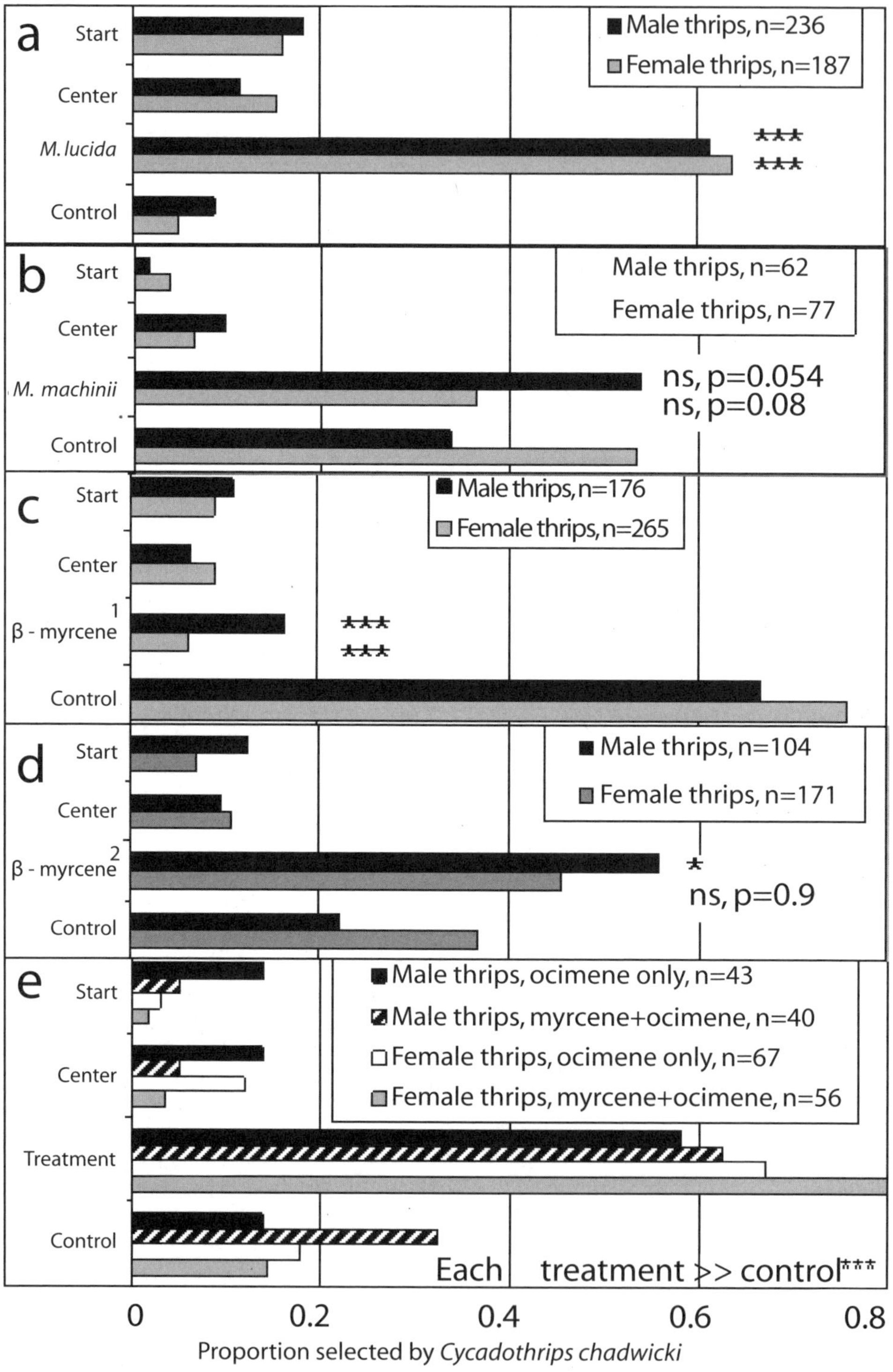

a
Start
Center
M. lucida
Control
Male thrips, n=236
Female thrips, n=187
***
***
b
Start
Center
M. machinii
Control
Male thrips, n=62
Female thrips, n=77
ns, p=0.054
ns, p=0.08
c
Start
Center
β - myrcene 1
Control
Male thrips, n=176
Female thrips, n=265
***
***
d
Start
Center
β - myrcene 2
Control
Male thrips, n=104
Female thrips, n=171
*
ns, p=0.9
e
Start
Center
Treatment
Control
Male thrips, ocimene only, n=43
Male thrips, myrcene+ocimene, n=40
Female thrips, ocimene only, n=67
Female thrips, myrcene+ocimene, n=56
Each treatment >> control***
0
0.2
0.4
0.6
0.8
Proportion selected by Cycadothrips chadwicki

(n=216, or 82%, of the females, and n=146, or 83%, of the males). Of these, a high percentage, 91% of males and 81% of females went into the control arm. Many individuals that first moved into the β-myrcene–loaded arm quickly moved away and into the control arm. At the highest level tested, 100 μg, several insects that did not move into the control arm either died or moved around erratically (twitching) during the observation period, suggesting that this level of β-myrcene may be toxic (total n=17 individuals, two trials, most insects did not move, and 6 died in less than five minutes). With lower levels of β-myrcene (0.5–2 μg in oil applied to filter paper), most females moved into the Y-tube arms, and there was no statistical difference in the proportion that moved into the control and the treatment arm. Males also responded well, and were slightly attracted to the arm containing the low concentration β-myrcene (Fig. 22-7D). Because a detectable but small amount of (E)-β- and (allo-)ocimene (<0.04% and <0.001%, respectively, of the β-myrcene levels) was present in the β-myrcene standard as determined from GC/GC-MS analysis, an attraction by the ocimene isomers in this sample cannot be ruled out without further testing. The reaction of *Tranes* to 2.5 μg of β-myrcene in oil on filter paper disk suggests repellency (Fig. 22-6D).

Both sexes of *Cycadothrips* responded positively to the ocimene standard containing the three isomers (Fig. 22-7E). Their response was about the same, in extent, as the response by these thrips to the *M. lucida* sporophyll odors. The response to this blend was positive even in the presence of relatively low but seemingly repellent levels (5 μg) of β-myrcene standard (Fig. 22-7E).

Overall, comparisons of distributions between male and female responses indicated both sexes of *Cycadothrips* responded similarly to most test components and treatments (not significant, $p>.11$), except for their response to *M. machinii* ($\chi^2=11.9$, df=3, $p=.007$) and to the low level of β-myrcene ($\chi^2=23.3$, df=3, $p=.0007$). In addition, the distribution of responses at the lower levels of β-myrcene was significantly different from that at higher levels for both males and females ($\chi^2>16.3$, df=3, $p<.001$).

Figure 22-7. Proportion of *Cycadothrips chadwicki* (from *Macrozamia lucida*) in Y-tube olfactometer tests counted as not moving (start arena); in the base of the Y-tube (center); responding to treatment (odor) arm; or control arm. Odors tested are: **a.** host (*M. lucida*) or **b.** nonhost (*M. machinii*) male cone sporophylls; **c.** β-myrcene,[1] standard=≥5 μg; **d.** β-myrcene,[2] standard=0.5–2 μg; and, **e.** Ocimene standard ((Z)-β-, (E)- β-, and (allo-) isomers), alone and in combination with 5 μg of β-myrcene standard. The significance of the statistical comparison, a two-sided binomial, between the treatment and control (n s= not significant) is $p>.05$; *= $p\leq.05$; **= $p\leq.01$; ***,=$p\leq.001$.

Discussion

The results reported above are interpreted here in relation to the species specificity of pollinators in these systems and the structure of the pollination syndrome of *Macrozamia*. Previous research on thermogenesis in *Macrozamia* cones is included because this is part of the pollination system in these plants, although, as yet, no conclusive interpretation is possible. Further avenues for research in all of these areas are suggested.

Cone Odors and Pollination

The combined results of physiological and behavioral tests suggest a difference in response by *C. chadwicki* obtained from *M. lucida* and by *Tranes* sp. obtained from *M. machinii* to cone volatile components that apparently reflect their specificity to their respective hosts.

Cycadothrips responded physiologically to the minor components (E)-β- and (allo-) ocimene found in *M. lucida* cone volatiles (but not in *M. machinii*), and they were attracted to a mixture of these components in olfactometer tests. *Cycadothrips* detected the major component of their host cone volatiles, β-myrcene, over about two orders of magnitude change in volume. In olfactometer experiments, *Cycadothrips* were strongly repelled by the higher levels of β-myrcene tested. Low levels of this component generated no responses in females but did attract males slightly. Combining a low but slightly repellent level of β-myrcene with the ocimene mix did not affect the attraction toward (E)-β- and (allo-)ocimene. Finally, *Cycadothrips* appear to be physiologically "blind" to linalool and to (Z)-β-ocimene based on both EAG and GC-EAD tests. There was no antennal response to CO_2 by either male or female *Cycadothrips*, although this is not proof that thrips do not detect CO_2 (Stange, 1992).

The *Tranes* sp. from *M. machinii* can detect linalool and β-myrcene, the two major components of their host cone volatiles (Terry et al., 2004b), according to the EAG results. β-Myrcene appears to be slightly repellent to *Tranes* based on our Y-tube olfactometer tests. More testing is needed on these weevils for their response to different levels of both β-myrcene and linalool, as well as to minor components of their host fragrance. The coning season was completed, and not enough weevils were found to test other odor components.

These cone odors alone may be responsible for mediating pollinator behavior in each of these systems. Indeed, although preliminary field data indicate that insects are attracted to the cone odors alone, other host-related factors, such as cone color and shape, or timing of thermogenesis, may play a role in the host selection process by these insects. Preliminary data (I. Terry et al., unpublished data) suggest that light levels affect both insects, but in opposite ways (photopositive versus photonegative by *Cycadothrips* and *Tranes*, respectively), and that thrips respond to specific colors. Overall, the mechanism of locating hosts is likely to comprise several steps, some of which may involve subtleties in chemical concentrations and combinations of factors. Further testing is required on other species of *Macrozamia* pollinators to determine how similar these responses are among the different species of *Cycadothrips* and *Tranes* found across Australia and to establish if different host-associated populations of these pollinators respond to different cues or combinations of cues. That this could be the case is suggested by the results from *Macrozamia communis,* which is dual-pollinated by both *Tranes lyterioides* and *C. chadwicki* (Terry, 2001), but which apparently has only low levels of β-myrcene relative to its numerous other components (R. Oberprieler, pers. comm.). Linalool is present, but isomers of ocimene have not been reported during the periods when samples were taken (late afternoon and evening). It is not uncommon to find different floral and cone odors among closely related plant species and genera (Pellmyr et al., 1991; Tang, 1993; Levin et al., 2003; Knudsen, 1999; Knudsen et al., 1999) for various reasons. These include homoplasy of enzymes and/or labile precursor enzymes and differences in specific evolutionary history with insect predators and pollinators (Levin et al. 2003), all of which increase the difficulty in finding an association between traits and plant phylogeny.

Thermogenesis and Pollination

The function of volatiles in attracting and dispersing insects may be intertwined with cone thermogenesis (and increased metabolism), because the emission of volatiles occurs in close temporal proximity to thermogenic temperature increases (Terry et al., 2004b; Roemer et al., 2005). The current study is beginning to elucidate the role of some of these volatiles, but the role of the elevated temperatures remains unknown. Thermogenesis has been detected in many species within basal angiosperm families as well as in many other cycads that have been studied (Tang, 1987b, Tang et al., 1987; Tang 1993; Azuma et al., 1999; Dieringer et al., 1999; Thien et al., 2000; Seymour et al., 2003a, 2003b). Several functions have been proposed (Seymour, 1997; Seymour & Schultze-Motel, 1998;

Ervik & Barfod, 1999; Patino et al., 2000; Seymour, 2001; Raguso, 2004; Terry et al., 2004b; Donaldson, 2007), including some that could affect pollinator insects and specificity in these cycad–pollinator mutualisms.

Given the temporal association between peaks in cone temperatures and insect arrivals (Roemer et al., 2005; unpublished data), elevated temperatures could play a primary role in attracting pollinators. However, several lines of evidence argue against this: (1) specific volatiles appear to be attractants in *M. lucida*; (2) receptive female cones of both *M. lucida* and *M. machinii* readily attract their specific insects even though their thermogenic temperature elevations are small or even negligible (Seymour et al., 2004; Terry et al., 2004b; Roemer et al., 2005); and (3) the elevated midday cone temperatures of *M. lucida* are comparable to those of nearby solar-heated objects, which do not attract thrips. Thus, elevated cone temperatures may only enhance an attracting signal (see below). For the pollinators of the thermogenic *Arum*, temperature alone (without odors) was not an attractant (Knoll, 1926).

Another proposed role for thermogenesis is the stimulation of insect dispersal from flowers or cones (Heinrich, 1974; Burquez, 1986; Meeuse & Raskin, 1988). For *Cycadothrips*, a threshold temperature could be a cue, because the insects disperse well before the cone temperature peak is reached (Roemer et al., 2005). However, experiments and observations, in both laboratory and field, suggest that a threshold temperature alone does not cause insects to leave their cones. In preliminary laboratory experiments, increasing temperatures from 13 to 35°C over a 2.5-hour interval caused *Cycadothrips* (placed in clear vials) to increase their activity only slightly (fewer than 20% of thrips were actively moving around at temperatures less than 32°C and up to 25% at greater than 33°C) or caused thrips inside a darkened tube to emerge (fewer than 10% at any temperature above) (R. Roemer, unpublished data). In field measurements, thrips first begin to emerge en masse from *M. lucida* cones when cone temperatures range from 22 to 32°C (I. Terry, unpublished data), suggesting that something other than a specific temperature is involved. *Tranes* on *M. machinii* have not been observed moving out of cones during the day because they avoid the light, even though cones may be as warm as thermogenic cones, but they do emerge during dusk and evening, and it could be the interaction between thermogenic temperatures and other changes (time of day, lighting, volatile components emitted and their concentration) that stimulates them to move to the cone surface at this time.

Other effects of plant thermogenesis could play a role in insect behavior and energetics. Energy savings for endothermic insects could be substantial. For example, *Cyclocephala colasi* beetles save up to 4.8 times their energy by being inside thermoregulating flowers of *Philodendron solimoesense* compared with endothermic energy expenditure under cooler ambient temperatures (Seymour et al., 2003b). However, *Tranes* and *Cycadothrips* likely have too small a mass to be endothermic (Heinrich, 1974; Seymour et al., 2004), in which case their body temperatures track the ambient temperature to the extent they would save no energy in thermogenic cones. Nonetheless, they could become more active under increasing or optimal temperatures (Seymour et al., 2004). Indeed, increased activity by *Tranes* weevils and *Cycadothrips* (mating and fighting by both insects, and oviposition by thrips) has been observed on male cones that are thermogenic (Seymour et al., 2004; J. A. Hall, unpublished data). Cone thermogenic temperatures could warm wing muscles for lift during take-off during cooler periods. However, many small insects cool rapidly to ambient if air temperatures are not enough for sustained wing beat frequency, thus grounding their flight (Heinrich, 1974). Exceptions to this are some small insects that fly at very low temperatures (0–15°C) (Heinrich, 1993; Heinrich, 1996; Seymour et al., 2003a). Further studies are needed to determine whether the pollinators of *Macrozamia* can fly when exposed to lower temperatures experienced during the coning season. Finally, *Cycadothrips* and *Tranes* larval development could be enhanced if the cone temperatures are kept in a more optimal range for development, as is common for most insects, but this has not been investigated on the pollinators of cycads.

In general, given the mixed evidence for temperature alone affecting insect movement between cones, it is more likely that temperature plays a role as part of a suite of cues (e.g., with volatiles as in Araceae; Moodie, 1976), as an attractant and/or a dispersant. It is not uncommon for flowers to use many types of advertisements to attract pollinators, and some combinations of signals may result in synergistic responses by pollinators compared with single signals (Raguso & Willis, 2002, 2005). A combination of odor plus heat is used by some flowers that mimic carrion to attract carrion feeding insects to flowers that offer no reward (Patino et al., 2000; Stensmyr et al., 2003). Experiments with the dead horse arum, *Helicodicerus muscivorous*, have demonstrated that heat plus odors attract more of the pollinating flies than do odors alone (Angioy et al., 2003). Another possibility is that thermogenesis could be related indirectly to the production or release of volatiles or volatile components (Skubatz et al., 1995, 1996) that attract and/or disperse insects, rather than being a cue in itself. The heat by-product of the increased metabolism also would increase the rate of diffusion of the released volatiles.

Pollination Syndrome of *Macrozamia*

Various strands of evidence suggest that many plant factors mediate insect behaviors at different steps in the interactions between these cycads and their pollinators. For *Tranes* on *M. machinii*, a single factor could be involved in stimulating them to venture outside their cones, such as: (1) the onset of dusk and night time alone (as the weevils crawl to the outside of cones when placed in the dark [see Materials and Methods] and do so in situ, unless temperatures drop rapidly during the evening); (2) heat alone (although this is not likely; *Tranes* does not leave warm/hot cones during the middle of the day because of their negative photoresponse); or (3) an elevated volatile component (β-myrcene and linalool both increase during thermogenesis [Terry et al., 2004b] around sunset). Alternatively, an interaction of the above factors could stimulate *Tranes* to exit cones. Once *Tranes* are outside the cone, they fly/walk to other cones if ambient conditions are warm enough; subsequently they are attracted by one or more of the volatile components that increase during the evening. In *Cycadothrips*, several of the above are also possibilities, including elevated levels of a repellent during thermogenesis or a combination of a repellent chemical and heat.

The interactions so far observed in the *Macrozamia* pollination mutualism suggest that the plant influences pollinator behavior more than is usual in plant–pollinator relationships. In addition to a kairomonal response by the insect toward a host, there appears to be an allomonal response, whereby the insect's aversive behavior is activated. This type of system could be characterized as a "pull-push" system. The evidence presented in the current study suggests that not only are there stimuli that attract pollinators, "pulling" insects toward cones, sometimes over several hours, but also some forces, possibly β-myrcene in combination with heat, that "push" or drive insects rapidly to the outside of cones and away from their safe feeding and breeding havens (even away from cones with plenty of resources remaining). A similar terminology, "push-pull" rather than "pull-push" (*Macrozamia* pollinators must initially be "pulled in"), has been used to describe pest control approaches that incorporate artificial repellents to push pests away from a protected resource and then pulled toward traps by attractants (Lindgren & Borden, 1993), and a type of "push-pull" phenomenon has been recognized as functioning in directing two different tick species toward their specific attachment sites within the same host animal (Wanzala et al., 2004). No such terminology has, however, been used in the context of pollination. In any case, this is a general phenomenon in which two antagonistic processes act in concert to achieve a particular end, be that location of a specific site (ticks), movement of pollinators (cycads), or the localization of pests (humans). This "pull-push" designation may be superficially ascribed to other plant–animal interactions, but they may lack the "push" part of the process. In summary, the studies reported here provide direct evidence that in both *Cycadothrips* and *Tranes*, specific cone fragrance components released by their *Macrozamia* hosts play a role in their rather complex interaction with their specific hosts.

Acknowledgments

We thank Adriana Najar, Lisa Anici, and Alex Yaku for help with laboratory equipment and tests, as well as Bruce Duncombe at Brisbane Forest Park for help with access to various field sites. We are also indebted to Robert Raguso for his many insights into the role of volatiles in this plant–pollinator mutualism and his stimulating thoughts about a "pull-push" pollinator system. As always, we thank Laurence Mound and Rolf Oberprieler for their valuable assistance with insect identifications.

Literature Cited

Angioy, A.-M., M. C. Stensmyr, I. Urru, M. Puliafito, I. Collu & B. S. Hansson. 2003. Function of the heater: the dead horse arum revisited. Proc. Roy. Soc. London, Ser. B (Suppl) 271: S13–S15.

Armbruster, W. S., J. J. Howard, T. P. Clausen, M. Debevec, J. C. Loquvam, M. Matsuki, B. Cerendolo & F. Andel. 1997. Do biochemical exaptations link evolution of plant defense and pollination systems? Historical hypothesis and experimental tests with *Daechampia* vines. Amer. Naturalist 148: 461–484.

Azuma, H., L. B. Thien & S. Kawano. 1999. Floral scents, leaf volatiles and thermogenic flowers in Magnoliaceae. Pl. Spec. Biol. 14: 121–127.

Burquez A., K. J. Sarukhan & L. A. Pedroza. 1986. Floral biology of a primary rain forest palm, *Astrcaryum mexicanum* Liebm. Bot. J. Linn. Soc. 94: 407–419.

Connell, S.W. & P.G. Ladd. 1993. Pollination biology of *Macrozamia riedlei*—the role of insects. Pp. 96–102 *in* D.W. Stevenson and K J. Norstog (eds.), Proceedings of CYCAD 90, the second international conference on cycad biology. Palm and Cycad Societies of Australia, Milton, Brisbane.

Dieringer, G., L. C. R., M. Lara, L. Loya & P. Reyes-Castillo. 1999. Beetle pollination and floral thermogenicity in *Magnolia tamaulipana* (Magnoliaceae). Int. J. Pl. Sci. 160: 64–71.

Donaldson, J. S. 1997. Is there a floral parasite mutualism in cycad pollination? The pollination biology of *Encephalartos villosus* (Zamiaceae). Amer. J. Bot. 84: 1398–1406.

———. 2007. Hot and smelly sex! Theoretical considerations of the adaptive significance of cone thermogenesis and volatile odors in cycad reproduction. Mem. New York Bot. Gar. 97: 308–325.

Ervik, F. & A. Barfod. 1999. Thermogenesis in palm inflorescences and its ecological significance. Acta Bot. Venez. 22: 195–212.

Forster, P. I. 2004. The cycads of Queensland—diversity and conservation. Palms and Cycads 82: 4–28.

———, P. J. Machin, L. A. Mound & G. W. Wilson. 1994. Insects associated with reproductive structures of cycads in Queensland and north-east New South Wales, Australia. Biotropica 26: 217–222.

Hall, J. A., G. H. Walter, D. M. Bergstrom & P. J. Machin. 2004. Pollination ecology of the Australian cycad *Lepidozamia peroffskyana* (Zamiaceae). Austral. J. Bot. 52: 333–343.

Heinrich, B. 1974. Thermoregulation in endothermic insects. Science 185:747–756.

———. 1993. The hot-blooded insects. Strategies and mechanisms of thermoregulation. Harvard University Press, Cambridge, MA.

———. 1996. The thermal warriors: strategies of insect survival. Harvard University Press, Cambridge, MA.

Jones, D. L. 2002. Cycads of the world. Reed New Holland, Sydney, Australia.

Knoll, F. 1926. Die Arum-Blutenstande und ihre Besucher Insekten und Blumen IV. Abhandlungen Zool. Bot. Gesellschaft Wien 12: 379–481.

Knudsen, J. T. 1999. Floral scent differentiation among coflowering, sympatric species of *Geonoma* (Arecaceae). Pl. Spec. Biol. 14: 137–142.

———, L. Tollsten & F. Ervik. 1999. Floral scent chemistry in geonomoid palms (Palmae: Geonomeae) and its importance in maintaining reproductive isolation. Mem. New York Bot. Gard. 83: 141–168.

Levin, R. A., L. A. McDade & R. A. Raguso. 2003. The systematic utility of floral and vegetative fragrance in two genera of Nyctaginaceae. Syst. Biol. 52: 334–351.

Lindgren, B. S. & J. H. Borden. 1993. Displacement and aggregation of mountain pine beetles, *Dendroctonus pondersae* (Coleoptera: Scoytidae), in response to their anti-aggregation and aggregation pheromones. Canad. J. Forest Res. 23: 286–290.

Meeuse, A. D. J. & I. Raskin. 1988. Sexual reproduction in the arum lily family with emphasis on thermogenicity. Sex. Pl. Reprod. 1: 3–15.

Moodie, G.E.E. 1976. Heat production and pollination in Araceae. Canad. J. Bot. 54: 545–546.

Mound, L. A. 1991. The first thrips species (Insecta, Thysanoptera) from cycad male cones, and its family level significance. J. Nat. Hist. 25: 647–652.

——— & I. Terry. 2001. Thrips pollination of the central Australian cycad, *Macrozamia macdonnellii* (Cycadales). Int. J. Pl. Sci. 162: 147–154.

———, E. den Hollander & L. den Hollander. 1998. Do thrips help pollinate *Macrozamia* cycads? Vict. Entomol. 28: 86–88.

Norstog, K., D. W. Stevenson & K. J. Niklas. 1986. The role of beetles in the pollination of *Zamia furfuracea* L.fil. (Zamiaceae). Biotropica 18: 300–306.

Oberprieler, R. G. 1995. The weevils (Coleoptera: Curculionoidea) associated with cycads. 2. Host specificity and implications for cycad taxonomy. Pp. 335–365 *in* P. Vorster (ed.), Conservation through cultivation, Proceedings of the third international conference of cycad biology. Cycad Society of South Africa, Stellenbosch, South Africa.

———. 2004. "Evil weevils"—the key to cycad survival and diversification? Pp. 170–194 *in* A. Lindstrom (ed.), Proceedings of the sixth international cycad conference on cycad biology. Nong Nooch Tropical Botanical Garden, Thailand.

Patino, S., J. Grace & H. Banziger. 2000. Endothermy by flowers of *Rhizanthes lowii* (Rafflesiaceae). Oecologia 124: 149–155.

Pellmyr, O., W. Tang, I. Groth, G. Bergstrom & L. B. Thien. 1991. Cycad cone and angiosperm floral volatiles: inferences for the evolution of insect pollination. Biochem. Syst. Ecol. 19: 623–627.

Raguso, R. A. 2004. Flowers as sensory billboards: progress towards and integrated understanding of floral advertisement. Curr. Opin. Pl. Biol. 7: 434–440.

——— & M. Willis. 2002. Synergy between visual and olfactory cues in nectar feeding by naïve hawkmoths, *Manduca sexta*. Anim. Behav. 64: 685–695.

——— & ———. 2005. Synergy between visual and olfactory cues in nectar feeding by wild hawkmoths, *Manduca sexta*. Anim. Behav. 64: 685–695.

Roemer, R., I. Terry, C. Chockley & J. Jacobsen. 2005. Experimental evaluation and thermophysical analysis of thermogenesis in male and female cycad cones. Oecologia DOI: 10.1007/s00442-005-0045-0.

Schiestl, F. P. & F. Marion-Poll. 2002. Detection of physiologically active flower volatiles using gas chromatography coupled with electroantennography. Pp. 173–198 *in* J.F. Jackson, H.F. Linskens, and R.S. Inman (eds.), Molecular methods of plant analysis. Springer-Verlag, Berlin.

Seymour, R. S. 1997. Plants that warm themselves. Sci. Amer. 276: 104–109.

———. 2001. Biophysics and physiology of temperature regulation in thermogenic flowers. Biosci. Rep. 21: 223–236.

——— & P. Schultze-Motel. 1997. Heat-producing flowers. Endeavor 21: 125–129.

——— & ———. 1998. Physiological temperature regulation by flowers of the sacred lotus. Nova Acta Regia. Soc. Sci. Upsal. 353: 935–943.

———, M. Gibernau & K. Ito. 2003a. Thermogenesis and respiration of inflorescences of the dead horse arum *Helicodiceros muscivorus*, a pseudo-thermoregulatory aroid associated with fly pollination. Funct. Ecol. 17: 886–894.

———, I. Terry & R. B. Roemer. 2004. Respiration and thermogenesis by cones of the Australian cycad, *Macrozamia machinii*. Funct. Ecol. 18: 925–930.

———, C. R. White & M. Gibernau. 2003b. Heat reward for insect pollinators. Nature 426: 243–244.

Skubatz, H., D. D. Kunkel, W. N. Howald, R. Trenkle & B. Mookherjee. 1996. The *Sauromatum guttatum* appendix as an osmophore: excretory pathways, composition of volatiles and attractiveness to insects. New Phytol. 34: 631–640.

———, ———, J. M. Patt, W. Howard, T. G. Hartman & B.J.D. Meeuse. 1995. Pathway of terpene excretion by the appendix of *Sauromatum guttatum*. Proc. Natl. Acad. Sci. U.S.A. 92: 10084–10088.

Stange, G. 1992. High resolution measurement of atmospheric carbon dioxide concentration changes by the labial palp organ of the moth *Heliothis armigera* (Lepidoptera: Noctuidae). J. Comp. Physiol. A 171: 317–324.

Stensmyr, M. C., I. Urru, I. Collu, M. Celander, B. S. Hansson & A.-M. Angioy. 2002. Rotting smell of dead-horse arum florets. Nature 420: 625–626.

Stevenson, D. W., K. J. Norstog & P.K.S. Fawcett. 1998. Pollination biology of cycads. Pp. 277–294 *in* S.J. Owens and P.J. Rudall. Reproductive biology, Royal Botanic Gardens, Kew, UK.

Tang, W. 1987a. Insect pollination in the cycad *Zamia pumila* (Zamiaceae). Amer. J. Bot. 74: 90–99.

———. 1987b. Heat production in cycad cones. Bot. Gaz. 148: 165–174.

———. 1993. Heat and odour production in cycad cones and their role in insect pollination. Pp. 140–147 *in* D.W. Stevenson and K.J. Norstog (eds.), Proceedings of CYCAD 90, the second international conference on cycad biology. Palm and Cycad Societies of Australia, Milton, Brisbane.

———, L. Sternberg & D. Price. 1987. Metabolic aspects of thermogenesis in male cones of five cycad species. Amer. J. Bot. 74: 1555–1559.

Terry, L. I. 2001. Thrips and weevils as dual, specialist pollinators of the Australian cycad *Macrozamia communis* (Zamiaceae). Int. J. Pl. Sci. 162: 1293–1305.

———, C. J. Moore, P. I. Forster, G. H. Walter, P. J. Machin & J. Donaldson. 2004a. Pollination ecology of the genus *Macrozamia*: cone volatiles and pollinator specificity. Pp.155–169 *in* A. Lindstrom (ed.), Proceedings of the sixth international cycad conference on cycad biology. Nong Nooch Tropical Botanical Garden, Thailand.

———, C. J. Moore, G. H. Walter, P. I. Forster, R. Roemer, J. S. Donaldson & P. J. Machin. 2004b. Association of cone thermogenesis and volatiles with pollinator specificity in *Macrozamia* cycads. Pl. Syst. Evol. 243: 233–247.

———, G. H. Walter, J. S. Donaldson, E. Snow, P. I. Forster & P. J. Machin. 2005. Pollination of Australian *Macrozamia* cycads: Effectiveness and behavior of specialist vectors in a dependent mutualism. Amer. J. Bot. 92: 931–940.

Thien, L. B., H. Azuma & S. Kawano. 2000. New perspectives on the pollination biology of basal angiosperms. Int. J. Pl. Sci. 161: S225–S235.

Wanzala, W., N. F. K. Sika, S. Gule & A. Hassanali. 2004. Attractive and repellent host odours guide ticks to their respective feeding sites. Chemoecology 14: 229–232.

Wilson, G. W. 2002. Insect pollination in the cycad genus *Bowenia* Hook. ex Hook. f. (Stangeriaceae). Biotropica 34: 438–441.

SYSTEMATICS
CHAPTER 23

Molecular Systematics in Amorphocerine Weevils on Encephalartos—Preliminary Results

Douglas A. Downie and John S. Donaldson

Abstract

Weevils in the tribe Amorphocerini have been implicated in pollination of *Encephalartos* species in southern Africa. The benefits they provide these plants and the unique attributes of the cycad–weevil interaction make them important from both conservation and evolutionary standpoints. Oberprieler, using morphological characters, proposed a tentative hypothesis of relationships among the Amorphocerini, but it is useful to estimate their phylogeny from molecular markers as well. Preliminary results from work in progress are presented comprising DNA sequence data from one mitochondrial and one nuclear gene. The results are reasonably consistent with the morphological hypothesis of relationships and species concepts, though important differences are observed in the placement of the more generalist species *Amorphocerus talpa* Schoenherr as a recent divergence and a more narrow delineation of a *Porthetes hispidus* Boheman group, which is indicated to have experienced recent divergences. This chapter represents the initial step of a more comprehensive study to (1) estimate relationships and the direction of evolution among species of *Amorphocerus* and *Porthetes*, (2) test the hypotheses that *Porthetes* species are host restricted and that speciation has been host plant mediated, (3) examine the extent of cophylogeny between the weevils and their hosts, and (4) estimate levels and patterns of genetic variation within species of Amorphocerini.

Resumen

Los gorgojos de la tribu Amorphocerini estuvieron implicados en la polinización de especies de *Encephalartos* en Sudáfrica. Los servicios que aportan a las plantas y los atributos únicos de la interacción cícada-gorgojo los hacen importantes desde ambos puntos de vista de conservación y evolución. Oberpreiler, usando caracteres morfológicos, ha propuesto una hipótesis tentativa de relaciones entre los Amorphocerini, también es útil en estimar su filogenia por marcadores moleculares. De los resultados preliminares del trabajo en camino se presenta datos de secuenciación de ADN de un gene mitocondrial y otro nuclear. Los resultados son razonablemente consistentes con la hipótesis morfológica de relaciones y conceptos de especie, aunque se observa diferencias importantes en la ubicación de la especie más generalista *Amorphocerus talpa* Schoenherr como una divergencia reciente y una delineación más estrecha de un grupo de *Porthetes hispidus* Boheman, del cual se ha indicado divergencias recientes. Esto representa un paso inicial de un estudio más comprensivo para: (1) estimar las relaciones y la dirección evolutiva entre especies de *Amorphocerus* y *Porthetes*, (2) probar la hipótesis que las especies de *Porthetes* son restringidas al anfitrión y que la especiación ha sido mediado por la planta anfitrión, (3) examinar el nivel de co-filogénia entre los gorgojos y sus anfitriones, y (4) estimar los niveles y patrones de variación genética dentro las especies de Amophocerini.

Introduction

Insects that feed on plants often specialize on one or a few plant taxa. The observation of closely related herbivorous insects on closely related host plant taxa (Ehrlich & Raven, 1964; Beccera, 1997; Futuyma et al., 1995; Funk et al., 2002) suggests a causal link between plant diversity and insect diversity. Evolutionary interactions between plants and the parasitic insects that feed on them may lead to tandem diversification, or cospeciation, but more often the pattern observed has been asymmetric, with insects diversifying independently of their host plants (Jaenike, 1990; Percy et al., 2004). Insect herbivores have often shifted onto closely related host plants, which may be a consequence of a lack of genetic variation to enable shifts to more distantly related hosts (Futuyma et al., 1995). Host shifts occur in a geographic context as well as a taxonomic one—shifts may occur onto sympatric hosts, an event that is considered to be a potential source of ecological speciation in the face of gene flow (Schluter, 2001, or onto allopatric hosts. Indeed, phytophagous insects continue to play a pivotal role in the controversy over the relative importance of sympatric versus allopatric speciation. Thus, the evolutionary history of specialized herbivorous insects and their host plant associations offers an attractive arena to explore the pattern and causes of diversification.

The majority of studies in evolution of plant–insect interactions have justifiably involved angiosperms, but a number of extremely interesting interactions exist between insects and gymnosperms. One of these is the interaction between cycads and the beetles (Coleoptera), particularly weevils (Curculionoidea), that are associated with their reproductive structures. Extant cycads (Cycadales) are long-lived woody plants and descendents of a lineage that extends well into the Mesozoic (Norstog & Nichols, 1997). Feeding on cycad reproductive structures by a relatively small group of cycad-specific beetles (Curculionoidea and Cucujoidea) in several different parts of the world, coupled with fossil evidence of insect damage in cycadaeoids, initially led to the conclusion that at least some of the current associations are derived from ancient cycad–insect interactions (Crowson, 1981; Norstog & Nichols, 1997). Diversification of beetles and cycads in different centers of cycad diversity after the breakup of the Gondwanan and Laurasian supercontinents was thought to be the result of coevolutionary processes acting on these ancient interactions. Subsequent studies of the beetles associated with cycads in the main centers of cycad diversity (South, Central, and North America; Africa; Southeast Asia; Australia) indicate that cycad beetles from different centers are not closely related and that novel cycad–insect interactions have evolved independently in each region (Oberprieler, 2004). Moreover, studies of beetles and thrips associated with Australian cycads show that different insect pollinator mutualisms have evolved among apparently closely related cycads (Terry et al., 2004), suggesting that cycad–insect interactions are more dynamic than was previously postulated.

This study focused on the evolution of the Amorphocerini, a tribe in the weevil family Curculionidae that is endemic to Africa and occurs exclusively on the cycad genus *Encephalartos*.

The genus *Encephalartos* (Zamiacea) is endemic to Africa, where it occurs mostly along the eastern escarpment from South Africa to Sudan but also ranges into central and west Africa. There are approximately 65 described species of *Encephalartos* in Africa, of which 37 species occur in South Africa (29 endemics), mostly in the provinces of the Eastern Cape and KwaZulu Natal (Norstog & Nichols, 1997; Donaldson, 2003; Hill, Stevenson & Osborne, 2007).

The tribe Amorphocerini consists of at least 23 species in two described genera, *Amorphocerus* Schoenherr and *Porthetes* Schoenherr (Oberprieler, 1995, 1996), and one undescribed, provisionally named *Protoporthetes* by Oberprieler (1996). The latter genus is monotypic, its single species having been collected rarely owing to its concealed location within old leaf stalks. Six species of *Amorphocerus* have been described, all but one of which have been associated most often with the female cones of *Encephalartos* species in the Eastern Cape (Oberprieler, 1996). *Amorphocerus setosus* Boheman feeds on old leaf bases that compose the outer trunk of most cycads. *Amorphocerus* species have been collected to the north of this province only in NW-KwaZulu Natal (one species) and in Zimbabwe (one undescribed species) (Oberprieler, 1996). *Porthetes* species generally develop in male cones, but visit female cones, and have been implicated in pollination of at least some *Encepharlatos* species (Donaldson et al., 1995; Donaldson, 1997). *Porthetes* comprises about 16 species and has been collected as far north as Kenya, though most species are known from South Africa, particularly the Eastern Cape (Fig. 23-1). This bias is almost certainly at least in part an artefact of sampling effort and has serious implications for understanding the systematics and ecology of the weevils and the pollination biology in *Encephalartos*. Alternative hypotheses to explain the disjunct distributions in Porthetes are that extant species are relictual and intervening populations have become extinct, which

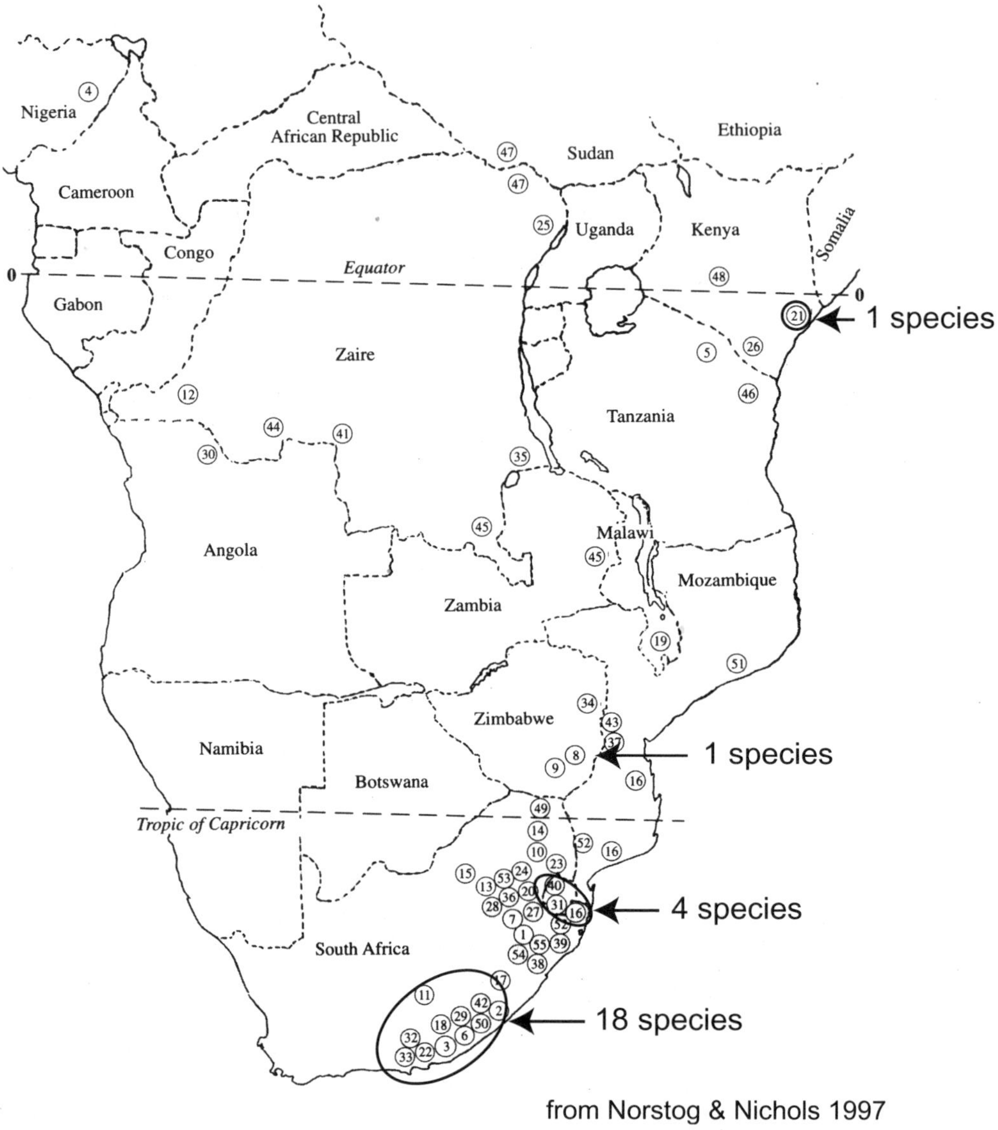

Figure 23-1. Map showing approximate distribution of amorphocerine weevils superimposed on distribution of *Encephalartos*. Representatives of 13 species from the Eastern Cape of South Africa and one species from Swaziland (*P. sp. n.* 1, sequenced for EF-1α only) were sampled for this study. The species from Zimbabwe is undescribed. For details on *Encephalartos* species numbering, see Norstog and Nichols (1997).

predicts substantial divergence between the surviving species, or that long-distance dispersal is involved.

Oberprieler (1996) presented a well-resolved phylogeny of the tribe based on morphological characters. The major features of this phylogeny with respect to host use are the deep divergence of *Amorphocerus* onto female cones of *Encephalartos* species and *Porthetes* onto male cones. Within *Amorphocerus*, *A. setosus*, which develops on trunks, was sister to all other species, all of which develop in cones. Two species on woolly coned *Encephalartos* (*E. friderici-guilielmii* and *E. cycadifolius*) were in a terminal position within the clade. Within *Porthetes,* divergence was indicated between a *Porthetes zamiae* group (*P. zamiae* Boheman, *P. dissimilis* Hesse, and an undescribed species [*Porthetes* sp. n. 1 of Oberprieler (1995)], *P. vorsteri* MS name of Oberprieler [1996], Vorster & Oberprieler [1999], plus two other undescribed species), and a group of highly host-specific species (*P. hispidus* Boheman group) onto male cones was indicated. There was some evidence for divergence of weevils at the level of species groups of the cycads, and some species pairs suggested cospeciation with their hosts, but finer resolution of the cophylogenetic pattern awaits a robust phylogeny for *Encephalartos*.

The tight association of these weevils with their host plants, as well as the possible link to the evolution of cycad pollination mutualisms, makes this a particularly interesting group for the study of speciation and diversification. Moreover, the unique history and role of cycads in plant evolutionary biology and the highly threatened status of many species of *Encephalartos* (Donaldson, 2003) and possible extinction of associated insects, makes a better understanding of the systematics and population genetics of the group a matter of some urgency. Molecular data will be useful in validating the species concepts and phylogenetic hypothesis of Oberprieler, as well as providing a framework for comparative study when a phylogeny for *Encephalartos* becomes available.

Here, a preliminary analysis from work in progress on two genomic regions is provided to address three compound questions: (1) Do molecular data support Oberprieler's species concepts and hypothesis of their relationships? (2) How much divergence is there between, and variation within, species? (3) What is the pattern of evolution with respect to host use?

Materials and Methods

Sampling

We collected weevils from 10 species of *Encephalartos* in the Eastern Cape Province. Effort was made to sample from multiple populations of each host species, though often this was not possible owing to absence of cones (Table 23-1). On morphological grounds, we included 13 species of weevils in the Amorphocerini from 21 populations of the 10 species of *Encephalartos* in the analysis. A number of species in this tribe have been described in a Ph.D. thesis (Oberprieler, 1996), but their names have not yet been validated in a formal publication (and thus remain nomenclaturally unavailable) and are referred to by numbers (see and continued from Oberprieler, 1995) or their host plant association, or both. Voucher specimens for the majority of samples are housed in the Department of Zoology and Entomology at Rhodes University, and at CSIRO, Canberra, Australia. Outgroup sequences from GenBank included *Rhynchophorus palmarum* and *Diocalandra frumenti* (Dryophthorinae)(O'Meara & Farrell, unpublished) for both *EF-1α* and *COI*, and *Araucarius minor* (Cossoninae) (Sequeira et al., 2000) for *COI* only.

Molecular Procedures

We extracted DNA from either whole individuals or thoraces via the Qiagen DNeasy Kit (Qiagen Inc., Valencia, CA) and diluted it in sterile water at half volume. Portions of one nuclear (elongation factor 1α [*EF-1α*]), and one mitochondrial (cytochrome oxidase I, *COI*) gene were amplified via polymerase chain reaction (PCR) under standard conditions, cleaned using the Qiagen MinElute Kit, and sequenced using Big Dye Terminator ver. 3.1 (Applied Biosystems, Foster City, CA). Sequencing reactions were separated on an ABI 3100 Analyzer. Primers used for PCR amplification were EF149B and EF1106A for *EF-1α* (Normark et al., 1999) and either C1-J-1718 (Simon et al., 1994) and BEN3R (Brady et al., 2000), or Zeus (Normark et al., 1999) and BEN3R for *COI*. Alignment of sequences was carried out via Clustal X ver. 1.83 (Thompson et al., 1997) and edited by eye.

Table 23-1 Collection details for Amorphocerini weevils on *Encephalartos* species in South Africa.

Weevil species (sample)	Host species	Location	Plant organ	Date
Amorphocerus				
A. setosus	*E. cycadifolius*	Winterberg Mountains	Trunk	30/11/2004
*A. rufipes*1	*E. friderici-guilielmi*	Cathcart	Female cone	12/11/2003
*A. rufipes*2	*E. friderici-guilielmi*	Thomas River	Female cone	01/02/2004
*A. rufipes*3	*E. friderici-guilielmi*	Quanti	Old male cone	07/11/2004
A. sp. n. 4.1	*E. cycadifolius*	Winterberg Mountains	Male cone	17/11/2004
A. sp. n. 4.2	*E. cycadifolius*	Winterberg Mountains	Male cone	17/11/2004
*A. talpa*1	*E. altensteinii*	Grahamstown	Male cone	Oct. 2003
*A. talpa*2	*E. trispinosus*	North Grahamstown	Old cone peduncle	01/04/2004
*A. talpa*3	*E. longifolius*	Addo NP, Zuurberg	Female cone, surface	08/04/2004
*A. talpa*4	*E. longifolius*	Addo NP, Kabouga	Female cone	07/04/2004
*A. talpa*5	*E. arenarius*	Woody Cape Preserve	Old axis female cone	13/04/2004
*A. talpa*6	*E. latifrons*	Kleinemonde	Male cone	15/07/2004
*A. talpa*7	*E. longifolius*	Riebeek-East	Female cone	22/07/3004
*A. talpa*8	*E. longifolius*	Groendal Wilderness	Male cone	23/07/2004
*A. talpa*9	*E. altensteinii*	Thomas Baines NR	Old female cone	11/10/2004
*A. talpa*10	*E. altensteinii*	King William's Town	Female cone	07/11/2004
Porthetes				
*P. dissimilis*1	*E. latifrons*	Kleinemonde	Male cone	15/07/2004
*P. dissimilis*2	*E. altensteinii*	Grahamstown	Female cone, dehiscence	22/10/2004
*P. dissimilis*3	*E. altensteinii*	King William's Town	Female cone	07/11/2004
P. sp. n. 1	*E. senticosus*	Siteki, Swaziland	Male cone	nd
P. sp. n. 2.1	*E. longifolius*	Riebeek-East	Male cone	22/07/2004
P. sp. n. 2.2	*E. longifolius*	Groendal Wilderness	Male cone	23/07/2004
P. sp. n. 2.3	*E. longifolius*	Riebeek-East	Male cone	22/07/2004
P. sp. n. 6.1	*E. villosus*	Northeast London	Old female cone	08/11/2004
P. sp. n. 6.2	*E. villosus*	Northeast London	Old female cone	08/11/2004
P. sp. n. 6.3	*E. villosus*	Umtiza Reserve, s.East London	Female cone	06/11/2004
P. sp. n. 10	*E. longifolius*	Groendal Wilderness	Male cone	23/07/2004
*P. hispidus*1	*E. friderici-guilielmi*	Cathcart	Male cone	12/11/2003
*P. hispidus*2	*E. friderici-guilielmi*	Thomas River	Male cone	07/11/2004
P. sp. n. 9.1	*E. altensteinii*	King William's Town	Male cone	11/10/2004
P. sp. n. 9.2	*E. altensteinii*	King William's Town	Male cone	11/10/2004
P. sp. n. 5.1	*E. horridus*	Uitenhage	Old male cone axis	21/11/2003

P. sp. n. 5.2	*E. horridus*	Uitenhage	Old male cone axis	21/11/2003
P. sp. n. 4.1	*E. trispinosus*	Kwandwe Game Reserve	Old male cone	10/30/2003
P. sp. n. 4.2	*E. trispinosus*	Manley's Flat	Old male cone	25/01/2004
P. sp. n. 4.3	*E. trispinosus*	North. Grahamstown	Old male cone	01/04/2004

In most cases conspecific samples were collected from different host populations, samples from the same populations were collected from different host individuals. Not all samples appear in figures because some were sequenced for either *EF-1α* or *COI* only. Species names are followed by a sample identifier. Species not formally described are designated by a species number and sample identifier as decimal (i.e., *A.* sp. n. 4.1).

Analysis

We estimated phylogeny via by maximum parsimony (MP) and Bayesian methods. For MP, heuristic searches excluding uninformative characters were used in 100 replications of random sequence addition during tree bisection reconnection. Support for nodes on the tree was obtained by nonparametric bootstrapping under the same search conditions as described in the preceding text. We found the model of sequence evolution that best fit the data using ModelTest ver. 3.06 (Posada & Crandall, 1998). Subsequently, we carried out Bayesian estimation via MrBayes (Huelsenbeck & Ronquist, 2001) with six substitution types and gamma distributed rates with shape parameter and proportion of invariant sites estimated from the data (General Time Reversible model), which was consistent with the ModelTest results. The analysis was run with 1,000,000 Markov Chain Monte Carlo (MCMC) generations sampling every 100 generations. Stationarity of likelihoods was reached rapidly in all cases, and a conservative burnin of 300,000 (300 trees) generations was applied before generating a consensus tree and posterior probabilities.

Results and Discussion

Sequence Details

The final alignment for EF-1α was 868 base pairs (bp). This includes an intron of variable length (99 bp of the alignment) inserted between positions 557 and 558 of the coding sequence. The intron presented no alignment problems, and indels in the intron were largely informative with respect to species groups. The final *COI* alignment was 746 bp.

Phylogenetic Analysis

EF-1α

There were 171 parsimony informative characters in the *EF-1α* data set with the intron included, and 114 when it was excluded. Excluding the intron from the analysis had no effect on tree topology, but did reduce subsequent bootstrap values somewhat. Though there were a very large number of most parsimonious trees (79,004 MPTs), the strict consensus tree is reasonably well resolved (L=376, CI=0.644, RI=0.897) (not shown), with most nodes supported by high bootstrap values. The samples from *Porthetes* sp. n. 9 from *E. altenstienii* were unresolved, one of the three *Amorphocerus rufipes* Boheman samples from *E. friderici-guilielmi* failed to be placed with the other two, and *Porthetes* sp. n. 1 was not separated from *P. dissimilis* in both MP and Bayesian analyses. The latter point is of interest in that these species are separated by a large geographic gap and might be expected to be more divergent if they have been separate for a long time. This is inconsistent with the hypothesis of a relictual distribution. Sequences of all other species for which multiple specimens were sequenced were monophyletic. Notable departures from Oberprieler's tree based on morphology was the placement of *A. setosus* as sister to *A. rufipes* and the species *Amorphocerus* sp. n. 4 on *E. cycadifolius*, rather than to all other *Amophocerus*, a grouping that is satisfying in the respect that a monophyletic clade on the woolly coned *Encephalartos* is formed but dissatisfying based on grounds that it requires a reversal to trunk feeding (*A. setosus* develops in old leaf bases on the trunk of *E. cycadifolius*, though it may not be restricted to this host), which is hypothesized to be ancestral in Amorphocerini (Oberprieler, 1996). The clade is not well supported by bootstrap or posterior probabilities. In addition, Oberprieler's *P. hispidus* group has *P.* sp. n. 10 sister to all other members of the clade, with *P.* sp. n. 6, followed by *P.* sp. n. 2, split from a small terminal clade consisting of *P.* sp. n. 4, *P.* sp. n. 5, *P.* sp. n. 9, and *P. hispidus*. With the exception of the branches subtending five of the six deepest nodes in the tree, all branches are quite short, suggesting little divergence between species or individuals from different populations within species, for this gene. An apparently new and previously uncollected species of *Porthetes* (here called *Porthetes* sp. n. 10), morphologically similar to a species on *E. caffer* (Oberprieler, pers. comm., 2004; not sampled for this study), was placed deep in the tree, after the split of the *P. zamiae* group in MP, and sister to all other *Porthetes* in Bayesian analysis.

COI

Of 746 bp of *COI*, 233 were parsimony informative. Problems with homoplasy were apparent for MP analysis of the *COI* data, with a number of relationships being difficult to accept based on both the morphological and *EF-1α* tree. Down-weighting of third codon positions did not improve this, but rather sacrificed phylogenetic signal. Bayesian analysis performed better for this data set, retrieving a well resolved and supported tree (not shown), differing from Oberprieler's tree, for *Porthetes*, in placing *Porthetes* sp. n. 2 on *E. longifolius* as a more basal split in the *P. hispidus* group (*Porthetes* sp. n. 10 was sister to all other species in this group). The relationships of the other species in this group were quite similar to the morphological tree, except that the relations of the three species on *E. altensteinii* (*P.* sp. n. 9), *E. trispinosus* (*P.* sp. n. 4), and *E. horridus* (*P.* sp. n. 5) were switched, so that *P.* sp. n. 4 and *P.* sp. n. 5 were sister taxa; with *P.* sp. n. 9 sister to them. For *Amorphocerus*, *A. talpa* was placed into a more derived position. *Amorphocerus* sp. n. 4 on *E. cycadifolius* was anomalously placed within *A. talpa*.

Combined Data

Combining the data into a matrix of 1616 bp (363 parsimony informative) led to a well resolved tree with high bootstrap and posterior probability values for nearly all nodes (Fig. 23-2), which is here taken to be the best estimate of amorphocerine phylogeny based on these data. Sequences of all species with multiple samples were monophyletic, excepting *Porthetes* sp. n. 4 on *E. trispinosus* for MP. As with the Bayesian analysis of the *COI* data, *P.* sp. n. 10 and *Porthetes* sp. n. 2 are the most basal splits in the *P. hispidus* group (and on relatively long branches, suggesting fairly old divergences). This topology suggests an early colonization of *E. longifolius*, with a somewhat later shift to *E. friderici-guilielmi* than implied by the morphological tree, though it should be noted that the records of adult beetles of the new species from three populations of *E. longifolius* have not been accompanied by observations of larvae in the plant tissues. Though only a single species of the *P. zamiae* group is represented for both genes (*P. dissimilis*), a deep divergence between this group and other *Porthetes* is supported by these data. *P. dissimilis* and two other members of this group develop in female cones, while members of the so-called *P. hispidus* clade develop in male cones (*Porthetes* sp. n. 6, in part, called *P. pearsoni* in Oberprieler

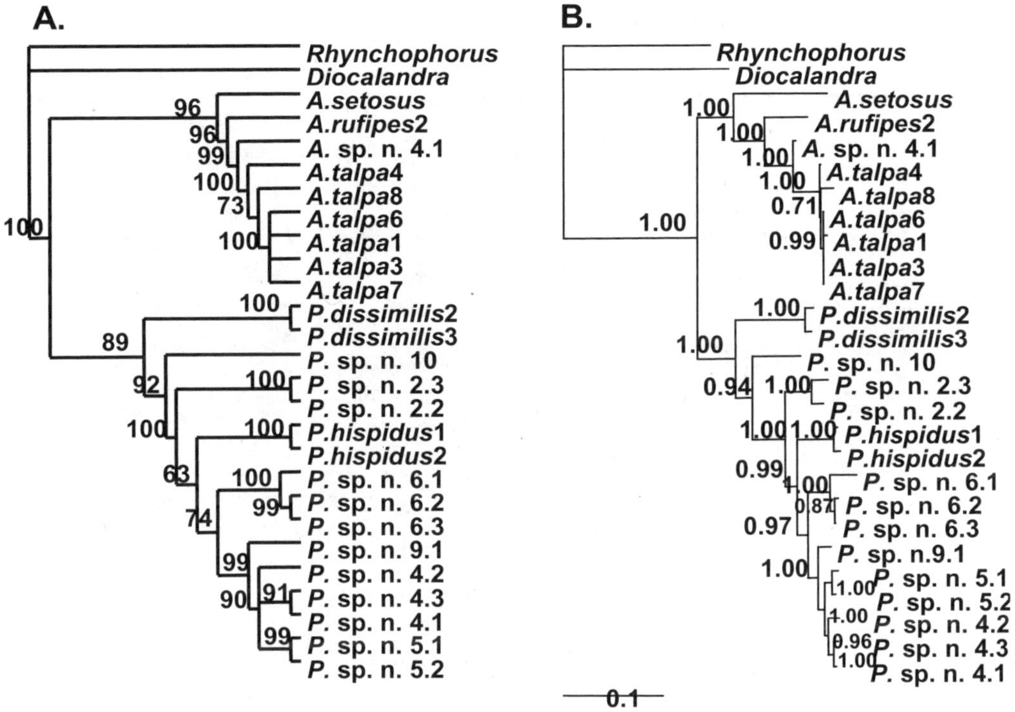

Figure 23-2. **A.** Strict consensus of two most parsimonious trees for combined *EF-1α* and *COI* data (L=963, CI=0.529, RI=0.779). Bootstrap values are based on 200 replications and are placed to the left of some subtending branches for clarity. **B.** Tree based on Bayesian analysis. Numbers are posterior probabilities. Samples sequenced for only one of the genes were omitted from the analysis.

[1996] and Vorster & Oberprieler [1999] apparently develops in both male and female cones). The shift to male cones in this clade may represent the onset of the evolution of pollinator mutualisms, but additional taxon sampling is necessary to locate the origin of this trait with any confidence.

The species in the clade of the *P. hispidus* group are distinguished by relatively few apomorphies and separated by low nucleotide divergences (branch lengths ranging from 0.004 to 0.038 and between species divergences from 0.010 to 0.072 substitutions per site; Fig. 23-2B), suggesting recent divergence, and by inference recent association of these species with *Encephalartos*. This is consistent with studies that suggest recent divergence among *Encephalartos* species (Treutlein & Wink, 2002; Treutlein et al., 2005, Vorster, 2007) and indicate that some of these weevil–cycad associations are not ancient and compose a

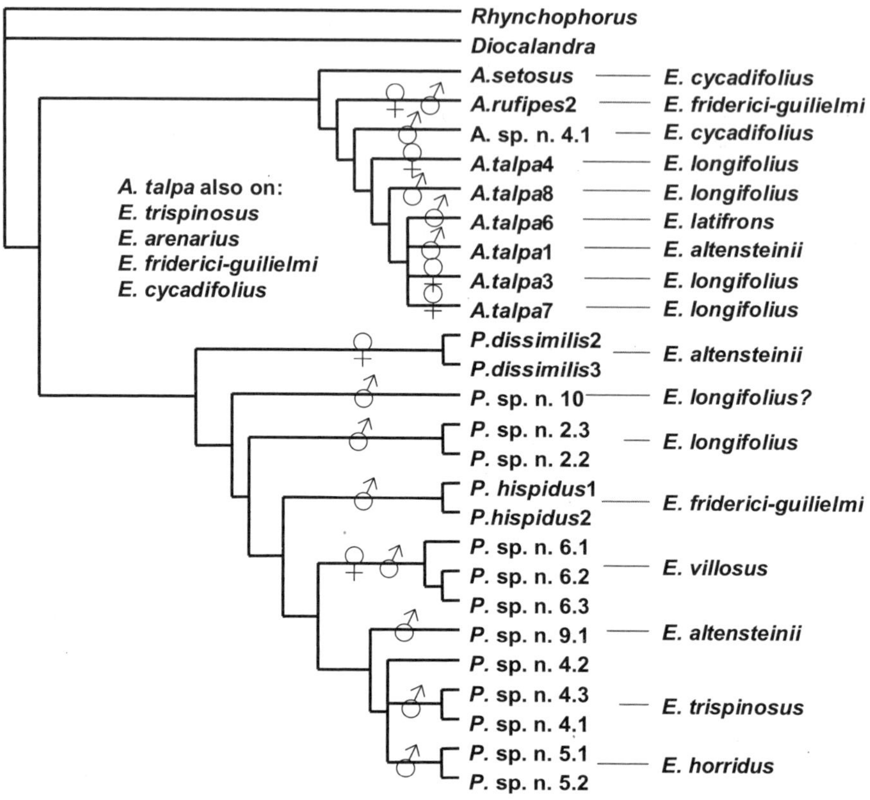

Figure 23-3. Strict consensus tree of combined *EF-1α* and *COI* data with host associations shown to the right. Development site (male or female cones) for each weevil species is indicated along subtending branch.

dynamic and actively evolving system. Even the colonization of male cones by *Porthetes* may be relatively recent.

Conclusions

The molecular analysis of a relatively small subset of Amorphocerini provides little evidence for sister species or clades occurring on sympatric hosts, though the incomplete sample precludes detailed analysis (Fig. 23-3). The current data suggest, however, that speciation in the Amorphocerini has been mediated by host shifts across geography, rather than host-driven divergence in sympatry, or cospeciation. Analyses of cophylogenetic patterns between amorphocerine weevils and their hosts can be undertaken only once there is a resolved phylogeny for *Encephalartos*. This work represents the early phases of a larger project to examine the Amorphocerini across their full range, and it is anticipated that more robust analyses of speciation and host use will be possible when this study is complete.

Acknowledgments

Rhodes University Joint Research Grants and a grant from the Bressler Foundation to John Donaldson provided funding for this research. We thank Rolf Oberprieler for identifications, helpful comments, and insight, and the farmers and rangers who graciously gave of their time and resources to assist in the collection of weevils. We also thank Andrew Vovides and Roy Osborne, and Andrea Cruz for endless helpful e-mails.

Literature Cited

Beccera, J. X. 1997. Insects on plants: macroevolutionary chemical trends in host use. Science 276: 253–256.

Brady, S. G., J. Gadau & P. S. Ward. 2000. Systematics of the ant genus *Camponotus* (Hymenoptera: Formicidae): a preliminary analysis using data from the mitochondrial gene cytochrome oxidase I. Pp. 131–139 *in* A. D. Austin & M. Dowton (eds.), Hymenoptera. Evolution, biodiversity, and biological control. CSIRO Publishing, Collingwood, Victoria, Australia.

Crowson, R. A. 1981. The biology of the Coleoptera. Academic Press, London.

Donaldson, J. S. 1997. Is there a floral parasite mutualism in cycad pollination? The pollination biology of *Encephalartos villosus*. Am. J. Bot. 84: 1398–1406.

———. 2003. Regional overview: Africa. Pp. 9–19. *in* J. S. Donaldson (ed.), Cycads: status survey and conservation action plan. IUCN—The World Conservation Union, Cambridge, UK.

———, I. Nänni & J. de Wet Bosenberg. 1995. The role of insects in the pollination of *Encephalartos cycadifolius*. Pp. 423–434 *in* P. Vorster (ed.), Proceedings of the third international conference on cycad biology. Cycad Society of South Africa, Stellenbosch.

Ehrlich, P. R. & P. H. Raven. 1964. Butterflies and plants: a study in coevolution. Evolution 18: 586–608.

Funk, D. J., K. E. Filchak, & J. L. Feder. 2002. Herbivorous insects: model systems for the comparative study of speciation ecology. Genetica 116: 251–267.

Futyuma, D. J., M. C. Keese, & D. J. Funk. 1995. Genetic constraints on macroevolution: the evolution of host affiliation in the leaf beetle genus *Ophraella*. Evolution 49: 797–809.

Hill, K. D., D. W. Stevenson & R. Osborne. 2007. The world list of cycads. Mem. New York Bot. Gard. 97: 454–483.

Huelsenbeck, J. P. & Ronquist, F. 2001. MrBayes: Bayesian inference of phylogeny. Bioinformatics 17: 754–755.

Jaenike, J. 1990. Host specialization in phytophagous insects. Annu. Rev. Ecol. Syst. 21: 243–273.

Normark, B. B., B. H. Jordal & B. D. Farrell. 1999. Origin of a haplodiploid beetle lineage. Proc. R. Soc. Lond. [Biol.] 266: 2253–2259.

Norstog, K. J. & T. J. Nichols. 1997. The biology of the cycads. Comstock Publishing Associates, Ithaca, NY.

Oberprieler, R. G. 1995. The weevils associated with cycads 1. classification, relationships, and biology. Pp. 335–378 *in* P. Vorster (ed.), Proceedings of the third international conference on cycad biology. Cycad Society of South Africa, Stellenbosch.

———. 1996. Systematics and evolution of the tribe Amorphocerini, with a review of the cycad weevils of the world. Ph.D. dissertation, University of the Free State, Bloemfontein, South Africa.

———. 2004. "Evil weevils"—the key to cycad survival and diversification? *in*: A. Lindstrom (ed.), Proceedings of the 6th international conference on cycad biology. Nong Nooch Tropical Garden, Chonburi, Thailand.

Percy, D. M., R. D. M. Page & Q. C. B. Cronk. 2004. Plant-insect interactions: double-dating associated insect and plant lineages reveals asynchronous radiations. Syst. Biol. 53: 120–127.

Posada D. & Crandell K. A. 1998 Modeltest: Testing the model of DNA substitution. Bioinformatics 14: 817–818.

Schluter, D. 2001. Ecology and the origin of species. Trends Ecol. Evol. 16: 372–380.

Sequeira, A. S., B. B. Normark & B. D. Farrell. 2000. Evolutionary assembly of the conifer fauna: distinguishing ancient from recent associations in bark beetles Proc. Roy. Soc. Lond. B 267: 2359–2366.

Simon, C., F. Frati, A. Beckenbach, B. Crespi, H. Liu & P. Flook. 1994. Evolution, weighting, and phylogenetic utility of mitochondrial gene sequences and a compilation of conserved polymerase chain reaction primers. Ann. Entomol. Soc. Am. 87: 651–701.

Terry, I. L., C. Moore, G. H. Walter, P. I. orster, R. B. Roemer, J. S., Donaldson & P. Machin, P. 2004. Association of cone thermogenesis and volatiles with pollinator specificity in *Macrozamia* cycads. Plant Syst. Evol. 243: 233–247.

Thompson, J. D., T. J. Gibson, F. Plewniak, F. Jeanmougin & D. G. Higgins. 1997. The CLUSTAL X windows interface: flexible strategies for multiple sequence alignment aided by quality analysis tools. Nucleic Acids Res. 25: 4876–4882.

Treutlein, J. & M. Wink. 2002. Molecular phylogeny of cycads inferred from *rbc*L sequences. Naturwissenschaften 89:221–225.

———, P. Vorster & M. Wink. 2005. Molecular relationships in *Encephalartos* (Zamiaceae, Cycadales) based on nucleotide sequences of nuclear ITS 1 & 2, rbcL, and genomic ISSR fingerprinting. Plant Biol. 7: 79–90.

Vorster, P. & R. G. Oberprieler. 1999. Entomological evidence for and against taxonomical decisions in *Encephalartos* sp. Pp. 198–207 *in* C. J. Chen (ed.), Biology and conservation of cycads—Proceedings of the fourth international conference on cycad biology, Panzhihua, China, International Academic Publishers, Beijing, China.

Vorster, P. 2007. Subdivision of *Encephalartos*. Mem. New York Bot. Gard. 97: 597–610.

SYSTEMATICS
CHAPTER 24

The Populations of Dioon caputoi in the Tehuacán-Cuicatlán Biosphere Reserve

Juan C. Flores-Vázquez, Silvia Salas-Morales, and Alfredo Saynes-Vázquez

Abstract

Dioon caputoi is a Mexican endemic species that is distributed along a very small area inside the Tehuacán-Cuicatlán Biosphere Reserve. Our main goal was to assess the basic demographic features of two populations and to verify the presence of other possible populations in the region. In both populations, the number of individuals is small, the population structure is an "inverse-j" type, the percentages of reproductive individuals are 30% and 40% respectively, and the sex ratio is 2.5:1 in favor of males. Besides these analyses, we carried out a survey of the surrounding region and found three new *D. caputoi* localities, which have larger population sizes than the original two. The natural vegetation at two of the five population sites presented a very advanced level of degradation.

Resumen

Dioon caputoi es una especie endémica de México que se distribuye en un área muy reducida dentro de la Reserva de la Biosfera Tehuacán-Cuicatlán. Nuestro objetivo fue registrar las características demográficas básicas en dos poblaciones y verificar la presencia de otras poblaciones en la región. Ambas poblaciones presentan una estructura tipo j-invertida, el porcentaje de individuos en estado reproductivo es 30% y 40 % y la proporción sexual es 2.5:1 siendo mayor el número de individuos masculinos. Además, realizamos recorridos de campo en la región encontrando tres nuevas localidades, las cuales tienen un mayor tamaño poblacional que las poblaciones anteriormente mencionadas. De cinco localidades en total, dos presentan un deterioro muy avanzado de la vegetación.

Introduction

The genus *Dioon* is the most endemic of the cycad genera present in Mexico (*Ceratozamina, Dioon*, and *Zamia)*, with 13 of its 14 existing species found only in that country. Although the genus is widely distributed in Mexico, from latitude 15° to 29°N, most of the species are highly localized (Sabato & De Luca, 1985). An exceptional example of this localized endemism in Mexico is in the region located in northwestern Oaxaca and southeastern Puebla that contains a significant number (five) of *Dioon* species (Gregory et al., 2003). This phytogeographical region, the Tehuacán-Cuicatlán valley, has a unique and outstanding flora, with many plant groups exhibiting high endemism (Villaseñor et al., 1990). In consideration of the great biological diversity in this region, the government of Mexico declared 149,187 ha in Oaxaca and Puebla a natural protected area in 1998, the Tehuacán-Cuicatlán Biosphere Reserve. All *Dioon* species existing in the Tehuacán-Cuicatlán Biosphere Reserve are considered endangered owing to their restricted geographic distribution and small population sizes. In addition, their slow rate of reproduction and growth places them in a situation of high vulnerability to any kind of extraction or disturbance. For these reasons, all are included in the IUCN Red List (Walter & Gillet, 1997), and also in the severe risk category of the Mexican Official Norm of Endangered Species (Diario Oficial de la Federación, 2002).

Despite the delicate situation of *Dioon* spp. and their potential as models for cycad conservation throughout the new world, and in Mexico in particular, studies that provide useful knowledge for their conservation have been scarce. If we look at the 13 currently known Mexican *Dioon* species, the only two on which demographic and genetic studies have been carried out are *D. edule* Lindl. (Vovides & Peters, 1987; Vovides, 1990; González-Astorga et al., 2003a, b) and *D. angustifolium* Miq. (González-Astorga et al., 2005). As part of defining the level of threat or extinction to a cycad species, a detailed analysis of its demographics, population genetics, and a systematic evaluation of the real and potential risk factors should be included (Osborne, 1995). We agree with the view of some authors that, in the short run, the study of demographics, instead of population genetics, will render more useful knowledge about the minimum size a population requires to maintain viability (Lande, 1988; Schemske et al., 1994). Therefore, we believe it is necessary to undertake demographic studies in order to understand the current situation of endangered species. Although it is clear that all 13 of the *Dioon* species are at some level of endangerment, some are more threatened than others. Hence our strategy is to begin by focusing on the species that are at greatest risk.

Dioon caputoi De Luca, Sabato & Vázq. Torres is a cycad distributed in a very small area of the Tehuacán-Cuicatlán Biosphere Reserve. Its populations are the smallest reported to date for a *Dioon* species. The IUCN places it at the maximum threat level (Critically Endangered) for species that are still located in a wild environment (Stevenson et al., 2003). Despite this, very little is known about the species. The only published works are focused on its taxonomic description (De Luca et al., 1980) and its evolutionary and phylogenetic relations to other *Dioon* species (Sabato & De Luca, 1985; Moretti et al., 1993). Therefore, we consider it important to begin to understand the situation with the populations of this species and to acquire information that could contribute to the implementation of a conservation action plan.

Methods

Study Site

We carried out the study in the central part of the Tehuacán-Cuicatlán Biosphere Reserve, in a portion of the territory of the state of Puebla (Fig. 24-1). Five populations of *D. caputoi* are located in this area. Two had already been reported (De Luca et al., 1980), and three previously unknown populations are reported here. The two populations previously reported are located in the rural community of San Luis Atolotitlán in two regions known by us as "Santuario" and "La Grana." The first site, Santuario, has been highly disturbed and thus it is very difficult to determine accurately the type of vegetation originally present there. Nevertheless, some evidence suggests that the site was located in a transition zone between oak forest and tropical deciduous forest. Among the tree species, we identified the following: *Quercus* sp., *Ipomoea arborescens* Sweet, *Bursera* sp., and the candelabra cacti *Polaskia chichipe* (Rol. Goss.) Backeb. and *P. chende* (Gossel.) A. C. Gibson & K. E. Horak. At La Grana, the second site, the type of vegetation is a tropical deciduous forest. Among the tree species there, we identified *Bursera* sp., *Lysiloma* sp., *Heliocarpus* sp., *Plumeria rubra* L., and the cacti *Polaskia chichipe*, and *P. chende*. We obtained climate data were obtained from the weather station in Caltepec, Puebla, and according to the Köppen climate classification system modified

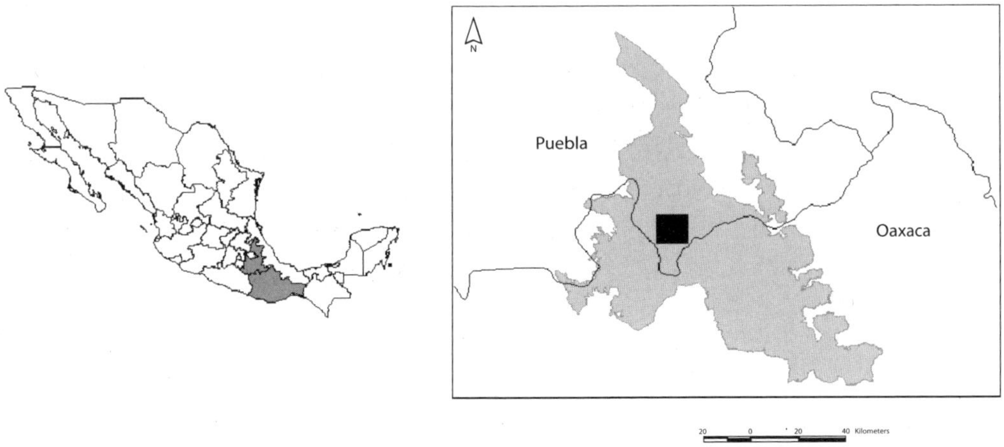

Figure 24-1. Location of the Tehuacán-Cuicatlán Biosphere Reserve (gray area), and the geographical distribution of the *D. caputoi* populations (black square).

by García (1988), this area has a predominantly dry climate with summer rains, with median annual temperature of 22°C, and a well defined "canícula" ("dog days" period) at the middle of the rainy season. The average annual rainfall varies from 400 mm to 450 mm.

Sampling of Populations

The number of individuals was small in both populations, occupying areas smaller than 1 ha each, so it was not possible to use an area delimitation method for sampling the populations. Therefore, all individuals in each population were marked with a consecutive number on an identification label. Stem height and three diameter measurements, number of leaves, sex, and the degree of maturity of the cones were recorded for every individual.

From the total sampled pool, five megastrobili were harvested, and their length and diameter measured. Also, the viable and nonviable seeds were separated by the floating method and then counted. We obtained the mean number, with standard deviation, of viable seeds per cone.

Trips were made with local guides to verify the existence of new populations. When found, their size was estimated, and the quality of the habitat recorded.

Statistical Analysis

We constructed contingency tables using female and male distributions in the two populations. Expected data were calculated based on a 1:1 sex ratio. We tested sex ratio differences between populations via a *G* test and compared *G*-values with χ^2 tables. Differences of $p < 0.05$ were considered significant.

Results

Population Structure

The sizes of the *D. caputoi* populations at the Santuario and La Grana sites are only 47 and 43 individuals respectively, so we pooled the data from the two populations for the analysis. In each population, the individuals were grouped into different size categories by their stem

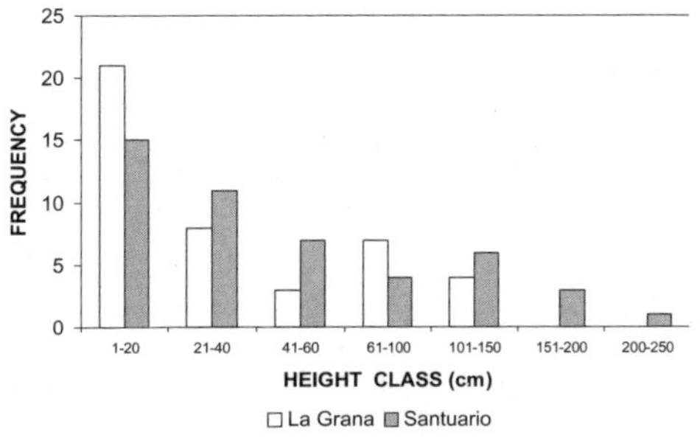

Figure 24-2. Frequency of individuals of *D. caputoi* for each height class in the Santuario and La Grana populations.

height. The first three, and smallest, categories differed by increments of 20 cm; the fourth by an increment of 40 cm; and the last three, and largest, differed by 50 cm. In the La Grana population, just the five first categories were present because the tallest individual was 143 cm (Fig. 24-2). Both population structures showed an inverted "j" type curve according to Bongers et al. (1988), where there a greater number of individuals in the initial, smaller, categories, and gradually decrease in the taller categories (Fig. 24-2). No recruitment of seedlings was recorded in either population, which may indicate that there are strong pressures against the establishment of seedlings and therefore regeneration by sexual reproduction.

In both populations, individuals exhibited vegetative growth, as evidenced by ramets growing from the base of large plants. In the Santuario population a larger percentage of individuals with more than one ramet was found, 23%, versus 16% at La Grana. The number of ramets per individual also differed between the populations. In the La Grana population, which is in a more preserved habitat, the individuals showed from 1 to 4 stems, whereas at Santuario, a more deteriorated habitat, 1 to 10 stems per individual were recorded.

Reproductive Parameters

Although the population sizes are small and the largest numbers of individuals are in the juvenile categories in both populations, the percentages of reproductive individuals were high, 30% in Santuario and 40% in La Grana. The number of male individuals was double that of female individuals (Table 24-1). The distribution of sexes in the two populations did not show significant differences ($G=0.22$; $p>0.05$, d.f. $=1$). This could mean that the males cone more frequently than the females, thus giving a bias toward males as occurs in some species of Zamia (Clark & Clark, 1987; Negrón-Ortíz, et al., 1996; but we do not have data to demonstrate if this ratio changes temporally or is due to environmental

Table 24-1 Reproductive parameters for La Grana and Santuario populations.

Variable	Population	
	La Grana	Santuario
Reproductive individuals	40%	30%
male	12	10
female	5	4
Sex ratio male/female	2.4	2.5

changes. The percentage of total viable seeds obtained from five megastrobili was relatively low (43%). The mean (± SD) of viable seeds per megastrobilus was 171.6 ± 32.4.

New Populations

During explorations in the zone of study, we located three new populations. Two of them are in the rural community of Coatepec and the third is in the territory of the community of Caltepec. The localities where the Coatepec populations are located are named Loma Pachona and Agua de Burro. The Caltepec locality is named Cerro Grande. These new populations are larger and cover larger surface areas than the other two populations previously reported. Of the five populations, two of them, Santuario and Cerro Grande, are located in highly disturbed sites; two populations, Loma Pachona and Agua de Burro, are located in less disturbed sites; and only one, La Grana, is in a preserved habitat (Table 24-2).

Discussion

Similar population structure exhibited in the two populations of *D. caputoi* in Atolotitlán has been reported for other Mexican cycads, *Ceratozamia matudae* Lundell (Pérez-Farrera et al., 2000) and *D. edule* (Vovides, 1990). It is important to note that the two *Dioon* species share similar patterns of population structure, despite differing in other population characteristics such as geographical distribution and population size. *D. edule* is the species

Table 24-2 Habitat quality, population size (no. individuals) and approximate area for each population of D. caputoi in the Tehuacán-Cuicatlán Biosphere Reserve.

Locality	Approximate area (ha)	Population size	Habitat condition
La Grana	1	43	Undisturbed
Agua de Burro	5	>100	Disturbed
Loma Pachona	2	<100	Disturbed
Santuario	1	47	Very disturbed
Cerro Grande	5	>100	Very disturbed

with the widest geographical distribution and the largest population sizes, while *D. caputoi* has the most restricted geographical distribution and the smallest population sizes.

Theoretically, the population structure exhibited by *D. caputoi* implies a good regenerative process because of the high frequency of juvenile individuals that could replace dead adults. Nevertheless, owing to the small size of both populations and the slow growth rate of this species, this conclusion must be carefully evaluated. No recent recruitment of seedlings was recorded. Also, no recent recruitments were observed in the three new populations recorded in Coatepec and Caltepec. This could mean that there are serious natural limitations for the establishment of seedlings. It could also be that advanced human disturbance of the habitat is affecting germination or seedling establishment. The extraction of firewood and land conversion to cultivated fields has decreased plant density and diversity. Soil erosion on the steep slopes where *D. caputoi* grows can have a negative effect on the availability of favorable sites for seed germination and seedling establishment. Another possibility is that even if the seedlings could become established, they are vulnerable to trampling by the abundant goats and cattle that graze in the area.

The discovery of three new populations slightly increases the geographical distribution and the total number of individuals in *D. caputoi*, but its ecological situation continues to be critical because, according to Stevenson et al. (2003), it still exhibits the lowest number of individuals of all the species in the genus. It must also be taken into account that this species is distributed among five fragmented populations, but low levels of genetic flow among them are occurring (D. Cabrera, unpubl. data).

A group of populations that maintain contact among themselves through the migration of individuals is known as metapopulation (Valverde, 1999). If we apply the concept of a metapopulation to *D. caputoi*, we would have a metapopulation composed of five subpopulations, supposing that migration between the subpopulations could occur, and this could be possible via seed dispersion between subpopulations or via pollinators carrying pollen between individuals in different subpopulations. According to the information collected by this study and evidence from other studies, the distances that separate different *D. caputoi* subpopulations could be large enough to prevent migration between them. The evidence recorded from other cycad species suggests that dispersers/predators do not move seeds more than 50 m (Burbidge & Whelan, 1982; Ballardie & Whelan, 1986). In the case of *D. caputoi*, the shortest distance between different populations is 741 m in a straight line between La Grana and Agua de Burro populations. The population at Cerro Grande is the most isolated of the group and is located 5 km in a straight line from the closest population. Lastly, the suspected pollinators of these plants are very small weevils and, up to now, the distance they can cover has not been recorded; nor has the possibility that they could move pollen among the populations been evaluated.

Conclusions

Obtaining demographic structure and searching for new populations are the first steps to understanding the status of *D. caputoi*. Ultimately a more detailed and long-term demographic study must be done (ca. 5 years), including all the five known populations of this species. Such a demographic study must consider the use of population projection matrix models in order to define the finite demographic increase rate, the main demographic variables, and the most sensitive steps in the life cycle. With this information, we will have a real understanding of the true status of the species.

Understanding the interactions that *D. caputoi* maintains with other organisms, mainly dispersers and pollinators, could be of great importance to managing and maintaining the populations effectively. Therefore it is necessary to evaluate the importance of these interactions as soon as possible. Likewise, it is important to know the genetic viability inside and among populations, and to use all of this knowledge in making management and preservation decisions. A global preservation plan for *D. caputoi* must integrate demographic, genetic, and biological interactions information. Only in this way will it be possible to manage effectively the preservation of this highly endangered species.

Acknowledgments

We acknowledge the personnel of the Tehuacán-Cuicatlán Biosphere Reserve, Juan Manuel Salazar, Fernando Reyes, and Mireya Hernández, for the logistical support they provided us and for introducing us to the people of the towns of Atolotitlán, Caltepec, and Coatepec. We also acknowledge our colleague at SERBO, A.C., Arturo Sánchez for field work. We also thank the people of San Luis Atolotitlán, Ciro Mendoza, Taurino López, and Andrés for their guiding and help with field work. We also thank two anonymous reviewers for their comments, which improved the manuscript. Special thanks are expressed to Tim Gregory for the manuscript translation, and to Gerardo Rodríguez for drawing the map. Lastly, we thank Irene Sánchez Gallén for the review and improvement of the entire document.

Literature Cited

Ballardie, R. T. & R. J. Whelan. 1986. Masting, seed dispersal and seed predation in the cycad *Macrozamia communis*. Oecologia 70: 100–105.

Bongers, F., J. Popma, J. Meave & J. Carabias. 1988. Structure and composition of the low rain forest of Los Tuxtlas, Mexico. Vegetatio 74: 55–88.

Burbidge, A. H. & R. J. Whelan. 1982. Seed dispersal in a cycad, *Macrozamia riedlei*. Aust. J. Ecol. 7: 63–67.

Clark, D. A. & D. B. Clark. 1987. Temporal and environmental patterns of reproduction in *Zamia skinneri*, a tropical rain forest cycad. J. Ecol. 75: 135–149.

De Luca, P., S. Sabato & M. Vázquez-Torres. 1980. *Dioon caputoi* (Zamiaceae), a new species from Mexico. Brittonia 32: 43–46.

Diario Oficial de la Federación. 2002. Norma Oficial Mexicana NOM-059-ECOL-2001. Pp. 1–56. Protección ambiental-especies nativas de México de flora y fauna silvestres-Categorías de riesgo y especificaciones para su inclusión, exclusión o cambio-lista de especies en riesgo. Segunda Sección, 6 de marzo de 2002. México, D.F.

García, E. 1988. Modificaciones al sistema de clasificación climática de Köppen. Instituto de Geografia, UNAM, México, D.F.

González-Astorga, J., A. P. Vovides, M. Ferrer & C. Iglesias. 2003a. Population genetics of *Dioon edule* Lindl. (Zamiaceae, Cycadales): biogeographical and evolutionary implications. Biol. J. Linn. Soc. 80: 457–467.

———, ——— & C. Iglesias. 2003b. Morphological and geographical variation of the cycad *Dioon edule* Lindl. (Zamiaceae): ecological and evolutionary implications. Bot. J. Linn. Soc. 141: 465–470.

———, ———, A. Cruz-Angón, P. Octavio-Aguilar & C. Iglesias. 2005. Allozyme variation in three extant populations of the narrowly endemic cycad *Dioon angustifolium* Miq. (Zamiaceae) from north-eastern Mexico. Ann. Bot. 95: 999–1007.

Gregory, T. J., J. Chemnick, S. Salas–Morales & A. P. Vovides. 2003. A new species in the genus *Dioon* (Zamiaceae) from north-central Oaxaca. Bot. J. Linn. Soc. 141: 471–476.

Lande, R. 1988. Genetics and demography in biological conservation. Science 241: 1455–1460.

Moretti, A., P. Caputo, S. Cozzolino, P. De Luca, L. Gaudio, G. Siniscalco-Gigliano & D. W. Stevenson. 1993. A phylogenetic analysis of *Dioon* (Zamiaceae). Amer. J. Bot. 80: 204–214.

Negrón-Ortiz, V., D. L. Gorchov & G. J. Breckon. 1996. Population structure in *Zamia* (Zamiaceae) in northern Puerto Rico. II. Seed germination and stage-structured population projection. Int. J. Plant. Sci. 157: 605–614.

Osborne, R. 1995. The world cycad census and proposed revision of the threatened species status for cycad taxa. Biol. Conser. 71: 1–12.

Pérez-Farrera, M. A., P. Quintana-Asencio, B. Salvatierra & A. P. Vovides. 2000. Population dynamics of *Ceratozamia matudai* Lundell (Zamiaceae) in El Triunfo Biosphere Reserve, Chiapas, México. J. Torr. Bot. 127: 291–299.

Sabato, S. & P. De Luca. 1985. Evolutionary trends in *Dioon* (Zamiaceae). Am. J. Bot. 72: 1353–1363.

Schemske, D. W., B. C. Husband., M.H. Ruckelhause, C. Goodwillie, I. M. Parker & J. G. Bishop. 1994. Evaluating approach to the conservation of rare and endangered plants. Ecology 75: 584–606.

Stevenson, D. W., A. P. Vovides & J. Chemnick. 2003. Regional overview: New World. Pp. 31–38 in J. D. Donaldson (ed.), Status survey and conservation action plan. Cycads. The World Conservation Union, Cambridge.

Valverde, T. 1999. Las metapoblaciones en la naturaleza, ¿realidad o fantasía?. Ciencias 53: 56–63.

Villaseñor J. L., P. Dávila & F. Chiang. 1990. Fitogeografía del Valle de Tehuacán-Cuicatlán. Bol. Soc. Bot. México 50: 135–149.

Vovides, A. P. 1990. Spatial distribution, survival, and fecundity of *Dioon edule* (Zamiaceae) in Veracruz, México, with notes on its habitat. Am. J. Bot. 77: 1532–1543.

——— & C. M. Peters. 1987. *Dioon edule*: la planta más antigua de México. Ciencia y Desarrollo XIII 73: 19–24.

Walter, K. S. & H. J. Gillet. 1997. IUCN. Red list of threatened Plants. Compiled by the World Conservation Monitoring Center. IUCN-The World Conservation Union, Gland, Switzerland and Cambridge, UK.

SYSTEMATICS
CHAPTER 25

Genetic Structure and Diversity of Dioon edule: A Comparison with Other Cycads

Jorge González-Astorga and Andrew P. Vovides

Abstract

We evaluated the diversity and genetic structure of *Dioon edule* throughout its known range in eastern Mexico, from the states of Nuevo León to Veracruz, north to south, respectively. We analyzed 14 allozyme loci in 8 populations of *D. edule*. The estimates of genetic diversity were: mean number of alleles per locus ($A=1.44$) and percentage of polymorphic loci ($P=54.8$), the mean observed ($H_O=0.27$ and expected heterozygosity

(H_E=0.24). In relationship to genetic structure, *F*-statistics revealed an excess of heterozygous genotypes, local (F_{is}=−0.27) and globally (F_{it}=−0.17). The F_{st} value was 0.75, which indicates that genetic variation explained by differences among populations is only 7.5%. Also, we detected a negative relationship between genetic diversity (i.e., *A*, *P*, and H_O) and latitude of the populations. Patterns of genetic diversity in *D. edule* appear to be associated with a post-Pleistocene spread of the species, from its southerly (origin) to its northern range (derived populations, including its central distribution). Also, in this study we detected that *D. edule* present relatively higher levels of genetic diversity, compared with other cycads and also compared with other long-lived narrow endemic perennial plant species and regionally distributed tropical long-lived trees. We discuss the results of this study in terms of biogeographical and evolutionary aspects and conservation biology.

Resumen

En este estudio evaluamos la diversidad y la estructura genética de *Dioon edule* en todo el rango de distribución de la especie, desde el Nuevo León (al norte) a Veracruz (al sur). Para esto, analizamos 19 loci aloenzimáticos de ocho poblaciones. Los estimadores de diversidad genética y sus valores fueron: número de alelos por locus (A=1.44), el porcentaje de loci polimórficos (P=54.8) y el promedio de las heterocigosis observadas (H_O=0.27) y esperadas (H_E=0.24). Los parámetros que evalúan la estructura genética de las poblaciones (i.e., *F*-estadísticos) revelaron exceso de heterocigotos, local (F_{is}=−0.27) y para el conjunto de todas las poblaciones (F_{it}=−0.17). El valor estimado de diferenciación genética entre poblaciones fue de F_{st}=0.75. También, detectamos una relación negativa y significativa entre los estimadores de diversidad genética (i.e., *A*, *P* y H_O) y la latitud de las poblaciones. Los patrones de diversidad genética detectados en las poblaciones de *D. edule*, parecen estar relacionados con la distribución de la especie después del Pleistoceno, desde las poblaciones del sur (originarias) al centro y norte, poblaciones que pudieron haber sido derivadas de las primeras. Colateralmente, detectamos que *D. edule* presenta niveles relativamente altos de diversidad genética comparada con otras cícadas, con especies perennes de larga vida, de distribución restringida y también con especies arbóreas del trópico. Finalmente, los resultados de este estudio se discuten en el contexto biogeográfico, de la evolución y la biología de la conservación.

Introduction

The integral study of geographic variation, diversity, and genetic structure of natural population is important for understanding the origin of geographic distribution of species, in terms of differential action of evolutionary process such as natural selection, genetic drift, inbreeding, mating system, mutation, and gene flow (Lewontin, 1974; Endler, 1986).

The genetic structure of populations (i.e., the distribution of genetic variation within and among populations) reflects the interactions of different processes including long-term evolutionary and short-term ecological history of species, including the effects of habitat fragmentation, shifts in distribution, and population isolation (Wright, 1978). In addition, several factors may influence the genetic structure of plant populations; among the more important are the demographic factors (Antonovics & Via, 1987), breeding systems (Lande & Schemske, 1985), and seed dispersal mechanisms (Hamrick et al., 1995; Hamrick & Godt, 1996a, b).

According to theoretical predictions, in small and restricted populations, inbreeding and genetic drift can lead to loss of genetic diversity (Wright, 1931, 1978). However, high genetic diversity has been reported for the following rare or endangered plant species in tropical regions, for example, the Hawaiian fern *Adenophorus periens* L.Bishop (Ranker, 1994), the endangered *Caesalpinia echinata* Lam. (Cardoso et al., 1998), the monoecious endemic shrub *Brongniartia vazquezii* O. Dorado of the tropical dry forests of Central Mexico (González-Astorga & Núñez-Farfán, 2001), the narrow endemic tree *Antirhea aromatica* G. Castillo-Campos & D. H. Lorence (González-Astorga & Castillo-Campos, 2004), and the epiphyte plant *Tillandsia achyrostachys* Ed. Morr. ex Baker (González-Astorga et al., 2004), among others.

In cycad species, reports indicated that diversity and genetic variability are relatively low, for example, *Macrozamia communis* L. A. S., Johnson (Ellstrand et al., 1990), *Cycas pectinata* Buch.-Ham. (Yang & Meerow, 1996), *M. parcifolia* P. I. Forst. & D. L. Jones and *M. pauli-guilielmi* W. Hill & F. Muell. (Sharma et al., 1998), *M. heteromera* C. Moore (Sharma et al., 1999), *C. panzhihuaensis* L. Zhou & S.Y. Yang (Chao-Luan et al., 1999), *C. taitungensis* C. F. Shen, K. D. Hill, C. H. Tsou & C. J. Chen (Lin et al., 2000), and *C. seemannii* A. Braun (Keppel, 2002; Keppel et al., 2002). The results for these species correspond to those for plants with restricted geographical range and relatively low population sizes (Hamrick & Godt, 1996a). However, there are cycads in which the genetic diversity is greater than expected, when we consider that the populations have relatively few individuals. This occurs, for example, in *Macrozamia riedlei* (Gaudich.) C. A. Gardner (Byrne & James,

1991), *Zamia loddigesii* Miq. (González-Astorga et al., in press), and three species of *Dioon*: *D. sonorense* (De Luca, Sabato & Vázq. Torres) Chemnick, T. J. Greg. & S. Salas-Mor., *D. tomasellii* De Luca, Sabato & Vázq. Torres (González-Astorga et al., unpublished data) and *D. caputoi* De Luca, Sabato & Vázq. Torres (Cabrera-Toledo et al., unpublished data). Historical effects (Lewis & Crawford, 1995; Irwin, 2002) and ecological factors (Hamrick & Godt, 1996a, b) may account for this greater diversity.

We chose *D. edule* for this study for two principal reasons: (1) *D. edule* is the only widely distributed *Dioon* on the Sierra Madre Oriental and (2) the presence of populations on the Trans-Mexican Neovolcanic Belt from sea level to mid-elevations (500–800 m).

Objectives

The following were the objectives of our investigations throughout the distribution range of *D. edule*:

- To analyze the genetic diversity of the populations.
- To estimate the genetic variation between and within populations.
- To search for patterns genetic variation with respect to latitudinal distribution and geographical distances between populations.
- To compare the genetic diversity of *D. edule* in relationship other cycads.

Materials and Methods

Target Species

D. edule is a long-lived arborescent cycad with trunks up to 5 m tall. The plants produce cones that are held for extended times on the plants throughout the year, and are found in ecotone zones between tropical deciduous forest and groves of evergreen oaks (Vovides et al., 1983). *D. edule* is distributed along the Sierra Madre Oriental, from Nuevo León, Tamaulipas to Veracruz (De Luca et al., 1982) (Fig. 25-1). The conservation status of populations of this species is rare (R) according to the IUCN (2003) category. It is also listed as a protected species and is incorporated, along with other Mexican cycads, within the Official Mexican Norm NOM-059-ECOL-1994-2000, which covers wild species of threatened, rare, and endangered flora and fauna that require special protection. Cycads are also considered a priority by the Mexican conservation authorities for conservation and study (INE-SEMARNAP, 2000).

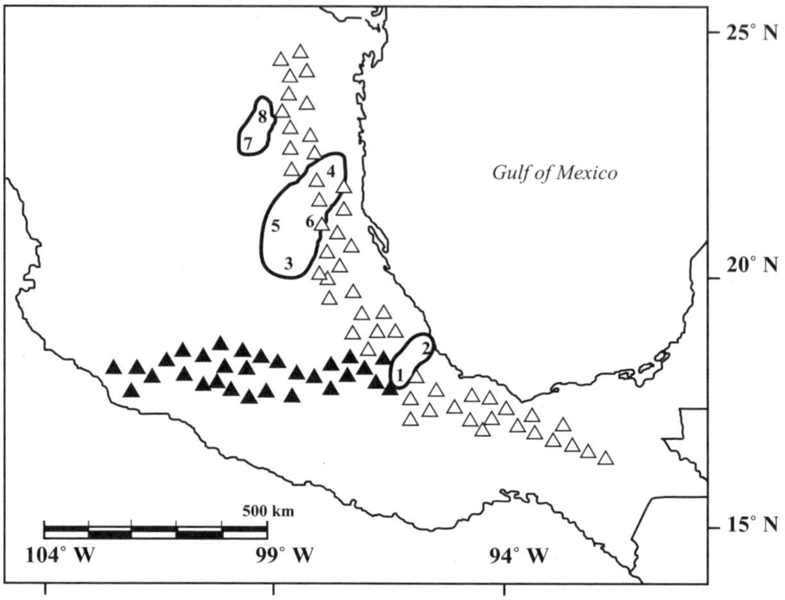

Figure 25-1. Geographical distribution of *D. edule* in Mexico. The open triangles indicate the Sierra Madre Oriental (ORI), and black triangles the Trans-Mexican Neovolcanic Belt. The numbers show the geographical position of each population evaluated. (1) Monte Oscuro, Veracruz, (2) Farallón, Veracruz, (3) Mazacintla, Querétaro, (4) Vado del Morro, Tamaulipas, (5) El Rascón, San Luis Potosí, (6) Tamuin, San Luis Potosí, (7) Alamos, Nuevo León, and (8) Iturbide-Linares, Nuevo León.

Sampling

We sampled tissues in the eight populations of *Dioon edule* throughout its distribution range. Populations are separated by distances that range from *ca.* 50 to 950 km (Fig. 25-1). We collected tissue samples of young leaves from 20 to 100 individual plants from each population, transported the tissue on ice, and then stored it in a REVCO freezer at −70°C until extraction for electrophoretic analysis.

Electrophoresis

Approximately 300 mg of fresh leaflet tissue was ground in liquid nitrogen. About 250 μl of extraction buffer (i.e., 0.1 M Tris-HCl, pH 7.5, 4% PVP-40, 0.001 M EDTA, 0.01 M $CaCl_2$, 0.01 M $MgCl_2$, and 0.1% β-mercaptoethanol; Chao-Luan et al., 1999; González-

Table 25-1 Allozyme systems resolved in this study.

Allozyme	Locus abbreviation	E.C. No.	Buffer[a]
Malate-dehydrogenase	*Mdh1* and *Mdh2*	1.1.1.37	R
Phosphogluco isomerase	*Pgi1* and *Pgi2*	5.3.1.9	R
Diaphorase	*Dia1* and *Dia2*	1.6.99.–	R
Isocitrate dehydrogenase	*Idh*	1.1.1.41	R
Esterase	*Est*	3.1.1	9
Leucine aminopeptidase	*Lap*	3.4.11.1	9
6–Phosphogluconate dehydrogenase	*6Pgd*	1.1.1.44	R
Peroxidase anodic	*Apx1, Apx2, Apx3,* and *Apx4*	1.11.1.7	9

Nomenclature and abbreviations follow Wendel and Weeden (1989), based on IUBNC. E.C. No. = Enzyme Commission number.

[a]R and 9 refer to system buffers (gel/electrode) of Chao-Luan et al. (1999) and Soltis et al. (1983), respectively.

Astorga et al., 2005) was added to dilute and stabilize the enzyme extracts, which were stored on filter paper wicks at −70°C until used for analyses. Multilocus genotypes of individuals from each population were obtained through horizontal starch gel electrophoresis (10% w/v). For each individual plant, allozymic variation was scored for 14 loci (Table 25-1). Electrophoresis was carried out at 4°C for 8 hours (constant current of 50 mA, and voltage of 80 V).

Data Analysis

We assigned the bands from each allozyme system to alleles and genotypes based on theoretical expectations and observed banding patterns. We used the *TFPGA* v.1.3 Program (Miller, 1997) to obtain the genetic estimators from the data analysis. The genotypic frequencies obtained were used to calculate observed mean heterozygosity (H_O), and allelic frequencies. Allelic frequencies at each population were used to estimate the mean number of alleles per locus (A), the average proportion of polymorphic loci (P), and expected mean heterozygosity (H_E), based on Hardy–Weinberg expectations (Hartl & Clark, 1997). We obtained significance of estimators by Monte Carlo methods (Weir, 1990).

We carried out partitioning of genetic variability via F statistics (Wright, 1978) and calculated the partitioning according to the Weir and Cockerham (1984) formula, which estimates genetic structure by partitioning variation in the same way as a regular analysis of variance. The F_{st} estimates populations' divergence through allele frequencies, whereas F_{is} and F_{it} estimate heterozygote excess (<0) and deficit (>0) relative to Hardy–Weinberg expectations in local populations and the total set of populations, respectively. To determine whether F_{is} and F_{it} estimations for each locus were significantly different from zero, we obtained χ^2 statistics ($\chi^2 = F[2N][k-1]$), with $k(k-1)/2$ degrees of freedom, where N is the sample size and k the number of alleles (Weir, 1990). To determine the significance of the F_{st} statistic per locus, we used the χ^2statistic: $\chi^2 = (2N)F_{st}(k-1)$, with $(k-1)(n-1)$ degrees of freedom, where n is the number of populations (Workman & Niswander, 1970). We obtained the confidence intervals (at 95%) of the F-statistics by bootstrapping over loci for the multilocus estimate, and jackknifing over populations for the single-locus estimates (Weir & Cockerham, 1984; Weir, 1990). Finally, we performed regression models (Sokal & Rohlf, 1995) for the latitude and genetic diversity estimates (A, P, and H_O) to explore the relationship between these variables.

Results

We detected genetic variation in 14 loci. The genetic diversity of *D. edule* was as follows: the mean of number of alleles per locus $A = 1.44 \pm 0.19$, the percentage mean of polymorphic loci $P = 54.8 \pm 10.9\%$; observed 0.273 ± 0.09 and expected mean heterozygosity 0.239 ± 0.0 (Table 25-2).

The regression models for the latitude and genetic diversity estimates (alleles per locus: $b = -0.086 \pm 0.03$; $R^2 = 0.523$; $p = 0.042$; polymorphic loci: $b = -4.081 \pm 1.4$; $R^2 = 0.585$; $p = 0.027$; and observed heterozygosity: $b = -0.037 \pm 0.01$; $R^2 = 0.637$; $p = 0.017$) were statistically significant. The negative b coefficients indicated that genetic diversity decreased as latitude increased (Fig. 25-2).

We evaluated the genetic structure via Wright's F-statistics (Table 25-3), where the means of F_{it} (-0.172 ± 0.16) and F_{is} (-0.270 ± 0.15) were negative and significantly different from zero, indicating an excess of heterozygotes. Finally, the mean genetic differentiation among populations (F_{st}) was 7.5% of the total variance.

Finally, we compared these results with those published for other cycads. Our findings are summarized in Tables 25-4 and 25-5.

Table 25-2 Genetic diversity for eight populations of D. edule in Mexico.

Population	A	P	H_O	H_E
(1) Monte Oscuro (Ver.)	1.86	71.43	0.4073	0.2982
(2) Farallón (Ver.)	1.79	71.43	0.4502	0.2893
(3) Mazacintla (Qro.)	1.36	54.54	0.2079	0.2053
(4) Vado del Morro (Tam.)	1.36	45.45	0.2283	0.2262
(5) El Rascón (SLP.)	1.21	50.00	0.2404	0.2393
(6) Tamuin (SLP.)	1.29	45.45	0.2321	0.2036
(7) Alamos (NL.)	1.36	54.54	0.2021	0.2585
(8) Iturbide-Linares (NL.)	1.29	45.45	0.2191	0.1935
Mean ± SD	1.44 ± 0.19	54.78 ± 10.94	0.2734 ± 0.0973	0.2392 ± 0.0397

A = number of alleles per loci; P = percentage of polymorphic loci; H_O and H_E = observed and expected mean heterozygosity, respectively.

Table 25-3 Wright's *F*-statistics for eight populations of *D. edule* in Mexico.

Loci	F_{it}	F_{st}	F_{is}
Mdh1	−0.1742*	0.0804*	−0.2768*
Mdh2	−0.1895**	0.0728**	−0.2829**
Pgi1	−0.1258*	0.0862*	−0.2320*
Pgi2	−0.1701**	0.0692**	−0.2571**
6Pgd	−0.1854**	0.0833*	−0.2932**
Dia1	−0.0998**	0.0797*	−0.1951**
Dia2	−0.0501*	0.0860*	−0.1489*
Idh	−0.2346**	0.0685**	−0.3253**
Est	−0.2076**	0.0660*	−0.2929**
Lap	−0.2143**	0.0637**	−0.2970**
Apx1	−0.1658	0.0755	−0.2611
Apx2	−0.1658	0.0755	−0.2611
Apx3	−0.1658	0.0755	−0.2611
Apx4	−0.1658	0.0755	−0.2611
Mean ± SD	−0.1724* ± 0.1647	0.0749* ± 0.0246	−0.2700* ± 0.1553
Confidence interval to 95%	−0.4624 − 0.1345	0.0292 − 0.1223	−0.5311 − 0.0292

* $p < 0.05$; ** $p < 0.01$.

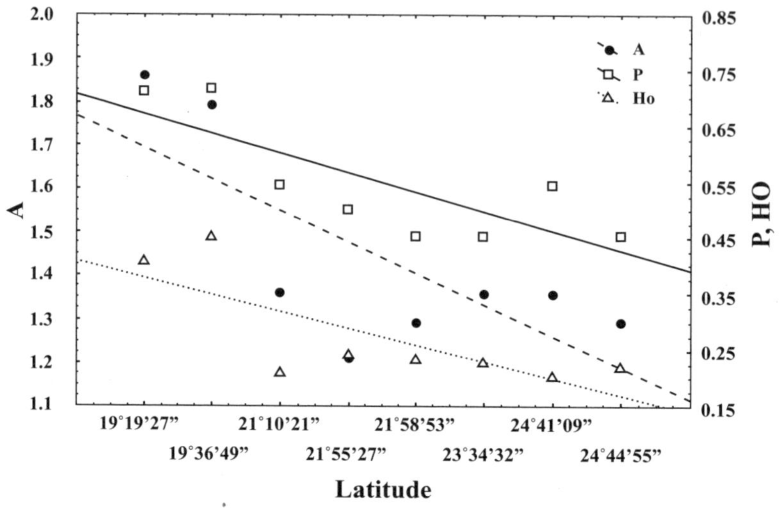

Figure 25-2. Lineal regression between diversity parameters and latitude in populations of *D. edule* in Mexico.

Discussion

D. edule has been found to have relatively higher levels of genetic diversity, compared with other cycads within the Zamiaceae such as *Macrozamia communis* (Ellstrand et al., 1990), *M. parcifolia* and *M. pauli-guilielmi* (Sharma et al., 1998), *M. heteromera* (Sharma et al., 1999), and also species in Cycadaceae such as *Cycas pectinata* (Yang & Meerow, 1996), *Cycas panzhihuaensis* (Chao-Luan et al., 1999), *C. taitungensis* (Lin et al., 2000), and *C. seemannii* (Keppel, 2002; Keppel et al., 2002) (see Table 25-4). On average, the *D. edule* populations show a higher genetic diversity than the mean of 12 species of Zamiaceae and 6 species of *Cycas* (Table 25-5). Nevertheless, within its genus *D. edule* shows relatively low values of genetic diversity (Table 25-4). Further, compared with other plant species, *D. edule* shows a higher percentage of polymorphic loci (>54%) than that reported for other long-lived narrow endemic perennials (40–45.1%, Hamrick & Godt, 1996a). The mean expected heterozygosity within populations of *D. edule* is also higher (0.24) than that reported for other regionally distributed tropical long-lived trees (Hamrick et al., 1994).

Genetic variability in *D. edule* indicated an excess of heterozygous genotypes, locally and globally (F_{it}=−0.172, and F_{is}=−0.270) (cf. Lin et al., 2000, *C. taitungensis*; Cabrera-Toledo et al., unpublished data, *D. caputoi*; Pinares de la Fe et al., unpublished data, *Microcycas*

Table 25-4 Summary of diversity and genetic structure of cycads plant species.

Species	N	A	P	H_E	F_{st}	Reference
Cycas pectinata Buch.-Ham.	11	1.82	58.5	0.076	0.387	Yang and Meerow (1996)
C. siamensis Miq.	13	1.48	58.9	0.134	0.291	Yang and Meerow (1996)
C. panzhihuaensis L. Zhou & S.Y. Yang	3	1.13	14.3	0.061	0.139	Chao-Luan et al. (1999)
C. guizhouensis	3	1.61	58.3	0.100	0.080	Yang and Meerow (1996)
C. taitungensis C.F. Shen, K. D. Hill, C. H. Tsou & C. J. Chen	2	1.07	2.5	0.013	0.034	Lin et al. (2000)
C. seemannii A. Braun	5	1.20	21.3	0.057	0.594	Keppel (2002); Keppel et al. (2002)
Macrozamia communis L. A. S. Johnson	5	1.61	50.0	0.045	0.270	Ellstrand et al. (1990)
M. riedlei (Gaudich.) C. A.	15	2.43	93.0	0.274	0.092	Byrne and James (1991)
M. parcifolia P. I. Forst. & D. L. Jones	2	1.20	17.6	0.037	0.090	Sharma et al. (1998)
M. pauli-guilielmi W. Hill & F. Muell.	3	1.30	31.3	0.081	0.030	Sharma et al. (1998)
M. heteromera C. Moore	5	1.30	26.0	0.077	0.100	Sharma et al. (1999)
Zamia loddigesii Miq.	4	1.80	66.6	0.266	0.790	González-Astorga et al. (unpublished data)
Microcycas calocoma Miq.	7	1.49	48.1	0.170	0.337	Pinares de la Fe et al. (unpublished data)
Dioon sonorense (De Luca, Sabato & Vázq. Torres) Chemnick, T. J. Greg. & S. Salas-Mor.	4	2.00	81.6	0.314	0.151	González-Astorga et al. (unpublished data)
D. tomasellii De Luca, Sabato & Vázq. Torres	5	1.96	83.1	0.309	0.295	González-Astorga *et al.* (unpublished data)
D. caputoi De Luca, Sabato & Vázq. Torres	4	1.91	79.0	0.350	0.099	Cabrera-Toledo et al. (unpublished data)
D. edule Lindl.	8	1.44	54.8	0.240	0.075	González-Astorga et al., 2003
D. angustifolium	3	1.67	52.4	0.218	0.167	González-Astorga et al., 2005

N = number of *populations*; A = mean alleles per locus; P = percentage of polymorphic loci; H_E = expected heterocigosis; F_{st} = genetic differentiation between populations.

Table 25-5 Summary of the genetic variation for cycads species.

	No. of species	A	P	H_E	F_{st}
D. edule	1	1.44	54.8	0.240	0.075
Dioon spp.	5	1.80±0.23	70.2±15.2	0.286±0.05	0.157±0.09
Zamiaceae species	12	1.68±0.36	57.0±24.3	0.198±0.11	0.208±0.21
Cycas species	6	1.4±0.3	35.6±25.8	0.074±0.04	0.254±0.21
All species of cycads	18	1.58±0.36	49.9±26.2	0.157±0.11	0.223±020

calocoma). Because dioecy favors Hardy–Weinberg equilibrium resulting in low positive or zero fixation indices (Hartl & Clark, 1997; Hedrick, 2000), we suggest that in *D. edule* the the negative values are probably due to natural selection that favors a higher percentage of heterozygous individuals than can adapt to environmental changes. Negative values of the indices have also been described for other plant species, mainly long-lived perennials (cf. Linhart et al., 1981; Smyth & Hamrick, 1984; Sampson et al., 1989; Eguiarte et al., 1992; Alvarez-Buylla et al., 1996), and the cycad species we tested, such as: *D. sonorense* and *D. tomasellii* (González-Astorga et al., unpublished data), *D. caputoi* (Cabrera-Toledo et al., unpublished data), and *Microcycas calocoma* (Pinares de la Fe et al., unpublished data) (Table 25-4). These results suggest that in the genus *Dioon* (see Table 25-5) excess of heterozygosis may be of phylogenetic importance (Harvey & Pagel, 1991); also, in the context of comparison between widespread rare plants in the same genus (Gitzendanner & Soltis, 2000).

On the other hand, only 7.5% of the variation in *D. edule* accounts for differences among its populations. The genetic differentiation between populations of this species is the lowest reported for cycads to date, with the exception of *C. taitungensis* (0.034) and *Macrozamia pauli-guilielmi* (0.03), which is higher (Table 25-4). This can be explained by the species' life history traits, which includes dioecy, pollination by specific insects *Rhopalotria* spp. and *Pharaxonotha* spp. (Tang, 1987; Norstog & Fawcett, 1989; Vovides, 1991), and the long life cycles of *D. edule* (Vovides, 1990; Norstog, 1987; Schneider et al., 2002). These appear to promote homogeneity or hinder divergence among populations (Loveless & Hamrick, 1984; Hamrick & Godt, 1996b).

Because they have the most genetic variation, we believe the southern populations of the species in Veracruz are precursory. These variations correspond to the base of the regression analysis between latitude versus estimates of genetic diversity (Fig. 25-2). In this context, we suspect that the northern populations are derived because genetic diversity is lower than the expected relative to the southern populations. Occurrence of high species diversity in *Dioon* in the northern region of Oaxaca State supports this proposition, and Oaxaca borders the States of Puebla, and south-central Veracruz (Fig. 25-1; cf. Vovides et al., 1983; Moretti et al., 1993; Vovides et al., 2004).

The present distribution of genetic diversity within populations of *D. edule* appears to correlate with factors such as life and biogeographic history (cf. González-Astorga et al., 2003) as well as the history of vegetation in Mexico, where the latter has been documented in the light of the floristic refuge hypothesis (Lundell, 1945; Toledo, 1982, 1988; Wendt, 1987). Based on this hypothesis, we envisage a scenario of southern immigration of warm tropical floras during the Pleistocene glaciations followed by a loss of biological diversity among northern populations of *D. edule*, thus reducing their population sizes (and genetic variability) to a minimum through climate change. Therefore a bottleneck effect could induce genetic drift and significant reduction in genetic diversity. Alternatively, a recent colonization in the northern range and associated founder effects could also explain this lower genetic variation (cf. González-Astorga et al., 2005). This colonization-and-spread scenario has also been reported for *Cycas seemanii* in the Vanatau archipelago by Keppel et al. (2002), explained by possible recent stepwise migration from Australasia into the Pacific (Hill, 1996). In addition, the northward migration of tropical floras, including *D. edule*, during the postglacial amelioration of climates occurred, and only genotypes adapted to these conditions are present today. Similar trends have been found in *Pinus flexilis* (Jørgensen et al., 2002) and *Lophocereus schottii* (Nason et al., 2002). DNA sequencing data for the genus *Ceratozamia* (González & Vovides, 2002) showed a similar trend, where the basal clades suggest a probable ancestral geographic area in southern Mexico with taxa associated to the aforementioned Pleistocene refuges. A single unresolved clade of *Ceratozamia* in northeastern Mexico north of the Trans Mexican Neovolcanic Mountain range suggested a recent speciation process in the genus.

The current pattern of genetic diversity throughout the geographical range of *D. edule* is complex. This can be attributed to the interaction of a multiplicity of factors such as the geographical distribution of Pleistocene and post-Pleistocene populations, the influence of corresponding climatic changes, as well as high gene flow among neighboring populations in this long-lived paleoendemic plant species. Finally, based on the studies of

González-Astorga et al. (2003), we recognize species status for the disjunct northern populations of *D. edule* var. *angustifolium* by referring to the basionym *D. angustifolium* Miq (González-Astorga et al., 2005).

Acknowledgments

A grant from CONACyT-SEMARNAT No. 2002-C01-0183 funded this research. We thank Carlos Iglesias, Pablo O. Aguilar, Miriam Ferrer, and Julia Hernández Villa for support in the laboratory and field work.

Literature Cited

Alvarez-Buylla, E. R., Chaos, A., Piñero, D. & A. A. Garay. 1996. Demographic genetics of a pioneer tropical tree species: patch dynamics, seed dispersal, and seed banks. Evolution 50: 1155–1166.

Antonovics J. & S. Via. 1987. Genetic influences on the distribution and abundance of plants. Pp. 185–203. *in* A. J. Devy, M. J. Hutchings & A. R. Watkinson (eds.), Perspectives on plant population biology. Sinauer Associates, Sunderland, MA.

Byrne, M. & S. H. James. 1991. Genetic diversity in the cycad *Macrozamia riedlei*. Heredity 67: 35–39.

Cardoso, M. A., J. Provan, W. Powell, P. C. G. Ferreira & D. E. de Oliveira. 1998. High genetic differentiation among remnant populations of the endangered *Caesalpinia echinata* Lam. (Leguminoseae-Caesalpinioideae). Mol. Ecol. 7: 601–608.

Chao-Luan, L., Q. Wang, S. Y. Jiang, S. Ge & K. E. Wang. 1999. Genetic diversity of allozymes in populations of *Cycas panzhihuaensis* L. Zhou and S. Y. Yang. Pp: 323–327 *in* C. J. Chen (ed.), Biology and conservation of cycads—Proceedings of the fourth international conference on cycad biology. International Academic Publishers Beijing, China.

De Luca, P., S. Sabato & M. Vázquez Torres. 1982. Distribution and variation of *Dioon edule* (Zamiaceae). Brittonia 34: 355–362.

Eguiarte, L. E., N. Pérez-Nasser & D. Piñero. 1992. Genetic structure, outcrossing rate and heterosis in *Astrocaryum mexicanum* (tropical palm): implications for evolution and conservation. Heredity 69: 217–228.

Ellstrand, N. C., R. Ornduff & J. M. Clegg. 1990. Genetic structure of the Australian cycad, *Macrozamia communis* (Zamiaceae). Am. J. Bot. 77: 677–681.

Endler, J. A. 1986. Natural selection in the wild. Princeton University Press, Princeton NJ.

Gitzendanner, M. A. & P. S. Soltis. 2000. Patterns of genetic variation in rare and widespread plant congeners. Am. J. Bot. 87: 783–792.

González, D. & A. P. Vovides. 2002. Low intralineage divergence in Ceratozamia (Zamiaceae) detected with nuclear ribosomal DNA ITS and chloroplast DNA trnL-F non-coding region. Syst. Bot. 27: 654–661.

González-Astorga, J. & J. Núñez-Farfán. 2001. Effect of habitat fragmentation on the genetic structure of the narrow endemic *Brongniartia vazquezii*. Evol. Ecol. Res. 3: 861–872.

——— & G. Castillo-Campos. 2004. Genetic variability of the narrow endemic tree *Antirhea aromatica* Castillo-Campos and Lorence (Rubiaceae, Guettardeae) in a tropical forest of Mexico. Ann. Bot. 93: 521–528.

———, A. P. Vovides, M. M. Ferrer & C. Iglesias. 2003. Population genetics of *Dioon edule* Lindl. (Zamiaceae, Cycadales): biogeographical and evolutionary implications. Biol. J. Linn. Soc. 80: 457–467.

———, A. Cruz-Angón, A. Flores-Palacios & A. P. Vovides. 2004. Diversity and genetic structure of the Mexican endemic epiphyte: *Tillandsia achyrostachys* E. Morr. ex Baker var. *achyrostachys* (Bromeliaceae). Ann. Bot. 94: 545–551.

———, A. P. Vovides, A. Cruz-Angón, P. Octavio-Aguilar & C. Iglesias. 2005. Allozyme variation in three extant populations of the narrowly endemic cycad *Dioon angustifolium* Miq. (Zamiaceae) from North-eastern Mexico. Ann. Bot. 95: 999–1007.

———, ———, P. Octavio-Aguilar, D. Aguirre-Fey D. & Iglesias C. 2006. Diversity and genetic structure of the Cycad *Zamia loddigesii* Miq. (Zamiaceae): implications for evolution and conservation. Bot. J. Linn. Soc. 152: 533–544.

Hamrick, J. L., A. Schnabel & P. W. Wells. 1994. Distribution of genetic diversity within and among populations of Great Basin conifers. Pp. 147–161 *in* K. T. Harper, L. L. St. Clair, K. H. Thorne & W. W. Hess (eds.), Natural history of the Colorado Plateau and Great Basin. University of Colorado Press, Niwot, CO.

———, M. J. W. Godt & S. L. Sherman-Broyles. 1995. Gene flow among plants populations: evidence from genetic markers. Pp. 215–232 *in* P. C. Hoch & A. G. Stephenson (eds.), Experimental and molecular approaches to plant biosystematics. Missouri Botanical Garden, St. Louis, MO.

——— & ———. 1996a. Conservation genetics of endemic plant species. Pp. 218–304. *in* J. C. Avise & J. L. Hamrick (eds.), Conservation genetics. Case histories from nature. Chapman and Hall, New York.

——— & ———. 1996b. Effects of the history traits on genetic diversity in plants. Philos. Trans. R. Soc. L. B. 351: 1291–1298.

Hartl, D. L. & A. G. Clark. 1997. Principles of population genetics. Ed. 3. Sinauer Associates, Sunderland MA.

Harvey, P. H. & M. D. Pagel. 1991. The comparative method in evolutionary biology. Oxford University Press, New York.

Hedrick, P. W. 2000. Genetics of populations.Ed. 2. Jones and Bartlett Publishers, Sudbury, MA.

Hill, K. D. 1996. Cycads in the Pacific. Pp. 267–274 *in* A. Keast & S. E. Miller (eds.), The origin and evolution of Pacific island biotas, New Guinea to eastern Polynesia. SPB, Amsterdam.

INE-SEMARNAP. 2000. Prep 6: Protección, conservación y recuperación de la familia Zamiaceae (Cycadales) de México. Instituto Nacional de Ecología, México D.F.

Irwin, D. E. 2002. Phylogeographic breaks without geographic barriers to gene flow. Evolution 56: 2383–2394.

IUCN. 2003. IUCN red list of threatened species. www.redlist.org.

Jørgensen, S., J. L. Hamrick & P. V. Wells. 2002. Regional patterns of genetic diversity in *Pinus flexilis* (Pinaceae) reveal complex species history. Am. J. Bot. 89: 792–800.

Keppel, G. 2002. Low genetic variation in a pacific cycad: conservation concerns for *Cycas seemannii* (Cycadaceae). Oryx 36: 41–49.

———, S. W. Lee & P. D. Hodgskiss. 2002. Evidence for long isolation among populations of a Pacific Cycad: genetic diversity and differentiation in *Cycas seemannii* A. Br. (Cycadaceae). J. Hered. 93: 133–139.

Lande, R. & D. W. Schemske. 1985. The evolution of self-fertilization and inbreeding depression in plants. I. Genetic models. Evolution 39: 24–40.

Lewis, P. O. & A. J. Crawford. 1995. Pleistocene refugium endemics exhibit greater allozymic diversity than widespread congeners in the genus *Polygonella* (Polygonaceae). Am. J. Bot. 82: 141–149.

Lewontin, R. C. 1974. The genetic basis of evolutionary change. Columbia University Press, New York.

Lin, T. P., Y. C. Sun, H. C. Lo & Y. P. Cheng. 2000. Low genetic diversity of *Cycas taitungensis* (Cycadaceae), an endemics species in Taiwan, revealed by allozyme analysis. Taiwan J. For. Sci. 15: 13–19.

Linhart, Y. B., J. B. Mitton, K. B. Sturgeon & M. L. Davis. 1981. Genetic variation in space and time in a population of ponderosa pine. Heredity 46: 407–426.

Loveless, M. D. & J. L. Hamrick. 1984. Ecological determinants of genetic structure in plant populations. Annu. Rev. Ecol. Syst. 15: 65–95.

Lundell, C. L. 1945. Vegetation and natural resources of British Honduras. Pp. 270–273 *in* V. F. Waltham (ed.), Plant and plant science in Latin America. Chronica Botanica, MA.

Miller, M. P. 1997. Tools for population genetic analyses (TFPGA) 1.3: A Windows program for the genetic data. Computer software distributed by author.

Moretti, A., P. Caputo, S. Cozzolino, P. De Luca, L. Gaudio, G. S. Gigliano & D. W. Stevenson. 1993. A phylogenetic analysis for *Dioon* (Zamiaceae). Am. J. Bot. 80: 204–214.

Nason, J. D., J. L. Hamrick & T. H. Fleming. 2002. Historical vicariance and postglacial colonization effect on the evolution of genetic structure in *Lophocereus*, a Sonoran desert columnar cactus. Evolution 56: 2214–2226.

Norstog, K. J. 1987. Cycads and the origin of insect pollination. Am. Sci. 75: 270–279.

——— & P. K. S. Fawcett. 1989. Insect-cycad symbiosis and its relation to the pollination of *Zamia furfuraceae* (Zamiaceae) by *Rhopalotria mollis* (Curculionidae). Am. J. Bot. 76: 1380–1394.

Ranker, T. A. 1994. Evolution of high genetic variability in the rare Hawaiian fern *Adenophorus periens* and implications for conservation management. Biol. Cons. 70: 19–24.

Sampson, J. E., S. D. Hopper & S. H. James. 1989. The mating system and population genetic structure in a bird-pollinated mallee, *Eucalyptus rhodantha*. Heredity 63: 389–393.

Schneider, D., M. Wink, F. Sporer & P. Lounibos. 2002. Cycads: their evolution, toxins, herbivores and insect pollinators. Naturwissenschaften 89: 281–294.

Sharma, I. K., D. L. Jones, P. I. Forster & A. G. Young. 1998. The extent and structure of variation in the *Macrozamia pauli-guilielmi* complex (Zamiaceae). Bioch. Syst. Ecol. 26: 45–54.

———, ———, ——— & ———. 1999. Low isozymic differentiation among five species of the *Macrozamia heteromera* group (Zamiaceae). Bioch. Syst. Ecol. 27: 67–77.

Smyth, C. A. & J. L. Hamrick. 1984. Measurement of outcrossing in natural population of musk thistle, *Carduus mutans* L. J. Hered. 75: 303–307.

Sokal, R. R. & F. J. Rohlf. 1995. Biometry. 3rd ed. W. H. Freeman, New York.

Soltis, D. E, C. H. Haufler, D. C. Darrow & G. J. Gastony. 1983. Starch gel electrophoresis of ferns: a compilation of grinding buffers, gel and electrode buffers, and staining schedules. Am. Fern J. 73: 9–27.

Tang, W. 1987. Insect pollination in the cycad *Zamia pumila* (Zamiaceae). Am. J. Bot. 74: 90–99.

Toledo V. M. 1982. Pleistocene changes of vegetation in tropical Mexico. Pp. 93–111 *in* G. T. Prance (ed.) Biological diversification in the tropics: Proceedings of the 5th international symposium of the Association for Tropical Biology, Caracas. Columbia University Press, New York.

———. 1988. La diversidad biológica de México. Ciencia y Desarrollo 14: 17–30.

Vovides, A. P. 1990. Spatial distribution, survival and fecundity of *Dioon edule* (Zamiaceae) in a tropical deciduous forest in Veracruz, Mexico, with notes on its habitat. Am. J. Bot. 77: 1532–1543.

———. 1991. Insect symbionts of some Mexican Cycads in their natural habitat. Biotropica 23: 102–104.

———, J. D. Rees & M. Vázquez Torres. 1983. Flora de Veracruz. Zamiaceae. Fascículo 26. INIREB, Xalapa, Veracruz, México.

———, M. A. Pérez-Farrera, D. González & S. Avendaño. 2004. Relationship and phytogeography in *Ceratozamia* (Zamiaceae). Pp. 109–125 *in* T. Walters & R. Osborne (ed.), Cycad classification: concepts and recommendations. CABI Publishing, Wallingford, Oxford, UK.

Weir, B. S. 1990. Genetic data analysis. Sinauer Associates, Sunderland, MA.

——— & C. C. Cockerham. 1984. Estimating *F*-statistics for the analysis of populations structure. Evolution 38: 1358–1370.

Wendt, T. 1987. Las selvas de Uxpanapa, Veracruz-Oaxaca, México: evidencia de refugios faunísticos Cenozoicos. Anales del Instituto de Biología UNAM (Ser. Bot.) 58: 29–54.

Workman, P. L. & J. D. Niswander. 1970. Population studies on southwestern Indian tribes II. Local differentiation in the Papago. Am. J. Hum. Gen. 22: 24–29.

Wright, S. 1931. Evolution in Mendelian populations. Genetics 16: 97–159.

——— . 1978. Evolution and the genetics of populations. Vol. 4. University of Chicago Press, Chicago and London.

Yang, S. L. & A. W. Meerow. 1996. The *Cycas pectinata* (Cycadaceae) complex: genetic structure and gene flow. Int. J. Plant Sci. 157: 468–483.

SYSTEMATICS
CHAPTER 26

An Emended Description of Dioon mejiae Standl. & L. O. Williams (Zamiaceae)

Jody L. Haynes and Mark A. Bonta

Abstract

Cycad taxonomy and systematics are plagued by insufficient and inconsistent species descriptions. Problems associated with the inadequate description of *Dioon mejiae* were perpetuated for more than 50 years. These include (1) unfounded and inaccurate reports of mature plant size, species distribution, and population numbers; (2) incomplete and/or erroneous descriptions of the type source locality; and (3) inaccurate designation of conservation status. In this chapter, the description of *D. mejiae* is emended following, to the extent possible, Walters and Osborne's (2004) taxonomic description guidelines. Included herein are (1) name, location, number of adult plants, area encompassed, habitat type, geology/soil type, and reproductive status of all known populations; (2) emended Latin description; (3) taxonomic key of the species and its closest relatives; (4) designation of a standard vernacular name; (5) brief overview of human uses; (6) estimates of the number of domestic plants in Honduras and Nicaragua and cultivated plants in ex situ collections worldwide; and (7) discussion of threats, conservation status, and in situ conservation recommendations.

Resumen

La taxonomía y la sistemática de las especies de Cycadales son plagadas por descripciones escasas y contrarias. Los problemas asociados a la descripción inadecuada de *Dioon mejiae* fueron perpetuados por más de 50 años. Éstos incluyen 1) los informes infundados e inexactos del tamaño de la planta adulta, de la distribución de la especie, y de los números de poblaciones; 2) descripciones incompletas y/o erróneas del lugar tipo; y 3) designación inexacta del estado de la conservación. En este artículo, la descripción de *D. mejiae* se enmienda, siguienodo al grado posible, las pautas taxonómicas de la descripción de Walters y de Osborne (2004). Se incluyen adjunto (1) el nombre, la localización, el número de plantas adultas, el área abarcada, el tipo de habitat, el tipo de geología/suelo, y el estado reproductivo de poblaciones conocidas; (2) descripción latina ampliada; (3) llave taxonómica de la especie y de sus parientes más cercanos; (4) designación de un nombre vernáculo estándar; (5) breve descripción de aplicaciones humanas; (6) estimaciones del número de plantas domésticas en Honduras y Nicaragua y plantas cultivadas en colecciones ex situ en el exterior; y (7) discusión de amenazas, del estado de la conservación, y de recomendaciones "in situ" de la conservación.

Introduction

Inconsistent and inadequate species descriptions have led to confusion in cycad taxonomy and systematics. To rectify these issues, a Cycad Classification Concepts (CCC) Workshop was held in 2002 at Montgomery Botanical Center (MBC), Miami, FL. The product of this workshop was a book (Walters & Osborne, 2004) containing the following prime tenets: (1) "species" is the smallest taxonomically useful entity in cycad systematics; (2) cycad species are best designated by a geomorphological species concept; and (3) a standard set of guidelines for taxonomic descriptions is necessary and timely. The book then introduced a set of guidelines for cycad taxonomists to follow.

Dioon mejiae Standl. & L. O. Williams is an example of a species that was inadequately described, and its meager description had far-reaching effects for more than half a century. Paul C. Standley and Louis O. Williams described *D. mejiae* in 1950 from a plant growing in the garden of Dr. Isidoro Mejía H., of Danlí, El Paraíso, Honduras. Mejía had "brought seed home 40 years earlier from a dry rocky canyon in the Department of

Olancho," and, although the authors never visited the population from which the seeds came, they reported its location as "on the road between Pueblo Nuevo and Olanchito" (Standley & Williams, 1950). They also reported that this population represented "the only locality at which [the species] has been found wild." These statements are problematic because there is no Pueblo Nuevo in the relevant area of Olancho and because references to this species in the literature mistakenly perpetuated the single-population concept without any direct supporting evidence.

This chapter provides a much-needed emended description of *D. mejiae* and is offered as a case study of and justification for a taxonomic description standard. It should be noted that this emended description does not affect the type specimen, taxonomy, or nomenclature of *D. mejiae* in any way. It is simply meant to supplement the original description by correcting some long-standing mistakes and clarifying some misconceptions about the species. Most of the data reported herein were gathered during an MBC-sponsored expedition to Honduras in 2003 (hereinafter referred to as HN03), and every effort has been made to conform to as many of the CCC taxonomic guidelines as possible.

Population Characteristics

Geographic Location

Bonta's (2003) report of widespread populations of a large cycad known locally as "teocinte" (or "tiusinte"; see later) in Olancho was the first indication that the range, abundance, and status of the species as reported by Standley and Williams (1950) were questionable at best. Bonta (2003) suggested that confusion over the common name may explain why botanists had been unaware of this cycad's true range for so long. In eastern Honduras—and Olancho in particular—"teocinte" and "tiusinte" refer to *D. mejiae*, whereas throughout the rest of Mesoamerica the names are applied to members of the grass genus *Zea*. Standley and Williams (1950) were aware of this confusion, but stated that it was "futile to speculate" how the same name became affixed to such dissimilar plants. Most populations of *D. mejiae* are referred to locally as "teocintales" or "tiusintales." Once botanists understood this, it was not difficult to pinpoint their locations.

Table 26-1 Name, locality, altitude, habitat type, geology/soil type, and reproductive status for all known *D. mejiae* populations in Olancho and Yoro, Honduras. Gray rows indicate superpopulations or components thereof.

Population information No	Name	Altitude (m)	Topo sheet	Munici pality	Habitat type	Geology/ soil type	Reprod. status	No. of adult plants	Area (ha)	TEFH specimens
1	Saguay/Río Grande	593	Gualaco	Gualaco	TDF	Hills along Río Grande and Quebrada de los Hornos composed of sandstone of the Agua Fría Formation, Jurassic Period; remainder of population occurs on first, second, and third terraces of Quaternary alluvium.	R: s+j+♀+♂	168,000	600	–
2	Quebrada Zarca	620–638	Gualaco	Gualaco	TDF	Quaternary alluvium	?	50	3	–
3	Teocintalito/ Quebrada del Quiscamote	700–750	Gualaco	Gualaco	POT	Cacaguapa Schist . Formation (Paleozoic)—schists, phyllite, gneiss, marble, quartzes; oldest basement rock in Honduras	R: s+j+♀+♂	250	10	39,075
4	Talquezate/Río El Riachuelo	638	Gualaco	Gualaco	TDF	Quaternary alluvium, along boundary with Chindona Batholith	NR?	50	3	–
5	El Nance/ Quebrada El Manchon	580	Gualaco	Gualaco	TDF	Quaternary alluvium	?	50	3	–

6	Camalotal/ Río Alao	698	Pueblo Viejo	Gualaco	TDF	Clastic sedimentary rocks of the	R: s+j+♂	1000	15	39,078
7	Los Encuentros/ Río de Oro	488	Pueblo Viejo	Gualaco	TDF	Valle de Angeles Formation,	R: j+♀+♂	12,000	200	–
8	Jacaleapa de Guata	671	Pueblo Viejo	Guata	TDF	Cretaceous; bedrock is group known as "redbeds;" along boundary with Chindona batholith.	R: s+j+♀+♂	500	30	39,079
9	Coamapa/ Río Negro 1	698	Pueblo Viejo	Guata	TRF	Cacaguapa schist, along boundary	R: s+♀	5000	?	39,082
10	Coamapa/ Río Negro 2	700–800	Pueblo Viejo	Guata	TRF	with Valle de Angeles Group	?	?	?	–
Mame Superpopulation: Sheets (S to N): Guata, Pueblo Viejo, Esquipulas del Norte, Conf. Ríos Aguán-Mame							250,000	2700		
11	Río Tezapa/ Río Mame (S end)	574	Guata	Guata	POT	River cuts through Cacaguapa schist; some populations may	R: s	200	10	39,093
12	Quebrada de Tiscan/Río Mame (middle)	432	Pueblo Viejo	Guata	POT	occur within Chindona batholith, which covers the middle course of	R: s+j	250	25	39,080
13	Quebrada Agua Caliente/Río Mame (N end)	657	Esquipulas del Norte	Esquipulas del Norte	POT	the river (between . Tezapa and Pueblo Viejo)	?	500	?	–
14	Las Labranzas/ Quebrada de Los Copantes	962	Guata	Jano	POT	Cacaguapa schist, along boundary with Chindona batholith	R: s+♀	58	10	39,085
15	Pueblo Viejo/ Quebrada del Carao	900–1000	Guata	Jano	?	Tertiary volcanics	?	200	?	–
16	El Quebrachal/ Río de Jano	704	Guata	Jano	TDF	Atima limestones of Yojoa Group, at edge of Chindona batholith	NR	200	2	39,086

Population information No	Name	Altitude (m)	Topo sheet	Munici pality	Habitat type	Geology/ soil type	Reprod. status	No. of adult plants	Area (ha)	TEFH specimens
17	La Laguna/ Quebrada El Teocinte	829	Esquipulas del Norte	Esquipulas del Norte	POT	Cacaguapa schist, at boundary with Chindona batholith	R: s+♀	200	10	39,092
18	Río de Arenas/ Quebrada del Camalote	690	Esquipulas del Norte	Esquipulas del Norte	TDF	Same as 17	R: s+♀	5,000	400	39,095
19	La Suncuya/ Quebrada del Repecho	500	Esquipulas del Norte	Esquipulas del Norte	TDF	Same as 17	R: ♀	3,000	300	39,098
20	Cerro El Gorrión/ Quebrada del Coyol	500	Esquipulas del Norte	Esquipulas del Norte	TDF	Cacaguapa schist	NR?	100	5	–
21	Filo Mulón-Palmira	420	Esquipulas del Norte	Esquipulas del Norte	POT	Valle de Angeles Group	NR?	500	10	–
22	"Río Chiquito o Esquipulas"	500–800	Esquipulas del Norte	Esquipulas del Norte	?	?	?	?	?	–
23	Palmira/ Quebrada del Teocintal	600	Esquipulas del Norte	Esquipulas del Norte	TDF	?	?	?	?	–
24	Cerro de la Cruz/Quebrada de la Cera	600–700	Olanchito	Esquipulas del Norte	?	?	NR?	?	?	–
25	Calpules/Río de Calpules	400+	Olanchito	Olanchito	?	Tertiary volcanics	?	?	?	–
Aguán Superpopulation: Sheets (W to E): Jimia, Arenal, Olanchito, Yaruca, Sonaguera								200,000	6000	
26	El Tamarindo/ Quebrada del Tamarindo (W end)	317	Arenal	Olanchito	TDF	In west is Cacaguapa schist; to east is Tertiary volcanics. Poorly mapped.	R: s+j	200	10	39,099
27	Agalteca/ Río de la Pimienta (middle)	380	Yaruca	Olanchito	TDF		R: s+j	1000	20	39,097

28	Río Lepaca (E end)	327	Sonaguera	Olanchito	TDF		R: s+j	1000	20	39,100
29	Cerro Teocinte/ Río San Pedro	120+	Saba-Tocoa	Tocoa, Colon	TRF	Probably Cacaguapa schist	?	500	50	–
30	El Teocinte/ La Soledad	600	Parumble	Manto	TDF	Same as 21	NR	6	?	–
E1	La Unión/ Cerro Teocinte (extinct)	900–1300	La Unión	La Unión	N/A	Cacaguapa schist	–	–	–	–
E2	La Crucita/ Quebrada del Teocinte (extinct)	900	La Unión	Jano	N/A	Same as E1	–	–	–	–
E3	Mocanquire/ Cerro del Teocintal (extinct)	850–950	Pueblo Viejo	Guata	N/A	Same as E1	–	–	–	–
Totals:							15 of 21 (71%) reprod. pops.	646,658 plants	10,351 ha	14 pop. vouchers

Most population names derive from a pairing of the name of the nearest village or human habitation and the river or stream with which the population is associated. "Habitat Type" codes: TDF = tropical deciduous forest, POT = pine/oak transition, TRF = tropical rain forest. "Reproductive Status" codes: R = reproductive; NR = non-reproductive; s = seedlings present; j = juveniles present; ♀ = female cone(s) present; ♂ = male cone(s) present. "TEFH Specimens" refer to herbarium vouchers deposited in the Herbario Nacional de Honduras, Universidad Nacional Autónoma de Honduras. (Modified from Tables 3 and 4 of Haynes & Bonta, 2003.)

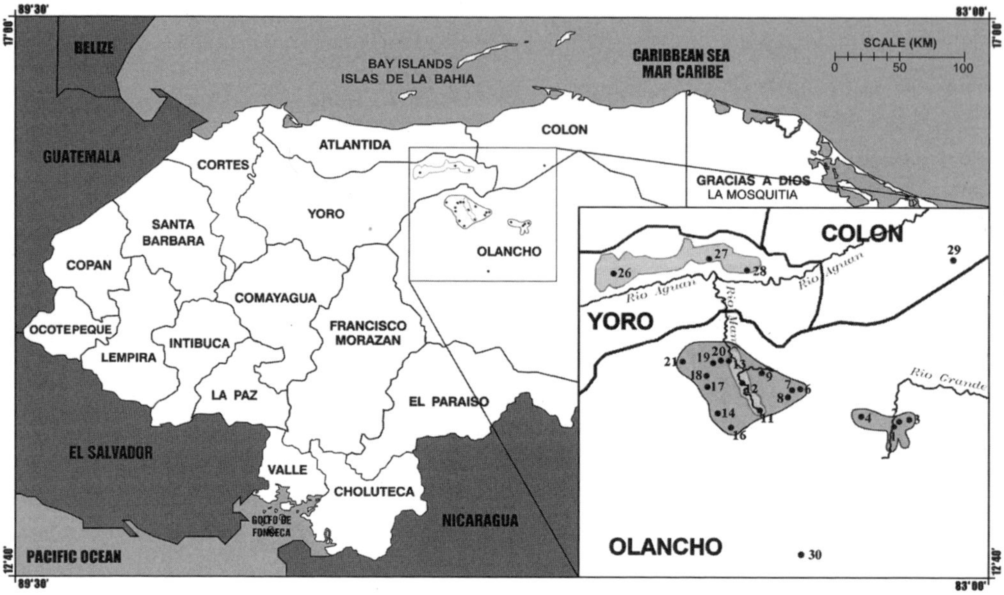

Figure 26-1. Location of known extant populations and three extinct populations of *dioon mejiae*.

Current Distribution

The distribution of the species, as it is known at present, extends from 14°48' to 15°32'N latitude and from 86°00 to 86°58'W longitude. A list of all known populations in Honduras is provided in Table 26-1, and the location of the known extant populations, in addition to three populations identified by map toponyms but confirmed to be extinct during HN03, is shown in Fig. 26-1. It is interesting to note that (1) most large populations are not reflected in place names on maps, (2) most extant high-altitude populations are quite small, and (3) three (of only seven) populations listed on topographical maps are extinct and all three occur at high elevations.

Most populations are found in the upper drainage basins of the Aguán and Sico rivers on the Caribbean slope in the departments of Olancho and Yoro (Fig. 26-1). In the Sico basin, the Saguay/Río Grande population (Pop. 1) and a few small satellite populations occur along the SW fringe of the Valle de Agalta, a Quaternary terrace plain. A large superpopulation—consisting of numerous small subpopulations distributed along smaller streams that empty into the larger river—skirts the northern edge of the Valle de Aguán

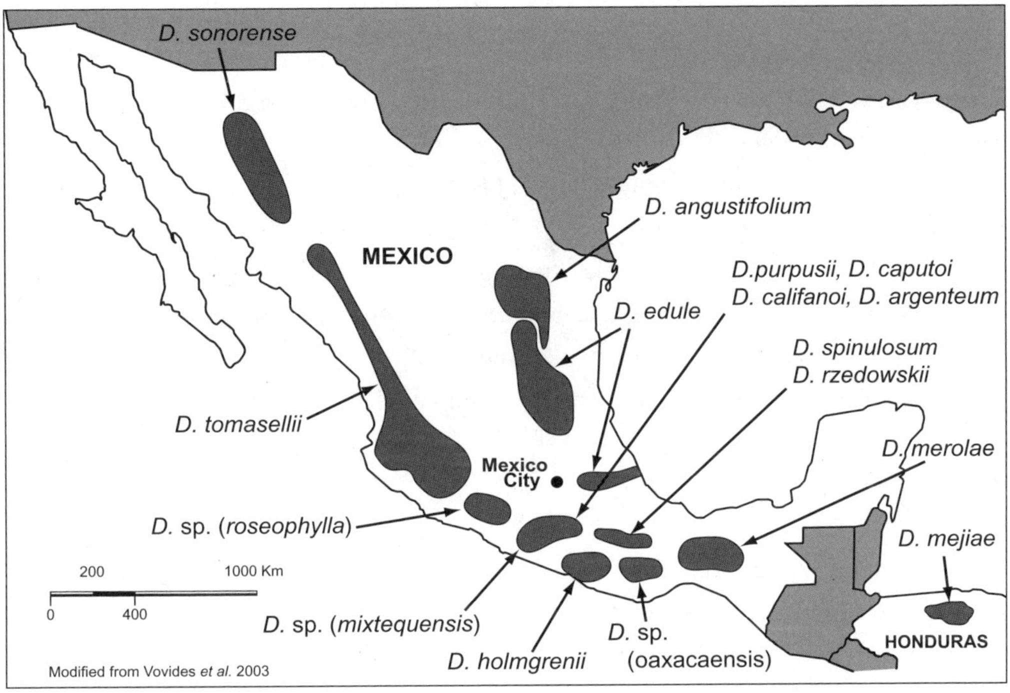

Figure 26-2. Distribution map of all known *Dioon* species.

terrace plain in Olanchito, Yoro. All but two of the remaining populations occur in the watershed of the Río Mame, a major tributary of the Aguán. Several of these are scattered along small streams entering the Mame—forming a second superpopulation—while the majority of the others occur in the watersheds of the Mame's major tributaries, the Chiquito and Negro rivers. Finally, one tiny disjunct population (Pop. 30) occurs in the upper Río Telica watershed (a tributary of the Río Patuca) in central western Olancho, while another disjunct population (Pop. 29) is found along the Río San Pedro, a stream that flows into the lower Río Aguán in the Department of Colón.

D. mejiae is the only Central American representative of this otherwise Mexican genus. Its distribution in relation to the other *Dioon* species is illustrated in Fig. 26-2. The nearest species is *D. merolae*, which occurs ca. 700 km to the west in Chiapas, while the other two tree dioons, *D. spinulosum* and *D. rzedowskii*, are greater than 1000 km disjunct in N. Oaxaca.

Other Locations

Anecdotal evidence and map toponyms suggest that *D. mejiae* may occur in central and west Honduras as well as north Nicaragua. An MBC-sponsored excursion by C. Nelson (Curator, Herbario Nacional de Honduras [TEFH], Universidad Nacional Autónoma de Honduras) to the San Juan de Opoa area of west Honduras in 2003 confirmed the existence of a *Zea* × *Tripsacum* hybrid called "teosinte" but no cycads (Haynes & Bonta, 2003). Though *D. mejiae* has been reported from gardens in Nicaragua and "teosintal" toponyms are also known from northern Nicaragua, whether these plants are descended from local wild populations or were brought there from Honduras is still unknown.

Type "Source" Locality

The exact locality/population from which the seed of the type specimen came remains a mystery. After much detective work, however, the most likely interpretation of Standley and Williams' (1950) locality reference is as follows:

1. Their report of the type source population being between Pueblo Nuevo and Olanchito is erroneous because (1) there is no Pueblo Nuevo in that part of Olancho, but there is a Pueblo Viejo and (2) the village name is written as "Pueblo Viejo" on the label of a herbarium voucher collected by Standley from plants in Mejía's garden around the same time the type specimen was collected—thus, it can safely be assumed that the name given in the formal description was a typographical error.
2. When Standley and Williams referred to a "road," if one assumes that they meant an automobile road, the La Suncuya/Quebrada del Repecho (Pop. 19) is the only population between Pueblo Viejo and Olanchito that is near such a road. However, no vehicular road existed in the region until after 1980; the main "road" was the "camino real," or old royal road. The main camino real route was preferred and was used during the dry season; the alternate route was used only during the rainy season.
3. The Cañón del Mame—the Río Mame Canyon—is one of only a handful of canyons in Olancho and is the only one that occurs along either the main or the

alternate camino real route from Pueblo Viejo to Olanchito. Because the main route runs along the Río Mame for much of its length, it is likely that this canyon is what Mejía (via Standley and Williams) was referring to. Also, because Standley and Williams referred to a "*dry* rocky canyon," one must assume that Mejía's trek occurred during the dry season, at which time the main route would have been preferred. Thus, the type source locality is most likely within the superpopulation that exists within the Río Mame Canyon, which extends for ca. 8 km in north-central Olancho. The precise locality will likely never be known.

Physical Environment

Habitat, Topography, and Climate

D. mejiae is typically an understory component of tropical broadleaf forests between 120 and 1000 masl and with average annual precipitation of 800–2,000 mm. Though it flourishes on steep slopes of 30–70 degrees, it also grows well on gentle and even flat terrain. Local habitat composition varies widely from lowland evergreen rain forest in Tocoa (Pop. 29), to deciduous mixed forests throughout NW Olancho, to arid thorn forest in Gualaco and Olanchito. It is rarely encountered in undisturbed habitat; rather, it is generally a hardy component of agro-pastoral habitats and modified forest zones, including pine/oak transition zones affected by anthropogenic fires. Habitat type and altitude for each population are listed in Table 26-1.

Geology and Soils

The soil type and underlying geology for each population are also provided in Table 26-1. Although many populations are restricted to rocky hillside terrain, the more intact populations spill down onto sandy Quaternary terraces above deeply incised watercourses. There is no single "preferred" soil type. Some populations thrive in sandy soil or sandy to clayey alluvial deposits, while others grow in loamy, limestone-derived soils; in soils weathered from metamorphics (schists, gneisses); etc. Most populations in Olancho lie along the edge of the Chindona batholith, a structurally homogeneous block of intrusive granite stretching across northwest Olancho (R. Rogers, pers. comm.). Analyses of

Table 26-2 Analyses of soil collected from eight *Dioon mejiae* populations across its range during the HN03 expedition.

Population information		pH	%	%	Available nutrients (ppm)			
No	Name	(H_20)	O.M.	N_{Total}	P	K	Ca	Mg
1	Saguay/Río Grande	6.71	2.26	0.11	29.58	402	1480	250
3	Teocintalito/Quebrada del Quiscamote	6.69	3.42	0.17	15.81	144	1330	280
6	Camalotal/Río Alao	8.05	18.94	0.95	11.69	360	14,500	450
8	Jacaleapa de Guata	7.64	9.76	0.49	123.77	610	6000	420
11	Río Tezapa/Río Mame	6.63	4.58	0.23	2.74	166	2720	430
12	Quebrada de Tiscan/Río Mame	7.77	4.17	0.21	55.43	240	4750	370
16	El Quebrachal/Río de Jano	7.95	15.74	0.79	5.86	410	9100	430
26	El Tamarindo/Quebrada del Tamarindo	6.32	4.94	0.25	14.00	168	1730	360

Analyses conducted at El Zamorano, the Panamerican School of Agriculture, Danlí, El Paraíso, Honduras.

soil collected from eight localities suggest a surprising adaptability with respect to soil physical properties and nutrient availabilities (Table 26-2).

Population Sizes and Densities

Estimates of the number of adult plants and geographic area encompassed for most populations are provided in Table 26-1. A conservative estimate of the total number of adult plants in Honduras is greater than 650,000 growing in greater than 30 populations distributed in three main sites in central and north Olancho and east Yoro and covering greater than 10,000 ha of land area (Table 26-1; Fig. 26-2), making it by far the most abundant cycad in the New World. The Saguay/Río Grande population, alone, contains more plants than most other cycad species (n=ca. 168,000 mature plants). Estimates for the other *Dioon* species range from 200 to 20,000 (Stevenson *et al.*, 2003), and the largest estimate for any species reported in the *IUCN/SSC Cycad Action Plan* is greater than 200,000 (*Cycas pectinata* and *C. panzhihuaensis*; Hill *et al.*, 2004. Density estimates averaged 62.5 plants/ha but varied greatly, ranging from 10 to 280 plants/ha.

Population Ecology

Life History

Reproductive Status and Phenology

More than 70% of the populations visited during HN03 were reproducing, as evidenced by the presence of seedlings and juveniles (Table 26-1). Seedlings were commonplace in non-fire-damaged populations but were not observed in populations most heavily impacted by human burning regimes. Female plants have an average rate of production of one cone every 2–3 years and appear to outnumber male plants in most populations. Coning occurs early in the rainy season (late May/early June), and female cones become receptive from June to August. People harvest cones (with immature seeds) from January to May the following year. Cones that are not harvested ripen and the seeds fall to the ground in September to December. As few as 10% of cones actually mature naturally, given the heavy demand on seeds as a subsistence food (see later).

Pollination

Very little is known about the pollination biology of this species. In 2002, S. Wells and the second author collected a weevil from a male cone at the Saguay/Río Grande population. It was later identified by C. O'Brien (former Professor Emeritus, Florida A&M University) as *Rhopalotria* sp. nov. All known *Rhopalotria* species are associated with cycad cones, and those with a known biology actually breed in the microsporophylls (Norstog & Nicholls, 1997).

Seed Dispersal

Potential mechanisms of natural seed dispersal include gravity, water, and animal transport. It is not clear to what extent the current distribution is also due to human agency, because seeds have been transported long distances by people for many centuries. It is also unclear whether the species originated in the lowlands and spread upslope, or vice versa.

A model wherein a historic climate shift had allowed species to expand gradually upslope (such as Gregory & Chemnick's [2004] "dynamic habitat" hypothesis for Mexican *Dioon* species) would necessitate either dispersal by animals or slow advance through seedling recruitment. Thus far, only the Central American agouti, *Dasyprocta punctata*, has been identified as a seed transporter, though it is possible that a now extinct specialized disperser existed in the past.

A downhill dispersal model would require only gravity or water transport, or both. The terrain of the region is extremely abrupt and subject to devastating flash floods that can rapidly transport seeds tens of kilometers. After the highest (i.e., 100+-year) floods, seeds might be safely deposited high above more frequent flood levels, allowing them time to germinate, mature, and gradually create new populations. More work needs to be done on seed dispersal in this and most other cycad species.

Growth Rate

Based on observations of garden plants in Juticalpa, Olancho, O. Flores Pinot estimated that *D. mejiae* grows only 70 cm per century at best (Bonta, 2003). If the assumption is made that plants in cultivation grow at least as fast as plants in habitat, Flores Pinot's growth rate would yield an estimated age of 1250–1400 years for the largest wild plants observed in Olancho—many of which had 8- to 10-m trunks (Fig. 26-3).

Predation

Insect Predators

During HN03, a brightly colored lepidopteran, was observed and photographed in every life stage exclusively on the leaves of *D. mejiae*. Newly emerging adults were collected at the Quebrada Agua Caliente/Río Mame population and were later identified as *Eumaeus childrenae* (R. Lehman, Butterfly & Insect Museum, La Ceiba, Atlántida). No record of this species using *D. mejiae* as a larval host plant is known.

A second caterpillar was observed burrowing into and feeding on the rachis of *D. mejiae* leaves at Talquezate/Río El Riachuelo. J. Heppner (Curator of Lepidoptera, Florida State Collection of Arthropods) identified it as an armyworm in the genus *Spodoptera*. Although armyworms are omnivorous, there is no record of a cycad being used as a larval host plant (J. Heppner, pers. comm.).

Non-insect Predators

At Saguay/Río Grande, cows were observed feeding on newly emerging leaves and tooth marks from an unidentified small mammal were observed on old sclerotestas. Domestic

Figure 26-3. Giant *dioon mejiae* at Teocintalito, Olancho; first author at bottom right indicates scale.

geese have been observed eating new leaves and small seedlings in Juticalpa (O. Flores Pinot, pers. comm.). In addition, D. Graham (pers. comm.) reported that the Central American agouti (known locally as "guatuza") is recognized as a tiusinte seed hoarder in the community of Limoncito, Gualaco.

Interpopulation Variation

Reports in the literature and discussions with experienced researchers suggest that interpopulation differences in vegetative characteristics are typical for *Dioon* species in Mexico (Gregory & Chemnick, 2004; T. Gregory, pers. comm.); thus, some level of variation was expected among populations in *D. mejiae* before HN03. Vegetative morphological variation was observed within each of the 20+ populations visited during HN03—and even within plants in any given population—but relatively little differentiation was noted between populations. A detailed analysis of intra- and interpopulation morphological variation is currently under way, and the results will be presented in a separate publication.

Species Characteristics

Nomenclature, Taxonomy, and Systematics

De Luca and Sabato (1978) are the only authors to address the nomenclature of *D. mejiae*, and they presented convincing arguments supporting specific rank for this taxon. With regard to taxonomy, Standley and Williams (1950) stated that, "on a geographic basis alone, it is quite safe to assume that the Honduran cycad is distinct" and that *D. mejiae* is "evidently related" to *D. spinulosum*. De Luca and Sabato (1978) echoed the latter testament, stating that *D. mejiae* is "probably more related to *D. spinulosum* Dyer (1883) than to *D. edule* Lindley (1843)." Sabato and De Luca (1985) supported these assumptions in their examination of evolutionary trends in *Dioon* and recognized the following three natural groups within the genus based on seed, seedling, and female cone morphology: (1) *D. mejiae*, (2) *D. spinulosum*, and *D. rzedowskii*, and (3) all other species.

With regard to systematics, the molecular work of Moretti et al. (1993) and De Luca et al. (1995) yielded two well-defined major clades within *Dioon*. The first consisted of the tree dioons (*D. mejiae* [*D. spinulosum* & *D. rzedowskii*] = "Spinulosum" clade) and the second of the remaining species ("Edule" clade). The resulting cladistic phylogeny was broadly congruent with a phenetic analysis on the same data, with morphological data, and with biogeography. A more recent study using analyses of different molecular sequences further supported the distinction between the "Spinulosum" clade and the "Edule" clade (Bogler & Francisco-Ortega, 2004).

Distinguishing Traits

The few authors who have reported vegetative or reproductive traits for *D mejiae*, or both, have done so primarily to distinguish it from the other two tree dioons. For example, Whitelock (2002) stated that, although *D. mejiae* is somewhat similar to *D. spinulosum* in outward appearance, adults can be separated by their "relatively stiff, straight, flat leaves, and leaflets without marginal spines." Whitelock (2002) further stated that *D. mejiae* seedlings have proximal leaflets reduced to prickles, whereas *D. spinulosum* seedlings have long, smooth petioles and lower leaflets that do not reduce to prickles. Sabato and De Luca (1985) reported that *D. mejiae* has the largest and most distinctive seeds, the most distinctive seedlings, and the largest male cones of any *Dioon*. The unpublished HN03 final report (Haynes & Bonta, 2003) provided extensive data on vegetative and reproductive traits for more than 200 plants from 14 populations. The distinguishing characteristics were used to develop an emended Latin description for this species as well as a diagnostic key that distinguishes the tree dioons from the more xeric *Dioon* species and from each other (see later). Vouchers from several populations were created during HN03 and deposited in the Herbario Nacional de Honduras [TEFH] at the Universidad Nacional Autónoma de Honduras, Tegucigalpa. They are listed in the far right column of Table 26-1.

Emended Latin Description

Planta arboriformis gigantea, trunco usque ad 6–10 m. vel ultra, saepe unius plantae multis magnis truncis erectis, his interdum crasse stipatis, folia numerosa rigida et coriacea, vel adscendentia vel declinantia, in statu juvenili tomentosa, cito glabrata, laete viridia, 87248 cm. longa vel ultra; rachis supra plana, subtus convexa, folii segmenta (pinnae) in uno latere rachidis 83–228 (i.e., omnino 163–456 in tota rachidi), segmenta media usque ad 8.5–30 cm. longa et 1.1–2.2 cm. lata, opposita, pungenti-attenuata, vel integra, vel cum usque ad 4 serraturis in segmento; distalia sensim decrescentia, inferiora sensimque decrescentia, pauca pungenti-serrata, infima elliptica 1–2 cm. tantum longa; strobili feminei vel erecti vel dependentes, ovoidei, usque ad 21–47 cm. longi atque 25–32 cm. lati, apice subabrupte acuminato, ubique tomento densissimo pallide brunneo molli obtecti; squamae stipitem obtegentes lineari-lanceolatae, usque ad 6–14.5 cm. longae, anguste attenuatae, extus dense lanosae; squamae strobili feminei deltoideo-ovatae usque ad 8–12 cm. longae et 7–9 cm. latae, anguste obtusae, imbricatae, extus dense lanosae; strobili masculi erecti, fusiformi, circa 53 cm. longi atque 12 cm. lati, apice rotundato, squamae stipitem obtegentes lineareae, 8 cm. tantum longae; squamae strobili masculi deltoideo-ovatae, usque ad 1–1.3 cm. longae, 1.7–2 cm. latae; semina usque ad 250 in uno strobilo, 3.7–6.0 cm. longa et 3.0–4.0 cm. lata; sarcotesta flava, sclerotesta fissurata, chalaza appendix plana.

Diagnostic Key

1A Median leaflets elliptic-acuminate, pungent; eophyll and seedling leaves resemble adult leaves, with petioles short or absent; female cones leaning to moderately pendulous at maturity; female cone peduncles ≤ 14.5 cm long; seeds with prominent chalazal appendix . . . . *D. mejiae*

1B Median leaflets lanceolate, not pungent; eophylls and seedling leaves with long, unarmed petioles; female cones completely pendulous at maturity, typically hanging below the leaves; female cone peduncles ≥ 17.5 cm long; seeds lacking chalazal appendix

2A Adult leaves lacking petioles (some or all leaves); leaflet margins serrated (some or all leaves); emerging leaves glabrous; female cone peduncles ≥ 30 cm long; megasporophylls appressed at maturity . . . . . . *D. spinulosum*

2B Adult leaves with unarmed petioles, 10–15 cm long; leaflet margins entire; emerging leaves pubescent; female cone peduncles ≤ 20 cm long; megasporophylls unappressed, becoming deflexed distally at maturity . . . . *D. rzedowskii*

Human Aspects

The Name "Tiusinte"

Derived from classic Nahuatl, "tiusinte'" (and its variants) means "sacred ear of maize." "Tiusinte" is herein designated as the standard vernacular name for *D. mejiae*. Although the authors had used "teocinte" in earlier writings (Bonta, 2003; Haynes & Bonta, 2003)—following the spelling on topographic maps—it is now important to standardize "tiusinte" (pronounced "tee-oo-SEEN-tay") over "teocinte" ("tay-oh-SEEN-tay"), not only because the former more closely reflects local pronunciation, but also because this should help avoid confusion between this cycad and other plants known as "teocinte."

Human Uses

Despite being little known outside its native range, *D. mejiae* is of great importance to the diets of 30–40,000 people in the municipalities of Gualaco, Guata, Jano, and Esquipulas del Norte in the Department of Olancho; Olanchito and Arenal in Yoro; and Tocoa in Colón. In some villages, up to 80% of the human population relies on tiusinte seeds as a starchy staple during times of maize scarcity. Cone harvests have taken on such importance that in some historical cases tiusintes received government protection. The average harvest is six cones per year for a household of six. Processed, one cone can feed one person for up to 2 months on a near-starvation diet; under better circumstances, tiusinte is used as a dietary supplement to maize and *Phaseolus* beans. Foods prepared from tiusinte seeds include tamales, tortillas, atol (a beverage), rosquillas (unleavened pastry), rosquetes (leavened pastry), and bread. The former three foods are widespread while the latter three are uncommon and the last two are restricted to Gualaco. At least two processes exist for the removal of toxins. Although no cases of progressive neurological damage in humans have ever been reported or observed, it is possible that isolated cases have occurred.

Tiusinte has a secondary importance within and outside its native range as a decorative species. The use of leaves for Catholic rites—particularly for Day of the Dead (Nov. 2), Christmas Eve, and Holy Week (Semana Santa)—is more widespread than seed consumption. Leaf use is uncommon in tiusinte-consuming villages and most common in larger towns and cities, where people grow tiusinte in dooryard gardens solely for the leaves—which are highly valued for their showiness, elasticity, and durability.

Tertiary uses are fading in importance, even while demand for tiusinte as a subsistence food is increasing in some areas. Early in the twentieth century, seeds were commonly made into whistles and "enchutes," or pin-and-target toys, while the mucilage from the cones was used to secure envelopes; mucilage and tomentum also had medicinal applications. (See Bonta, 2007 for a detailed discussion of human uses of this and other Honduran cycads.)

Conservation Status and Recommendations

Threats

Though *D. mejiae* faces numerous threats, its worth in local culture is unprecedented in areas where it is valued for cone production. As a common-property resource, tiusinte cannot, in theory, be owned by an individual (in contrast with most other trees). One rarely finds tiusintes chopped down by harvesters; damage is done primarily through conversion of habitat to farmland and ranchland, and, to a lesser extent, by habitat damage through road-building and logging. Roads are harmful to tiusintales, both through direct damage and through the introduction of new threats—such as conversion to pasture—that are brought about by the improved access. These changes are largely a result of post-1980 economic modernization.

Destruction of tiusintales is often an unintended result of illegal (but widely tolerated) enclosure of public lands. Subsequent treatment of tiusinte on private farmland ranges from protection to intentional destruction, with the most common case being the progressive but nonmalicious degradation of tiusinte plants due to excessive use of fire. In one village, small farmers have devised a management scheme whereby they balance the requirements of tiusintes and maize growing in the same field. In another case, a cattle rancher provides strict protection to part of the Saguay/Río Grande population at the village of Río Grande, Gualaco. But such examples are not commonplace.

Conservation Assessments

Osborne (1995) reported estimates of the number of plants in the wild, in public gardens, and in private collections for 200 cycad species. Osborne's estimates for *D. mejiae* were 5,000, 33, and 52 plants, respectively. As a result of these estimates, this species was considered "Rare" or "Vulnerable" based on the IUCN "Red Data" categories of Lucas and Synge (1978) and Mace et al. (1992), respectively (Osborne, 1995). The 1997 IUCN Red List of Threatened Plants listed *D. mejiae* as "Vulnerable" (Walter & Gillett, 1998), whereas Stevenson et al. (2003) treated it as "Data Deficient." The conservative estimate of greater than 650,000 wild adult plants reported herein effectively elevates this species to "Least Concern," (IUCN, 2001). Even though *D. mejiae* may very well be the most abundant cycad in the New World, many populations have been seriously degraded and the number of adult plants has diminished by ca. 50% over the past 25 years. As a CITES Appendix II species, it still merits a protection program. More than anything, it is a marvelous plant, not only in the context of its beauty and its importance to local diets, but also because it is the only locally abundant cycad in Honduras.

Representation in Ex Situ Collections

In May 2004, the first author conducted a survey in an attempt to obtain a more accurate and current estimate of the representation of this species in public and private gardens and parks, private collections and seed colonies, and nurseries around the world. Survey responses indicated the existence of 371 plants in 18 public or private gardens or parks in seven countries, 319 plants in 32 private collections or seed colonies in six countries, and 4,963 plants in 21 nurseries in six countries. Reported age of plants in cultivation ranged from 0 from 100+ years, and reported size ranged from 0 to 250 cm trunk height. Overall maturity ratio was 199:1 (immature:mature), overall sex ratio of mature plants was 1.25:1 (male:female), and age at first coning was 7–10 years, with some 30-year-old plants reported to have never coned. Although the number of cultivated plants in Honduras and Nicaragua remains unknown, it is important to include an estimate here because *D. mejiae* is a popular garden plant in many areas of these countries. It is reasonable to assert that the number is in the low thousands. Therefore, a conservative estimate of the total number of plants in cultivation worldwide exceeds 8000.

In Situ Conservation Recommendations

The following recommended actions would guarantee a safe future for *D. mejiae* and would assist the slow process of recovery in damaged tiusintales, together with a generalized "rescue" of the cultural traditions associated with the plant:

1. Immediate protection of tiusintales by the Protected Areas and Wildlife Department of the Administración Forestal del Estado-Corporación Hondureña de Desarrollo Forestal (AFE-COHDEFOR) by any means possible in the short term. This could include training of AFE-COHDEFOR personnel in recognition of the plant, where it grows, monitoring it, and trying to protect it against damage without prohibiting traditional extraction of tiusinte products.
2. Training of schoolteachers in tiusinte zones so that they can, in turn, educate the human populations who affect the plants.
3. Promotion of tiusinte in the media, and by private as well as state-run cultural centers ("casas de cultura").
4. In addition to botanical studies, studies of the nutritional and commercial values of tiusinte should be carried out.
5. Delimitation of tiusintales and zoning into the micro-watersheds of villages and towns, with their protection turned over to the communities.
6. Incorporation of tiusintales into actual and proposed protected areas, as mentioned I the preceding text.
7. Legal declaration of tiusintales as reserves, preferably as extractive reserves.
8. Development of the ecotourism potential of tiusintales.
9. Promotion of the hacienda of María Sarmiento Colindres, Río Grande, Saguay, Gualaco, as a special protection and study zone, because, thanks to family tradition, a great effort has been made there toward the protection of tiusinte—and as a result of HN03, it is now the most studied population of this species.
10. Reestablishment of municipal protection (the "embargo" of the "finca") in Gualaco and mechanisms for protection in other municipalities, perhaps by trying to have them protected as municipal reserves.
11. Presidential decree for protection of tiusintales throughout Honduras, as well as all Zamiaceae.

12. Support of legislation in Honduran Congress for the long-term protection of tiusinte and all Zamiaceae in Honduras.
13. Strict control and regulation of traffic in the species outside its local range, whether to domestic nurseries and cities or to the exterior.
14. Revival of the use of tiusinte foods (given adequate knowledge of toxin removal).
15. Recovery of the traditions of using leaves for religious and decorative purposes, so that domestic plants will be better appreciated. Also, commerce could be developed in leaves, from tiusinte populations to the urban population centers, where they are needed for both religious and secular purposes.
16. Study of the problems associated with tiusinte seed storage (pests, for example).
17. Study of the effects of the *Eumaeus* caterpillar on the health of the plant and populations.
18. Complete inventory of populations.
19. Collection of living material, for ex situ preservation, from populations not covered during HN03.
20. Inclusion of tiusinte in sustainable agriculture projects and biodiversity protection projects.
21. Promotion of knowledge of tiusinte by way of school texts, stamps, websites, and other means, to give it a higher national profile as a recognized part of Honduran natural and cultural heritage.
22. With the Instituto Hondureño de Antropología y Historia (IHAH), archaeological study of zones where tiusinte grows.
23. Incorporation of tiusinte into socioeconomic development projects that affect the zones where it grows.

Acknowledgments

We thank all members of the HN03 expedition party; Dr. Cyril Hardy Nelson-Sutherland, Universidad Nacional Autónoma de Honduras; Vince Murphy and the Instituto Hondureño de Turismo; and Marta Moreno and Manuel Rey Figueroa, AFE-COHDEFOR. Special thanks go to Dr. Barry Osborne, retired Latin professor in Santa Barbara, CA, for translating the emended description into Latin.

Literature Cited

Bogler, D. J. & J. Francisco-Ortega. 2004. Molecular systematic studies in cycads: evidence from *trn*L intron and ITS2 rDNA sequences. Bot. Rev. 70: 260–273.

Bonta, M. 2003. *Teocinte,* "Ear of God." Cycad Newsletter 26(1): 7–12.

———. 2007. Ethnobotany of Honduran cycads. Mem. New York Bot. Gard. 97: 120–142.

De Luca, P., A. Moretti, G. Siniscalco Gigliano, P. Caputo, S. Cozzolino, L. Gaudio, D. W. Stevenson, E. T. Wurtzel & R. Osborne. 1995. Molecular systematics of cycads. Pp. 131–137 *in* P. Vorster (ed.), *Proceedings of the third international conference on cycad biology.* Cycad Society of South Africa, Stellenbosch.

——— & S. Sabato. 1978. Nomenclature and taxonomy of *Dioon mejiae* Standl. & Williams (Zamiaceae). Delpinoa, n.s. 20: 45–50.

Gregory, T. J. & J. Chemnick. 2004. Hypotheses on the relationship between biogeography and speciation in *Dioon* (Zamiaceae). Pp. 137–148 in T. W. Walters & R. Osborne (eds.), Cycad classification: concepts and recommendations. CABI Publishing, Wallingford, UK.

Hill, K. D., D. W. Stevenson & R. Osborne. 2004. The world list of cycads. Appendix I in T. W. Walters and R. Osborne (eds.), Cycad classification: concepts and recommendations. CAB International Publishing, Wellingford, UK; available at http://plantnet.rbgsyd.gov.au/PlantNet/cycad/wlist.html.

IUCN. 2001. IUCN Red List categories and criteria version *3.1.* IUCN—The World Conservation Union, Gland, Switzerland, and Cambridge, UK; available at www.iucnredlist.org/info/categories_criteria.html.

Lucas, G. L. & H. Synge. 1978. The IUCN plant red data book. IUCN Threatened Plants Committee, Kew, UK.

Mace, G., N. Collar & J. Cooke. 1992. The development of new criteria for listing species on the IUCN Red List. Species 19: 16–22.

Moretti, A., P. Caputo, S. Cozzolino, L. Gaudio, G. Siniscalco-Gigliano & D. W. Stevenson. 1993. A phylogenetic analysis of *Dioon* (Zamiaceae). Am. J. Bot. 80: 204–214.

Norstog, K. J. & T. J. Nichols. 1997. The biology of the cycads. Cornell University Press, Ithaca, NY.

Osborne, R. 1995. The world cycad census and a proposed revision of the threatened species status for cycad taxa. *Biol. Cons.* 71: 1–12.

Sabato, S. & P. De Luca. 1985. Evolutionary trends in *Dion* (Zamiaceae). Am. J. Bot. 72: 1353–1363.

Standley, P. C. & L. O. Williams. 1950. *Dioon mejiae*, a new cycad from Honduras. Ceiba 1: 36–38.

Stevenson, D. W., A. Vovides & J. Chemnick. 2003. Regional overview: New world. Chapter 6 in J. S. Donaldson (ed.), Cycads: status survey and conservation action plan, IUCN—The World Conservation Union, Gland, Switzerland; available at www.iucn.org/themes/ssc/news/cycadsap.htm.

Walter, K. S. & H. J. Gillett (eds.). 1998. 1997 IUCN Red List of threatened plants. Compiled by the World Conservation Monitoring Centre. IUCN—The World Conservation Union, Gland, Switzerland and Cambridge, UK; available at www.wcmc.org.uk/species/plants/red_list.htm.

Walters, T. W. & R. Osborne. 2004. Cycad classification: concepts and recommendations, CAB International Publishing, Wellingford, UK.

Whitelock, L. M. 2002. The cycads. Timber Press, Portland, OR; available at www.timberpress.com/books/ index.cfm?do=detailsandID=519.

SYSTEMATICS
CHAPTER 27

Overview of the Cycad Flora of Honduras

Jody L. Haynes

Abstract

The cycad flora of Honduras includes two fairly well known species, *Dioon mejiae* and *Zamia standleyi*, and at least four undescribed species. The current state of knowledge of the distribution, taxonomy, and conservation of these cycads is reviewed.

Resumen

La flora de cícadas de Honduras incluye dos especies, *Dioon mejiae* y *Zamia standleyi*, bastante bien conocidas, y por lo menos cuatro especies no descritas. El estado actual del conocimiento de la distribución, taxonomía, y conservación de estas cícadas se repasa.

Introduction

The most recent *World List of Cycads* (Hill et al., 2007 lists three species as occurring in Honduras: *Dioon mejiae* Standl. & L. O. Williams, *Zamia standleyi* Schutzman, and *Z. herrerae* Calderón & Standl. No verifiable records seem to exist for *Z. herrerae* in Honduras, however, and, given the small amount of habitat available for it on Honduras's Pacific slope, it is highly unlikely that it even exists in the country. Therefore, one should not consider *Z. herrerae* a member of Honduras's cycad flora until indisputable proof is obtained. The status of the former two species in the wild was, until recently, not well known (Haynes & Bonta, 2003)—owing partly to a paucity of botanical exploration in the region and, in the case of *D. mejiae*, to an inadequate and partially erroneous species description (see Haynes & Bonta, 2007. Four new species have been discovered in Honduras in recent years, thus tripling the total cycad flora to six species. Because one of the new species is in the genus *Ceratozamia*, Honduras is now only the second country in the New World known to possess representatives of three cycad genera (the other is Mexico).

This chapter represents the published form of the first of three presentations given during a Workshop on Honduran Cycads at CYCAD 2005, organized and led by Mark Bonta and the current author. General ranges are provided for the two described species, but specific locality information has been intentionally omitted for all species to protect the plants from poaching. Notes regarding known vernacular names, human uses, associated insects, and herbarium specimens—including those collected by the author and deposited in the Herbario Nacional de Honduras (TEFH), Universidad Nacional Autónoma de Honduras, Tegucigalpa, as well as those examined in the Herbario Paul C. Standley, Escuela Agrícola Panamericana (EAP), Danlí, Honduras—are included at the end of each respective species summary.

Described Species

Dioon mejiae Standl. & L. O. Williams

The renowned Mesoamerican botanists Paul C. Standley and Louis O. Williams described *D. mejiae* Standl. & L. O. Williams in 1950 from a cultivated specimen growing in the garden of Dr. Isidoro Mejía H. of Danlí, El Paraíso, Honduras. *Dioon mejiae* has long been an enigma owing, in part, to the inadequate and partially incorrect description, which, unfortunately, resulted in misinformation being perpetuated about the species for more than half a century. No reproductive structures were available when the species was described, and hence this information was not included in the description. In addition, the description misnamed one of the towns in the location reference of the population from which Mejía collected his seeds 40 years earlier, indicated that only a single population occurred in the wild, and grossly underestimated the maximum size of the plants.

Based on Bonta's (2003) earlier work and the present study (HN03), it is now known that *D. mejiae* is a giant "tree" cycad with multiple trunks reaching greater than 10 m in height. It is common and widespread, occurring in at least 30 populations, encompassing 10,000 ha in the departments of Olancho and Yoro, and totaling in excess of 650,000 mature plants—thus making it the most abundant cycad in the New World (Haynes & Bonta, 2003). Population sizes and densities vary widely (from 6 to 150,000 plants/population and from 10 to 240 plants/ha, respectively), as do the habitats and soil conditions in which it grows (lowland evergreen rain forest to deciduous mixed forest to arid thorn forest; 120–1000 m above sea level [masl]; 800–2000 mm annual precipitation; and soils characterized by 6.3–8.0 pH, 2.2–18.9% organic matter, 2.7–123 ppm P, 166–610 ppm K, and 1330–14,500 ppm Ca). It is the southern- and easternmost *Dioon* species of an otherwise Mexican genus, occurring greater than 700 km east of its nearest congener. There are anecdotal reports of domestic plants in Nicaragua (Stevenson, 2001), but its presence in the wild there remains unsubstantiated (see Haynes & Bonta, 2007 for an emended description of this species.)

Dioon mejiae was listed as "Rare" or "Vulnerable" based on the IUCN Red Data categories set forth in Lucas and Synge (1978) and Mace et al. (1992), respectively (Osborne, 1995). The 1997 IUCN Red List of Threatened Plants also considered it "Rare" (Walter & Gillett, 1998), whereas it was more appropriately deemed "Data Deficient" in the IUCN Cycad Action Plan (Stevenson et al., 2003). Although much of its preferred habitat is affected by human burning regimens—its most serious threat—the plants can survive such fires. Therefore, it is recommended that the IUCN category be changed from "Data Deficient" to "Least Concern," based on the IUCN Red List Categories and Criteria v. 3.1 (IUCN, 2001).

Notes:

- Known locally as "tiusinte" (Spanish) or "tiñúk" (Tol) (Bonta & Osborne, 2007; Haynes & Bonta, 2003).
- The seeds are an important food source for 30–40,000 people (see Bonta, 2007 for details on human uses of this and other Honduran cycads).
- It appears to be an exclusive host species for the "tiusinte" hairstreak butterfly, *Eumaeus childrenae* (Haynes & Bonta, 2003).
- A new species of *Rhopalotria* weevil is thought to be the pollinator (Haynes & Bonta, 2003).
- TEFH specimens collected: 39075, 39078–39080, 39082, 39085, 39086, 39092, 39093, 39095, 39097–39100.
- EAP specimens examined: Merrill 15717, Standley 16522 (isotype), Standley & Williams 16756, Valerio 3805 (labeled *Cycas circinalis*).

Zamia standleyi Schutzman

In his description of *Z. standleyi* Schutzman, Schutzman (1989) indicated that it had first been reported from Honduras (as *Z. furfuracea*) by Standley (1931), who claimed that its range extended from the Lancetilla Valley, Atlántida, into southern Mexico. Work conducted since then separated out several Mexican species that were poorly known at that time (Schutzman, 1989) and eliminated all evidence of *Z. standleyi* occurring outside of Honduras. Although the true extent of its range is still uncertain, it is now known that the species is common and widely distributed in Honduras; in fact, its range may encompass as many as 500,000 ha in the northern river valleys of the departments of Colón, Olancho, and Yoro alone (Haynes & Bonta, 2003). Schutzman (1989) stated that the type specimen was collected in Atlántida, that he had also examined specimens from the departments of Cortés and Santa Barbara, and that the species had been reported as far east as the Río Plátano in the Department of Gracias a Dios (also known as "La Mosquitia"). Although some authors have suggested that it may also extend into Nicaragua and/or Guatemala (Whitelock, 2002), this claim is unsubstantiated at present, and most recent references refer to *Z. standleyi* as a Honduran endemic (Hill & Stevenson, 2007; Hill et al.,2007; Stevenson et al., 2003).

It typically grows in colonies of only 10–20 plants (populations are difficult to circumscribe), sometimes growing in sympatry with *D. mejiae*. It generally prefers moist hillsides under dense tree canopy below 650 masl. Plant density is fairly consistent from place to place, reaching only ca. 0.10 plants per hectare. A conservative estimate of the number of wild plants in Honduras is 50,000 (Haynes & Bonta, 2003). An earlier estimate of 6500 wild plants resulted in this species being considered "Rare" or "Vulnerable" based on the IUCN Red Data categories set forth in Lucas and Synge (1978) and Mace et al. (1992), respectively (Osborne, 1995). The 1997 IUCN Red List of Threatened Plants considered it "Rare" (Walter & Gillett, 1998), and the more recent estimate of Stevenson et al. (2003) of less than 1000 wild plants suggested a "Vulnerable" listing. However, because the previous population estimates were at least an order of magnitude too low, much of the preferred habitat is not easily converted to farmland, and several well-protected national parks and biosphere reserves are found within its range, it is recommended that the IUCN category be downgraded from "Vulnerable" to "Near Threatened," based on the IUCN Red List Categories and Criteria v. 3.1 (IUCN, 2001).

Notes:

- *Zamia standleyi* Schutzman is known locally as "camotillo" or "yuca de ratón" (Bonta & Osborne, 2007 Haynes & Bonta, 2003).
- It is widely regarded for its poisonous properties and has been used in Pech, Ladino, and Tol ethnic communities to kill both rodent pests and human enemies (Bonta, 2007; Haynes & Bonta, 2003).
- Several preserved specimens were collected during the HN03 expedition; unfortunately, none of them survived in good enough condition to be catalogued into the herbarium.

Undescribed Species

Zamia Species 1

Zamia species 1 is a beautiful arborescent plant with 2 m or greater trunks and 16 cm diameter. It carries 30 or more dark green, glossy leaves measuring to 1.5 m in length, and its long-lanceolate, basally subfalcate, apically acute, papyraceous leaflets measure up to

36 cm long and 4.5 cm wide (Haynes & Bonta, 2003). Its large seed cones (to 30 cm long and 11 cm in diameter) and pollen cones (to 36 cm long and 4 cm in diameter) most closely resemble those of *Z. tuerckheimii* Donn. Sm., although the cones of the latter are significantly smaller (Haynes & Bonta, 2003; Hill & Stevenson, nd; pers. obs.). Highly ornamental, even in habitat, the species also resembles *Z. tuerckheimii* in leaf, leaflet, and trunk characteristics, but it differs from the latter by carrying many more, slightly shorter leaves with numerous stout prickles along the petiole and extending up onto the rachis, and by having many more, longer but narrower leaflets.

Plants grow on steep slopes of 700–1300 masl in deep humus within dense tropical deciduous forest understory. In the higher elevations, the plants actually grow within a cloud forest transition zone. As many as 10,000 plants of every size, from seedling to large adult, were observed during HN03 in the only known population. Even though the population occurs within a protected area and the estimated total number of mature plants in the wild is relatively high, the threat of over-collecting is very real. Therefore, it is recommended that the species be considered "Critically Endangered" at present (IUCN Red List Criteria = CR B1ab(iv)+2ab(iv) [IUCN, 2001]).

Notes:

- *Zamia* species 1 is known locally as "camotillo" (Bonta & Osborne, 2007; Haynes & Bonta, 2003), like many of the smaller cycads in Honduras (as well as some ferns and even angiosperms [(Bonta, 2007).
- *Zamia* species 1 has been in cultivation since the 1990s, and seeds are being collected and grown in a local nursery as well as exported to the United States.
- A new species of *Pharaxonotha* beetle—thought to be the pollinator—was collected from a dehiscing pollen cone during HN03; it is currently being described.
- TEFH specimens collected: 39182–39189.

ZAMIA SPECIES 2

Zamia species 2 is known as the Honduran "cliff-dwelling" *Zamia* because of its tendency to grow on steep hillsides and road cuts. It has an acaulescent habit and holds one or two pendant leaves that can reach 2.5 m in length. It is sometimes informally referred to as

Z. "*lacanophila*" because of its morphological resemblance to *Z. lacandona* Schutzman & Vovides and its pendant leaves and cliff-dwelling habit reminiscent of *Z. cremnophila* Vovides, Schutzman & Dehgan (R. Adams, pers. comm.). It resembles the former species in having linear-lanceolate, falcate leaflets, and similarly shaped, although slightly larger seed cones. However, it differs from it in having much longer, green-emergent leaves, a four-sided petiole and rachis, and truly coriaceous leaflets.

It typically occurs in small colonies of 6–12 plants on steep, rocky hillsides that are not likely to be converted to farmland. Its distribution and range are not well known, but it seems to grow in isolated, disjunct populations along the Atlantic slope. Although the total number of plants and populations is unknown at present, it is recommended that the species be considered "Endangered" at present, owing to the threat of over-collecting (tentative IUCN Red List Criteria=EN B1ac(i,iii)+2ac(i,iii); C1+2a(i) [IUCN, 2001]).

Notes:

- *Zamia* species 2 is known locally as "camotillo" and "yuca de ratón" (Bonta & Osborne, this volume; Haynes & Bonta, 2003), and it is reportedly the most cold-sensitive of all the zamias in cultivation.
- Zamia species 2 has been in cultivation since the 1990s; it exists in several private collections and at least one ex situ research collection (MBC).
- Larvae of a second hairstreak butterfly, *Eumaeus toxea*, were observed feeding on the leaves in habitat.
- TEFH specimens collected: 39155–39159, 39190–39193.

Zamia Species 3

Zamia species 3 is the least known undescribed species in Honduras. At one time, a single acaulescent plant with red-emergent leaves, numerous narrow and variable leaflets, and resembling a dwarf version of *Z. chigua* Seem. existed in the New York Botanical Garden's living collection (D. Stevenson, pers. comm.). Although B. Schutzman (pers. comm.) suggested that this plant may have been nothing more than a red-emergent form of *Z. herrerae*, it differs from this species in having smaller, red-emergent leaves and smaller, thinner leaflets (D. Stevenson, pers. comm.). Moreover, the presence of *Z. herrerae* on the Atlantic coast of Honduras, in a highly distinct biogeographical province and across the Isthmus from the known Pacific slope range of the species in El Salvador, is extremely unlikely (M. Bonta, pers. comm.). This species was not observed during HN03, but plants fitting this general description were photographed on the Atlantic coast by R. Steiner in January 2005. Nothing more can be said about this species until it is better understood or formally described. It is recommended that this species be listed as "Data Deficient" based on the IUCN Red List Categories and Criteria v. 3.1 (IUCN, 2001).

Ceratozamia Species

The undescribed species of *Ceratozamia* is acaulescent and highly ornamental, holding up to 20 or more elegantly arching leaves measuring 2.5 m or more in length. Its broad, papyraceous leaflets (up to 13 cm wide and 30 cm long) suggest that it belongs to the *C. miqueliana* Miq. species complex, and it most closely resembles *C. euryphyllidia* Vázq. Torres, Sabato & D.W. Stev. However, it differs from the latter in having many more, shorter, shiny lime-green emergent leaves and dark green, glossy, nearly symmetrically obovate leaflets with acuminate tips that are only slightly off-center phyllodistally. It should be noted that various herbarium specimens of this species have been mislabeled *Z. tuerckheimii* (see later), with which it shares no affinities.

Although its range and distribution are not well known, it is most likely a Honduran endemic. It occurs in isolated, widely disjunct populations along the Atlantic slope. It grows in small to medium-sized colonies (from 6 to 24 or more plants), generally on moist hillsides along streams and under dense canopy in tropical lowland forest. Two of only three populations confirmed before and during the HN03 expedition are now extinct: one has been converted to farmland and the other was wiped out by collectors in 2003 (pers. obs.). The third confirmed population occurs in a protected area, but its size and extent are unknown. Because of its apparent scarcity and the fact that at least one entire population has been extirpated by collectors, it is recommended that this species be considered "Endangered" at present (tentative IUCN Red List Criteria = EN A2acd; B1ac(i,ii,iii,iv,v)+2ac(i,ii,iii,iv,v); C1+2a(i)b [IUCN, 2001]).

Notes:

- The undescribed *Ceratozamia* species is the southeasternmost *Ceratozamia* species known to date.

- The only known vernacular name is "camotillo" (Bonta & Osborne, 2007; Haynes & Bonta, 2003).
- It has been in cultivation since the 1990s, and it exists in only a handful of private collections.
- TEFH specimens collected: 39165, 39177–39180.
- EAP specimens examined (all incorrectly labeled *Z. tuerckheimii*): A. E. Brant & R. Zúniga 2830; D. L. Hazlett, A. E. Brant & R. Zúniga 8036; R. L. Leisner & D. Mejía 26238; P. C. Standley 7593.

Acknowledgments

Special thanks go to Mark Bonta, whose broad and extensive knowledge and experience in Honduras, exceptional planning and organizational skills, and steadfast friendship not only made the HN03 expedition possible, but also allowed me to establish myself as a field cycad botanist. Mark and the two reviewers also provided valuable comments on the final draft of the manuscript. Thanks also go to MBC Member Loran Whitelock, Bart Schutzman, Russell Adams, and Dennis Wm. Stevenson for graciously sharing their knowledge of population localities and Honduran cycads in general. Finally, I thank all of the wonderful people who participated in the 2003 Honduras expedition and the 2005 Honduran Cycad Study Tour, as well as the individuals, agencies, and nongovernmental organizations in Honduras that have supported MBC's cycad research and conservation efforts in their wonderful country. This study was sponsored by the Montgomery Botanical Center (2003 expedition targeting cycads).

Literature Cited

Bonta, M. 2003. Teocinte, "ear of God." The Cycad Newsletter 26(1): 7–12.

———. 2007. Ethnobotany of Honduran cycads. Mem. New York Bot. 97: 120–142.

——— & R. Osborne. 2007. Cycads in the vernacular: A compendium of local names. Mem. New York Bot. Gard. 97: 143–175.

Haynes, J. & M. Bonta. 2003. Montgomery Botanical Center Honduras 2003 cycad expedition final report. Unpublished report submitted to AFE-COHDEFOR in Tegucigalpa, Honduras.

——— & ———. 2007. An emended description of *Dioon mejiae* Standl. & L.O. Williams (Zamiaceae). Mem. New York Bot. Gard. 97: 418–443.

Hill, K. D. & D. W. Stevenson. 2007. *The* cycad pages website: http://plantnet.rbgsyd.nsw.gov.au/cgi-bin/cycadpg?taxname=Zamia+herrerae. Royal Botanic Gardens Sydney, Australia.

———, ——— & R. Osborne. 2007. The world list of cycads. Mem. New York. Bot Gard. 97: 454–483.

IUCN. 2001. IUCN Red list categories and criteria version 3.1. IUCN—The World Conservation Union, Gland, Switzerland, and Cambridge, UK: http://www.iucnredlist.org/info/categories_criteria.html.

Lucas, G. L. & H. Synge. 1978. The IUCN plant red data book. IUCN Threatened Plants Committee, Kew, UK.

Mace, G., N. Collar & J. Cooke. 1992. The development of new criteria for listing species on the IUCN Red List. Species 19: 16–22.

Osborne, R. 1995. The world cycad census and a proposed revision of the threatened species status for cycad taxa. Biol. Cons. 71: 1–12.

Schutzman, B. 1989. A new species of *Zamia* from Honduras. Syst. Bot. 14: 214–219.

Standley, P. C. 1931. Pp. 84–85 *in* Flora of Lancetilla Valley, Honduras. Field Mus. Nat. Hist. Bot. Ser., vol. 10, Chicago.

——— & L. O. Williams. 1950. *Dioon mejiae*, a new cycad from Honduras. Ceiba 1: 36–38.

Stevenson, D. W. 2001. Zamiaceae. Flora de Nicaragua 1: 6–7.

——— A. Vovides & J. Chemnick. 2003. Regional overview: New World. Pp. 31–38 *in* J. S. Donaldson (ed.), Cycads: status survey and conservation action plan. IUCN—The World Conservation Union, Gland, Switzerland: http://www.iucn.org/themes/ssc/news/cycadsap.htm.

Walter, K. S. & H. J. Gillett (eds.). 1998. 1997 IUCN Red list of threatened plants. Compiled by the World Conservation Monitoring Centre. IUCN—The World Conservation Union, Gland, Switzerland, and Cambridge, UK: http://www.wcmc.org.uk/species/plants/red_list.htm.

Whitelock, L. 2002. The cycads. Timber Press, Portland, OR: http://-www.timberpress.com/books/ index.cfm?do=detailsandID=519.

SYSTEMATICS
CHAPTER 28

The World List of Cycads/[La Lista Mundial de Cícadas]

Ken D. Hill, Dennis Wm. Stevenson, and Roy Osborne

Abstract

The authors present an update of the World List of Cycads, which is a list of the valid names of all extant cycads, together with details of their countries of occurrence, the primary publication references, and herbarium localities for the types and details of their conservation status. The data given for each genus is followed by a list of synonyms. The number of species on the current list is as follows: *Bowenia* (2), *Ceratozamia* (21), *Chigua* (2), *Cycas* (98), *Dioon* (13), *Encephalartos* (65), *Lepidozamia* (2), *Macrozamia* (41), *Microcycas* (1), *Stangeria* (1), and *Zamia* (57) to give a total of 303 species.

Resumen

Los autores presentan una actualización de la Lista Mudial de las Cícadas. Este consiste de una lista de los nombres válidos de todas las cícadas vivientes, con detalles de los países de procedencia, citas originales de los nombres, referencias, ubicación de los herbarios para los tipos y detalles del estatus de conservación de las especies. Los datos de cada género están seguidos por una lista de sinónimos. El número de las especies en la lista actual es lo siguiente: *Bowenia* (2), *Ceratozamia* (21), *Chigua* (2), *Cycas* (98), *Dioon* (13), *Encephalartos* (65), *Lepidozamia* (2), *Macrozamia* (41), *Microcycas* (1), *Stangeria* (1) y *Zamia* (57), sumando un total de 303 especies.

Introduction

The first World List of Cycads was published in *Encephalartos* (*Journal of the Cycad Society of South Africa*) (Osborne & Hendricks, 1985) with minor amendments in a supplementary list in a subsequent issue of the same journal (Osborne & Hendricks, 1986). Several updates followed as changes in taxonomy and in outlook, especially in the genera *Cycas, Encephalartos, Macrozamia,* and *Zamia,* made the previous lists obsolete. These have been presented at the various international conferences on cycad biology (Stevenson et al., 1990, 1995; Stevenson & Osborne, 1993a; Osborne et al., 1999; Hill et al., 2004a, c) an-delsewhere (Stevenson & Osborne, 1993b; Hill et al., 2004b). The list is also now available as a continuously updated searchable database on the Internet (Hill & Stevenson, 1998–).

The list that follows gives the valid names of all known extant cycads at the time of final editing of this text, i.e., April 2007 (column 1), and details of their countries of appearance (column 2) (states, provinces, departments, or districts in parentheses). Compilation is alphabetical by genus/species and author's names are abbreviated in accordance with Brummit and Powell (1992). The type species for each genus is indicated by an asterisk. Dates (column 3) refer to the year of publication of the name. Journals (column 4) are abbreviated according to the *Botanico-Periodicum-Huntianum* standard (Lawrence et al., 1968) while other reference material follows the style of Stafleu and Cowan (1976–1983). Column 5 gives details of the nature and herbarium locality of the type. The following abbreviations are used: H = holotype; I = isotype; L = lectotype; N = neotype; SYN = syntype and BAS = under basionym. Abbreviations for herbaria follow those of *Index Herbariorum* (Holmgren et al., 1990) and at the website: http://sciweb.nybg.org/science2/IndexHerbariorum.asp. The final column (column 6) gives details of the threatened status as listed in Donaldson et al. (2003) or, where the taxon is not listed in that work, from other authoritative sources. We use the abbreviations: EW = extinct in the wild; CR = critically endangered; EN = endangered; VU = vulnerable; NT = near threatened; LC = least concern; DD = data deficient, and NE = not evaluated.

After each genus we have given a partial list of synonomy; the synonyms given are all validly published legitimate names.

Taxon	Location	Date	Original Source	Type	Status
BOWENIA Hook. ex Hook. F., 1863 (2 species, Australia)			t. 5398	Bot. Mag. 89: sub	
B. serrulata (W. Bull) Chamb.	Australia (Qld)	1912	Bot. Gaz. 54: 419	t. 5	LC
⋆*B. spectabilis* Hook. ex Hook. f.	Australia (Qld)	1863	Bot. Mag. 89: sub t. 5398	t. 5398	LC
Synonyms:.					
B. spectabilis var. *serrata* Bailey	= *B. serrulata*	1883	Queensland Flora, p. 501	I – K	
B. spectabilis var. *serrulata* W. Bull	= *B. serrulata*	1878	Retail List 143: 4, t. 5	t. 5	
CERATOZAMIA Brongn., 1846 (20 species; Belize, Guamemala, Mexico)			Ann. Sci. Nat. Bot. Ser. 3, 5: 7–9, t. 1		
C. alvarezii Pérez-Farr., Vovides & Iglesias	Mexico (Chiapas)	1999	Novon 9(3) 410–413	H—CHIP	EN
C. becerrae Pérez-Farr., Vovides & Schutzman	Mexico (Chiapas, Tabasco)	2004	Bot. J. Linn. Soc.146: 123–128	H—XAL	EN
C. euryphyllidia Vázq. Torres, Sabato & D.W. Stev.	Guatemala, Mexico (Oaxaca, Veracruz)	1986	Brittonia 38(1): 17–26	H—NY	CR
C. hildae G. P. Landry & M. C. Wilson	Mexico (Querétaro, San Luis Potosí)	1979	Brittonia 31(3): 422–424, fig. 1	H—GH	EN
C. huastecorum S. Avendaño, Vovides & Cast.-Campos	Mexico (Veracruz)	2003	Bot. J. Linn. Soc. 141: 395–398	H—XAL	DD
C. kuesteriana Regel	Mexico (Tamaulipas)	1857	Bull. Soc. Imp. Naturalistes Moscou 30: 186–188, t. 3, fig. 6; t. 4, fig. 22	H—LE	CR
C. latifolia Miq.	Mexico (Hidalgo, Querétaro, San Luis Potosí, Veracruz)	1848	Tijdschr. Wis-Natuurk. Wetensch. Eerste Kl. Kon. Ned. Inst. Wetensch. 1: 206	N—NY	VU
C. matudae Lundell	Guatemala, Mexico (Chiapas)	1939	Lloydia 2(2): 75–76	H—MICH	EN
⋆*C. mexicana* Brongn.	Mexico (Hidalgo, Puebla, Veracruz)	1846	Ann. Sci. Nat. Bot. ser. 3, 5: 7–8, t. 1	H—P	VU
C. microstrobila Vovides & J. D. Rees	Mexico (San Luis Potosí)	1983	Madroño 30(1): 38–41	HXAL	VU
C. miqueliana H. Wendl.	Mexico (Chiapas, Veracruz)	1854	Index Palm.: 68	N—NY	VU

C. mirandae Vovides, Pérez-Farr. & Iglesias	Mexico (Chiapas)	2001	Bot. J. Linn. Soc. 137(1): 81–85, fig. 1–2 as *C. mirandai*	H—CHIP	VU
C. mixeorum Chemnick, T. J. Greg. & S. Salas-Mor.	Mexico (Oaxaca)	1997	Phytologia 83(1): 47–52 (issued 1998)	H—HNT	DD
C. morettii Vázq. Torres & Vovides	Mexico (Veracruz)	1998	Novon 8(1): 87–90, fig. 1	H— CIB	EN
C. norstogii D.W. Stev.	Mexico (Chiapas, Oaxaca)	1982	Brittonia 34: 181–184, figs 1–2	H—NY	CR
C. robusta Miq.	Belize, Guatemala, Mexico (Chiapas, Oaxaca, Veracruz)	1847	Tijdschr. Wis-Natuurk. Wetensch. Eerste Kl. Kon. Ned. Inst. Wetensch. 1: 42–43	N—NY	VU
C. sabatoi Vovides, Vázq. Torres, Schutzman & Iglesias	Mexico (Hidalgo, Querétaro)	1993	Novon 3(4): 502–504	H—XAL	EN
C. vovidesii Pérez-Farr. & Iglesias	Mexico (Chiapas)	2007	Bot. J. Linn. Soc. 153: 393–400, figs 1–6	H—HEM	VU
C. whitelockiana Chemnick & T.J. Greg.	Mexico (Oaxaca)	1995	Phytologia 79(1): 51–57 (issued 1996)	H—HNT	EN
C. zaragozae Medellin-Leal	Mexico (San Luis Potosí)	1963	Brittonia 15: 175–176	H—SLPM	CR
C. zoquorum Pérez-Farr., Vovides & Iglesias	Mexico (Chiapas)	2001	Bot. J. Linn. Soc. 137(1): 77–80, fig.1	H—CHIP	CR
Synonyms:					
DIPSACOZAMIA Lehm.	= *CERATOZAMIA*	1846	*in* Lindl., Veg. Kingd.: 225	C	
C. boliviana Brongn.	= *Zamia boliviana*	1846	Ann. Sci. Nat. Bot. ser. 3, 5: 9	L—P	
C. brevifrons Miq.	= *C. mexicana*	1847	Tijdschr. Wis-Natuurk. Wetensch. Eerste Kl. Kon. Ned. Inst. Wetensch. 1: 41–42	?	
C. intermedia Miq.	= *C. mexicana*	1847	Tijdschr. Wis-Natuurk. Wetensch. Eerste Kl. Kon. Ned. Inst. Wetensch. 1: 40–41	?	
C. katzeriana Regel	= *Zamia katzeriana*	1876	Trudy Imp. S. Petersburgsk. Bot. Sada 4: 298	L—LE	
C. longifolia Miq.	= *C. mexicana*	1847	Tijdschr. Wis-Natuurk. Wetensch. Eerste Kl. Kon. Ned. Inst. Wetensch. 1: 40	?	
CHIGUA D.W. Stev., 1990 (2 species, South America)			Mem. New York Bot. Gard. 57: 169–172		
C. bernalii D.W. Stev.	Colombia (Cordoba)	1990	Mem. New York Bot. Gard. 57: 169–172	H—COL	CR

Taxon	Location	Date	Original Source	Type	Status
C. restrepoi D. W. Stev.	Colombia (Cordoba)	1990	Mem. New York Bot. Gard. 57: 169–172	H—HUA	CR
CYCAS L., 1753 (98, Asia, Australia, Indian and SW Pacific ocean countries)			Sp. Pl.: 1188.		
C. aculeata K. D. Hill & H. T. Nguyen	Vietnam (Da Nang)	2004	Bot. Rev. 70(2): 149–150	H—HN	EN
C. annaikalensis Rita Singh & P. Radha	India (Western Ghats)	2006	Brittonia 58: 119–123	H—IUPH	DD
C. angulata R. Br.	Australia (NT & Qld)	1810	Prodr. 1: 348	H—BM	LC
C. apoa K. D. Hill	New Guinea	1994	Austral. Syst. Bot. 7: 553–554, fig. 9	H—CANB	NT
C. arenicola K. D. Hill	Australia (NT)	1993	Telopea 5(2): 419–422	H—NSW	NT
C. armstrongii Miq.	Australia (NT)	1868	Arch. Néerl. Sci. Exact. Nat. 3(5): 235–236	H—U	LC
C. arnhemica K. D. Hill	Australia (NT)	1994	Telopea 5(4): 693–696, fig. 1	H—NSW	LC
subsp. *muninga* Chirgwin & K. D. Hill	Australia (NT)	1996	Telopea 7(1): 44–46, fig. 20	H—NSW	LC
subsp. *natja* K. D. Hill	Australia (NT)	1996	Telopea 7(1): 46–47, fig. 21	H—DNA	LC
C. badensis K. D. Hill	Australia (Qld)	1996	Telopea 7(1): 20–21, fig. 9	H—BRI	NT
C. balansae Warb.	China (Guangxi), Vietnam (Lang Son, Quang Ninh, Thai Nguyen, Vinh Phuc)	1900	Monsunia 1: 179	H—P	NT
C. basaltica C. A. Gardner	Australia (WA)	1923	Bull. Woods Forests Dept., Western Australia 32: 31	H—PERTH	LC
C. beddomei Dyer	India (Andhra Pradesh)	1885	Trans. Linn. Soc. London, Bot. ser. 2, 5: 85–86, pl. 17	H—K	CR
C. bifida (Dyer) K. D. Hill	China (Guangxi, Yunnan), Vietnam (Cao Bang, Lang Son, Tuyen Quang)	2004	Bot. Rev. 70(2): 161–163	H—K	VU
C. bougainvilleana K. D. Hill	Bougainville, New Britain, Solomon Islands	1994	Austral. Syst. Bot. 7: 557–560, fig. 11	H—NSW	NT
C. brachycantha K. D. Hill, H. T. Nguyen, & Phan K. Loc	Vietnam (Bac Kan)	2004	Bot. Rev. 70(2): 155–157, fig. 6	H—HN	NT
C. brunnea K. D. Hill	Australia (Qld)	1992	Telopea 5(1): 200–201, fig. 15	H—DNA	NT
C. cairnsiana F. Muell.	Australia (Qld)	1876	Fragm. 10: 63, 121	H—MEL	NT

C. calcicola Maconochie	Australia (NT)	1978	J. Adelaide Bot. Gard. 1(3): 175, fig. 1	H—DNA	LC
C. campestris K. D. Hill	New Guinea	1994	Austral. Syst. Bot. 7: 538–540	H—NSW	NT
C. canalis K. D. Hill	Australia (NT)	1994	Telopea 5(4): 698–700, fig. 4a–d	H—NSW	EN
C. candida K. D. Hill	Australia (Qld)	2004	Telopea 10(2): 607–611	H—NSW	NT
C. chamaoensis K. D. Hill	Thailand (Chantaburi)	1999	Brittonia 51(1): 58, fig. 6	H—NSW	CR
C. changjiangensis N. Liu	China (Hainan)	1998	Acta Phytotax. Sin. 36(6): 552–554, fig. 1	H—IBSC	EN
C. chevalieri Leandri	Vietnam (Ha Tinh, Nghe An, Quang Binh, Quang Tri)	1931	In Lecomte, Fl. Indo–Chine 5(10): 1092	I—P	NT
**C. circinalis* L.	India (Andra Pradesh, Karnataka, Kerala, Maharashtra, Tamil Nadu)	1753	Sp. Pl.: 1188	t. 19, *in* Rheede, Hort. Malab, 1682	DD
C. clivicola K. D. Hill	Malaysia, Thailand (Chumphon, Narathiwat, Phang Nga, Phuket, Ranong, Trang)	1999	Brittonia 51(1): 62–63, fig. 8a–d, g–h	H—NSW	NT
subsp. *lutea* K. D. Hill	Cambodia, Thailand (Prachinburi, Rayong), Vietnam (An Giang, Kien Giang)	1999	Brittonia 51(1): 64, fig. 8e,f,I	H—NSW	DD
C. collina K. D. Hill, H. T. Nguyen, & Phan K. Loc	Vietnam (Son La)	2004	Bot. Rev. 70(2): 142–144	H—HN	VU
C. condaoensis K. D. Hill & S. L. Yang	Vietnam (Ba Ria-Vung Tau)	2004	Bot. Rev. 70(2): 178–179, fig. 14	H—K	VU
C. conferta Chirgwin ex Chirgwin & Wigston	Australia (NT)	1993	J. Adelaide Bot. Gard. 15(2): 147	H—DNA	VU
C. couttsiana K. D. Hill	Australia (Qld)	1992	Telopea 5(1): 197–198, fig. 13	H—NSW	NT
C. cupida P. I. Forst.	Australia (Qld)	2001	Austrobaileya 6: 153	H—BRI	VU
C. curranii (J. Schust.) K. D. Hil	Philippines (Palawan)	1995	Proc. Third Int. Conf. Cycad Biol.: 150	L—K	DD
C. debaoensis Y. C. Zhong & C. J. Chen	China (Guangxi)	1997	Acta Phytotax. Sin. 35(6): 571	H—PE	CR
C. desolata P. I. Forst.	Australia (Qld)	1995	Austrobaileya 4(3): 345–352, figs 1–5	H—BRI	VU
C. diannanensis Z. T. Guan & G. D. Tao	China (Yunnan), ?N Vietnam	1995	Sichuan Forestry & Design 1995: 1	?H—I	VU

Taxon	Location	Date	Original Source	Type	Status
C. dolichophylla K. D. Hill, H. T. Nguyen, & Phan K. Loc	China (Yunnan), Vietnam (Bac Kan, Cao Bang, Ha Giang, Lai Chau, Lao Cai, Ninh Binh, Son La, Thai Nguyen, Thanh Hoa, Tuyen Quang)	2004	Bot Rev. 70(2): 157–160, fig. 7	H—HN	VU
C. edentata de Laub.	Philippines (Sulu)	1998	*in* De Laub. & Adema, Blumea 43: 372	H—L	DD
C. elephantipes A. Lindstr. & K. D. Hill	Thailand (Chaiyaphum)	2002	Brittonia 54(4): 301. Fig. 3 (issued 2003)	H—BKF	EN
C. elongata (Leandri) D.Yue Wang	Vietnam (Binh Dinh, Khanh Hoa, Ninh Thuan, Phu Yen, Quang Ngai)	1996	Cycads in China: 51, as *C. elonga*	L—P	VU
C. fairylakea D.Yue Wang	China (Guangdong)	1996	Cycads in China: 54	H—SZG	EN
C. falcata K. D. Hill	Indonesia	1999	Kew Bull. 54(1): 209	H—K	DD
C. ferruginea F. N. Wei	China (Guangxi), Vietnam (Lang Son, Thai Nguyen)	1994	Guihaia 14: 300	H—IBK	NT
C. fugax K. D. Hill, H. T. Nguyen, & Phan K. Loc	Vietnam (Phu Tho)	2004	Bot. Rev. 70(2): 145–146, fig. 3	H—HN	CR
C. furfuracea W. Fitzg.	Australia (WA)	1918	J. & Proc. R. Soc. West. Austral. 3: 108	L—NSW	LC
C. glauca hort. ex Miq.	East Timor, Indonesia	1840–41	Comm. Phytog.: 127	L—U	DD
C. guizhouensis K. M. Lan & R. F. Zou	China (Guangxi, Guizhou, Yunnan)	1983	Acta Phytotax. Sin. 21(2): 209–210	H—PE	NT
C. hainanensis C. J. Chen	China (Hainan)	1975	Acta Phytotax. Sin. 13(4): 82, t. 2, figs 5 & 6	H—IBSC	EN
C. hoabinhensis Phan K. Loc & H. T. Nguyen	Vietnam (Ha Nam, Ha Tay, Hoa Binh, Ninh Binh)	2004	Bot. Rev. 70(2): 153–155, fig. 5	H—HNU	EN
C. hongheensis S. Y. Yang & S. L. Yang	China (Yunnan)	1996	*in* D. Y. Wang, Cycads in China: 62	H—Panzhihua Inst. Hort	CR
C. inermis Lour.	Vietnam (Da Nang, Dong Nai, Khanh Hoa, Quang Nam)	1793	Fl. Cochinch. Ed. 2, 2: 776–777	H—BM	VU
C. javana (Miq.) de Laub.	Indonesia	1996	*in* D. Y. Wang, Cycads in China: 65	H—L	DD
C. lane-poolei C. A. Gardner	Australia (WA)	1923	Bull. Woods Forests Dept., West. Austral. 32: 30–32, fig. e	H—PERTH	LC

C. lindstromii S. L. Yang, K. D. Hill, & N. T Hiêp	Vietnam (Ba Ria-Vung Tau, Binh Thuan, Khanh Hoa, Ninh Thuan)	1997	Novon 7: 213–215, fig. 1	H—FTG	VU
C. litoralis K. D. Hill	Indonesia (Sumatra), Malaysia, Myanmar, Thailand (Chumphon, Narathiwat, Phang Nga, Trat), Vietnam (Kien Giang)	1999	Brittonia 51(1): 70–72, fig. 11	H—NSW	NT
C. maconochiei Chirgwin & K. D. Hill	Australia (NT)	1996	Telopea 7(1): 48–49, fig. 22	H—NSW	NT
subsp. *lanata* K. D. Hill	Australia (NT)	1996	Telopea 7(1): 49–51	H—DNA	LC
subsp. *viridis* K. D. Hill	Australia (NT)	1996	Telopea 7(1) : 51–52, fig. 23	H—NSW	LC
C. macrocarpa Griff.	Malaysia, Thailand (Chantaburi, Chumphon, Narathiwat, Ranong)	1854	Not. Pl. Asiat. 4: 11–14, pl. 362, fig. 2	H—K	VU
C. media R. Br.	Australia (Qld)	1810	Prodr. 1: 348	H—BM	LC
subsp. *banksii* K. D. Hill	Australia (Qld)	1996	Telopea 7(1): 15–16	H—NSW	LC
subsp. *ensata* K. D. Hill	Australia (Qld)	1996	Telopea 7(1): 14–15, fig. 6	H—NSW	LC
C. megacarpa K. D. Hill	Australia (Qld)	1992	Telopea 5(1): 188–189, fig. 5	H—NSW	EN
C. micholitzii Dyer	E Laos, Vietnam (Dac Lak, Gia Lai, Kon Tum)	1905	Gard. Chron. 38: 142–144, figs. 48 & 49	H—K	VU
C. micronesica K. D. Hill	Micronesia (Marianas Islands)	1994	Austral. Syst. Bot. 7: 554–556, fig. 10	H—NSW	NT
C. multipinnata C. J. Chen & S. Y. Yang	China (Yunnan), Vietnam (Yen Bai)	1994	Acta Phytotax. Sin. 32: 239, 480–481	H—PE	EN
C. nathorstii J. Schust.	N Sri Lanka	1932	Pflanzenr. 99: 76, fig. 10e	L—G	DD
C. nongnoochiae K. D. Hill	Thailand (Nakhon Sawan)	1999	Brittonia 51(1): 60–62, fig. 7	H—NSW	VU
C. ophiolitica K. D. Hill	Australia (Qld)	1992	Telopea 5(1): 190–191, fig. 7	H—NSW	NT
C. orientis K. D. Hill	Australia (NT)	1994	Telopea 5(4): 696–697, fig. 3	H—NSW	LC
C. pachypoda K. D. Hill	Vietnam (Binh Thuan, Ninh Thuan)	2004	Bot. Rev. 70(2): 176–178, fig. 13	H—HN	VU
C. panzhihuaensis L. Zhou & S. Y. Yang	China (Sichuan, Yunnan)	1981	Acta Phytotax. Sin. 19(3): 335, t. 10, figs. 1–6; t. 11, figs. 1–10	H—PE	NT
C. papuana F. Muell.	New Guinea	1876	Descr. Notes Papuan Pl. 4: 71–72	L—MEL	NT

Taxon	Location	Date	Original Source	Type	Status
C. pectinata Buch.-Ham.	Bangladesh, Bhutan, China (Yunnan), NE India, Laos, Myanmar, Nepal, Thailand (Chiang Mai, Kanchanaburi, Mae Hong Son, Phetchabun, Phrae, Sukhothai), Vietnam (Gia Lai, Kon Tum, Lam Dong, Quang Ngai)	1826	Mem. Wern. Nat. Hist. Soc. 5(2): 322–323, figs. 3 & 5	L—K	VU
C. petraea A. Lindstr. & K. D. Hill	Thailand (Loei)	2002	Brittonia 54 (4): 299. fig. 1 (issued 2003)	H—BKF	NT
C. platyphylla K. D. Hill	Australia (Qld)	1992	Telopea 5(1): 193–194, fig. 9	H—NSW	EN
C. pranburiensis S.L. Yang, W. Tang, K. D. Hill, & P. Vatcharakorn	Thailand (Prachuap Khiri Khan)	1999	Brittonia 51(1): 44–47, fig. 1	H—K	VU
C. pruinosa Maconochie	Australia (WA)	1978	J. Adelaide Bot. Gard. 1(3): 177–178, fig. 2	H—DNA	LC
C. revoluta Thunb.	Japan (Ryukyu Islands)	1782	Verh. Holl. Maatsch. Weetensch. Haarlem 20(2): 424, 426–427	H—UPS	NT
C. riuminiana Porte ex Regel	Philippines (Luzon, Mindoro)	1863	Gartenflora 12: 16–17	H—LE	DD
C. rumphii Miq.	S Borneo, E Indonesia, NE Java, Moluccan Islands, New Guinea, Sulawesi	1839	Bull. Sci. Phys. Nat. Néerl. 2: 45	t. 23 *in* Rumphius, Herb. Amb., 1749	NT
C. schumanniana Lauterb.	New Guinea	1900	Fl. Schutzgeb. Sudsee: 154–155	L—WRSL	NT
C. scratchleyana F. Muell.	New Guinea, Australia (Qld)	1885	Victorian Naturalist 2(2): 18–19	H—MEL	NT
C. seemannii A. Braun	SW Pacific islands under the jurisdiction of Fiji, New Caledonia, Tonga, Vanuatu	1876	Sitzungsber. Ges. Naturf. Freunde Berlin : 114–115	L—K	VU
C. segmentifida D. Yue Wang & C.Y. Deng	China (Guangxi, Guizhou, Yunnan), ?N Vietnam	1995	Encephalartos 43: 11–14	H—SZG	VU
C. semota K. D. Hill	Australia (Qld)	1996	Telopea 7(1): 23–26, fig. 11	H—NSW	NT
C. sexseminifera F. N. Wei	China (Guangxi), Vietnam (Cao Bang, Ninh Binh, Thanh Hoa)	1996	Guihaia 16: 1	H—IBK	NT

C. shanyaensis G. A. Fu	China (Hainan Island)	2006	Bull. Bot. Res. 26(1): 2–3	?	DD
C. siamensis Miq.	Cambodia, Laos, Myanmar, Thailand (Chachoengsao, Chaiyaphum, Chonburi, Kanchanaburi, Lampang, Nakhon Ratchasima, Phetchabun, Ratchaburi, Sakon Nakhon, Tak, Uthai Thani, Uttaridit), Vietnam (Dac Lak, Gia Lai, Kon Tum, Nghe An, Thanh Hoa)	1863	Bot. Zeitung Berlin 21: 334	H—U	VU
C. silvestris K. D. Hill	Australia (Qld)	1992	Telopea 5(1): 181–182, fig. 1	H—NSW	VU
C. simplicipinna (Smitinand) K. D. Hill	Laos, Myanmar, Thailand (Chiang Mai, Loei, Mae Hong Song, Phrae), Vietnam (Quang Tri)	1995	Proc. Third Int. Conf. Cycad Biol.: 150	L—BKF	NT
C. spherica Roxb.	India (Orissa)	1832	Fl. Ind.: 747	?	DD
C. szechuanensis C.Y. Cheng, W. C. Cheng, & L. K. Fu	China (Fujian, Guangdong)	1975	Acta Phytotax. Sin. 13(4): 81, t. 1, figs. 7 & 8	H—PE	CR
subsp. *fairylakea* (D.Yue Wang) N. Liu	China (Guangdong)	1996	Proc. Sixth Int. Conf. Cycad Biol.: 1–4	H—SZG	EN
C. taitungensis C.F. Shen, K. D. Hill, C.H. Tsou, & C. J. Chen	Taiwan	1994	Bot. Bull. Acad. Sin. 35: 135–138	H—HAST	VU
C. taiwaniana Carruth.	China (Guangdong)	1893	J. Bot. 31: 1–3, pl. 331	H—BM	EN
C. tanqingii D.Yue Wang	China (Yunnan), Vietnam (Lai Chau)	1996	Cycads of China: 134, fig. 4	H—SZG	NT
C. tansachana K. D. Hill & S. L. Yang	Thailand (Saraburi)	1999	Brittonia 51(1): 65–66	H—NSW	CR
C. thouarsii R. Br. ex Gaudich.	E Africa, Indian Ocean Islands	1826	Voy. Uranie, Bot.: 434	?H—P	LC
C. tropophylla K. D. Hill & Phan K. Loc	Vietnam (Hai Phong, Quang Ninh)	2004	Bot. Rev. 70(2): 168–170, fig. 10	H—HN	NT
C. tuckeri K. D. Hill	Australia (Qld)	1996	Telopea 7(1): 20–23, fig. 10	H—NSW	NT
C. wadei Merrill	Philippines (Culion)	1936	Philipp. J. Sci. 60(3): 234–236, pls. 1–4	L—NY	DD

Taxon	Location	Date	Original Source	Type	Status
C. xipholepis K. D. Hill	Australia (Qld)	1996	Telopea 7(1): 32–33, fig. 15	H—NSW	LC
C. yorkiana K. D. Hill	Australia (Qld)	1996	Telopea 7(1): 18–19, fig. 8	H—NSW	NT
C. zambalensis Madulid & Agoo	Philippines (Luzon)	2005	Blumea 50: 519–522. fig. 1.	H—PNH	DD
C. zeylanica (J.Schust.) A. Lindstr. & K. D. Hill	Andaman Islands, S. Sri Lanka	2002	Novon 12(2): 238	L—G	DD
Synonyms:					
C. acuminatissima H. T. Chang, Y. C. Zhong, & Z. F. Lu	= *C. sexseminifera*	1998	Acta Sci. Nat. Univ. Sunyatseni 37: 6	H—SYS	
C. baguanheensis L. K. Fu & S. Z. Cheng	= *C. panzhihuaensis*	1981	Acta Phytotax. Sin. 19(3): 337	H—PE	
C. bellefontii Linden & Rodigas	Nomen dubium	1886	Ill. Hort. 33: 27, pl. 586 as *C. bellefonti*	?	
C. brevipinnata H. T. Chang, Y. Y. Huang, & Y. C. Zhong	= *C. sexseminifera*	1998	Acta Sci. Nat. Univ. Sunyatseni 37: 8	H—SYS	
C. canalis var. *carinata* K. D. Hill	= *C. canalis*	1994	Telopea 5(4): 699–700, fig. 4e–g	H—NSW	
C. celebica Miq.	= *C. rumphii*	1839	Bull. Sci. Phys. Nat. Néerl. 1839: 45	Tab. 20 & 21 *in* Rumphius, Herb. Amb., 1749	
C. chamberlainii W. H. Brown & Kienholz	= *C. riuminiana*	1925	Philipp. J. Sci. 26: 47–48, pls 1-2, fig. 1	H—NY	
C. circinalis var. *javana* Miq.	= *C. javana*	1842	Monogr. Cycad.: 28. t. 1, figs t, u, t II, fig. E1	H—L	
C. circinalis var. *orixensis* Haines	= *C. spherica*	1924	Bot. Bihar Orissa 6: 1228	?	
C. circinalis var. *swamyii* Pant	Nomen nudum ~ *C. circinalis*	2002	Intro. to Gymnosperms, Cycas and the Cycadales	?	
C. gracilis Miq.	= *C. media*	1863	Versl. Meded. Afd. Natuurk. Kon. Acad. Wetensch. ser. 1, 15: 366–367	H—U	
C. gracilis var. *glauca* Regel	= *C. media*	1876	Trudy Imp. S. Petersburgsk. Bot. Sada 4(4): 282	H—LE	
C. gracilis var. *viridis* Regel	= *C. media*	1876	Trudy Imp. S. Petersburgsk. Bot. Sada 4(4): 282	L—LE	
C. hypoleuca Presl	= nomen dubium	1849	Epimeliae Botanicae: 238	?	
C. immersa Craib	= *C. siamensis*	1912	Bull. Misc. Inform. 1912: 434–435	H—K	

C. jenkinsiana Griff.	= *C. pectinata*	1854	Not. Pl. Asiat. 4: 9–10, pl. 360, figs 1-2, pl. 362, fig. 1	H—K
C. kennedyana F. Muell.	= *C. media*	1882	Australas. Chem. Drugg. : 16	H—MEL
C. lingshuigensis G. A. Fu	= *C. hainanensis*	2004	Bull. Bot. Res. 24: 387–388	
C. longiconifera H.T. Chang, Y. C. Zhong, & Y.Y. Huang	= *C. segmentifida*	1998	Acta Sci. Nat. Univ. Sunyatseni 37: 6	H—SYS
C. longipetiolula D.Yue Wang	= *C. multipinnata*	1996	Cycads in China: 68	H—SZG
C. longisporophylla F. N. Wei	= *C. sexseminifera*	1997	Guihaia 17: 209	H—IBK
C. longlinensis H.T. Chang & Y. C. Zhong	= *C. guizhouensis*	1997	Acta Sci. Nat. Univ. Sunyatseni 36: 68	H—SYS
C. madagascariensis Miq.	= *C. thouarsii*	1840–41	Comm. Phytog.: 127	t. 2a–e *in* Petit-Thouars, Hist .Veg., 1804
C. micholitzii var. *simplicipinna* Smitinand	= *C. simplicipinna*	1971	Nat. Hist. Bull. Siam Soc. 24: 164, figs. 2–3, 4f	L—BKF
C. miquelii Warb.	= *C. revoluta*	1900	Monsunia 1: 179, 184	SYN—H. Amstelod. U028127, U208129
C. multifida H.T. Chang & Y. C. Zhong	= *C. segmentifida*	1997	Acta Sci. Nat. Univ. Sunyatseni 36: 70	H—SYS
C. multifrondis D.Yue Wang	= *C. bifida* x ?*segmentifida*	1996	Cycads in China: 80	H—SZG
C. multiovula D.Yue Wang	= *C. guizhouensis*	1996	Cycads in China: 83	H—SZG
C. neocaledonica Linden	Nomen nudum ~ *C. seemannii*	1881	Ill. Hort. 28: 32	?
C. normanbyana F. Muell.	= *C. media*	1874	Fragm. 8: 169–171, Icon. 65–66	H—MEL
C. palmatifida H.T. Chang, Y.Y. Huang, & Y. C. Zhong	= *C. balansae*	1998	Acta Sci. Nat. Univ. Sunyatseni 37: 7	H—SYS
C. parvulus S. L.Yang	= *C. diannanensis*	1996	*in* D.Y. Wang, Cycads in China: 93	H—HWA
C. pectinata var. *elongata* Leandri	= *C. elongata*	1931	*in* Lecompte, Fl. Indo-Chine 5(10): 1091	L—P
C. pectinata var. *manhaoensis* C. J. Chen & P. Yun	= *C. diannanensis*	1995	Acta Bot.Yunnan. 17: 400	H—YUN
C. pygmaea Blume	Nomen dubium	1848	Rumphia 4: 16	?
C. rumphii var. *bifida* Dyer	= *C. multifrondis*	1902	J. Linn. Soc., Bot. 26: 560	H—K

Taxon	Location	Date	Original Source	Type	Status
C. rumphii var. *timorensis* Miq.	= *C. glauca*	1840–41	Comm. Phytolog.: 125–126	H—L	
C. septemsperma H. T. Chang, Y.Y. Huang, & H. X. Zheng	= *C. sexseminifera*	1998	Acta Sci. Nat. Univ. Sunyatseni 37: 8	H—SYS	
C. shiwandashanica H. T. Chang & Y. C. Zhong	= *C. balansae*	1995	Chinese Bull. Bot. 12: 12	H—SYS	
C. spiniformis J. Y. Liang	= *C. sexseminifera*	1997	*in* F. N. Wei, Guihaia 17: 211	H—IBK	
C. tonkinensis (Linden & Rodigas) Rodigas	Nomen dubium	1886	Ill. Hort. 33: 27	pl. 547 *in* Ill. Hort. 32, 1885	
C. undulata Desf. ex Gaudich.	= *C. circinalis*	1829	Voy. Uranie, Bot.: 434	H—P	
C. wallichii Miq.	= *C. circinalis*	1842	Monogr. Cycad.: 32	H—U	
C. xilingensis H. T. Chang & Y. C. Zhong	= *C. segmentifida*	1997	Acta Sci. Nat. Univ. Sunyatseni 36: 69	H—SYS	
DYEROCYCAS Nakai	= *CYCAS*	1943	Ordines, familiae, tribi, ... et combinationes a Prof. Nakai-Takenosin adhuc novis edita : 208		
D. micholitzii (Dyer) Nakai	= *C. micholitzii*	1943	Ordines, familiae, tribi, ... et combinationes a Prof. Nakai-Takenosin adhuc novis edita : 208	BAS	
EPICYCAS de Laub.	= *CYCAS*	1998	*in* De Laub. & Adema, Blumea 43: 388		
E. elongata (Leandri) S. L. Yang	= *C. elongata*	1998	*in* De Laub. & Adema, Blumea 43: 393	BAS	
E. lindstromii (S. L. Yang, K. D. Hill, & H. T. Nguyen) de Laub.	= *C. lindstromii*	1998	*in* De Laub. & Adema, Blumea 43: 395	BAS	
⋆*E. micholitzii* (Dyer) de Laub.	= *C. micholitzii*	1998	*in* De Laub. & Adema, Blumea 43: 389	BAS	
E. miquelii (Warb.) de Laub.	= *C. revoluta*	1998	*in* De Laub. & Adema, Blumea 43: 393	BAS	
E. multipinnata (Chen . & Yang) de Laub	= *C. multipinnata*	1998	*in* De Laub. & Adema, Blumea 43: 391	BAS	
E. siamensis (Miq.) de Laub.	= *C. siamensis*	1998	in De Laub. & Adema, Blumea 43: 393	BAS	
E. tonkinensis (Linden & Rodigas) de Laub.	= *C. tonkinensis* nomen dubium	1998	*in* De Laub. & Adema, Blumea 43: 391	BAS	

DIOON Lindl., 1843 (13 species; Honduras, Mexico, Nicaragua)			Edward's Bot. Reg. 29: misc. 59–60		
D. angustifolium Miq.	Mexico (Nuevo Leon, Tamaulipas)	1848	Tijdschr. Wis-Natuurk. Wetensch. Eerste Kl. Kon. Ned. Inst. Wetensch. 1: 37	L—K	DD
D. argenteum T. J. Greg, Chemnick, S. Salas-Mor. & Vovides	Mexico (Oaxaca)	2003	Bot. J. Linn. Soc. 141: 471–476	H—XAL	VU
D. califanoi de Luca & Sabato	Mexico (Puebla)	1979	Brittonia 31(1): 170–172	H—NAP	VU
D. caputoi de Luca, Sabato & Vázq. Torres	Mexico (Puebla)	1980	Brittonia 32(1): 44–46	H—XALUV	CR
**D. edule* Lindl.	Mexico (Hidalgo, Querétaro, San Luis Potosi, Veracruz)	1843	Edward's Bot. Reg. 29: misc. 59–60	H—CGE	NT
D. holmgrenii de Luca, Sabato & Vázq. Torres	Mexico (Oaxaca)	1981	Brittonia 33(4): 552–554	H—XALUV	EN
D. mejiae Standl. & L. O. Williams	Honduras (Colón, Olancho, Yoro)	1951	Ceiba 1: 36–38	H—US	DD
D. merolae de Luca, Sabato & Vázq. Torres	Mexico (Chiapas, Oaxaca)	1981	Brittonia 33(2): 180–184	H—NAP	VU
D. purpusii Rose	Mexico (Oaxaca)	1909	Contr. U.S. Natl. Herb. 12(7): 260–261	H—US	VU
D. rzedowskii de Luca, A. Moretti, Sabato & Vázq. Torres	Mexico (Oaxaca)	1980	Brittonia 32(2): 225–229	H—XALUV	VU
D. sonorense (de Luca, Sabato & Vázq. Torres) Chemnick, T. J. Greg. & S. Salas-Mor.	Mexico (Sinaloa, Sonora)	1997	Phytologia 83(1): 1–6 (issued 1998)	BAS	EN
D. spinulosum Dyer & Eichler	Mexico (Oaxaca, Veracruz)	1883	Gart. Zeit. Berlin. 2: 411–413, fig. 80	H—K	VU
D. tomasellii de Luca, Sabato & Vázq. Torres	Mexico (Durango, Guerrero, Jalisco, Nayarit, Michoacán)	1984	Brittonia 36: 225–227	H—NAP	EN
Synonyms:					
PLATYZAMIA Zuccarini	= *DIOON*	1845	Abh. Math. Phys. Cl. Königl. Bayer. Akad. Wiss. 4: 23		
P. rigida Zuccarini	= *D. edule*	1845	Abh. Math. Phys. Cl. Königl. Bayer. Akad. Wiss. 4: 23	H—M	

Taxon	Location	Date	Original Source	Type	Status
D. aculeatum Lem.	= *D. angustifolium*.	1846	*in* Lem., Catalog. Vanhoutt. 27: 45	?	
D. edule f. *angustifolium* (Miq.) Miq.	= *D. angustifolium*	1861	Prodr. Syst. Cycad.: 10, 22	BAS	
D. edule subsp. *angustifolium* (Miq.) E. Murray	= *D. angustifolium*	1983	Kalmia 13: 5	BAS	
D. edule var. *angustifolium* (Miq.) Miq.	= *D. angustifolium*	1868	Arch. Néerl. Sci. Exact. Nat. 3: 427	BAS	
D. edule var. *lanuginosum* Wittmack	= *D. edule*	1899	Gartenflora 48: 153	L—K	
D. edule var. *latipinnium* Dyer	= *D. mejiae*	1884	*in* Hemsl., Biol. Cent. Am., Bot. 3: 191, t. 81, figs. 1–3	H—K	
D. edule var. *sonorense* (de Luca, Sabato, & Vázq. Torres) R. McVaugh & Perez de la Rosa	= *D. sonorense*	1992	*in* R. McVaugh, Fl. Novo-Galiciana 17: 112	BAS	
D. imbricatum Miq.	= *D. edule*	1848	Tijdschr. Wis-Natuurk. Wetensch. Eerste Kl. Kon. Ned. Inst. Wetensch. 1: 36	t. 4d & e *in* Verh. Eerste Kl. Kon. Ned. Inst. Wetensch 3, 1851	
D. strobilaceum Lem. ex A. DC	= *D. edule*	1868	Prodr. 16(2): 537	?	
D. tomasellii var. *sonorense* de Luca, Sabato, & Vázq. Torres	= *D. sonorense*	1984	Brittonia 36: 226–227	H—NAP	
ENCEPHALARTOS Lehm., 1834 (65 species, Africa)			Pugill. 6: 14		
E. aemulans Vorster	South Africa (kwaZulu-Natal)	1990	S. African J. Bot. 56(2): 239–243	H—PRE	CR
E. altensteinii Lehm.	South Africa (E Cape)	1834	Pugill. 6: 11, t. 4–5	tt. 4 & 5	VU
E. aplanatus Vorster	Swaziland	1996	S. African J. Bot. 62: 57–60	H—PRE	VU
E. arenarius R. A. Dyer	South Africa (E Cape)	1956	J. S. African Bot. 22(1): 1	H—PRE	EN
E. barteri Carruth. ex Miq.	Benin (Bergu), Ghana, Nigeria	1868	Arch. Néerl. Sci. Exact. Nat. 3: 243	I—K	VU
subsp. *allochrous* L. E. Newton	Nigeria (Plateau)	1978	J. Linn. Soc., Bot. 77: 125–129	H—K	EN
E. brevifoliolatus Vorster	South Africa (Limpopo)	1996	S. African J. Bot. 62(1): 61–64	H—PRE	EW

E. bubalinus Melville	Kenya, Tanzania (Masai)	1957	Kew Bull. : 252	H—K	LC
⋆*E. caffer* (Thunb.) Lehm.	South Africa (E Cape)	1834	Pugill. 6: 14	BAS	NT
E. cerinus Lavranos & D. L. Goode	South Africa (kwaZulu-Natal)	1989	Durban Mus. Novit. 14: 153–156	H—NH	CR
E. chimanimaniensis R. A. Dyer & I. Verd.	Mozambique, Zimbabwe	1969	Kirkia 7: 147–158	H—PRE	EN
E. concinnus R. A. Dyer & I. Verd.	Zimbabwe (Mberwenga, Runde)	1969	Kirkia 7: 147–158	H—PRE	EN
E. cupidus R. A. Dyer	South Africa (Mpumalanga)	1971	Bothalia 10(2): 379–383	H—PRE	CR
E. cycadifolius (Jacq.) Lehm.	South Africa (E Cape)	1834	Pugill. 6: 13	?	LC
E. delucanus Malaisse, Sclavo, & Crosiers	Tanzania (Mapanda)	1992	Ann. Gembloux 98: 153	H—BR	VU
E. dolomiticus Lavranos & D. L. Goode	South Africa (Limpopo)	1988	Bull. Jard. Bot. Belg. 58: 219–224	H—NH	CR
E. dyerianus Lavranos & D. L. Goode	South Africa (Limpopo)	1988	Bull. Jard. Bot. Belg. 58: 219–224	H—PRE	CR
E. equatorialis P. J. H. Hurter	Uganda	1995	S. African J. Bot. 61(4): 226–229	H—PRE	CR
E. eugene-maraisii I. Verd.	South Africa (Limpopo)	1945	J. S. African Bot. 11(1): 1–3	H—PRE	EN
E. ferox Bertol. f.	Mozambique, South Africa (kwaZulu-Natal)	1851	Mem. Reale Accad. Sci. Ist. Bologna 3: 264	?	LC
E. friderici-guilielmi Lehm.	South Africa (E Cape, kwaZulu-Natal)	1834	Pugill. 6: 8, tt. 1–3	tt. 1–3	NT
E. ghellinckii Lem.	South Africa (E Cape, kwaZulu-Natal)	1867	Ill. Hort. 14, misc.: 80, pl. 567	?	VU
E. gratus Prain	Malawi, Mozambique	1916	Bull. Misc. Inform.: 181	L—K	VU
E. heenanii R. A. Dyer	South Africa (Mpumalanga), Swaziland	1972	Bothalia 10(4): 539–546	H—PRE	CR
E. hildebrandtii A. Braun & C. D. Bouché	Kenya (Kilifi, Lamu), Tanzania (Lushoto, Tanga), Zanzibar Island	1874	Ind. Sem. Hort. Berol. 18	I—K	NT
E. hirsutus P. J. H. Hurter	South Africa (Limpopo)	1996	S. African J. Bot. 62(1): 46–48, fig. 1	H—PRE	CR
E. horridus (Jacq.) Lehm.	South Africa (E Cape)	1834	Pugill. 6: 14	BAS	EN
E. humilis I. Verd.	South Africa (Mpumalanga)	1951	Bothalia 6: 220–241, pl. 3	?H—PRE	VU
E. inopinus R. A. Dyer	South Africa (Limpopo)	1964	Bothalia 8(2): 169–170	H—PRE	CR
E. ituriensis Bamps & Lisowski	Dem. Rep. Of Congo, Uganda	1990	Mem. New York Bot. Gard. 57: 152–155	H—BR	NT
E. kisambo Faden & Beentje	Kenya (Taita-Taveta)	1989	Utafiti 2: 7–10	H—K	EN

Taxon	Location	Date	Original Source	Type	Status
E. laevifolius Stapf & Burtt Davy	South Africa (E Cape, kwaZulu-Natal, Limpopo, Mpumalanga), Swaziland	1926	Fl. Transvaal & Sw. 1: 40 & 99	H—K	CR
E. lanatus Stapf & Burtt Davy	South Africa (Mpumalanga)	1926	Fl. Transvaal & Sw. 1: 40, 90, fig. 4d	H—PRE	NT
E. latifrons Lehm.	South Africa (E Cape)	1837	Tijdschr. Natuurl. Gesch. Physiol. 4: 424, t. 9	L—t. 9	CR
E. laurentianus De Wild.	Angola, Dem. Rep. Of Congo	1903	Etudes Fl. Bas et Moy – Congo 1: 10, t. 25	H—BR	DD
E. lebomboensis I. Verd.	Mozambique, South Africa (kwaZulu-Natal), Swaziland	1949	Fl. Pl. Africa 27: pls 1078–1079	H—PRE	EN
E. lehmannii Lehm.	South Africa (E Cape)	1834	Pugill. 6: 14	?	NT
E. longifolius (Jacq.) Lehm.	South Africa (E Cape)	1834	Pugill. 6: 14	?	NT
E. mackenziei L. E. Newton	Sudan	2002	Bot. J. Linn. Soc. 140: 187–192, figs 1–6	H—EA	DD
E. macrostrobilus S. Jones & J. Wynants	N Uganda	1997	Encephalartos 50: 13–17, figs 1–2	H—BR	VU
E. manikensis (Gilliland) Gilliland	Mozambique (Manica), Zimbabwe	1939	Proc. Rhodesia Sci. Assoc. 37: 133–134	H—BM	VU
E. marunguensis Devred	Dem. Rep. Of Congo (Shaba)	1958	Bull. Soc. R. Bot. Belgique 91: 104–110	H—BM	NT
E. middelburgensis Vorster	South Africa (Mpumalanga)	1989	S. African J. Bot. 55: 122–126	H—PRE	CR
E. msinganus Vorster	South Africa (kwaZulu-Natal)	1996	S. African J. Bot. 62(2): 67–70	H—PRE	CR
E. munchii R. A. Dyer & I. Verd.	Mozambique (Manica)	1969	Kirkia 7: 147–158	H—PRE	CR
E. natalensis R. A. Dyer & I. Verd.	South Africa (kwaZulu-Natal)	1951	Bothalia 6(1): 205–211, pls 1–3	H—PRE	NT
E. ngoyanus I. Verd.	South Africa (kwaZulu-Natal), Swaziland	1949	Fl. Pl. Africa 27: pls 1053–1054	?H—PRE	VU
E. nubimontanus P. J. H. Hurter	South Africa (Limpopo)	1995	Phytologia 78(6): 409–410, fig. 1	H—PRE	EW
E. paucidentatus Stapf & Burtt Davy	South Africa (Mpumalanga), Swaziland	1926	Fl. Transvaal & Sw. 1: 40 & 99, fig. 4a	H—PRE	VU
E. poggei Asch.	Dem. Rep. Of Congo (Bas-Katanga, Lualaba-Kasai)	1878	Verh. Bot. Vereins Prov. Brandenburg 20: 35–36	?	LC
E. princeps R. A. Dyer	South Africa (E Cape)	1965	J. S. African Bot. 31(2): 111–112, pl. 19	H—PRE	VU
E. pterogonus R. A. Dyer & I. Verd.	Mozambique (Manica)	1969	Kirkia 7: 147–158	H—PRE	CR
E. relictus P. J. H. Hurter	Swaziland	2001	Bothalia 31(2): 197–199	H—PRE	EW
E. schaijesii Malaisse, Sclavo & Crosiers	Dem. Rep. Of Congo (Shaba)	1993	Bull. Jard. Bot. Belg. 62: 215–219	H—BR	VU

E. schmitzii Malaisse	Dem. Rep. Of Congo (Shaba), Zambia	1969	Bull. Jard. Bot. Belg. 39: 401–406	H—BR	NT
E. sclavoi De Luca, D.W. Stev., & A. Moretti	Tanzania (Tanga)	1989	Delpinoa 29–30: 3–5	H—K	VU
E. senticosus Vorster	South Africa (kwaZulu-Natal), Swaziland	1996	S. African J. Bot 62(2): 76–79	H—PRE	VU
E. septentrionalis Schweinf.	Sudan, Uganda	1871	Bot. Zeitung Berlin 29: 334	I—K	DD
E. tegulaneus Melville	Kenya	1957	Kew Bull.: 249	H—K	LC
subsp. *powysii* Mirungu & Beentje	Kenya	1999	J. East African. Nat. Hist. 88:35	H—EA	CR
E. transvenosus Stapf & Burtt Davy	South Africa (Limpopo)	1926	Fl. Transvaal & Sw. 1: 40 & 99, fig. 4b	H—PRE	LC
E. trispinosus (Hook.) R. A. Dyer	South Africa (E Cape)	1965	J. S. African Bot. 31(2): 112–116, pl. 20	BAS	VU
E. turneri Lavranos & D. L. Goode	Mozambique (Nampula)	1985	Garcia de Orta, ser. Bot. 7: 11–14	H—LISC	LC
E. umbeluziensis R. A. Dyer	Mozambique, Swaziland	1951	Fl. Pl. Africa 28: pl. 1100	H—PRE	VU
E. villosus Lem.	South Africa (E Cape, kwaZulu-Natal), Swaziland	1867	Ill. Hort. 14, misc.: 79; 15: t. 557	?	LC
E. whitelockii P. J. H. Hurter	Uganda	1995	Phytologia 78(6): 410–411, fig. 3	H—PRE	VU
E. woodii Sander	South Africa (kwaZulu-Natal), extinct in the wild	1908	Gard. Chron.: 257	?	EW
Synonyms:					
Cycas caffra Thunb.	= *E. caffer / E. longifolius*	1775	Nov. Acta Soc. Sc. Uppsala 2: 286, t. 5	t. 5	
E. brachyphyllus Lehm. & De Vriese	= *E. caffer*	1838	*in* Hoeven & De Vriese, Tijdschr. Natuurl. Gesch. Physiol. 4: 414, tt. 6–7	?tt. 6 & 7	
E. eugene-maraisii subsp. *middelburgensis* Lavranos & D. L. Goode	= *E. middelburgensis*	1988	Bull. Jard. Bot. Belg. 58: 219–224	H—PRE	
E. eximius I. Verd.	= *E. cycadifolius*	1934	Bothalia 6: 426	H—PRE	
E. flavostrobilus I. Turner, Sclavo, & Malaisse	= *E. schmitzii*	2006	Biotechnol. Agron. Soc. Environ. 10(3): 181–183	H—BR	

Taxon	Location	Date	Original Source	Type	Status
E. graniticolus Robbertse, Vorster & S. van der Westhuizen	= *E. dyerianus*	1988	S. African J. Bot. 54: 363–366, figs 1 & 2	H—PRE	
E. hildebrandtii var. *dentatus* Melville	= ?*E. hildebrandtii*	1957	Kew Bull.: 248	H—K	
E. horridus var. *trispinosus* Hook.	= *E. trispinosus*	1863	Bot. Mag. 89: t. 5371	?t.. 5371	
E. imbricans Vorster	= *E. equatorialis*	1995	Novon 5: 388, figs 2–5	H—PRE	
E. kosiensis Hutch.	= *E. ferox*	1932	Bull. Misc. Inform.: 512	?	
E. lanuginosus (Jacq.) Lehm.	= *E. longifolius*	1834	Pugill. 6: 14	BAS	
E. latifolius Sweet	= *Zamia latifolia* = *Z. muricata*	1839	Hort. Brit. Ed. 3: 626	?	
E. lemarinellianus De Wild. & Th. Dur.	= *E. poggei*	1900	Compt. Rend. Soc. Bot. Belg. 39: 80	?	
E. marumii De Vreise	= *E. altensteinii*	1839	*in* Hoeven & De Vriese, Tijdschr. Natuurl. Gesch. Physiol. 5: 187	?	
E. mauritianus Miq.	= *E. longifolius*	1842	Monogr. Cycad. 48	?	
E. nanus Lehm. in Hoeven & De Vriese	= *E. horridus*	1837	Tijdschr. Natuurl. Gesch. Physiol. 4: 421, t. 8c	?t. 8C	
E. spinulosus Lehm.	= *E. lehmannii*	1837–38	*in* Hoeven & De Vriese, Tijdschr. Natuurl. Gesch. Physiol. 4: 419, t. 8c	?t. 8C	
E. striatus Stapf & Burtt Davy	nomen dubium	1926	*in* Burtt-Davy, Fl. Transvaal & Sw. 1: 40 & 99	?	
E. successibus Vorster	= *E. whitelockii*	1995	S. African J. Bot. 61: 347	H—K	
E. tridentatus (Willd.) Lehm.	nomen dubium, *Macrozamia* or *E.* sp.	1834	Pugill. 6: 13	?	
E. venetus Vorster	= *E. nubimontanus*	1996	S. African J. Bot. 62: 71·	H—PRE	
E. verrucosus Robbertse, Vorster & S. van der Westhuizen	= *E. dolomiticus*	1988	S. African J. Bot. 54: 487–490, figs 1–3	H—PRE	
E. voiensis A. Moretti, D. W. Stev., & J. P. Sclavo	= *E. kisambo*	1989	Ann. Missouri Bot. Gard. 76: 935	H—K	
Zamia cycadifolia Jacq.	=*E. cycadifolius*	1801	Fragm. Bot. 1: 27, tt. 25 & 26	tt. 25 & 26	
Z. cycadis L. f.	= *E. caffer*	1782	Suppl. Pl.: 443	?	
Z. horrida Jacq.	= *E. horridus*	1801	Fragm. Bot. 1: 27, t. 28	t. 28	
Z. lanuginosa Jacq.	= *E. longifolius*	1801	Fragm. Bot. 1: 28, tt. 30 & 31	tt. 30 & 31	
Z. longifolia Jacq.	= *E. longifolius*	1801	Fragm. Bot. 1: 28, t. 29	t. 29	

LEPIDOZAMIA Regel, 1857 (2 species, Australia)			Bull. Soc. Imp. Naturalistes Moscou 30: 182		
L. hopei Regel	Australia (Qld)	1876	Gartenflora 25: 6	?	LC
★*L. peroffskyana* Regel	Australia (NSW, Qld)	1857	Bull. Soc. Imp. Naturalistes Moscou 30: 184	H—LE	LC
Synonyms:					
CATAKIDOZAMIA W. Hill	= *LEPIDOZAMIA*	1865	*in* Hook., Gard. Chron.: 1107		
C. hopei W. Hill	= *L. hopei*	1865	*in* Hook., Gard. Chron.: 1107	?	
Macrozamia denisonii C. Moore & F. Muell.	= *L. peroffskyana*	1858	*in* F. Muell., Fragm. 1: 41	L—MEL	
M. hopei C. Moore	= *L. hopei*	1883	J. & Proc. R. Soc. New South Wales 17: 116	BAS	
M. peroffskyana (Regel) Miq.	= *L. peroffskyana*	1868	Arch. Néerl. Sci. Exact. Nat. 3: 252	BAS	
MACROZAMIA Miq., 1842 (41 species, Australia)			Monogr. Cycad.: 35		
M. cardiacensis P. I. Forst. & D. L. Jones	Australia (Qld)	1998	Fl. Australia 48: 717	H—BRI	LC
M. communis L. A. S. Johnson	Australia (NSW)	1959	Proc. Linn. Soc. New South Wales 84(1): 98	H—NSW	LC
M. concinna D. L. Jones	Australia (NSW)	1998	Fl. Australia 48: 718	H—BRI	LC
M. conferta D. L. Jones & P. I. Forst.	Australia (Qld)	1994	Austrobaileya 4(2): 271–273, fig. 1	H—BRI	EN
M. cranei D. L. Jones & P. I. Forst.	Australia (Qld)	1994	Austrobaileya 4(2): 273–275, fig. 2	H—BRI	VU
M. crassifolia P. I. Forst. & D. L. Jones	Australia (Qld)	1994	Austrobaileya 4(2): 275–276, fig. 3	H—BRI	VU
M. diplomera (F. Muell.) L. A. S. Johnson	Australia (NSW)	1959	Proc. Linn. Soc. New South Wales 84(1): 98	L—MEL	LC
M. douglasii W. Hill ex F. M. Bailey	Australia (Qld)	1883	Syn. Queensland Fl. 500	?H—MEL	LC
M. dyeri (F. Muell.) C. A. Gardner	Australia (WA)	1930	Enum. Pl. Austral. Occid. 3	H—MEL	LC
M. elegans K. D. Hill & D. L. Jones	Australia (NSW)	1998	Fl. Australia 48: 718	H—NSW	EN
M. fawcettii C. Moore	Australia (NSW)	1884	J. & Proc. R. Soc. New South Wales 17: 120	H—NSW	LC

Taxon	Location	Date	Original Source	Type	Status
M. fearnsidei D. L. Jones	Australia (Qld)	1991	Austrobaileya 3(3): 481	H—CANB	LC
M. flexuosa C. Moore	Australia (NSW)	1884	J. & Proc. R. Soc. New South Wales 17: 121	H—NSW	EN
M. fraseri Miq.	Australia (WA)	1842	Monogr. Cycad.: 37	?H—MEL	LC
M. glaucophylla D. L. Jones	Australia (NSW)	1998	Fl. Australia 48: 718	H—CANB	LC
M. heteromera C. Moore	Australia (NSW)	1884	J. & Proc. R. Soc. New South Wales 17: 122	L—NSW	LC
M. humilis D. L. Jones	Australia (NSW)	1998	Fl. Australia 48: 719	H—CANB	VU
M. johnsonii D .L. Jones & K. D. Hill	Australia (NSW)	1992	Telopea 5(1): 31	H—CANB	LC
M. lomandroides D. L. Jones	Australia (Qld)	1991	Austrobaileya 3(3): 483	H—CANB	VU
M. longispina P. I. Forst. & D. L. Jones	Australia (Qld)	1998	Fl. Australia 48: 717	H—BRI	LC
M. lucida L. A. S. Johnson	Australia (NSW, Qld)	1959	Proc. Linn. Soc. New South Wales 84(1): 102	H—NSW	LC
M. macdonnellii (F. Muell. ex Miq.) A. DC.	Australia (NT)	1868	Prodr. 16(2): 537	H—MEL	LC
M. macleayi Miq.	Australia (Qld)	1868	Arch. Néerl. Sci. Exact. Nat. 3(5): 250	?H—U	LC
M. machinii P. I. Forst. & D. L. Jones	Australia (Qld)	1994	Austrobaileya 4(2): 276	H—BRI	VU
M. miquelii (F. Muell.) A. DC.	Australia (Qld)	1868	Prodr. 16(2): 535	L—MEL	LC
M. montana K. D. Hill	Australia (NSW)	1998	Fl. Australia 48: 717	H—NSW	LC
M. moorei F. Muell.	Australia (Qld)	1881	Australas. Chem. Drugg. 4: 84	?H—MEL	LC
M. mountperriensis F. M. Bailey	Australia (Qld)	1886	Syn. Queensland Fl., suppl. 1, 50	N—CANB	LC
M. occidua D. L. Jones & P. I. Forst.	Australia (Qld)	1994	Austrobaileya 4(2): 278, fig. 5	H—BRI	VU
M. parcifolia P. I. Forst. & D. L. Jones	Australia (Qld)	1994	Austrobaileya 4(2): 279–281, fig. 6	H—BRI	VU
M. pauli-guilielmi W. Hill & F. Muell.	Australia (Qld)	1859	Fragm. 1: 86	H—MEL	EN
M. platyrhachis F. M. Bailey	Australia (Qld)	1898	Queensland Agric. J. 3: 356	H—BRI	LC
M. plurinervia (L. A. S. Johnson) D. L. Jones	Australia (NSW, Qld)	1991	Austrobaileya 3(3): 484	H—NSW	NT
M. polymorpha D. L. Jones	Australia (NSW)	1998	Fl. Australia 48: 718	H—CANB	LC
M. reducta K. D. Hill & D. L. Jones	Australia (NSW)	1998	Fl. Australia 48: 718	H—NSW	LC

**M. riedlei* (Gaudich.) C. A. Gardner	Australia (WA)	1930	Enum. Pl. Austral. Occid. 3	H—P	LC
M. secunda C. Moore	Australia (NSW)	1884	J. & Proc. R. Soc. New South Wales 17: 122	L—MEL	VU
M. serpentina D. L. Jones & P. I. Forst.	Australia (Qld)	2001	Austrobaileya 6(1): 90–92, figs 20 & 21	H—BRI	EN
M. spiralis (Salisb.) Miq.	Australia (NSW)	1842	Monogr. Cycad.: 36	?	EN
M. stenomera L. A. S. Johnson	Australia (NSW)	1959	Proc. Linn. Soc. New South Wales 84(1): 106	H—NSW	LC
M. viridis D. L. Jones & P. I. Forst.	Australia (Qld)	1994	Austrobaileya 4(2): 281–283, fig. 7	H—BRI	VU
Synonyms:					
M. corallipes Hook. f.	= *M. spiralis*	1872	Bot. Mag 98: sub t. 5943	t. 5943	
M. cylindrica C. Moore	= ?*M. miquelii* or ?*M. johnsonii*	1884	J. & Proc. R. Soc. New South Wales 17: 119	H—NSW	
M. mackenziei hort. ex Mast.	= *M. miquelii*	1877	Gard. Chron. n.s. 7: 665 (as M. mackenzii)	H—BRI	
M. oldfieldii (Miq.) A. DC.	= *M. fraseri*	1868	Prodr. 16(2): 535	H—MEL	
M. pauli-guilielmi subsp. *flexuosa* (C. Moore) L. A. S. Johnson	= *M. flexuosa*	1959	Proc. Linn. Soc. New South Wales 84(1): 109	H—NSW	
M. pauli-guilielmi subsp. *plurinervia* L. A. S. Johnson	= *M. plurinervia* and *M. concinna*	1959	Proc. Linn. Soc. New South Wales 84(1): 108	H—NSW	
M. preissii Lehm.	= *M. fraseri*	1844	Pugill. 8: 31	?H—HBG	
MICROCYCAS (Miq.) A. DC., 1868 (1 species, Cuba)			Prodr. 16(2): 538		
**Microcycas calocoma* (Miq.) A. DC.	W Cuba	1868	Prodr. 16(2): 538	N—G-DC	CR
Synonyms:					
Zamia calocoma Miq.	= *Microcycas calocoma*	1852	*in* Van Houtte, Fl. Serres 7(6): 141	N—G-DC	
STANGERIA T. Moore, 1853 (1 species, Southern Africa)			Hook. Kew J. 5: 228		
**Stangeria eriopus* (Kunze) Baill.	South Africa (E Cape, kwaZulu-Natal)	1892	Hist. Pl. 12: 68	?	NT

Taxon	Location	Date	Original Source	Type	Status
Synonyms:					
Lomaria eriopus Kunze	= *S. eriopus*	1839	Linnaea 13: 152	?	
S. katzeri Regel	= *S. eriopus*	1874	Gartenflora 23: 163, t. 798	H—LE	
S. paradoxa T. Moore	= *S. eriopus*	1853	Hook. Kew J. 5: 228	?	
S. schizodon Bull	= *S. eriopus*	1872	Catalog 72: 8	?H—K	
ZAMIA L., 1763 (57 species; South, Central & North America)			Sp. Pl. Ed. 2, 2: 1659		
Z. acuminata Oers. ex Dyer	Costa Rica, Nicaragua, Panama (Coclé, Panamá)	1884	*in* Hemsl., Biol. Cent. Am., Bot. 3: 190–195	H—C	NT
Z. amazonum D. W. Stev.	Brazil (Amazonas), Colombia, (Amazonas, Chocó, Vaupés), (Morona-Santiago, Napo, Ecuador Sucumbíos), Peru (Loreto), S Venezuela	2001	Fl. Colombia 21: 33, fig. 3	H—INPA	NT
Z. amblyphyllidia D. W. Stev.	Cuba, Jamaica, Puerto Rico	1987	Fairchild Trop. Gard. Bull. 42(3): 26	H—NY	VU
Z. amplifolia hort. Bull ex Mast.	Colombia (Valle del Cauca)	1878	Gard. Chron. ser. 2, 10(261): 810	N—FTG	CR
Z. angustifolia Jacq.	Bahamas, Cuba	1789	Collectanea 3: 263, pl. 636	pl. 636	DD
Z. boliviana (Brongn.) A. DC.	N Bolivia, Brazil	1868	Prodr. 16(2): 540	L—P	NT
Z. chigua Seem.	Colombia (Chocó, Valle del Cauca), Panama (Chiriquí)	1854	Bot. Voy. Herald 6: 201–203, t. 43	t. 43	NT
Z. cremnophila Vovides, Schutzman, & Dehgan	Mexico (Tabasco)	1988	Bot. Gaz. 149: 351	H—MEXU	EN
Z. cunaria Dressler & D. W. Stev.	Panama (Colón, Panamá)	1993	Brittonia 45(1): 5–6	H—NY	NT
Z. disodon D. W. Stev. & Sabato	Colombia (Antioquia), Peru (Huánuco)	2001	Fl. Colombia 21: 38–40, fig. 4	H—COL	CR
Z. dressleri D. W. Stev.	Panama (Colón)	1993	Brittonia 45(1): 6	H—NY	EN
Z. elegantissima Schutzman, Vovides & Adams	Panama (Colón)	1998	Phytologia 85(3): 137–145	H—MO	DD
Z. encephalartoides D. W. Stev.	Colombia (Santander)	2001	Fl. Colombia 21: 40–42, fig. 5	H—COL	VU
Z. fairchildiana L. D. Gómez	Costa Rica, Panama (Chiriquí)	1982	Phytologia 50: 401–404	H—CR	NT
Z. fischeri Miq.	Mexico (Querétaro, San Luis Potosí)	1845	*in* Lem., Hort. Vanhoutt. 1: 20	N—U	EN
Z. furfuracea L. f.	Mexico (Veracruz)	1789	*in* Aiton, Hortus Kew. 3: 477	pl. 210 in Herm. Parad. Bat., 1698	VU
Z. gentryi Dodson	Ecuador (Carchi, Esmeraldas)	1998	Novon 8(1): 12–14, fig. 1a–f	H—QCNE	VU

Z. herrerae Calderón & Standl.	El Salvador, Guatemala, Honduras, Mexico (Chiapas)	1924	Proc. Wash. Acad. Sci. 14(4): 93–99, fig. 1	H—US	VU
Z. hymenophyllidia D. W. Stev.	Colombia (Amazonas), Peru (Loreto)	2001	Fl. Colombia 21: 43–44, fig. 6	H—COAH	CR
Z. inermis Vovides, J. D. Rees & Vázq. Torres	Mexico (Veracruz)	1983	Flora de Veracruz 26: 22–24	H—XAL	CR
Z. integrifolia L. f.	Bahamas, Cayman Islands, Cuba, USA (Florida, Georgia)	1789	*in* Aiton, Hortus Kew. 3: 477–479	H—BM	NT
Z. ipetiensis D. W. Stev.	Panama (Panamá)	1993	Brittonia 45(1): 7–9	H—NY	VU
Z. katzeriana (Regel) E. Rettig	Mexico (Chiapas, Tabasco, Veracruz)	1896	Gartenflora 45: 148–149.	L—NY	VU
Z. lacandona Schutzman & Vovides	Mexico (Chiapas)	1998	Novon 8(4): 441–446, figs 1–3	H—FLAS	EN
Z. lecointei Ducke	Brazil (Pará), Colombia (Antioquia, Amazonas), Peru (Loreto), Venezuela (Amazonas)	1915	Arch. Jard. Bot. Rio de Janeiro 1: 9–10, pls 1–2	H—MG	NT
Z. loddigesii Miq.	Guatemala, Mexico (Chiapas, Hidalgo, Oaxaca, Tabasco, Tamaulipas, Veracruz)	1843	Tijdschr. Natuurl. Gesch. Physiol. 10: 72–73	?	NT
Z. lucayana Britton	Bahamas	1907	Bull. NY Bot . Gard. 5(18): 311–318	H—NY	
Z. macrochiera D. W. Stev.	Peru (Loreto)	2004	Cycad Classification: Concepts & Recommendations: 185–186, fig. 14.7	H—NY	CR
Z. manicata Linden ex Regel	Colombia (Antioquia, Chocó), Panama (Darién)	1876	Trudy Imp. S. Petersburgsk. Bot. Sada 4: 310, t. 926, fig. E	t. 926, fig. e	NT
Z. melanorrhachis D. W. Stev.	Colombia (Amazonas, Córdoba, Meta, Santander)	2001	Fl. Colombia 21: 55, fig. 11	H—COL	EN
Z. montana A. Braun	Colombia (Antioquia)	1875	Monatsber. Königl. Preuss. Akad. Wiss. Berlin : 376	N—NY	CR
Z. monticola Chamb.	Guatemala	1926	Bot. Gaz. 81: 219, 223, figs. 1–3	H—MO	CR
Z. muricata Willd.	Colombia (Boyacá, La Guajira, Meta, Santander), Venezuela (Guarico)	1806	Sp. Pl. Ed. 4, 4: 847–848	H—B-W	NT
Z. neurophyllidia D. W. Stev.	Costa Rica, S Nicaragua, Panama (Bocas del Toro)	1993	Brittonia 45(1): 10, fig. 5	H—PMA	NT

Taxon	Location	Date	Original Source	Type	Status
Z. obliqua A. Braun	Colombia (Antioquia, Chocó, Valle del Cauca), Panama (Darién, Panamá)	1875	Monatsber. Königl. Preuss. Akad. Wiss. Berlin: 376	N—BM	NT
Z. oligodonta E. Calderón-Sáenz & D. W. Stev.	Colombia (Risaralda)	2003	Rev. Acad. Colomb. Cienc. 27: 485–490	H—FMB	VU
Z. paucijuga Wieland	Mexico (Chiapas, Colima, Guerrero, Jalisco, Michoacan, Nayarit, Oaxaca)	1916	American Fossil Cycads 2: 212, fig. 86	fig. 86	NT
Z. poeppigiana Mart. & Eichler	Brazil, Colombia (Antioquia, Huila, Meta, Tolima), Ecuador (Azuay, Bolivar, Chimborazo, El Oro, Esmeraldas, Guayas, Los Ríos, Manabi, Pichincha), Peru (Loreto, Madre de Dios, San Martin, Tumbes)	1863	Fl. Bras. 4(1): 414–416, pls. 108–109	L—F	NT
Z. polymorpha D. W. Stev., A. Moretti, & Vázq. Torres	Belize, Mexico (Campeche, Quintana Roo, Yucatán)	1995–96	Delpinoa n.s. 37–38: 3–8) (issued 1998	H—NY	NT
Z. portoricensis Urban	Puerto Rico	1899	Symb. Antill. 1: 291	I—NY	CR
Z. prasina W. Bull	Belize	1881	Retail List: 20	H—K	CR
Z. pseudomonticola L. D. Gómez	Costa Rica (Chiriquí)	1982	Phytologia 50(6): 401–404	H—CR	NT
Z. pseudoparasitica Yates in Seem.	Panama (Bocas del Toro, Coclé, Veraguas)	1854	Bot. Voy. Herald 6:201–203, 253	H—BM	NT
⋆*Z. pumila* L.	Cuba, Domican Republic, Puerto Rico	1763	Sp. Pl. Ed. 2, 2: 1659	t. 58 *in* Hort. Med. Amstelod 1, 1697	NT
Z. purpurea Vovides, J. D. Rees & Vázq. Torres	Mexico (Oaxaca, Veracruz)	1983	Flora de Veracruz 26: 28–31	H—XAL	EN
Z. pygmaea Sims	Cuba (W Cuba, Isla de la Juventud	1815	Bot. Mag. 43: t.1741	H—BM	DD
Z. roezlii Linden	Colombia (Amazonas, Chocó, Nariño, Valle del Cauca), Ecuador (Esmeraldas, Imbabura)	1873	Catalogue des Plantes Nouvelles 9: 10	pl. 133–134 *in* Ill. Hort. 20, 1873	NT
Z. skinneri Warsz. Ex A. Dietrich	Panama (Coclé, Veraguas)	1851	Allg. Gartenzeitung 19: 146, fig. s.n.	fig. s.n.	EN
Z. soconuscensis Schutzman, Vovides, & Dehgan	Mexico (Chiapas)	1988	Bot. Gaz. 149(3): 347–351	H—F	VU
Z. spartea A. DC.	Mexico (Oaxaca)	1868	Prodr. 16(2): 539	H—G-DC	CR
Z. standleyi Schutzman	Honduras (Colón, Olancho, Yoro)	1989	Syst. Bot. 14(2): 214–219	H—FLAS	VU

Z. stricta Miq.	Cuba (Oriente)	1851	Verh. Nederl. Inst. 3(4): 183	H—U	EN
Z. tuerckheimii Donn. Sm.	Guatemala	1903	Bot. Gaz. (Crawfordsville) 35(1): 8, pl. 1	H—US	NT
Z. ulei U. Dammer	W Brazil, Colombia (Amazonas, Guainía), Ecuador (Napo, Pastaza), Peru (Loreto, Madre de Dios)	1907	Verh. Bot. Vereins Prov. Brandenburg 47: 117–118	L—HBG	NT
Z. urep B. Walln.	Peru (Huánuco)	1996	Linz. Biol. Beitr. 28(2): 1056	H—W	CR
Z. variegata Warsz.	Guatemala, Mexico (Chiapas)	1845	Allg. Gartenzeitung 32: 252–253	N—NY	EN
Z. vazquezii D.W. Stev., Sabato, A. Moretti, & de Luca	Mexico (Veracruz)	1995–96	Delpinoa n.s. 37–38: 9–17 (issued 1998)	H—NY	CR
Z. verschaffeltii Miq. *(speies dubium)*	Mexico (Chiapas, Tabasco)	1870	Verh. Kon. Ned. Akad. Wetensch., Afd. Natuurk. 2(4): 31	H—U	EW
Z. wallisii A. Braun	Colombia (Antioquia)	1875	Monatsber. Königl. Preuss. Akad. Wiss. Berlin: 376	N—NY	CR
Synonyms:					
AULACOPHYLLUM Regel	= *ZAMIA*	1876	Gartenflora 25: 141		
A. lindeni (Regel ex André) Regel	= *Z. poeppigiana*	1876	Gartenflora 25: 141	BAS	
A. montanum (A. Braun) Regel	= *Z. montana*	1876	Gartenflora 25: 141	BAS	
A. ortgiesi Regel	= *Z. chigua*	1876	Gartenflora 25: 141	H—LE	
A. roezli (Linden) Regel	= *Z. roezlii*	1876	Gartenflora 25: 141	BAS	
**A. skinneri* (Warsz. ex A. Dietrich) Regel	= *Z. skinneri*	1876	Gartenflora 25: 143	BAS	
A. wallisii (hort. Veitch ex A. Braun) Regel	= *Z. wallisii*	1876	Gartenflora 25: 143	BAS	
Ceratozamia boliviana Brongn.	= *Z. boliviana*	1846	Ann. Sci. Nat. ser. 3, 5: 9	BAS	
C. katzeriana Regel	= *Z. verschaffeltii*	1876	Gartenflora 25: 261	L—LE	
Z. allison-armourii Millsp.	= *Z. pumila*	1900	Publ. Field Columbian Mus., Bot. ser. 2: 23	H—F	
Z. angustissima Miq.	= *Z. angustifolia*	1851	Verh. Nederl. Inst. 3(4): 184	H—U	
Z. baraquiniana hort. ex Regel	= *Z. poeppigiana*	1876	Trudy Imp. S. Petersburgsk. Bot. Sada 4: 7	H—LE	
Z. chamberlainii J. Schust.	= *Z. pygmaea*	1932	*in* Engl., Pflanzenr. 4(1): 153	?	

Taxon	Location	Date	Original Source	Type	Status
Z. cupatiensis Ducke	= *Z. ulei*	1922	Arch. Jard. Bot. Rio de Janeiro 3: 20, pl. 1	pl. 1	
Z. cycadifolia Dyer	= *Z. loddigesii*	1884	*in* Hemsl., Biol. Cent.-Am., Bot. 3: 195	H—C	
Z. debilis L. f.	= *Z. pumila*	1789	*in* Aiton, Hortus Kew. Ed. 1, 3: 478	t. 58 *in* Hort.Med. Amstelod 1, 1697	
Z. dentata Voigt	= ?*Z. integrifolia*	1828	Syll. Ratisb. 2: 53	?	
Z. erosa Cook & Collins	= *Z. integrifolia*	1903	Contr. U. S. Natl. Herb. 8: 267	L—NY	
Z. floridana A. DC.	= *Z. integrifolia*	1868	Prodr. 16(2): 544	H—G-DC	
Z. furfuracea var. *trewii* A. DC	= *Z. furfuracea*	1868	Prodr. 16(2): 541	H—G	
Z. galeotti De Vriese	= *Z. loddigesii*	1845	*in* Hoeven & De Vriese, Tijdschr. Natuurl. Gesch. Physiol. 12: 24	?	
Z. guggenheimiana Carabia	= *Z. angustifolia*	1941	Caribbean Forest. 2: 89	H—NY	
Z. gutierrezi Sauv.	= *Z. muricata*	1868	Anal. Acad. Cienc. Nat. Haban. (reimp. 2)	H—HAC	
Z. humilis Salisb.	= *Z. pumila*	1796	Prodr. Stirp.: 400	?	
Z. jirijirimensis R. E. Schult.	= *Z. lecointei*	1953	Mutisia 15: 2, fig. p. 5 s.n.	H—GH	
Z. katzeriana (Regel) Rettig	= *Z. verschaffeltii*	1896	Gartenflora 45: 148–149	BAS	
Z. kickxii Miq.	= *Z. pygmaea*	1842	Monogr. Cycad.: 71, t. 8	H—U	
Z. laeta Salisb.	nomen dubium	1796	Prodr. Stirp. 400	?	
Z. latifolia Lodd. ex Miq.	= *Z. muricata*	1849	Tijdschr. Wis-Natuurk. Wetensch. Eerste Kl. Kon. Ned. Inst. Wetensch. 2: 298	t. 7, fig. a *in* Linnaea 19, 1847	
Z. latifoliolata Prenl.	= *Z. pumila*	1872	Bull. Soc. Vaud. Sci. Nat. 11: 278	H—LAU	
Z. lawsoniana Dyer	= *Z. loddigesii*	1884	*in* Hemsl., Biol. Cent.-Am., Bot. 3: 195	H—OX	
Z. leiboldii Miq.	= *Z. loddigesii*	1847	Linnaea 19: 425	H—U	
Z. leiboldii var. *latifolia* Regel	= *Z. loddigesii*	1876	Trudy Imp. S. Petersburgsk. Bot. Sada 4: 307	H—LE	
Z. lindenii Regel ex André	= *Z. poeppigiana*	1875	Ill. Hort. 22: 23, pl. 195	pl. 195	
Z. lindleyi Warsz. Ex A. Dietrich	nomen dubium	1851	Allg. Gartenzeitung 19: 146, fig. s.n.	fig. s.n.	
Z. loddigesii var. *angustifolia* Regel	= *Z. loddigesii*	1857	Bull. Soc. Imp. Naturalistes Moscou 30: 190	H—LE	
Z. loddigesii var. *longifolia* J. Schust.	= *Z. loddigesii*	1932	*in* Engl., Pflanzenr. 4(1): 147	L—LE	

Z. loddigesii var. *obtusifolia* Regel	= *Z. loddigesii*	1857	Bull. Soc. Imp. Naturalistes Moscou 30: 190	t. 186, figs. 27 &28 *in* Gartenflora 6, 1857
Z. madida R. E. Schult.	= *Z. manicata*	1958	Bot. Mus. Leafl. 18: 114, pl. 18	H—GH
Z. media Jacq.	= *Z. integrifolia*	1798	Hort. Schoenb. 3: 77, tt. 397, 398	t. 398
Z. mexicana Miq.	= *Z. loddigesii*	1861	Prodr. Syst. Cycad.: 13, 26	H—U
Z. multifoliolata A. DC.	= *Z. angustifolia*	1868	Prodr. 16(2): 545	H—G-DC
Z. muricata var. *angustifolia* Miq.	= *Z. muricata*	1842	Monogr. Cycad.: 66	H—U
Z. muricata var. *obtustifolia* Miq.	= *Z. muricata*	1843	Tijdschr. Natuurl. Gesch. Physiol. 10: 71–72	t. 7, fig. a *in* Linnaea 19, 1847
Z. noeffiana Linden ex Wittmack	nomen dubium	1897	Gartenflora 46: 358	?
Z. obidensis Ducke	= *Z. lecointei*	1922	Arch. Jard. Bot. Rio de Janeiro 3: 19	H—MG
Z. obliqua Regel ex A. Ducos	= *Z. obliqua*	1877	Ill. Hort. 24: 140, pl. 289	pl. 289
Z. ottonis Miq.	= *Z. pygmaea*	1843	Linnaea 17: 740	H—U
Z. pallida Salisb.	nomen dubium	1796	Prodr. Stirp., 401	?
Z. picta Dyer	= *Z. variegata*	1884	*in* Hemsl., Biol. Cent.-Am., Bot. 3: 194	H—U
Z. potemkini hort. Ex Miq.	nomen dubium	1870	Verh. Kon. Ned. Akad. Wetensch., Afd. Natuurk. 2(4): 30	?
Z. silicea Britton	= *Z. pygmaea*	1916	Bull. Torrey Bot. Club 43: 462	H—NY
Z. silvicola Small	= *Z. integrifolia*	1926	J. NY Bot. Gard. 17: 128	H—NY
Z. splendens Schutzman	= *Z.katzeriana*	1984	Phytologia 55: 299–303	H—FTG
Z. sylvatica Chamb.	= *Z. loddigesii*	1926	Bot. Gaz. 81: 223, 225	L—NY
Z. tenuis Willd.	= *Z. integrifolia*	1806	Sp. Pl. Ed. 4, 4(2): 846.	H—B-W
Z. umbrosa Small	= *Z. integrifolia*	1921	J. NY Bot. Gard. 22: 136, in adnot.	L—NY
Z. yatesii Miq.	= *Z. angustifolia*	1851	Verh. Nederl. Inst. 3(4): 182	H—U

The total number of species, validly published or known to be in press, in our list, not including synonyms, *species dubia*, or taxa below specific rank, is 303.

Acknowledgments

The World List of Cycads is published under the auspices of the IUCN/SSC Cycad Specialist Group. We thank our cycad colleagues past and present for their for useful input to this edition of the World List.

Literature Cited

Brummit, R. K. & C. E. Powell. 1992. Authors of plant names. Royal Botanic Gardens, Kew, England.

Donaldson, J. S. (ed.). 2003. Cycads. Status survey and conservation action plan. IUCN/SSC Cycad Specialist Group. IUCN, Gland, Switzerland and Cambridge, UK.

Hill, K. D. & D. W. Stevenson. 1998–. The cycad pages: http://plantnet.rbgsyd.nsw.gov.au/PlantNet/cycad/.

———, ——— & R. Osborne. 2004a. The world list of cycads. Proceedings of the fifth international conference on cycad biology. Bot. Rev. 70(2): 274–298.

———, ——— & ———. 2004b. The world list of cycads. Pp. 219–235 (Appendix 1) *in* T. Walters and R. Osborne (eds.), Cycad classification: concepts and recommendations. CABI, Wallingford, UK.

———, ——— & ——. 2004c. The world list of cycads. Pp. 195–212 *in* A. J. Lindström (ed.), Proceedings of the sixth international conference on cycad niology. Nong Nooch Tropical Botanical Garden, Thailand.

———, ——— & ——. 2007. The world list of cycads / [La lista mundial de cícades]. Mem. New York. Bot Gard. 97: 454–483; available at: http://www.iucn.org/themes/ssc/sgs/csg/pages/worldlist.htm.

Holmgren, P. K., N. H. Holmgren & L. C. Barnett. 1990. Index herbariorum. Part I: The Herbaria of the world. Ed. 8. Regum vegetabile 120. New York Botanical Garden, Bronx, NY.

Lawrence, G. H. M., A. F. G. Buchheim, G. S. Daniels & H. Dolezal (eds.). 1968. Botanico-periodicum-huntianum. Hunt Botanical Library, Pittsburgh, PA, 1063 pp.

Osborne, R. & J. Hendricks. 1985. A world list of cycads. Encephalartos 3: 13–17.

——— & ———. 1986. A world list of cycads—supplement. *Encephalartos* 5: 27.

———, D. W. Stevenson & K. D. Hill. 1999. The world list of cycads. Pp. 224–239 *in* C. J. Chen (ed.), Proceedings of the fourth international conference on cycad biology. International Academic Publishers, Beijing, China.

Stafleu, F. A. & R. S. Cowan. 1976, 1979, 1981, 1983. Taxonomic literature. Ed. 2. Vols 1–4 (Regum vegetabile 94, 98, 105, 110).

Stevenson, D. W. & R. Osborne. 1993a. The world list of cycads. Pp. 354–364 *in* D. W. Stevenson & K. J. Norstog (eds.), Proceedings of the second international conference on cycad biology. Palm and Cycad Societies of Australia Ltd, Milton, Queensland, Australia.

——— & ———. 1993b. The world list of cycads. Encephalartos 33: 19–25.

———, ——— & J. Hendricks. 1990. A world list of cycads. Mem. New York Bot Gard. 57: 200–206.

———, ——— & K. D. Hill. 1995. The world list of cycads. Pp. 55–56 *in* P. Vorster (ed.), Proceedings of the third international conference on cycad biology. Cycad Society of South Africa, Stellenbosch, South Africa.

SYSTEMATICS
CHAPTER 29

Unlocking the Coontie Conundrum: The Potential of Microsatellite DNA Studies in the Caribbean Zamia pumila Complex (Zamiaceae)

Alan W. Meerow, Dennis Wm. Stevenson, Jeremy Moynihan, and Javier Francisco-Ortega

Abstract

The *Zamia pumila* complex (Cycadales: Zamiaceae) is a monophyletic, diploid ($2n=16$) and distinctive assemblage of cycad populations restricted to the West Indies and southeastern United States that has been treated as comprising as many as 14 or as few as one species. We are in the process of developing a microsatellite (simple sequence repeat, SSR) DNA library from *Z. integrifolia* L. fil. that we will use to study variation in the complex and apply these data to various evolutionary and biogeographic questions. Our overall goal is to document the patterns of microsatellite DNA variation across populations of the *Z. pumila* L. complex throughout its range, infer population structure and biogeographic history of the complex, and attempt to understand the processes of speciation within the group. We hypothesize that *Zamia* in the Caribbean will exhibit conformance to the stepping stone model, given the limited gene flow exhibited by most cycad species. We further hypothesize that we will uncover genetic evidence of bottlenecks within some populations, apropos to the history of the Greater Antilles. Finally, we intend to address the hypotheses that the *Z. pumila* complex represents a minium of five species, that pollinator mediated isolation restricts gene flow between species in the complex, and that every species in the *Z. pumila* complex will have a unique pollinator association. We have already designed seven SSR primer pairs that have successfully captured polymorphism in *Zamia*, and have screened only 15% of our clones.

Resumen

Zamia pumila (Cycadales: Zamiaceae) forma un grupo monofiletico, diploide (2n=16) constituido por poblaciones restringidas a las Indias Occidentales y el sureste de Estados

Unidos. Se ha sugerido que este grupo está formado por una especies o por al menos 14 de ellas. En estos momentos estamos desarrollando una biblioteca genómica para microsatelites (SSR) a partir de Z integrifolia L. f. que se utilizará para estudiar la variación existente en el complejo así como para resolver algunas preguntas de índole biogeográfico y evolutivo. Nuestro objetivo global es el documentar las pautas de variación de SSRs a lo largo del ambito geográfico de las poblaciones del complejo Z. pumila L. Estos datos nos ayudarán a entender la estructura genetica de sus poblaciones y nos daran información para comprender los procesos de especiación dentro de este grupo. Nuestra hipotesis de trabajo es que Zamia en el Caribe tiene un módelo de saltos entre islas dado los nivels limitados de flujo genético que tienen la gran mayoría de las especies de cicadinas. Tambien tenemos la hipótesis de que para varias poblaciones encontraremos evidencia de cuello de botella genético. Finalmente intentaremos de comprobar la hipótesis de que el complejo Z. pumila esta compuesto por un mínimo de de cinco especies y que aislamiento a traves de polinizadores específicos restringe en gran manera el flujo genético dentro del complejo y que cada una de las especies del complejo tiene asociado un único polinizador. Hasta el momento hemos diseñado cebadores para siete loci y hemos sido capaces de detectar polimorfismos a pesar de que solamente hemos muestreado un 15% de los clones obtenidos.

Zamia in the West Indies and Florida

Although the taxonomy of *Zamia* spp. in the West Indies has, in the past, been at best confusing, the work of Eckenwalder (1980a), Stevenson (1987a, b), and Stevenson and Sabato (1987) has clarified and resolved the nomenclatural aspects of the genus as well as identified the types for all 26 valid names. The taxonomic extremes range from the recognition of more than 14 species (De Candolle, 1868) to 1 species (*Z. pumila* L.) with 2 subspecies (Eckenwalder, 1980a), subsp. *pumila* and subsp. *pygmaea* (Sims) Eckenwalder, based on general plant size and leaflet shape. Stevenson (1987a, b) and Sabato (1990) later recognized 6 and 5 spp., respectively, for the West Indies. More recently, Géigel (2003) recognized 2 additional spp. in Cuba, the endemics *Z. stricta* Miq. and *Z. ottonis* Miq. Thus, 9 spp. are currently recognized for the West Indies. They are *Z. amblyphyllidia* D.W. Stev., *Z. angustifolia* Jacq., *Z. integrifolia* L. fil., Z. lucayana Britton, *Z. ottonis*, *Z. portoricensis* Urb., *Z. pumila*, *Z. pygmaea* Sims (including *Z. kickxii* Miq.), and *Z. stricta*.

Only *Z. angustifolia* was found to be distinct from all others in allozyme patterns

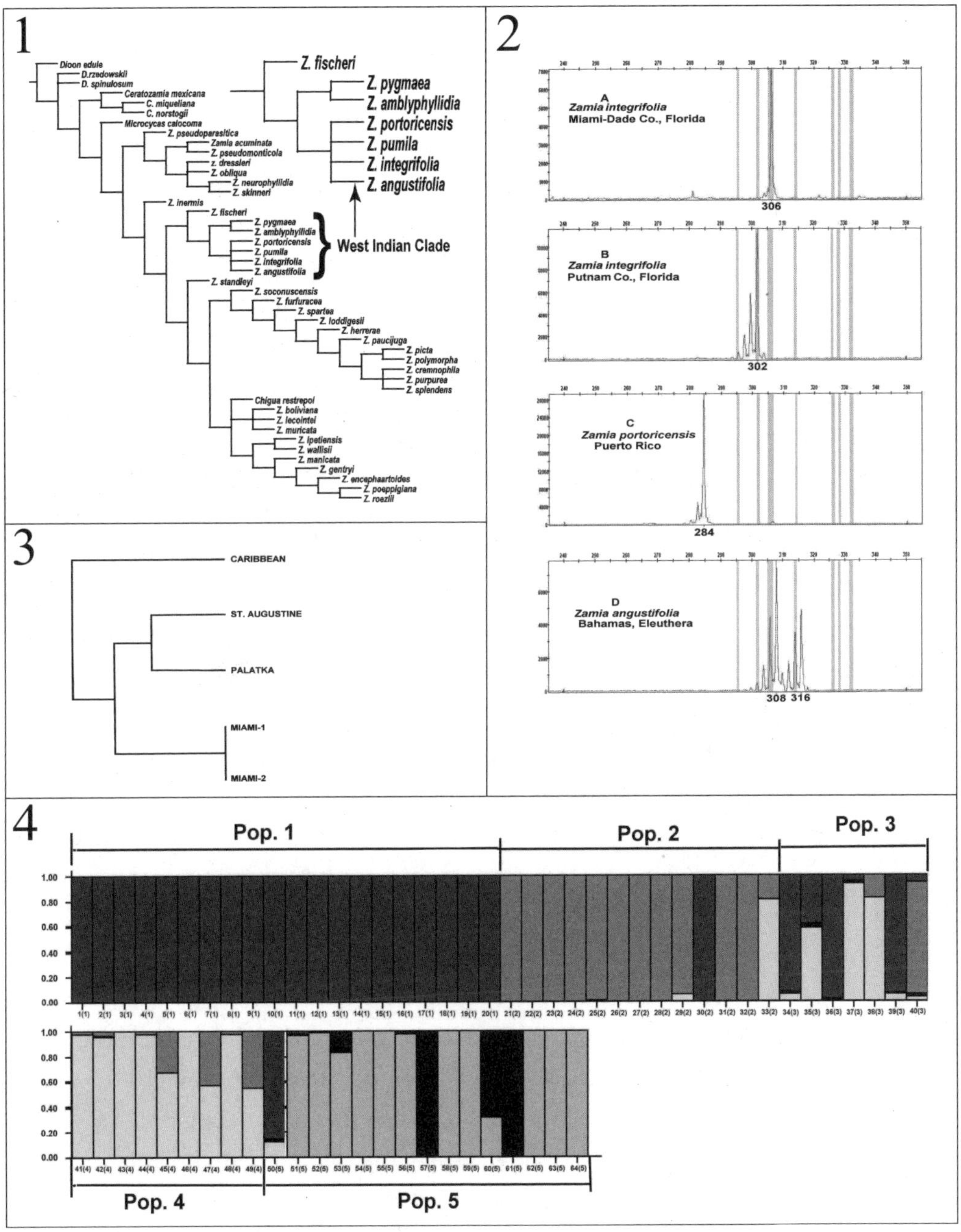

Figure 29-1. Strict consensus of 12 most parsimonious trees from cladistic analysis of *Zamia* using ITS2 sequences and morphological characters. (After Caputo et al., 2004.)

Figure 29-2. Representative chromatograms at a single SSR locus (GA repeat) from four populations of the *Zamia pumila* complex. Number below peak is the called allele size. D is a heterozygote.

Figure 29-3. UPGMA dendrogram of the five *Zamia* populations listed in Table 29-2 based on coancestry coefficient.

Figure 29-4. Clustering of 64 individuals representing the 5 populations of the *Zamia pumila* complex in Table 29-2.

(Walters & Decker-Walters, 1991). Similarly, recent phylogenetic analyses based on the gene region ITS2 and morphology (Caputo et al., 2004; Fig. 29-1) have failed to provide significant resolution within the clade (Fig. 29-1). However, the monophyly of the West Indian zamias was demonstrated and their sister taxon, *Z. fischeri* Miq., was identified. Consequently, a multilocus molecular marker approach at the populational level has the greatest potential for unraveling the relationships within this complex.

Biogeography and Dispersal

The distribution of West Indian and southeast U.S. species of *Zamia* shows little discernible pattern and no sympatry among the currently recognized species. Eckenwalder (1980b) and Tang (1989) found that nearly 50% of *Z. integrifolia* seeds were within 1 m of the parent plant (many within 15 cm), with only about 5% at distances greater than 4 m (Tang, 1989). In fact, one often finds many seeds actually germinating within the crown of leaves. Because a distance of 4 m from the parent plant is beyond the range of gravitational dispersal for the ground-level cones in an area of flat terrain, some animal agent would seem to be involved in these cases (Eckenwalder, 1980b). Indirect evidence in the form of marks on the seeds and the affinity of mockingbirds for the seeds of related species in cultivation indicates that birds and small mammals are dispersal agents (Eckenwalder, 1980b; Tang, 1989). Tang (1989) observed that most dispersed seeds are found under shrubs, suggestive of bird-landing sites and areas of shelter for small mammals, and noted that most potential bird dispersers had a mouth-gape either less than the minimum diameter of the seeds or less than the median diameter. Thus, birds could transport seeds only a short distance before dropping them, which is consistent with the observed distribution of seeds. This type of dispersal cannot explain the present distribution of species.

In the southeastern United States, only one species is recognized, *Z. integrifolia* (Stevenson, 1991). Although there is some local and ecotypic variation in vegetative morphology, it is not possible to correlate this variation with variable characters of reproductive morphology. *Zamia integrifolia* is known from extreme southeastern Georgia southward through peninsular Florida (including the Florida Keys) and sporadically in the Bahamas, where it occurs on Andros, Grand Bahama, Great Abaco, Long, and New Providence islands. It has been collected in coastal areas of north-central Cuba, the Cayman Islands, and south-central Puerto Rico (where it may now be extinct). Habitats vary from open coastal areas and sand dunes to pinelands and closed-canopy oak hammocks to tropical forest. This cycad is most commonly found in soil over limestone and in sand near sea level or in pinelands subjected to periodic wildfires. *Zamia portoricensis* is endemic to very dry areas of western Puerto Rico, where it is known from only three extant populations. *Zamia angustifolia* is confined to Eleuthera in the Bahamas and extreme southeastern Cuba in semideciduous dry forest. *Zamia pumila* is quite common in coastal areas of the Dominican Republic and rare in extreme eastern Cuba along the coast and appears to be extinct in southern Puerto Rico. *Zamia amblyphyllidia* occurs in mesic forests in north-central Puerto Rico, in western and eastern Cuba, and in Jamaica. *Zamia pygmaea* and *Z. ottonis* are both endemic to Cuba with the former in western Cuba and the latter in central and eastern Cuba without any sympatry. *Zamia stricta* is endemic to a very small area in southeastern Cuba in dry, evergreen microphyllous forest. *Zamia lucayana* is known only from the vicinity of Clarence Town on Long Island in the Bahamas. Plants of these populations appear to be *Z. integrifolia* except for their extremely wide leaflets. Perhaps the most enigmatic populations are those assignable to *Z. kickxii* Miq. (not currently recognized as a distinct species from *Z. pygmaea*). These plants differ much more dramatically from *Z. pygmaea* than does *Z. ottonis*. Of particular interest is the similarity of leaflet shape, texture, and marginal denticulation to that of the sister taxon to the West Indian Clade, *Z. fischeri* Miq., from San Luis Potosi, Mexico. Miquel, who described both species, thought that *Z. fischeri* might also be found in the West Indies. This broad Caribbean distribution suggests that long-distance dispersal has played at least an occasional role in the phytogeography of *Zamia*. Indeed, Eckenwalder (1980b) demonstrated potential for inter-island dispersal. After 2 weeks in seawater, 92% of seeds germinated and grew normally.

Pollination Biology

Pollination biology is poorly known for most species of *Zamia* (Stevenson et al., 1998), except *Z. integrifolia* in Florida (Tang, 1987) and *Z. furfuracea* L. fil. in Mexico (Norstog et al., 1986). *Zamia integrifolia* is pollinated by a snout weevil, *Rhopalotria slossonii* (Schäffer), whereas *Z. furfuracea* is pollinated by *R. mollis* (Sharp). When the two species of *Zamia* are grown together with both pollinator species present, there is fidelity between the pollinator species and the *Zamia* species. This is also true when both species of *Zamia* are grown together and only one pollinator is present. Although pollination studies on other West Indian species of *Zamia* have not been conducted, it is known that *R. dimidiata* Chevrolat is associated with pollen cones of *Z. pygmaea* in Cuba (Muñiz & Barrera, 1969). Further, one author (D. W. S.) has collected *R. slossonii* from pollen cones of *Z. integrifolia* in the Bahamas, and, while in Mexico, an unidentified new species of *Rhopalotria* on cones of *Z. fischeri* and another on *Z. inermis*, the sister taxa to the West Indian Clade (Fig. 29-1). This suggests a long evolutionary association between *Rhopalotria* and *Zamia*.

Population Genetics of Cycadales

All cycads are dioecious. Their pollination and dispersal biology (Tang, 1987, 1989; Donaldson et al., 1995; Norstog & Nicholls, 1997; Stevenson et al., 1998) constrains both gene flow and population recruitment to relatively small geographic areas. Thus, the expectation would be high genetic differentiation between populations, directly proportional to the distance among them, and low intrapopulation differentiation. This is what the majority of genetic studies of cycad populations have found (Ellstrand et al., 1990; Walters & Decker-Walters, 1991; Yang & Meerow, 1996; Lin et al., 2000; Huang et al., 2001, 2004). Yang and Meerow (1996) estimated gene flow at only 2–7 km between neighboring populations of Chinese *Cycas* species. However, Byrne and James (1991) reported much greater distances for *Macrozamia riedlei* (Gaudich.) C. A. Gardner in southern Australia. They attributed this disparity to the more continuous distribution of *M. riedlei* across a fairly homogeneous area of low relief, thereby promoting high levels of gene flow, relative to *M. communis* L. A. S. Johnson, which represents an isolated element of a long-term fragmentation process initiated by geologic activity.

More alarming, from the perspective of conservation of these ancient plants, are the low levels of heterozygosity that have been reported for some populations (or entire species). Walters and Decker-Walters (1991) reported observed heterozygosity (H_o) of only 0.047 in Caribbean *Zamia*. Ellstrand et al. (1990) found total heterozygosity (H_T) among Australian *Macrozamia* to be 0.08. Yang and Meerow (1999) reported $H_T = 0.049$ for the Chinese *Cycas panzhihuaensis* L. Zhou & S. Y. Yang and Lin et al. (2000) and Huang et al. (2004) found low levels of allozyme variability both within and among *C. taitungensis* C. F. Shen, K. D. Hill, C. H. Tsou, & C. J. Chen populations in Taiwan. This has dire consequences as habitats are destroyed or populations are reduced by over-collecting.

All of these studies utilized allozymes to assess genetic variation, the conserved nature of which may have biased the estimates (Brossart & Prowell, 1998). To date, the only other marker system that has been applied to cycad population genetics has been sequence variation in the chloroplast *atpB–rbcL* intergenic spacer, and the mtDNA internal transcribed spacer (ITS) region (Huang et al., 2001). In contrast to allozyme results of populations of the same species (Lin et al., 2000; Huang et al., 2004), there was a great deal of haplotype diversity in these two sequences. The authors attributed this to the long evolutionary history of *Cycas* coupled with the low level of functional constraints on spacer regions and the possibility of frequent intramolecular recombination. They also attributed the lack of congruence between the chloroplast and mitochondrial phylogenies to this frequent recombination.

Caribbean Phytogeography

The West Indies have a flora with a high proportion of endemic species, and it is estimated that approximately 60% of West Indian species are restricted to these islands (Myers et al., 2000). The Antilles and Bahamas have a very complex geological history and their ecosystems have been severely affected by several major environmental changes (reviewed by Iturralde-Vinent & MacPhee, 1999; Fritsch & McDowell, 2003; Graham, 2003a, b; Santiago-Valentín & Olmstead, 2004). These geological and environmental histories have had an important influence on the unique plant biodiversity patterns of the Antilles (Borhidi, 1996; Adams, 1997; Fritsch & McDowell, 2003).

Geologic and Environmental History of the West Indies and Southeastern United States

The geological history of the West Indies dates back to the Early Cretaceous and has involved mainly volcanism and tectonic processes. In addition, sedimentary depositions caused mostly by fluctuations in eustatic sea level changes have also influenced the historical geology of the West Indies (Pindell & Barrett, 1990; Sealey, 1992; Draper & Barros, 1994; Iturralde-Vinent & MacPhee, 1999). The impact of a giant meteor in the Yucatan area approximately 65,000,000 years ago marked a major planetary boundary between the Cretaceous and Tertiary geological periods, producing major extinction events in the Antilles (Hedges et al., 1992; Hedges, 2001) and the whole of North America (Sweet & Bramen, 1992; Schultz & D'Hondt, 1996; Vajda et al., 2001).

Major environmental changes in the Caribbean Basin have been linked to the closure of the Panama Isthmus between the Middle Miocene and the Late Pliocene (Coates, 1997; Droxler et al., 1998) and the Tertiary and Quaternary glacial and interglacial phases (Iturralde-Vinent & MacPhee, 1999). The formation of the Panama Isthmus established the Gulf Stream as a major thermal regulator for the West Indies and Florida (Collins, 1996; Collins et al., 1996; Droxler et al., 1998). Glacial and interglacial phase sea level changes have had a major effect on the shore lines of the West Indies since the formation of the proto-Antillean arc in the Pacific Ocean (Nunn, 1994; Aubrey et al., 1988). Island size and inter-island distances have varied dramatically during the geo-environmental history of the West Indies.

The Great Bahamas Bank once formed a large island during the last glacial maximum (approximately 18,000 years ago; Andersen & Borns, 1994; Kindler & Hearty, 1997; Hearty & Kaufman, 2000). This large paleo-island would have been approximately the same size as the current island of Hispaniola. During the last interglacial maximum (approximately 120,000 years ago; Andersen & Borns, 1994), the majority of the Great Bahamas Bank was totally below sea-level with little island growth during this phase (Kindler & Hearty, 1997; Hearty & Kaufman, 2000). At present, approximately 20% of the Great Bahamas Bank is below sea level and it is fragmented into six small archipelagos.

It has been suggested that most of Jamaica was below sea level between the Middle Eocene and the Middle Miocene and that tectonic uplifts, in the Late Miocene, have contributed to the current high elevations of this island (Lewis & Draper, 1990; Robinson, 1994, 1998). Therefore the current Jamaican biota is relatively recent, and reached this island by dispersal events in the last 10,000,000 million years (Buskirk, 1985; Graham, 2003a). Cuba was fragmented into various archipelagos several times in the last 65 million years (Gutiérrez-Domech & Rivero-Glean, 1997). Hispaniola also underwent similar fragmentation and is thought to be formed by at least four tectonically independent blocks (Lewis & Draper, 1990). Northern Hispaniola was connected to eastern Cuba and Puerto Rico (Graham, 2003a), and southwestern Hispaniola was connected to Jamaica until the Early Miocene. The two fragments united in the Middle Miocene. This collision caused a major tectonic uplift in the central parts of the island (Lewis & Draper, 1990; Iturralde-Vinent & MacPhee, 1999; Graham, 2003a). The fossil record indicates that *Zamia* achieved its modern distribution by the beginning of the Cenozoic (Balduzi et al., 1981–1982; Sabato, 1990; Walters & Decker-Walters, 1991), based on Eocene-Oligocene remains from Puerto Rico referable to *Z. pumila* (*sensu* Eckenwalder, 1980a).

Eustatic sea level changes have been one of the most important factors in determining the extent of the landmass of Florida (including the Florida Keys) and Georgia in the Tertiary and Quaternary (Schmidt, 1997; Scott, 1997), especially in Florida because of its low elevation. Pleistocene seas covered most of Florida and the Atlantic coast counties of Georgia during the interglacial phases (Randazzo, 1997), though high ridges such at the Lake Wales Ridge likely remained above water throughout these periods (Webb, 1990). During the glacial maxima the Florida platform was almost entirely above sea level and had at least three times its current dry land mass area (Scott, 1997).

Floristic Links between the West Indies and Southeastern United States

Despite these sea level fluctuations, there is agreement among geologists that no land bridges have existed between the dry land areas of the Bahamas, Florida, and Cuba, at least in the last 35 million years (Draper & Barros, 1994; Gutiérrez-Domech & Rivero-Glean, 1997; Iturralde-Vinent & MacPhee, 1999). South Florida, the Bahamas, and the Antilles form the "Caribbean Biodiversity Hotspot" (Myers et al., 2000). From a biogeographical perspective, South Florida belongs to the neotropical floral kingdom (reviewed by Cox, 2001), and has always been placed within the Caribbean biogeographical region (Good, 1947; Takhtajan , 1988; Rivas-Martínez et al., 1999). Ecologically, South Florida has been classified within the "Humid Tropical Domain," and it is the only area of the continental United States that belongs to this domain (Bailey, 1995). Long and Lakela (1971) reported that about 91% of the tropical plants native to South Florida are taxonomically related to Caribbean Basin endemics.

Microsatellite DNA

Microsatellites or simple sequence repeats (SSRs) are DNA sequences of one to four bases (mono-, di-, tri-, or tetranucleotide repeats, respectively) that are repeated in tandem arrays. Such microsatellite sequences have been found randomly distributed throughout the genomes of all eukaryotes assayed (Tautz & Renz, 1984; Hearne et al., 1992; Wang et al., 1994; Li et al., 2002), including gymnosperms, angiosperms, monocots, and dicots (Morgante & Olivieri, 1993). SSR loci tend to be imprecisely replicated during DNA synthesis, thus generating new alleles with different numbers of repeating units (Tautz et al., 1986; Harding et al., 1992; Tachida & Iizuka, 1992). This length variation is a source of polymorphisms even between closely related individuals.

Microsatellites provide an ideal tool for assessing the genetic diversity of plants at various levels of inquiry (Chase et al., 1996; Powell et al., 1996a; Zhang & Hewitt, 2003), owing to their high information content, ease of genotyping, codominant and multiallelic nature, high discriminatory power, and reproducibility (Jones et al., 1997; Morgante & Olivieri, 1993; Powell et al., 1996a, b; Russell et al., 1997).

Models of Microsatellite Mutation

Two very different models of mutation have been applied to microsatellite markers, the infinite allele model (IAM; Kimura & Crow, 1964) and the stepwise mutation model (SMM; Kimura & Otha, 1978). These two models differ in their basic assumptions. The IAM assumes that novel alleles are created by each mutation at a given rate with no allowance for homoplasy and with equal probability (Balloux & Lugon-Moulin, 2002). In the SMM, novel alleles are created by the addition or subtraction of a single repeat unit of the SSR. Thus it is assumed that alleles are similar size are more closely related than alleles that are much different in length. A variant of the SMM, the two-phase model (TPM; Valdés et al., 1993; Di Rienzo et al., 1994) allows for multiunit additions or deletions. Many empirical studies support SMM and its variants as the most likely model estimations for SSR mutational processes (Estoup & Angers, 1998; Estoup & Cornuet, 1999; Ellegren, 2000; Makova et al., 2000; Schlötterer, 2000). However, Nielsen and Palsbøl (1999) performed computer simulations of both multi-step mutations and constraints in allele size and compared them to results obtained with data from baleen whale populations. They concluded "that it may be futile to develop realistic and applicable models of microsatellite evolution based on an exact description of the mutational model" (Nielsen & Palsbøl, 1999).

Homoplasy

Size homoplasy of SSR electromorphs undoubtedly occurs in nature (Orti et al., 1997; Angers et al., 2000; Makova et al., 2000), but has generally not been addressed in most population level studies. Estoup et al. (2002) concluded that size homoplasy presents the greatest problems where high mutation rates and strong constraints in allele size are coupled with large population size, and that the large variability inherent in SSR loci compensates for their homoplasious evolution. The fact that SSR loci in *Zamia* are so well conserved across the genus (see later) makes it much less likely that size homoplasy will be a factor in using them interspecifically (Petren et al., 1999), especially among such closely related taxa as the *Z. pumila* complex.

Statistical Analysis of Microsatellite Data

Genetic Distance

The traditional distance measures developed originally for classical genetic markers such as isozymes assume the IAM, which has been shown to be inaccurately applied to the mutational process of microsatellite loci (Takezaki & Nei, 1996; Goldstein et al., 1995). Thus distance measures such as Rogers' (1972) distance, Nei's (1972) standard genetic distance, and minimum genetic distance (Nei, 1973) cannot be applied accurately where the SMM or one of its variants is in force with SSR loci (Goldstein & Pollack, 1997; Estoup & Angers, 1998; Estoup & Cornuet, 1999; Ellegren, 2000; Schlötterer, 2000). The distance measure of Goldstein et al. (1995) $(\partial\mu)^2$ or that of Shriver et al. (1995), D_{SW}, which is linear relative to time when applied to loci with a strict stepwise mutation model, is a preferred distance coefficient for genealogical trees of populations with SSR data. Accuracy is relative, however, depending on the fidelity of any given locus to the SMM, and is also extremely sensitive to the number of loci used to estimate distance (Zhivotovsky & Feldman, 1995; Zhivotovsky et al., 2001). Cavalli-Sforza and Edwards' (1967) chord distance (D_c), the distance (D_a), of Nei et al. (1983), and allele sharing distance (DAS) of Stephens et al. (1992) all make use of the product of allele frequencies shared between populations, do not incorporate any particular mutational model, and will reconstruct closely related genealogies better than SSM-based distances (Goldstein et al. 1995; Takezaki and Nei, 1996). These distances are much less accurate when population bottlenecks have occurred (Takezaki and Nei, 1996) or when populations are more distantly related (Goldstein and Pollack, 1997).

Population Structure

Populations of most species demonstrate some level of genetic structure, the consequence of isolation, breeding system, and evolutionary history (Donnelly & Townson, 2000; Gerlach & Musolf, 2000; Pálsson, 2000; Tiedemann et al., 2000; Balloux & Lugon-Loulin, 2002). Inextricably related to genetic structuring is an understanding of gene flow, which will have enormous influence over the allele composition of populations and, ultimately the process of speciation (Barton & Hewitt, 1985; Balloux & Lugon-Loulin, 2002). Estimates of population differentiation are exceedingly important in conservation biology, where it is essential to understand the extent to which populations are genetically isolated from each other (Rossiter et al., 2000; Eizirik et al., 2001). Genetic drift will affect small, isolated populations to a much greater extent than populations exchanging genes (Wright, 1977; Madsen et al., 1996; Frankham & Ralls, 1998; Saccheri et al., 1998; Eldridge et al., 1999; Higgins & Lynch, 2001).

Estimation of population structure and gene flow using codominant markers has largely relied on indirect estimates such as Wright's (1951) F_{ST} (which assumes the IAM) and/or Slatkin's (1995) $R_{ST,}$ which was developed with SMM as its core assumption. The drawback of F_{ST} when applied to SSRs is the fact that microsatellites may not fit the infinite allele model. The chief disadvantage of R_{ST} is its high variance (Balloux & Lugon-Moulin, 2002). Brossart and Prowell (1998) summarized the inherent problems associated with such estimates, and strongly cautioned against drawing explicit conclusions from their use. Balloux and Goudet (2002) concluded that R_{ST} better reflects differentiation of populations with low levels of gene exchange, while F_{ST} is more accurate when gene flow is high. Obtaining values for both statistics and making comparisons has been suggested for determination of whether mutation or migration rates has had the greatest effect on genetic differentiation of populations (Michalakis & Veuille, 1996; Estoup et al., 2002). A higher R_{ST} favors mutation (Estoup & Angers, 1998), while with low differentiation the two statistics will converge (Lugon-Moulin et al., 1999). Pritchard et al. (2000) introduced a model-based clustering method that does not assume a specific mutation process and that incorporates a Bayesian clustering approach to inferring population structure.

Methods of Analysis

There is no consensus on the best approaches for analyzing microsatellite data (Brossart & Prowell, 1998). Ultimately, some type of clustering method is applied, and these fall into two classes: (1) distance based and (2) model based. Distance methods calculate a pairwise distance matrix between individuals or populations and construct a visual plot of the relationships based on the matrix. Most commonly, a neighbor-joining algorithm is applied to the data. Before the mid-1990s, this was the typical approach for analyzing multilocus data (e.g., Bowcock et al., 1994). It is difficult to assign confidence to the clusters resolved by these methods (though bootstrapping [Efron, 1979] was an often ignored starting point), and their resolution may be inextricably related to the algorithm and type of graphical representation used (Pritchard et al., 2000). Model-based methods assume that data points represent random observations following a parametric model. The basis for overall clustering is determined in concert with the parameters by which each individual or population is assigned to a cluster. The problems with model-based approaches are the inherent limitations of conventional population genetic models themselves, such that inferences drawn from them may lack "biological realism" (Brossart & Prowell, 1998).

Three recent approaches to SSR data analysis have improved the accuracy of inferences from population genetic data (Luikart & England, 1999). Maximum Likelihood (ML; Griffiths & Tavare, 1994; Nielsen, 1997; Nielsen et al., 1998) uses raw data (versus summary statistics) to identify the parameters that maximize the probability of obtaining the observed allele frequencies under a particular model. ML is able to distinguish between isolated subpopulations undergoing genetic drift from those undergoing gene flow, even if their F_{ST} values are equal (Beaumont & Bruford, 1998). Coalescent methods (Tavare, 1984; Kingman, 1982, 2000; Kaplan et al., 1988; Hudson, 1990; Rosenberg & Nordborg, 2002) allow an understanding of the patterns of shared ancestry among alleles. They describe the shape of a genealogy based on the data that contain information about effective population size and demographic history, including rates and putative dates of recent changes in population size. Bayesian methods using Monte Carlo Markov chain (MCMC) simulations (Hastings, 1970; Gamerman, 1997; Nielsen & Wakely, 2001; Wilson & Rannala, 2003) can be applied to any number of genetic parameters, for example, effective population size, level of inbreeding, patterns of dispersal and migration. Bayesian methods develop a probability distribution (posterior probability) for the parameter being tested using the raw data and a prior probability of one to several model parameters (e.g., a prior distribution of mutation rates). Though sometimes criticized as subjective, they have proven to be very effective in estimating a wide range of genetic parameters (Luikart & England, 1999; Nielsen & Wakely, 2001; Wilson & Rannala, 2003).

Interspecific SSR Phylogenies

In contrast to the vast body of literature that has accumulated on interpopulational genetic variation using microsatellite DNA, very few studies have attempted to reconstruct interspecific SSR phylogenies (Pollack et al., 1998; Primmer & Ellegren, 1998; Petren et al., 1999), despite suggestions that allele length polymorphisms could be useful for resolving interspecific relationships (Takezaki & Nei, 1996; McDonald & Potts, 1997). The obvious concern in doing so is the ever greater likelihood of allele size homoplasy as one crosses species boundaries, and the possibility for "developmental bias," that is, when primers designed from genomic DNA of one species fail to produce an amplicon in other species or show much lower levels of polymorphism (Ellegren et al., 1995). However, where radiation has occurred rapidly and recently, homoplasy is expected to be low (Nauta & Weissing, 1996; Takezaki & Nei, 1996; Petren et al., 1999). The use of microsatellite data in this context can be complementary to a well resolved and well supported DNA sequence phylogeny for the taxa of interest.

SSR Library Construction and Preliminary Data Analyses

We have constructed a *Zamia* SSR library from the genomic DNA of *Z. pumila*, using a modification of the enrichment method of SSR marker development of Edwards et al. (1996). Genomic DNA was restricted, ligated to adaptors, and amplified via polymerase chain reaction (PCR). The amplicons were hybridized twice with biotin-labeled synthetic SSRs and isolated using streptavidin-coated beads (Dynal, Oslo, Norway) in conjunction with a Dynal Magnetic Particle Concentrator. The eluted fragments were size separated using Sepharose CL-4B SizeSep 400 Spin Columns (Amersham Pharmacia Biotech, Piscataway, NJ), amplified, and cloned (using phage and plasmid vectors with M13 priming sites), and the clones screened by sequencing. Approximately 20% of the enriched fragments have been cloned to date, yielding 672 clones (it will yield more than 3000 clones when completely plated out). Two rounds of sequence screening were performed, each screening 192 clones. SSR loci were identified as dinucleotide repeats with 9 or more repeat units, and trinucleotide repeats with 6 or more repeat units. The first round yielded 41 potential SSR loci for primer design, 28 simple SSRs, and 13 complex/compound SSRs. The majority were AG and AC dinucleotide repeats. We designed PCR primers for 18 loci from the original 41 potential loci, 7 of which are working unambiguously and capturing polymorphism in *Zamia* populations (Table 29-1; Fig. 29-2). The remaining 11 need further optimization for PCR conditions, or a different primer combination, but because of the large number of clones in our library we opted to screen these further before returning to problematic primer pairs. A second round of screening yielded 34 potential SSR loci. Only 5 loci from round 2 matched loci isolated previously, resulting in 29 new potential loci for primer design, 20 simple SSRs, and 9 complex/compound SSRs, again with the majority AG or AC dinucleotide repeats. We are currently testing these primers. Initial tests suggest that we have another 15–17 useful loci. We are confident that we will have no problem reaching our goal of a minimum of 25 SSRs. The larger the number of loci, the greater the likelihood that we can find loci that are linked to phenotypic characters of interest, such as gender expression. Differences in allele size were detected on an ABI 3730 Genetic Analyzer (Applied Biosystems, Foster City, CA) via capillary gel electrophoresis. Preliminary analysis of raw microsatellite data was performed using Genemapper 3.5 (Applied Biosystems). The PCR reaction mix for Zam5, 10, 14, and 15 was 1.0 μL of 10× buffer (with 15 mM $MgCl_2$), 0.2 μL of 2.5 mM dNTP mix, 0.25 μL of 10 mM forward primer, 0.25 μL of 10 mM reverse primer, 0.05 μL of *Taq* polymerase, 7.25 μL of dH_20, and 1.0 μL of template. Amplification parameters were 5 min at 94°C, 35 cycles (45 s at 94°C, 45 s at 55°C, 1 min at 72°C), and 7 min at 72°C. For Zam1, 16, and 17, 1.0 μL of 5 M betaine was substituted for 1 μL of the dH_20. A touchdown amplification procedure was used for these three loci: 5 min at 94° C, 14 cycles (45 s at 94°C, 45 s at 60°C [–1°C per cycle], 1 min at 72°C), 25 cycles (45 s at 94°C, 45 s at 46°C, 1 min at 72°C), and 7 min at 72°C.

We assessed four populations of Florida *Zamia*, in cultivation at the Montgomery Botanical Center (MBC), and 15 Caribbean *Zamia*, also at MBC, representing populations from Puerto Rico, Dominican Republic, Cuba, Jamaica, and the Bahamas (Table 29-2) with primers for 7 SSR loci (Table 29-1). We generated summary statistics and a

Table 29-1 Characteristics of Seven SSR Loci Used in Preliminary Analyses of Genetic Variation in *Zamia*.

Locus	GenBank Accession No.	T_a (°C)	Repeat Type	Expected Product Length
Zam4	DG230525	61	$(AG)_{13}(AC)_{11}$	147
Zam5	DG230526	61	$(GA)_{13}$	198
Zam10	DG230527	61	$(GA)_{11}$	284
Zam14	DG230528	61	$(CT)_8(GT)_2(AT)_3(GT)(AT)(GT)_6$	195
Zam15	DG230529	60	$(AG)_9$	162
Zam16	DG230530	63	$(GA)_{13}$	151
Zam17	DG230531	63	$(GAA)_{12}$	170
			Forward primer (5' to 3')	Reverse primer (5' to 3')
Zam4			TGGAAGGAAGAGATGTTCCAC	CCATGATTGAAAGAGGGCAAG
Zam5			GAAGAGTGGATGGAGAAAAGTG	CTAAGCACCAGCAAGAGATAAC
Zam10			TGCCCTTCAACTCTTGTTTC	CAAACTGACATATCCTGTTCCC
Zam14			GGATATGCCAAAACCTTACCC	ACTAAGTCCTCTGACTAGCAC
Zam15			CCTCAACCCTATTAGCATGAAC	TGAGCCATTAGATGCCTCC
Zam16			AGCCCAAAGTAGGACCTCAC	GCTTAACTCATTGGCGCTCTC
Zam17			TGCTCCTTCATAGGCAAGTC	CCAGTTCTTACCATCCTCCATC

Table 29-2 Populations of *Zamia pumila* Complex Used in Preliminary Analyses of SSR Variation.

Locality	n	P	A	H_o	H_e	f	No. of Unique Alleles
Florida							
1. St. Augustine, St. John's County, Florida	20	1.00	3.429	0.516	0.446	0.139	2
2. Miami, Dade County, Florida	13	1.00	4.000	0.567	0.438	0.233	3
3. Palatka, Putnam Co. FL	7	1.00	3.857	0.696	0.446	0.379	1
4. Miami, Dade County, Florida	9	1.00	3.286	0.551	0.307	0.459	0
5. Caribbean	15	1.00	11.000	0.794	0.543	0.324	53

n = sample size; P = proportion of polymorphic loci; A = alleles per locus; H_o = observed heterozygosity, H_e = expected heterozygosity, f = inbreeding coefficient.

preliminary distance base UPGMA tree (Fig. 29-3) using the coancestry coefficient (θ of Reynolds et al., 1983) with the program GDA (Lewis & Zaykin, 2001). All of the Caribbean collections were treated (artificially) as a single population (5) for this purpose. Populations 1, 2, and 4 represent the only "real" populations; population 3 (Palatka) was supposedly generated from hand pollination of several plants collected from the wild (but see later). All 7 primers successfully amplified across the 4 Florida populations and all of the Caribbean collections, and all 7 are polymorphic across all of the test populations. We also tested our primers once on single samples of several *Zamia* spp. from outside of the Caribbean. Six out of 7 successfully amplified the loci from *Z. fairchildiana* L. D. Gómez (Panama), 4/7 from *Z. poeppigiana* Mart. & Eichler (Ecuador), and 3/7 from *Z. neurophyllidia* D.W. Stev. (Panama). At least a subset of our primers will be able to amplify loci from *Zamia* spp. outside of the *Z. pumila* complex.

The Caribbean collections as a group showed a much higher level of heterozygosity than any of the Florida populations, a higher number of alleles per locus, and an extraordinary number of unique alleles (Table 29-2). Inbreeding was lowest in the St. Augustine population, and highest in the smaller of the two Miami populations. It is premature to assign too much significance to these statistics, as this time as sample size was small except in the St. Augustine population.

We also performed a preliminary assessment of population structure (Fig. 29-4) using STRUCTURE 2.1 (Pritchard et al., 2000). We ran 1 million MCMC simulations after a burn in of 100,000 generations with the following parameters: $K=5$, GENS-BACK=2, MIGRPRIOR=0.05, an admixture model, with α values inferred from each population, and a constant λ (1.0) for the allele frequencies parameter. All individuals from population 1 from St. Augustine had greater than 98% posterior probabilities (PP) of belonging to that population, as did all but one individual from population 2 (Miami). The St. Augustine population was so homogeneous that we suspected it might represent half-siblings from a single female cone (and, in fact, we later determined that they were). None of the plants from the Palatka population (3) had significant PP of membership in that population, and this group was probably generated with pollen from diverse male plants, most not originating from that population.

PP values of the second Miami population (4) ranged from 0.541 to 0.987. The artificially assembled Caribbean "population" (5) is clearly quite distinct from the Florida material, except for one individual from the Dominican Republic that had a high PP of belonging to Miami population 1, and three other individuals that likely represent another cluster not accounted for in our prior assumptions. Even with small sample sizes and a low number of loci, the resolving power of SSRs in *Zamia* appears quite substantial (Figs. 29-3 and 29-4).

Application of Microsatellite DNA to *Zamia*

Our overall goal is to document the patterns of microsatellite DNA variation across populations of the *Z. pumila* complex throughout its range, infer population structure and biogeographic and demographic history of the complex, and attempt to understand the processes of speciation within the group. Our specific goals target three broad categories: biogeography, cycad biology, and conservation biology–ethnobotany.

Biogeographical Hypotheses

We aim to investigate if major environmental events in the Caribbean have left traces in the genetic structure of populations of *Zamia*. Our two major hypotheses are: (1) There is congruence between patterns of microsatellite variation and sea level history in the West Indies and southeastern United States. We expect that the high-elevation islands of Cuba, Hispaniola, and Puerto Rico will harbor most of the genetic variation in the complex. Jamaican populations should have limited genetic variation, as most of this island was submerged during the Miocene. (2) There is congruence between population genetic structure and the geology of the Antilles. For instance, eastern Cuba, north-central Hispaniola, and Puerto Rico were connected until the Early Miocene. We therefore expect that *Zamia* populations occurring in these areas will have greater genetic identity and similar population structure, compared to populations that developed in isolation. We also will investigate if populations of the two species that occur on serpentine soils are genetically related. Correlation of allele frequencies with geographic patterns will be assessed with the program SPAIDA (Pálsson, 2004). SPAIDA calculates two estimates of spatial autocorrelation, Moran's *I* and Geary's *c* (Bertorelle & Barbujani, 1995), first by assuming the infinite allele model, and second by assuming a stepwise mutational model.

Hypotheses Relating to *Zamia* Biology

We aim to investigate if patterns of genetic variation among populations of *Zamia* included in the proposal are concordant with their known pollination and dispersal strategies, as well as the described patterns of morphological variation among them. We have two specific hypotheses. First, we expect patterns of microsatellite variation to confirm a low level of gene flow between populations. We hypothesize that *Zamia* in the Caribbean will exhibit conformance to the stepping stone model, given the limited gene flow exhibited by most cycad species (Ellstrand et al., 1990; Walters & Decker-Walters, 1991; Yang & Meerow, 1996; Lin et al., 2000; Huang et al., 2001, 2004). The program BAYESASS+ (Wilson & Rannalla, 2003), which uses the Monte Carlo Markov Chain (MCMC) algorithm to estimate recent migration rates between populations, each individual's immigrant ancestry, and the generation in which immigration occurred. Our second hypothesis is that morphological differences between population groups will be congruent with patterns of genetic variation. Evidence in support of this hypothesis will be the presence of private alleles across sets of morphologically coherent populations as well as genetic distance clustering tree topologies that support a priori morphological groupings. These groups are effectively the taxa recognized by Caputo et al. (2004), Géigel (2003), and Stevenson (1987a, b). An accessory expectation is that populations on different islands that are assignable to the same phenotypic group based on morphology will have higher genetic identity than populations on the same island with divergent morphology. The same data (private alleles and distance based clustering) will address this point as well. The null hypothesis is that the *Z. pumila* complex has evolved convergently on each island in response to ecological selection pressures, or that morphologically diverse populations on a single island represent independent introductions (see previous and following sections).

Hypotheses Germane to the Conservation Biology of and Anthropogenic Influences on the *Z. pumila* Complex

Patterns of genetic variation from our study need to be interpreted not only within natural history and environmental biology frameworks but also bearing in mind the great influence that humans have had on Caribbean ecosystems and on *Zamia* spp. itself. Historically, *Zamia* has been used throughout the West Indies as a source of starch by the Arawak and other Amerindians, and there is good evidence that *Zamia* was used in Florida both before and after the arrival of the Seminoles (Sturtevant, 1969). This could explain why *Z. integrifolia* (the Florida component of the *Z. pumila* complex), particularly in western Florida, is often found growing on shell middens (Small, 1921). The Arawaks traveled from island to island, perhaps transporting plants to use as food source. Within this context, we expect to find evidence of genetic erosion and bottlenecks in populations of *Zamia* that occur in habitats fragmented by human activity, and in areas of Florida and Hispaniola where there was industrial/commercial use of plants for starch exploitation and the horticulture trade. We also expect to find this pattern in the few remaining populations of the three rarest species of the complex. A drastic bottleneck results in a period of enhanced genetic drift (Charlesworth et al., 2003), and should be evident in a pattern of lineage coalescence at the bottleneck. We might thus expect to see an excess of rare alleles in populations that experienced the bottleneck (Tajima, 1989). We already see this with the Caribbean taxa (Table 29-1; though this number is inflated by the low subpopulation sampling). Fluctuations in population size and patterns of gene flow (as may well have occurred in Florida and the Caribbean during glacial and interglacial periods) can obscure the discrimination of potential bottlenecks due to human activities or natural processes. Bottlenecks can also result in high levels of linkage disequilibrium (Stumpf & Goldstein, 2003), another means of testing this hypothesis. The likelihood of population bottlenecks will be tested with the program BOTTLENECK 1.2.02 (Cornuet & Luikart, 1996). The program computes for each population sample and for each locus the distribution of the gene diversity expected from the observed number of alleles (*k*), given the sample size (*n*) under the assumption of mutation-drift equilibrium. This distribution is obtained through simulating the coalescent process of *n* genes under three possible mutation models—the IAM, the TPM, and the SMM. This enables the computation of the average (Hexp) which is compared to the observed gene diversity (Nei, 1987) to establish whether there is a gene diversity excess or deficit at this locus. In addition, the standard deviation (SD) of the mutation-drift equilibrium distribution of the gene diversity is used to compute the standardized difference for each locus (Hobs-Hexp)/SD). The distribution obtained through simulation enables also the computation of a *P*-value for the observed gene diversity. Once all loci available in a population sample have been processed, the three statistical tests are performed for each mutation model as explained in Cornuet and Luikart (1996) and the allele frequency distribution is established to see whether it is approximately L-shaped (as expected under mutation-drift equilibrium) or not (recent bottlenecks provoke a mode shift).

The related hypothesis that Native American migrations among islands and mainland Florida influenced the current patterns of genetic variation for the *Z. pumila* complex is more difficult to address, given that little is known about the routes, dates, and inventory of these putative movements. Bayesian methodologies that test the goodness of fit of each individual to their assigned population based on allele frequencies are excellent tools to identify introgression of alleles originating from distant populations. STRUCTURE 2.1 (Pritchard et al., 2000) and GeneClass 2.0 (Piry et al., 2004) will be used for this purpose.

Wright (1943) coined the term "isolation by distance" to describe the phenomenon of local genetic differences due to dispersal restricted by geographic isolation. Two models have been applied to these patterns of population subdivision. The island model (Wright, 1943) describes a population of *s* colonies of equal size with equal migration probabilities for each pair of colonies (Hudson, 1990; Nei & Takahata, 1993). In an island model, migration occurs at random among a group of equivalent populations; individuals migrate from population to population at some constant rate. Allele frequencies should converge on the average frequency of the set.

Kimura (1953) formulated the stepping stone model, with subsequent refinement (Kimura & Weiss, 1964; Weiss & Kimura, 1965; Malécot, 1967, 1969, 1975; Maruyama,

1970, 1971, 1972; Nagylaki, 1974; Crow & Aoki, 1984; Strobeck, 1987; Slatkin & Barton, 1989; Slatkin, 1991, 1993; Hey, 1991). The stepping-stone model describes island migration along a linear set of populations, where each population receives migrants only from neighboring populations. In the stepping stone pattern, closer populations would be more similar genetically because they can move only to close populations (Takahata, 1983). In an island model, all populations would be equally similar, while pairwise differentiation would be observed in a stepping-stone pattern. We hypothesize that *Zamia* in the Caribbean will exhibit conformance to the stepping stone model, given the limited gene flow exhibited by most cycad species (Ellstrand et al., 1990; Walters & Decker-Walters, 1991; Yang & Meerow, 1996; Lin et al., 2000; Huang et al., 2001, 2004).

Cycads are considered threatened and endangered plants wherever they occur in the world (all are listed on CITES Appendix II). We believe that the results of the proposed research will have important application to the conservation biology of the genus, and will be extremely useful for developing conservation priorities for *Zamia* in the Caribbean and Florida. It is also likely than a sizable number of our primers will be able to be used for such purposes in other regions of diversity for the genus in America. Cycads have fascinated the plant-oriented general public for years because of their antiquity, novelty, and association with dinosaurs. They are also valued ornamentals in the subtropics, tropics, and in conservatories. As the first extensive SSR project on a Caribbean plant species, our project may become a model for future plant biodiversity and conservation biology studies in the West Indies.

Acknowledgments

This chapter is contribution number 90 from the Tropical Biology Program of Florida International University. We thank Timothy Motley for his careful reading of the manuscript.

Literature Cited

Adams, C. D. 1997. Regional overview: Caribbean islands. Pp. 233–258 *in* S. D. Davis, V. H. Heywood, O. Herrera-MacBryde, J. Villa-Lobos, & A. C. Hamilton (eds.), Centres of plant diversity, vol. 3: The Americas. IUCN Publication Unit, Cambridge.

Andersen, B. G. & H. W. Borns. 1994. The ice age world. Scandinavian University Press, Oslo.

Angers, B., A. Estoup & P. Jarne. 2000. Microsatellite size homoplasy, SSCP, and population structure: a case study in the freshwater snail *Bulinus truncates.* Mol. Biol. Evol. 17: 1926–1932.

Aubrey, D. G., K. O. Emery & E. Uchipi. 1988. Changing coastal level of South America and the Caribbean region from tide-gauge records. Tectonophysics 154: 269–284.

Bailey, R. G. 1995. Description of the ecoregions of the United States. Ed. 2. USDA Forest Service, Washington, DC.

Balduzi, A., P. Deluca & S. Sabato. 1981–1982. A phytogeographical approach to the New world cycads. Delphinoa 23–24: 185–202.

Balloux, F. & J. Goudet. 2002. Statistical properties of population differentiation estimators under stepwise mutation in a finite island model. Mol. Ecol. 11: 771–783.

——— & N. Lugon-Moulin. 2002. The estimation of population differentiation with microsatellite markers. Mol. Ecol. 11: 155–165.

Barton, N. H. & G. M. Hewitt. 1985. Analysis of hybrid zones. Annu. Rev. Ecol. Syst. 16: 113–148.

Beaumont, M. A. & M. W. Bruford. 1998. Microsatellites in conservation genetics. Pp. 165–182 *in* D. B. Goldstein & C. Schlötterer (eds.), Microsatellites: evolution and applications. Oxford University Press, New York.

Bertorelle, G. & G. Barbujani. 1995. Analysis of DNA diversity by spatial autocorrelation. Genetics 140: 811–819.

Borhidi, A. 1996. Phytogeography and vegetation ecology of Cuba. Revised Ed. 2. Akademiai Kiado, Budapest.

Bowcock, A. M., A. Ruiz-Linares, J. Tomfohrde, E. Minch, J. Kidd & L. L. Cavalli-Sforza. 1994. High resolution of human evolutionary trees with polymorphic microsatellites. Nature 368: 455–457.

Brossart, J. L. & D. P. Prowell. 1998. Genetic estimates of population structure and gene flow: limitations, lessons and new directions. TREE 13: 202–206.

Buskirk, R. E. 1985. Zoogeographic patterns and tectonic history of Jamaica and the northern Caribbean. J. Biog. 12: 445–461.

Byrne, M. & S. H. James. 1991. Genetic diversity in the cycad *Macrozamia riedlei.* Heredity 87: 35–39.

Caputo, P., S. Cozzolino, P. De Luca, A. Moretti & D. Stevenson. 2004. Molecular phylogeny of *Zamia*. Pp. 149–158 *in* T. Walters & R. Osborne (eds.), Cycad classification: concepts and recommendations. CABI, Oxford.

Cavalli-Sforza, L. L. & A. W. F. Edwards. 1967. Phylogenetic analysis: models and estimation procedures. Am. J. Hum. Genet. 19: 233–257.

Charlesworth, B., D. Charlesworth & N. H. Barton. 2003. The effects of genetic and geographic structure on neutral variation. Annu. Rev. Ecol. Syst. 34: 99–125.

Chase, M., R. Kesseli & K. Bawa. 1996. Microsatellite markers for population and conservation genetics of tropical trees. Am. J. Bot. 83: 51–57.

Coates, A. G. 1997. The forging of Central America. Pp. 1–37 *in* A. G. Coates (ed.), Central America, a natural & cultural history. Yale University Press, New Haven, CT.

Collins, L. S. 1996. Environmental changes in Caribbean shallow waters relative to the closing of the Tropical American seaway. Pp. 130–167 *in* J. B. C. Jackson, A. F. Budd, & A. G. Coates (eds.), Evolution and environment in tropical America. The University of Chicago Press, Chicago, IL.

———, A. F. Budd & A. G. Coates. 1996. Earliest evolution associated with closure of the Tropical American Seaway. Proc. Natl. Acad. Sci. USA 93: 6969–6072.

Cornuet, J. M. & G. Luikart. 1996. Description and power analysis of two tests for detecting recent population bottlenecks from allele frequency data. Genetics 144: 2001–2014.

Cox, C. B. 2001. The biogeographical regions reconsidered. J. Biogeogr. 28: 511–523.

Crow, J. F. & K. Aoki. 1984. Group selection for a polygenic trait: estimating the degree of population subdivision. Proc. Natl. Acad. Sci. USA 81: 6073–6077.

De Candolle, A. P. 1868. Cycadeae. Pp. 361–521 *in* Prodromus systema natura et regnum vegetabile 16(2). Icteor Massen, Paris.

Di Rienzo, A, A. C. Peterson, J. C. Garza, A. M. Valdès, M. Slatkin & N. B. Freimer 1994. Mutational processes of simple-sequence repeat loci in human populations. Proc. Natl. Acad. Sci. USA 91: 3166–3170.

Donaldson, J. S., I. Nanni & B. de Wet. 1995. The role of insects in the pollination of *Encephalartos cycadifolius*. Pp. 423–434 *in* P. Voster (ed.), Proceedings of the thirdinternational conference of cycad biology. Cycad Society of South Africa, Stellenbosch.

Donnelly, M. J. & H. Townson 2000. Evidence for extensive genetic differentiation among populations of the malaria vector *Anopheles arabiensis*. Eastern Africa. Insect Mol. Biol. 9: 357–367.

Draper, G. & A. Barros. 1994. Cuba. Pp. 65–86 *in* S. K. Donovan & T. A. Jackson (eds.), Caribbean geology, an introduction. The University of the West Indies Publishers Association, Kingston, Jamaica.

Droxler A. W., K. C. Burke, A. C. Cunningham, A. C. Hine, E. Rosencrantz, D. S. Duncan, P. Hallock & E. Robinson. 1998. Caribbean constrains on circulation between Atlantic and Pacific Oceans over the past 40 million years. Pp. 160–191 *in* T. J. Crowley & K. C. Burke (eds.), Tectonic boundary conditions for climatic reconstructions. Oxford University Change, New York.

Eckenwalder, J. 1980a. Taxonomy of the West Indian cycads. J. Arnold Arb. 61: 701–702.

———. 1980b. Divergent dispersal methods in the West Indian cycad, *Zamia pumila* L. Biotropica 12: 79–80.

Edwards K. J., J. H. A. Baker, A. Daly, C. Jones & A Karp. 1996. Microsatellite libraries enriched for several microsatellite sequences in plants. BioTechniques 20: 758–760.

Efron, B. 1979. Bootstrap methods: another look at the jackknife. Ann. Statist. 7: 1–26.

Eizirik, E., J. H. Kim, M. Menotti-Raymond, P. G. Crawshaw, S. J. O'Brien, & W. E. Johnson. 2001. Phylogeography, population history and conservation genetics of jaguars (*Panthera onca*, Mammalia, Felidae). Mol. Ecol. 10: 65–79.

Eldridge, M. D. B., J. M. King, A. K. Loupis, P. B. S. Spencer, A. C. Taylor, L. C. Pope & G. P. Hall. 1999. Unprecedented low levels of genetic variation and inbreeding depression in an island population of the black-footed rock-wallaby. Cons. Biol. 13: 531–541.

Ellegren, H. 2000. Microsatellite mutations in the germline: implications for evolutionary inference. Trends Genet. 16: 551–558.

———, C. R. Primmer & B. C. Sheldon. 1995. Microsatellite "evolution": directionality or bias? Nat. Genet. 11: 360–362.

Ellstrand, N., R. Ornduff & J. M Clegg. 1990. Genetic structure of the Australian cycad, *Macrozamia communis* (Zamiaceae). Am. J. Bot. 77: 677–681.

Estoup, A. & B. Angers. 1998. Microsatellites and minisatellites for molecular ecology: theoretical and empirical considerations. Pp. 55–86 *in* G. Carvalho (ed.), Advances in molecular ecology. NATO Press, Amsterdam.

——— & J.-M. Cornuet. 1999. Microsatellite evolution: inferences from population data. Pp. 49–65 *in* D. Goldstein & C. Schlötterer (eds.), Microsatellites: evolution and applications. Oxford University Press, Oxford.

———, P. Jarne & J-M. Cornuet. 2002. Homoplasy and mutation model at microsatellite loci and their consequences for population genetics analysis. Mol. Ecol. 11: 1591–1604.

Frankham, R. & K. Ralls. 1998. Conservation biology: inbreeding leads to extinction. Nature 392: 441–442.

Fritsch, P. W. & T. D. McDowell. 2003. Biogeography and phylogeny of Caribbean plant - Introduction. Syst. Bot. 28: 376–377.

Gamerman, D. 1997. Markov Chain Monte Carlo: stochastic simulation for Bayesian Inference. Chapman and Hall, New York.

Géigel, L. G. 2003. Zamiaceae. Flora de Cuba 8(4): 3–22.

Gerlach, G. & K. F. Musolf 2000. Fragmentation of landscape as a cause for genetic subdivision in bank voles. Cons. Biol. 14: 1066–1074.

Goldstein, D. B., A. Ruiz Linares, L. L. Cavalli-Sforza & M. W. Feldman. 1995. An evaluation of genetic distances for use with microsatellite loci. Genetics 139: 463–471.

——— & D. D. Pollock 1997. Launching microsatellites: a review of mutation processes and methods of phylogenetic inference. J. Hered. 88: 335–342.

Good, R. 1947. The geography of flowering plants. Longmans, London.

Graham, A. 2003a. Geohistory models and Cenozoic paleoenvironments of the Caribbean region. Syst. Bot. 28: 378–386.

———. 2003b. Historical phytogeography of the Greater Antilles. Brittonia 55: 357–383.

Griffiths, R. C. & S. Tavare 1994. Simulating probability distributions. Theor. Pop. Biol. 46: 131–159.

Gutiérrez-Domech, R. & M. Rivero-Glean. 1997. Minigeografía de Cuba. Editorial Científico-Técnica, Havana.

Harding, R. M., A. J. Boyce & J. B. Clegg. 1992. The evolution of tandemly repetitive DNA: recombination rules. Genetics 132: 847–859.

Hastings, W. K. 1970. Monte Carlo sampling methods using Markov chains and their applications. Biometrika 57: 97–109.

Hearne, C. M., S. Ghosh & J. A. Todd. 1992. Microsatellites for linkage analysis of genetic traits. Trends Genet. 8: 288–294.

Hearty, P. J. & D. S. Kaufman. 2000. Whole-rock aminostratigraphy and Quaternary sea-level history of the Bahamas. Quaternary Res. 54: 163–173.

Hedges, S. B. 2001. Biogeography of the West Indies: an overview. Pp. 15–33 *in* C. A. Woods & F. E. Sergile (eds.), Biogeography of the West Indies: patterns and perspectives. 2nd ed. CRC Press, Boca Raton, FL.

Hedges, S. B., C. A. Hass & L. R. Maxson. 1992. Caribbean biogeography: molecular evidence for dispersal in West Indian terrestrial vertebrates. Proc. Natl. Acad. Sci. USA 89: 1909–1913.

Hey, J. 1991. A multi-dimensional coalescent process applied to multi-allelic selection models and migration models. Theor. Pop. Biol. 39: 30–48.

Higgins, K. & M. Lynch. 2001. Metapopulation extinction caused by mutation accumulation. Proc. Natl. Acad. Sci. USA 98: 2928–2933.

Huang, S., Y. C. Chiang, B. A. Schaal, C. H. Chou & T.Y. Chaing. 2001. Organelle DNA phylogeography of *Cycas taitungensis*, a relict species in Taiwan. Mol. Ecol. 10: 2669–2681.

———, H-T. Hsieh, K. Fang & Y-C. Chiang. 2004. Patterns of genetic variation and demography of *Cycas taitungensis* in Taiwan. Bot. Rev. 70: 86–92.

Hudson, R. R. 1990. Gene genealogies and the coalescent process. Oxford Surv. Evol. Biol. 7: 1–44.

Iturralde-Vinent, M. A. & R.D.E. MacPhee. 1999. Paleogeography of the Caribbean region: implications for Cenozoic biogeography. Bull. Am. Mus. Nat. Hist. 238: 1–95.

Jones, C. J., K. J. Edwards, S. Castaglione, M. O. Windield, F. Sala, C. van de Wiel, G. Bredemeijer, B. Vosman, M. Mathes, A. Daly, R. Brettschneider, P. Bettini, M. Buiatti, E. Maestri, A. Malcevschi, N. Marmiroli, R. Aert, G. Volckaert, J. Rueda, R. Linacero, A. Vasquez & A. Karp. 1997. Reproducibility testing of RAPD, AFLP and SSR markers in plants by a network of European laboratories. Mol. Breed. 3: 381–390.

Kaplan, N. L., T. Darden & R. R. Hudson. 1988. The coalescent process in models with selection. Genetics 120: 819–829.

Kimura, M. 1953. "Stepping stone" model of population. Ann. Rep. Nat. Inst. Genet. Japan 3: 62–63.

——— & G. H. Weiss. 1964. The stepping stone model of population structure and the decrease of genetic correlation with distance. Genetics 49: 561–576.

——— & J. F. Crow. 1964. The number of alleles that can be maintained in a finite populations. Genetics 49: 725–738.

——— & T. Otha. 1978. Stepwise mutation model and distribution of allelic frequencies in a finite populations. Proc. Natl. Acad. Sci. USA 75: 2868–2872.

Kindler, P. & P. J. Hearty. 1997. Geology of the Bahamas: architecture of Bahamian Islands. Dev. Sediment. 54: 141–160.

Kingman, J.F.C. 1982. The coalescent. Stoch. Proc. Appl. 13: 235–248.

———. 2000. Origins of the coalescent: 1974–1982. Genetics 156: 1461–1463.

Lewis, J. F. & G. Draper. 1990. Geology and tectonic evolution of the northern Caribbean margin. Pp. 77–140 *in* G. Dengo & J. E. Case (eds.), The geology of North America, vol. H: the Caribbean region. Geological Society of America, Boulder, CO.

Lewis, P. O. & D. Zaykin. 2001. Genetic data analysis: computer program for the analysis of allelic data. Version 1.0 (d16c). Free program distributed by the authors over the Internet from http://hydrodictyon.eeb.uconn.edu/people/plewis/software.php.

Li, Y. C., A. B. Korol, T. Fahima, A. Beiles & E. Nevo. 2002. Microsatellites: genomic distribution, putative functions and mutational mechanisms: a review. Mol. Ecol. 11: 2453–2465.

Lin, T.-P., Y.-C. Sun, H.-C. Lo & Y.-P. Cheng. 2000. Low genetic diversity of *Cycas taitungensis* (Cycadaceae), an endemic species in Taiwan, revealed by allozyme analysis. Taiwan J. Forest Sci. 15: 13–19.

Long, R. W. & O. Lakela. 1971. A flora of tropical Florida, a manual of the seed plants and ferns of Southern Peninsular Florida. University of Miami Press, Miami, FL.

Lugon-Moulin, N., H. Brunner, A. Wyttenbach, J. Hausser & J. Goudet. 1999. Hierarchical analyses of genetic differentiation in a hybrid zone of *Sorex araneus* (Instectivora: Soricidae). Mol. Ecol. 8: 419–431.

Luikart, G. & P. R. England. 1999. Statistical analysis of microsatellite DNA. TREE 14: 253–256.

Madsen, T., B. Stille & R. Shine. 1996. Inbreeding depression in an isolated population of adders *Vipera berus*. Biol. Cons. 75: 113–118.

Makova, K. D., A. Nekrutenko & R. J. Baker. 2000. Evolution of microsatellite alleles in four species of mice (genus Apodemus). J. Mol. Evol. 51: 166–172.

Malécot, G. 1967. Identical loci and relationship. Proc. Fifth Berkeley Symp. Math. Statist. Probab. 4: 317–332. University of California Press, Berkeley.

———. 1969. The mathematics of heredity. Freeman, San Francisco, CA.

———. 1975. Heterozygosity and relationship in regularly subdivided populations. Theor. Pop. Biol. 8: 212–241.

Maruyama, T. 1970. Effective number of alleles in a subdivided population. Theor. Pop. Biol. 1: 273–306.

———. 1971. Analysis of population structure. II. Two dimensional stepping stone models of finite length and other geographically structured populations. Ann. Hum. Genet. Lond. 35: 179–196.

———. 1972. The rate of decrease of genetic variability in a two dimensional continuous population of finite size. Genetics 70: 639–651.

McDonald, D. B. & W. K. Potts. 1997. DNA microsatellites as genetic markers at several scales. Pp. 29–49 *in* D. P. Mindell (ed.), Avian molecular evolution and systematics. Academic Press, New York.

Michalakis, Y. & M. Veuille. 1996. Length variation of CAG/CAA trinucleotide repeats in natural populations of *Drosophila melanogaster* and its relation to recombination rate. Genetics 143: 1713–1725.

Morgante, M. & A. M. Olivieri. 1993. PCR-amplified microsatellites as markers in plant genetics. Plant J. 3: 175–182.

Muñiz, R. & A. Barrera. 1969. *Rhopalotria dimidiate* Chevrolat, 1878: studio morfologico del adulto y descripción de la larva (Ins. Col. Curcul.: Oxycoryninae). Rev. Soc. Mex. His. Nat. 30: 205–222.

Myers, N., R. A. Mittermeier, C. G. Mittermeier, G. A. B. de Fonseca & J. Kent. 2000. Biodiversity hotspots for conservation priorities. Nature 403: 853–858.

Nagylaki, T. 1974. The decay of genetic variability in geographically structured populations. Proc. Natl. Acad. Sci., USA 71: 2932–2936.

Nauta, M. J. & F. J. Weissing. 1996. Constraints on allele size at microsatellite loci, implications for genetic differentiation. Genetics 143: 1021–1032.

Nei, M. 1972. Genetic distance between populations. Am. Nat. 106: 283–291.

———. 1973. Analysis of gene diversity in subdivided populations. Proc. Natl. Acad. Sci. USA 70: 3321–3323.

———. 1987. Molecular evolutionary genetics. Columbia University Press, New York.

——— & N. Takahata. 1993. Effective population size, genetic diversity, and coalescence time in subdivided populations. J. Mol. Evol. 37: 240–244.

Nielsen, R. 1997. A likelihood approach to population samples of microsatellite alleles. Genetics 146: 711–716.

——— & P. J. Palsbøll. 1999. Single-locus tests of microsatellite evolution: multi-step mutations and constraints on allele size. Mol. Phylogen. Evol. 11: 477–484.

——— & J. Wakely. 2001. Distinguishing migration from isolation: a Markov chain Monte Carlo approach. Genetics 158: 885–896.

———, J. L. Mountain, J. P. Huelsenbeck & M. Slatkin. 1998. Maximum-Likelihood estimation of population divergence times and population phylogeny in models without mutation. Evolution 52: 669–677.

Norstog, K., D. Stevenson & K. Niklas. 1986. The role of weevils in the pollination of *Zamia furfuracea* L.f. (Zamiaceae). Biotropica 18: 300–306.

——— & T. J. Nicholls. 1997. The biology of the cycads. Cornell University Press, Ithaca, NY.

Nunn, P. D. 1994. Oceanic islands. Blackwell, London.

Orti, G., D. E. Pears & J. C. Avise. 1997. Phylogenetic assessment of length variation at a microsatellite locus. Proc. Natl. Acad. Sci. USA 94: 10745–10749.

Pálsson, S. 2000. Microsatellite variation in *Daphnia pulex* from both sides of the Baltic Sea. Mol. Ecol. 9: 1075–1088.

———. 2004. Isolation by distance, based on microsatellite data, tested with spatial autocorrelation (SPAIDA) and assignment test (SPASSIGN). Mol. Ecol. Notes 4: 143–145.

Petren, K., B. R. Grant & P. R. Grant. 1999. A phylogeny of Darwin's finches based on microsatellite DNA length variation. Proc. R. Soc. Lond. B 266: 321–329.

Pindell, J. L. & S. F. Barrett. 1990. Geological evolution of the Caribbean region: a plate-tectonic perspective. Pp. 405–432 in G. Dengo & J. E. Case (eds.), The geology of North America, vol. H: the Caribbean region. Geological Society of America, Boulder, CO.

Piry, S, A. Alapetite, J. M. Cornuet, D. Paetkau, L. Baudouin & A. Estoup. 2004. GeneClass2: a software for genetic assignment and first-generation migrant detection. J. Hered. 95: 536–539.

Pollack, D. D., A. Bergman, M. W. Feldman & D. B. Goldstein. 1998. Microsatellite behavior with range constraints: parameter estimation and improved distances for use in phylogenetic reconstruction. Theor. Pop. Biol. 53: 256–271.

Powell, W., G. C. Machray & J. A. Provan. 1996a. Polymorphism revealed by simple sequence repeats. Trends Pl. Sci. 1: 215–222.

———, M. Morgante, C. Andre, M. Hanafey, J. Vogel, S. Tingey & A. Rafalski. 1996b. The comparison of RFLP, RAPD, AFLP and SSR (microsatellite) markers for germplasm analysis. Mol. Breed. 2: 225–238.

Primmer, C. R. & H. Ellegren. 1998. Patterns of molecular evolution in avian microsatellites. Mol. Biol. Evol. 15: 997–1008.

Pritchard, J. K., M. Stephens & P. Donnelly. 2000. Inference of population structure using multilocus genotype data. Genetics 155: 945–959.

Randazzo, A. F. 1997. The sedumentary platform of Florida: Mesozoic to Cenozoic. Pp. 39–56 *in* A. F. Randazzo & D. S. Jones (eds.), The geology of Florida. University Press of Florida, Gainesville, FL.

Reynolds, J., B. S. Weir & C. C. Cockerham. 1983. Estimation of the coancestry coefficient: basis for a short-term genetic distance. Genetics 105: 767–779.

Rivas-Martínez, S., D. Sánchez-Mata & M. Costa. 1999 North American boreal and western temperate forest vegetation. Itinera Geobot. 12: 5–316.

Robinson, E. 1994. Jamaica. Pp. 111–127 in S. K. Donovan & T. A. Jackson (eds.), Caribbean geology, an introduction. The University of the West Indies Publishers Association, Kingston, Jamaica.

———. 1998. Late Cretaceous and Early Tertiary sedimentary rocks of the Central Inlier, Jamaica. J. Geol. Soc. Jamaica 24: 49–67.

Rogers, J. S. 1972. Measures of genetic similarity and genetic distance. Pp. 145–153 *in* Studies in genetics VII. University of Texas Publication 7213, Austin.

Rosenberg, N. A. & M. Nordborg. 2002. Genealogical trees, coalescent theory and the analysis of genetic polymorphisms. Nat. Rev. Genet. 3: 380–390.

Rossiter, S. J., G. Jones, R. D. Ransome & E. M. Barratt. 2000. Genetic variation and population structure in the endangered greater horseshoe bat *Rhinolophus ferrumequinum*. Mol. Ecol. 9: 1131–1135.

Russell, J. R., J. D. Fuller, M. Macaulay, B. G. Hatz, A. Jahoor, W. Powell & R. Waugh. 1997. Direct comparison of levels of genetic variation among barely accessions detected by RFLPs, AFLPs, SSRs and RAPDs. Theor. Appl. Genet. 95: 714–722.

Sabato, S. 1990. West Indian and South American cycads. Mem. NY Bot. Gard. 57: 173–185.

Saccheri, I., M. Kuussaari, M. Kankare, P. Vikman, W. Fortelius & I. Hanski. 1998. Inbreeding and extinction in a butterfly metapopulation. Nature 392: 491–494.

Santiago-Valentín, E. & R. G. Olsmtead. 2004. Historical biogeography of Caribbean plants: introduction to current knowledge and possibilities from a phylogenetic perspective. Taxon 53: 299–320.

Schlötterer, C. 2000. Evolutionary dynamics of microsatellite DNA. Chromosoma 109: 365–371.

Schmidt, W. 1997. Geomorphology and physiography of Florida. Pp. 1–12 *in* A. F. Randazzo & DS Jones (eds.), The geology of Florida. University Press of Florida, Gainesville.

Schultz, P. H. & S. D'Hondt. 1996. Cretaceous-Tertiary (Chicxulub) impact angle and its consequences. Geology 24: 963–967.

Scott, T. M. 1997. Miocene to Holocene history of Florida. Pp. 57–67 *in* A. F. Randazzo & DS Jones (eds.), The geology of Florida. University Press of Florida, Gainesville.

Sealey, N. 1992. Caribbean world, a complete geography. Cambridge University Press, Cambridge, UK.

Shriver, M. D., L. Jin, E. Boerwinkle, R. Deka, R. E. Ferrell & R. Chakraborty. 1995. A novel measure of genetic distance for highly polymorphic tandem repeat loci. Mol. Biol. Evol. 12: 914–920.

Slatkin, M. 1991. Inbreeding coefficients and coalescence times. Genet. Res. Cambridge 58: 167–175.

———. 1993. Isolation by distance in equilibrium and non-equilibrium populations. Evolution 47: 264–279.

———. 1995. A measure of population subdivision based on microsatellite allele frequencies. Genetics 139: 457–462.

——— . & N. H. Barton. 1989. A comparison of three indirect methods of estimating average levels of gene flow. Evolution 43: 1349–1368.

Small, J. 1921. Seminole bread-the conti, a history of the genus *Zamia* in Florida. J. NY Bot. Gard. 22: 121–137.

Stephens, J. C., D. A. Bilbert, N. Yuhki, & S. J. O'Brien. 1992. Estimation of heterozygosity for single-probe multilocus DNA fingerprints. Mol. Biol. Evol. 9:729–743.

Stevenson, D. 1987a. Comments on character distribution, taxonomy, and nomenclature with respect to the genus *Zamia* L. in the West Indies and Mexico. Encephalartos 9: 3–7.

———. 1987b. Again, the West Indian zamias. Fairchild Trop. Gard. Bull. 42(3): 23–27.

———. 1991. The Zamiaceae in the Southeastern United States. J. Arnold Arb. Suppl. Ser. 1: 367–383.

——— & S. Sabato. 1987. Typification of names in *Zamia* L. and *Aulacophyllum* Regel (Zamiaceae). Taxon 35: 134–144.

———, K. Norstog & P. Fawcett. 1998. Pollination biology of cycads. Pp. 277–294 *in* S. Owens & P. Rudall (eds.), Reproductive biology. Royal Botanic Gardens, Kew, UK.

Strobeck, C. 1987. Average number of nucleotide differences in a sample from a single subpopulation: a test for population subdivision. Genetics 117: 149–153.

Stumpf, M.P.H & D. B. Goldstein. 2003. Demography, recombination hotspot intensity, and the block structure of linkage disequilibrium. Curr. Biol. 13: 1–8.

Sturtevant, W. 1969. History and ethnography of some West Indian starches. Pp. 177–199 *in* P. Ucko & G. Dimbleby (eds.), The domestication and exploitation of plants and animals. The University of Chicago Press, Chicago, IL.

Sweet, A. R. & D. R. Bramen. 1992. The K-T boundary and contiguous strata in western Canada: interactions between paleoenvironments and palynological assemblages. Cretaceous Res. 13: 31–79.

Tachida, H. & M. Iizuka 1992. Persistence of repeated sequences that evolve by replication slippage. Genetics 131: 471–478.

Tajima, F. 1989. The effect of change in population size on DNA polymorphism. Genetics 123: 597–601.

Takahata, N. 1983. Gene identity and genetic differentiation of populations in the finite island model. Genetics 104: 497–512.

Takezaki, N. & M. Nei. 1996. Genetic distances and reconstruction of phylogenetic trees from microsatellite DNA. Genetics 144: 389–399.

Takhtajan, A. 1988. Floristic regions of the world (translated by Theodore J. Crovello with the assistance of the author and under the editorship of Arthur Cronquist). University of California Press, Berkeley.

Tang, W. 1987. Insect pollination in the cycad *Zamia pumila* (Zamiaceae). Am. J. Bot. 74: 90–99.

———. 1989. Seed dispersal in the cycad *Zamia pumila* in Florida. Can. J. Bot. 67: 2066–2070.

Tautz, D. & M. Renz. 1984. Simple sequences are ubiquitous repetitive components of eukaryotic genomes. Nucl. Acids Res. 12: 4127–4138.

———, M. Trick & G. A. Dover. 1986. Cryptic simplicity in DNA is a major source of genetic variation. Nature 322: 652–656.

Tavare, S. 1984. Line of descent and genealogical processes, and their application in population genetics models. Theor. Pop. Biol. 26: 119–164.

Tiedemann, R., O. Hardy, X. Vekemans & M. C. Milinkovitch. 2000. Higher impact of female than male migration on population structure in large mammals. Mol. Ecol. 9: 1159–1163.

Vajda. V., J. I. Raine & C. J. Hollis. 2001. Indication of global deforestation at the Cretaceous-Tertiary boundary by New Zealand fern spike. Science 294: 1700–1702.

Valdès, A. M., M. Slatkin & N. B. Freimer. 1993. Allele frequencies at microsatellite loci: the stepwise mutation model revisited. Genetics 133: 737–749.

Walters, T. & D. Decker-Walters. 1991. Patterns of allozyme diversity in the West Indian cycad *Zamia pumila* (Zamiaceae). Am. J. Bot. 78: 436–449.

Wang, Z., J. L. Weber, G. Zhong & S. D. Tanksley. 1994. Survey of plant short tandem DNA repeats. Theor. Appl. Genet. 88: 1–6.

Webb, S. D. 1990. Historical biogeography. Pp. 70–100 *in* Myers, R. L. & J. J. Ewel (eds.), Ecosystems of Florida, University of Central Florida Press, Orlando.

Weiss, G. H. & Kimura, M. 1965. A mathematical analysis of the stepping stone model of genetic correlation. J. Appl. Prob. 2: 129–149.

Wilson, G. A. & B. Rannala. 2003. Bayesian inference of recent migration rates using multilocus genotypes. Genetics 163: 1177–1191.

Wright, S. 1943. Isolation by distance. Genetics 28: 114–156.

———. 1951. The genetical structure of populations. Ann. Eugenics 15: 323–354.

———. 1977. Evolution and the genetics of population: experimental results and evolutionary deductions. The University of Chicago Press, Chicago.

Yang, S. L. & A. W. Meerow. 1996. The *Cycas pectinata* (Cycadaceae) complex: genetic structure and gene flow. Int. J. Plant Sci. 157: 468–483.

——— & ———. 1999. Genetic variation in Chinese cycad population. Pp. 175–186 *in* C-J. Chen (ed.), Biology and conservation of cycads. International Academic Publishers, Beijing.

Zhang, D. & C. M. Hewitt. 2003. Nuclear DNA analyses in genetic studies of populations: practice, problems and prospects. Mol. Ecol. 12: 563–584.

Zhivotovsky, L. A. & M. W. Feldman. 1995. Microsatellite variability and genetic distances. Proc. Natl. Acad. Sci. USA 92: 11549–11552.

———, D. B. Goldstein & M. W. Feldman. 2001. Genetic sampling error of distance $(\partial\mu)^2$ and variation in mutation rate among microsatelllite loci. Mol. Biol. Evol. 18: 2141–2145.

SYSTEMATICS
CHAPTER 30

Variación morfológica y genética del complejo Ceratozamia miqueliana H.A. Wendl. (Cycadales, Zamiaceae)

Miguel A. Pérez-Farrera, Andrew P. Vovides,
Dolores González, Luis Hernández Sandoval
y Mahinda Martínez

Abstract

The morphological and genetic variation of species in *Ceratozamia miqueliana* complex (*C. becerrae, C. miqueliana, C. zoquorum* and *C. euryphyllidia*), from southern Mexico, were analyzed and evaluated using morphometric and genetic population methods. Seventeen vegetative and six reproductive parameters of 48 individuals from four populations were analyzed by discriminant analysis. A total of 58 leaflets were collected from four species out of five populations and analyzed using a dominant marker (RAPD's). The morphometric results show clear differences between the four species (Wilks-$\lambda^2 = 373$, $p < 0.001$). However, the Nei genetic distance shows that the *C. miqueliana* species complex is clustered into three groups. Low levels of genetic diversity within and among populations ($H_s = 0.19$; $G_{st} = 0.12$) were found in this complex. The high genetic value of identity (0.95–0.97) and a correlation between genetic distance and altitude (Mantel test $r = 0.67$, $p < 0.05$) appear to show a recent divergence, probably late Miocene to Pleistocene, and a possible ecological speciation or allele differences not yet fixed between taxa.

Resumen

La variación morfológica y genética de las especies del complejo *Ceratozamia miqueliana* (*C. becerrae, C. miqueliana, C. zoquorum* y *C. euryphyllidia*) del sur mexicano fueron analizadas y evaluadas por medio de técnicas morfométricas y de genética de poblaciones para analizar su variación. Diecisiete parámetros vegetativos y seis reproductivos de 48 individuos de cuatro poblaciones fueron analizados por análisis de discriminantes. Un total de 58 foliolos fueron colectados de cuatro especies de cinco poblaciones y analizados mediante un marcador dominante (RAPD's). Los resultados mostraron diferencias claras entre las cuatro especies (Wilks-$\lambda^2 = 373$, $p < 0.001$). Sin embargo, la distancia genética de Nei muestra que el complejo *C. miqueliana* se agrega en tres grupos. Se encontró un nivel bajo de diversidad genética dentro y entre las poblaciones de este complejo ($H_s = 0.19$; $G_{st} = 0.12$). El índice elevado de identidad genética (0.95–0.97) y una correlación entre distancia genética y altitud (prueba de Mantel $r = 0.67$, $p < 0.05$) parecen mostrar una divergencia reciente, probablemente Mioceno tardío al Pleistoceno, y una posible especiación ecológica o diferencias alélicas todavía no fijadas entre los taxa.

Introducción

Ceratozamia es un género neotropical con alrededor de 21 especies, la mayoría se distribuyen al sur de México. Sin embargo, existe un amplio debate entre los taxónomos sobre el reconocimiento de algunas de especies (Whitelock, 2004). Esto se debe, en gran parte, porque las especies de este género están estrechamente relacionadas y tienen atributos morfológicos que en muchos casos no permiten distinguirlas (Whitelock, 2004). Adicionalmente algunos caracteres se pierden dentro de los ejemplares de herbario, aunado a que muchos de éstas son colectas estériles o colectas de plantas inmaduras (Vovides et al., 1983), lo que ha hecho difícil su estudio. Sumado a esto, algunas especies del género presentan ciclos reproductivos irregulares y largos, lo que ha provocado que su estudio sea muy lento.

El primero en reconocer complejos de especies dentro del género *Ceratozamia* fue Moretti et al. (1980). Recientemente se han propuesto siete grupos o complejos de especies basado en su morfología (Vovides et al., 2004a), dentro de los cuales se ha propuesto el complejo *Ceratozamia miqueliana* H.Wendl., que se caracterizan por tener foliolos anchos (ovado-oblanceolado) y que comprende a *Ceratozamia euryphyllidia, Ceratozamia miqueliana* y *Ceratozamia zoquorum.*

Ceratozamia miqueliana es una especie que fue descrita en 1854 por H.W. Wendland de una localidad no precisa en el sur de México (Vovides et al., 1983; Vovides et al., 2001). Mientras *Ceratozamia euryphyllidia* fue descrita del sur de Veracruz por Stevenson et al. (1986). *Ceratozamia zoquorum* fue descrita de las montañas del norte de Chiapas por Pérez-Farrera et al. (2001) y recientemente *C. becerrae* (Vovides et al., 2004b), de las montañas del sur de Tabasco y Norte de Chiapas. Desafortunadamente, el rango de la variación morfológica y genética de las especies del complejo no está bien comprendida.

La escasez de conocimiento sobre su variación ha contribuido a la confusión taxonómica dentro del género por la dificultad de encontrar caracteres morfológicos vegetativos, así como atributos reproductivos, estos últimos han sido difíciles de encontrarse en las poblaciones naturales, y también porque muchas poblaciones de estas especies se localizan en áreas inaccesibles que hacen difícil su estudio.

En este contexto, el objetivo general de este trabajo fue examinar la variación morfológica y genética dentro del complejo *Ceratozamia miqueliana*, utilizando estadística multivariada para probar la hipótesis de que el complejo *C. miqueliana* está compuesto por especies o entidades discretas de reciente formación y no por una sola especie polimórfica.

Material y Método

Variación Morfológica

Se analizaron cuatro poblaciones naturales en el sur de México, de cuatro especies del complejo *C. miqueliana*, dentro de su rango de distribución que comprende: las montañas

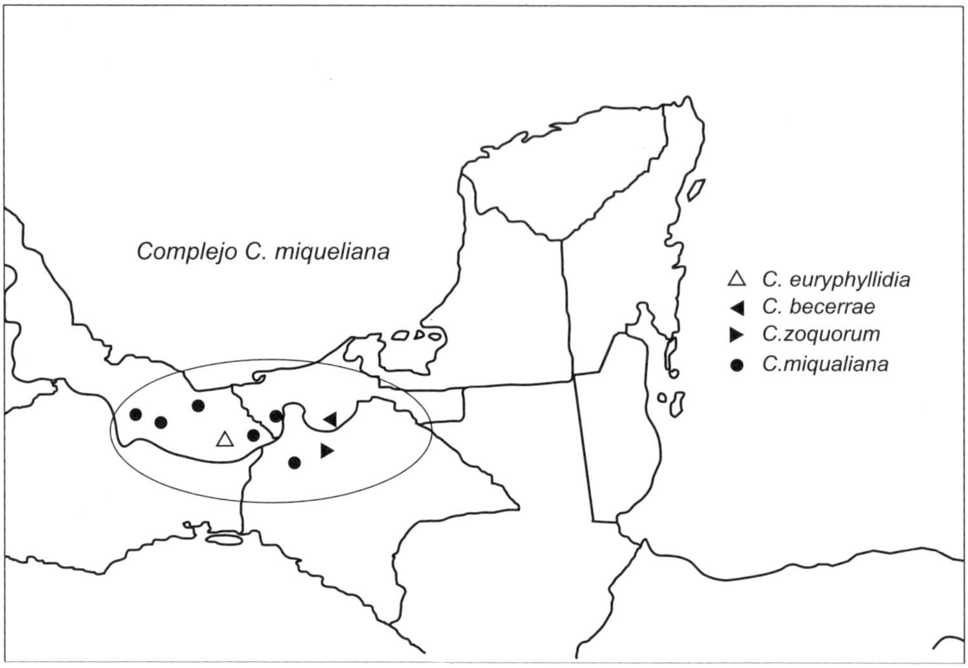

Figure 30-1. Distribución geográfica del complejo *Ceratozamia miqueliana*.

del norte de Chiapas, las montañas del sur de Tabasco, y las montañas del sur de Veracruz (Fig. 30-1). Se analizaron, 15 individuos de *C. miqueliana,* 10 de *C. euryphyllida,* 10 de *C. zoquorum y* 13 individuos maduros de *C. becerrae* y 23 variables morfológicas: 17 vegetativas y seis reproductivas (Tabla 30-1). Para cada medición se utilizó un flexómetro de 3 m y un vernier digital Trupper con resolución de 0.01 mm. Los datos fueron capturados en una hoja de cálculo Excel para Windows 2000.

Variación Genética

Colecta y selección de material de ejemplares

Se colectó foliolos de 10 individuos de la única población conocida de *Ceratozamia euryphyllidia* de Oaxaca, foliolos de 15 individuos de la única población de *C. becerrae* de Tabasco, foliolos de 15 individuos de una población de *C. miqueliana* de Chiapas y foliolos de 6 individuos de otra población de *C. miqueliana* en Veracruz y foliolos de 12 individuos

Tabla 30-1 Variables Morfométricas analizadas.

No	Abreviatura	Unidades y Unidades de Medición
1	LARGTR	Largo del tronco (cm)
2	DIAMET	Diámetro del tronco (cm)
3	NHOJA	Número de hojas por tronco
4	LARHOJ	Largo de hojas (cm)
5	ANCHHOJ	Ancho de hoja (cm)
6	LARGOPEC	Largo del pecíolo (cm)
7	DIAMPECI	Diámetro del pecíolo (mm)
8	LARGORAQ	Largo del raquis (cm)
9	NFOLIOL	Número de folíolos por fronda
10	LARGOFOL	Largo de folíolos (cm)
11	ANCHOFOL	Ancho de folíolos (cm)
12	NVENAS	Número de venas presentes en los folíolos
13	DISTVEN	Distancia entre venas (mm)
14	GROSORBASE	Es el grosor de la base de inserción del foliolo (mm)
15	GROSORMARG	Es el grosor del margen superior del foliolo (mm)
16	GOSORAPIC	Es el grosor de la parte apical del foliolo (mm)
17	ANCHOBASE	Es el ancho de inserción del foliolo (mm)
18	LARGOMICRO	Largo de microsporófilo (mm)
19	ANCHOMICRO	Ancho del microsporófilo (mm)
20	ANCHOMEGAS	Ancho de la cara distal megasporófilo (mm)
21	LARGOMEGAS	Largo de la cara distal megasporófilo (mm)
22	DIAMSEMI	Diámetro de la semilla (mm)
23	LARGSEMI	Largo de la semilla (mm)

de la única población de *C. zoquorum* en Chiapas. Se colectaron foliolos de ejemplares adultos, se tomaron de la parte media de la hoja y se verificó que los mismos estuvieran libres de manchas, líquenes, hongos u otros organismos.

Extracción ADN

El método de extracción y purificación de ADN que se usó fųé de acuerdo a González y Vovides (2002). Sin embargo, debido a las características coriáceas de los foliolos fue necesario eliminar la cutícula antes de la extracción. Una vez eliminada la cutícula, el tejido se mezcló con un buffer de extracción que contiene CTAB. El ADN se extrajo con una mezcla de cloroformo-alcohol isoamílico (24:1) y se precipitó con alcohol etílico absoluto. Por último, el precipitado se lavó con alcohol al 70% y se resuspendió en un buffer de TE (10 mM Tris-Cl, pH 8.0 1.0 mM EDTA).

Verificación y Purificación del ADN

La concentración de ADN de cada muestra fue determinada sobre un gel de agarosa al 1% en base a la concentración de una λ ADN de una concentración conocida. Previo a las amplificaciones, el ADN se purificó en un gel de bajo punto de fusión. La banda del ADN se cortó del gel y se extrajo con Geneclean (BIO 101) de acuerdo a las instrucciones del proveedor.

Amplificación

Para la amplificación se utilizó Pure Taq ™ Ready to go TM PCR Bead (Amersham Biosciences). En cada tubo se colocaron 5 µl de ADN de cada muestra con una concentración de 2.5 ηg/µl, 3 µl de primer a 10 ηM y 17 µl de agua destilada esterilizada para hacer un volumen total de la reacción de 25 µl. Se utilizó un termociclador Perkin-Elmer Thermal Cycler 480 (Norwalk, CT). Los parámetros del ciclo fueron: una desnaturalización inicial a 94° C por 2.5 minutos, seguido por la hibridación del primer al templado a 37 °C por 1 minuto y la extensión a 72°C por 2 minutos durante 44 ciclos para finalizar con una extensión por 7 minutos a 72 °C.

Se probaron 28 primers de los "kits" B y C de "Operon Technologies" (alameda, CA), de los cuales 5 fueron polimórficos. Los primers utilizados en la amplificación fueron: **B7** (5'-G-G-T-G-A-C-G-C-A-G-3'); **B11** (5'-G-T-A-G-A-C-C-C-G-T-3'); **C2** (5'-G-T-G-A-G-G-C-G-T-C-3'); **C3** (5'-G-G-G-G-G-T-C-T-T-T-3') y **C6** (5'-G-A-A-C-G-G-A-C-T-C-3'). Los productos amplificados fueron corridos en un gel de agarosa al 1% junto con el marcador de tamaño de 123 pb "ADN Ladder" manufacturado por Sigma.

Análisis

Variación Morfológica

La variación morfológica, se analizó mediante: a) análisis univariado de varianza. Este análisis se realizó para determinar el grado de variación morfológica existente entre las especies del complejo. Los análisis se realizaron utilizando un software estadístico JMP v. 3.2 y b) análisis de discriminantes. Este análisis se aplicó para separar dos o más grupos de individuos, basados en medidas morfométricas de cada una de las especies, sobre las variables estudiadas, misma que fue realizado con Statgraphics v. 2.0 para Windows 2000. Los datos se transformaron a $\log_{10}$ y las distancias de Mahalanobis fueron calculadas utilizando Statistica 5.5 para Windows 2000.

Variación genética

Los geles fueron fotografiados utilizando una cámara digital Kodak modelo DC 120 Zoom. Las fotos digitales fueron analizadas utilizando el software Kodak digital Science ds 1D V. 2.0.3. para Windows (Eastman Kodak Company). Los datos obtenidos fueron transferidos posteriormente a una hoja de cálculo utilizando el software Excel 2000 V 9.0 para Windows con los que se construyó una matriz de presencia (1) o ausencia de bandas (0).

La diversidad y la estructura genética de las poblaciones se analizó utilizando POPGENE para Windows V 1.32 (Yeh & Boyle, 1996). En este análisis se utilizó las opciones de alelos dominantes diploídes con 1,000 iteraciones sugeridas por el mismo programa, con la cual se obtuvieron los siguientes parámetros: Alelos por locus, número de alelos efectivo (estimación de la homocigocidad reciproca), porcentaje de loci polimórficos, y la heterocigocidad esperada (índice de similitud Shannon de acuerdo a Yeh & Boyle, 1996), la identidad genética (Nei, 1978), la diversidad genética (Nei, 1973), la diversidad genética total (H_t), la diversidad genética dentro de las poblaciones (H_s), coeficiente de diversidad genética entre poblaciones (G_{st}) y el flujo genético (Nm) (McDermott & McDonald, 1993). También se efectuó un análisis de conglomerado utilizando el método de agrupación UPGMA y utilizando las distancias genéticas. Adicionalmente se llevo a cabo una prueba estadística de χ^2 (chi-cuadrada) para ver la significancia estadística en los loci. Esta fue una prueba de homogeneidad y fue construido a partir de cuadros de contingencia bi-direccionales. Las pruebas se llevan a cabo para grupos o las poblaciones múltiples según (Yeh & Boyle, 1996).

Una prueba de Mantel (Sokal & Rohlf, 1995) para ver la significancia estadística en la asociación entre matrices. Adicionalmente fue efectuado un análisis de regresión lineal para observar el grado de relación entre distancias genéticas y geográficas; flujo genético y distancias geográficas y distancia genética y diferencia altitudinal, utilizando PC-ORD V. 3.17 para Windows 2000 (McCune & Mefford, 1997). Adicionalmente se llevo a cabo un análisis de Coordenadas Principales (PCO) basado en la matriz de presencia y ausencia de bandas utilizando MVSP V 3.0 para Windows para determinar las relaciones espaciales entre las especies.

Resultados

Analisis Univariado

En la Tabla 30-2 se muestra el resumen de los resultados del análisis de varianza. Todas las variables se traslapan excepto largo y ancho de la semilla y largo del megasporofilo que sólo se traslapa ligeramente. *Ceratozamia euryphyllidia* tiene los troncos (diámetro y largo); hojas (largo); pecíolos, ráquis (largo); foliolos (ancho); semillas (largo y ancho); megasporofilos (ancho y largo) y ancho de la base del foliolo más grandes del complejo. *C. becerrae* es

Tabla 30-2 Medias y desviaciones estándares no transformadas a $\log_{10}$ del análisis de poblaciones en el complejo *Ceratozamia miqueliana*.

Especie	*C.eury phyllidia*		*C.be cerrae*		*C.miq ueliana*		*C.zoq* uorum	
N	10		13		15		10	
Variables	Media	Desv. Stand	Media	Desv. Stand	Media	Desv. Stand	Media	Desv. Stand
LARGTRONCO	192.00	34.58	147.99	57.91	216.67	76.13	223.00	78.18
DIAMTRONCO	892.99	145.26	76.11	12.46	119.29	28.19	107.27	28.19
NOHOJA	3.60	0.97	2.85	1.07	9.33	4.86	3.30	1.06
LARGOHOJA	2456.50	690.00	1092.85	364.81	1843.80	303.30	1987.00	398.61
ANCHOHOJA	471.20	52.38	354.62	70.64	585.40	60.88	585.50	125.71
LARGOPECIO	948.80	227.02	497.00	175.71	831.87	297.42	888.50	224.48
DIAMPECIOL	11.46	2.67	8.04	1.84	11.16	1.63	11.40	1.67
RAQLAR	1436.50	497.86	417.62	203.16	957.80	191.13	812.00	208.32
NFOLIOLO	10.80	2.78	6.23	2.42	14.73	3.41	10.30	3.27
LARGOFOLIO	260.90	33.49	231.92	38.86	305.20	33.39	319.00	45.81
ANCHOFOLIO	140.30	33.06	68.92	15.31	54.00	7.28	57.70	5.17
NOVENAS	64.50	17.95	38.38	6.64	31.47	4.90	36.40	4.27
DISTAVENAS	2.36	0.38	2.09	0.17	1.79	0.23	1.64	0.15
GROSORBASE	1.96	0.23	1.74	0.39	1.88	0.63	1.70	0.30
GROSORMARG	0.27	0.03	0.46	0.13	0.37	0.07	0.45	0.16
GROSORAPIC	0.18	0.05	0.36	0.06	0.21	0.04	0.37	0.14
ANCHOBASE	11.60	3.62	8.09	2.21	9.81	1.91	6.68	1.89
LARGOMICRO	7.46	0.49	11.32	0.99	13.38	1.99	10.48	0.54
ANCHOMICRO	5.76	0.41	4.95	0.54	9.53	1.24	5.85	0.59
LARGOSEMIL	25.56	1.17	18.19	0.99	31.09	1.21	24.94	1.13
ANCHOSEMIL	17.09	0.69	12.58	0.77	15.55	0.65	14.91	1.69
LARGOMEGAS	42.91	3.43	26.52	3.43	54.95	2.16	27.11	2.55
ANCHOMEGAS	19.82	6.80	10.53	2.16	24.34	0.86	10.09	1.26

una especie menos robusta con troncos muy pequeños, con pocas hojas y foliolos pero con el largo de los microsporofilos más grandes del complejo. En términos generales las especies más robustas son: *C. euryphyllidia* y *C. miqueliana*. De las 23 variables analizadas, 22 exhibieron diferencias significativas ($p < 0.0001$) en el valor de las medias entre las especies (Tabla 30-3), excepto el grosor de la base del foliolo que no difirió significativamente entre las especies.

Tabla 30-3 Resumen del análisis de varianza de las 23 variables morfométricas usadas en el análisis de poblaciones en el complejo *C. miqueliana* en donde la especie fue utilizada como variable de clasificación.

Variable	R^2	F	*p*
LARGOTRO	0.30	8.91	< 0.0001
DIATRONC	0.92	238.86	< 0.0001
NHOJAS	0.55	25.00	< 0.0001
LARGOHOJ	0.47	17.97	< 0.0001
ANCHOHOJ	0.64	36.11	< 0.0001
LARGPECI	0.31	9.09	< 0.0001
DIAMPECI	0.21	5.63	0.0018
LARGRAQ	0.58	28.03	< 0.0001
NOFOLIO	0.54	24.3	< 0.0001
LARGOFOL	0.39	12.92	< 0.0001
ANCHOFOL	0.79	77.16	< 0.0001
NOVENAS	0.68	44.26	< 0.0001
DISTVENA	0.54	24.25	< 0.0001
GROSBASE	0.06	1.29	0.28
GROSMARG	0.14	3.5	0.02
GROSAPIC	0.58	27.79	< 0.0001
ANCHOBAS	0.44	15.73	< 0.0001
LARMICRO	0.81	70.22	< 0.0001
ANCHMICR	0.87	106.79	< 0.0001
LARGSEMI	0.94	283.41	< 0.0001
ANCHOSEM	0.73	44.49	< 0.0001
LARGOMEG	0.9	175.44	< 0.0001
ANCHMEGA	0.84	96.9	< 0.0001

R^2 coeficiente de correlación, F = valor de F, p = nivel de significancia.

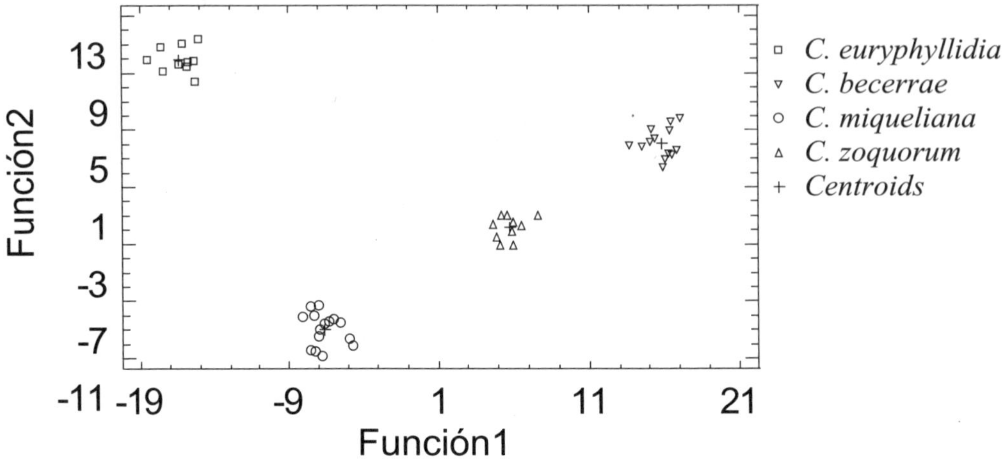

Figure 30-2. Diagrama de las dos primeras funciones discriminantes de 23 variables morfométricas del complejo *C. miqueliana*.

Análisis de Discriminantes

La Figura 30-2 presenta los resultados de los datos de dispersión derivados de las funciones discriminantes. 48 casos fueron usados para desarrollar un modelo discriminante entre las cuatro especies. 23 variables predictivas fueron usadas. Las tres funciones discriminantes fueron estadísticamente significativas ($p<0.001$) (Tabla 30-4). Las distancias de

Tabla 30-4 Resumen de los resultados del análisis de discriminantes de poblaciones en el complejo *C. miqueliana*.

Funciones Derivadas	Wilks-λ	χ^2	gl	*p*
1	7.7E-06	394.586	69	< 0.001
2	0.00121	225.036	44	< 0.001
3	0.06846	89.8296	21	< 0.001

Mahalanobis fueron altamente significativas entre las cuatro especies $p<0.0001$. Los resultados demuestran que *C. euryphyllidia* y *C. zoquorum* son especies muy distantes, mientras que *C. zoquorum* y *C. becerrae* son especies muy cercanas desde el punto de vista morfométrico (Tabla 30-5).

Las cuatro especies se separan ordenadamente en el espacio bidimensional y no presentan ninguna sobre posición entre los grupos. De las 23 variables que incluyeron en las funciones discriminantes estandarizadas, solamente seis variables tuvieron el valor más alto en el factor 1: diámetro del tronco, ancho de hoja, número de foliolo, largo y ancho del foliolo y ancho de semilla, mientras que en factor 2, las variables con mayor variación fueron: largo y ancho del foliolo. La primera variable demuestra que el 69% de la variación es en gran medida función de las variables morfológicas vegetativas. Este análisis de discriminantes refleja que existen características morfométricas distintivas entre las especies. Todas las especies fueron clasificadas correctamente sin sobrelapamiento la prueba de Wilks-λ fue altamente significativa ($p<0.0001$) para los tres factores. Las especies expresan su máxima variación en: largo del foliolo, largo del microsporofilo, largo y ancho de semilla y largo del megasporofilo (Tabla 30-6).

Tabla 30-5 Distancia cuadrática de Mahalanobis (D^2).

	C. becerrae	*C. zoquorum*	*C. miqueliana*	*C. euryphyllidia*
C. becerrae	****	48.7532	230.725	403.325104
C. zoquorum	48.7532	****	178.547	362.075684
C. miqueliana	230.725	178.547	****	379.094849
C. euryphyllidia	403.325	362.076	379.095	****
F-valores; gl = 23,22				
C. becerrae	****	12.0823	50.0322	71.3960571
C. zoquorum	12.0823	****	38.7174	64.0941467
C. miqueliana	50.0322	38.7174	****	60.8932571
C. euryphyllidia	71.3961	64.0941	60.8933	****
p-niveles				
C. becerrae	****	< 0.0001	< 0.0001	< 0.0001
C. zoquorum	< 0.0001	****	< 0.0001	< 0.0001
C. miqueliana	< 0.0001	< 0.0001	****	< 0.0001
C. euryphyllidia	< 0.0001	< 0.0001	< 0.0001	****

Tabla 30-6 Coeficientes de funciones de clasificación por especie.

Variables	*C. becerrae*	*C. zoquorum*	*C. miqueliana*	*C. euryphillidia*
LARGTRONCO	195.101	212.603	196.347	230.02
DIAMTRONCO	413.037	436.307	438.509	589.14
NOHOJA	14.4926	−5.83139	40.7621	36.75
LARGOHOJA	76.9525	82.6152	−22.4197	−63.79
ANCHOHOJA	−259.067	−205.295	−252.318	−156.31
LARGOPECIO	35.2207	55.7443	80.8407	118.64
DIAMPECIOL	−2.12485	3.7607	74.2798	57.82
RAQLAR	−630.622	−651.157	−619.41	−587.01
NFOLIOLO	173.178	182.759	139.733	172.25
LARGOFOLIO	1113.8	1128.93	1083.89	777.38
ANCHOFOLIO	−120.494	−185.349	−140.363	63.60
NOVENAS	392.623	415.062	313.54	357.02
DISTAVENAS	146.281	105.68	128.787	83.43
GROSORBASE	−31.0315	−41.8367	−27.4494	1.53
GROSORMARG	32.0869	34.7411	21.4571	59.01
GROSORAPIC	125.612	140.27	77.8415	66.54
ANCHOBASE	−11.2039	−21.4776	15.224	8.62
LARGOMICRO	1336.32	1267.28	1420.31	1144.67
ANCHOMICRO	−538.446	−494.216	−473.543	−522.73
LARGOSEMIL	1847.04	1863.12	2085.54	1834.06
ANCHOSEMIL	1407.54	1481.36	1580.1	1807.15
LARGOMEGAS	797.845	852.879	1024.91	945.31
ANCHOMEGAS	305.884	300.552	366.146	321.21
CONSTANT	−4380.42	−4699.31	−5284.83	−5179.59

Análisis Genético: Diversidad, Distancia, Identidad Genética y Flujo Genético

La Tabla 30-7 presenta los promedios de índices de diversidad de Nei (1973) para las especies del complejo *C. miqueliana*. Las cuatro especies no difieren mucho incluso entre poblaciones de las misma especies (*C. miqueliana*), siendo C. *becerrae* la más alta (0.20) con un promedio total de 0.36±0.19 para todas las poblaciones. También el índice varió entre los primers dentro de las poblaciones, siendo los promedios más altos con OPC02 y OPC06. Las bandas polimórficas por población variaron entre 41 a 99 %. De acuerdo a las pruebas estadísticas de homogeneidad de las frecuencias genéticas de las poblaciones (prueba de χ^2), 15 loci de los 119 fueron estadísticamente significativas.

Tabla 30-7 Índices de diversidad para las especies del complejo *Ceratozamia miqueliana.*

		na		A		h		Ho		%P
Sp/Población	n	M	D.S	M	D.S	M	D.S	M	D.S	
C. zoquorum	12	1.76	0.43	1.30	0.31	0.19	0.16	0.33	0.23	76.47
C. becerrae	15	1.79	0.41	1.31	0.31	0.20	0.16	0.32	0.23	78.99
C. euryphyllidia	10	1.68	0.47	1.30	0.31	0.19	0.17	0.30	0.25	68.01
C. miqueliana Veracruz	6	1.66	0.48	1.34	0.33	0.21	0.18	0.32	0.26	65.55
C. miqueliana Chiapas	15	1.83	0.38	1.32	0.29	0.21	0.16	0.34	0.22	83.19
Total	58	2	0	1.33	0.27	0.22	0.14	0.36	0.19	

M = Media; D.S.= desviación estándar; na = Número de álelos observados; A = Número de alelos efectivos (Kimura & Crow, 1964); h = Diversidad genética de Nei (1973); H_o= Información del índice de Shannon's (Lewontin, 1972) (Heterocigocidad observada) ; % P = porcentaje de loci polimorficos, n = número de individuos analizados.

El promedio de la diversidad total (H_t) en los loci polimórficos fue 0.22±0.01, mientras que el promedio de la diversidad dentro de las poblaciones (H_s) fue 0.19±0.01 indicando bajos niveles de variabilidad dentro de cada especie. La diferenciación de la diversidad genética entre poblaciones (G_{st}) vario entre 0.008 y 0.58 con una media de 0.11, lo cual sugiere que cerca del 90% de la variación esta presente dentro de cada especie.

El análisis de cluster UPGMA basado en la distancia genética (Fig. 30-3) demuestra al menos que *C. becerrae* esta estrechamente relacionado con *C. miqueliana,* lo cual concuerda con sus similitudes morfológicas (Fig. 30-2). En la Tabla 30-8 se presenta la distancia e identidad genética entre poblaciones de las especies del complejo *C. miqueliana,* siendo las especies *C. miqueliana* y C. *zoquorum* donde existe mayor distancia genética (0.0408) *y* en *C. miqueliana* de Chiapas y *C. becerrae* donde existe mayor identidad genética (0.9911). Es importante recalcar que entre todas las especies la identidad genética es superior al 90%, lo cual sugiere una baja o reciente divergencia entre las especies de *Ceratozamia.*

El promedio de Nm o migrantes por generación entre poblaciones de *Ceratozamia* fue de 4.03. Los valores estimados mas bajos fueron en poblaciones de *Ceratozamia*

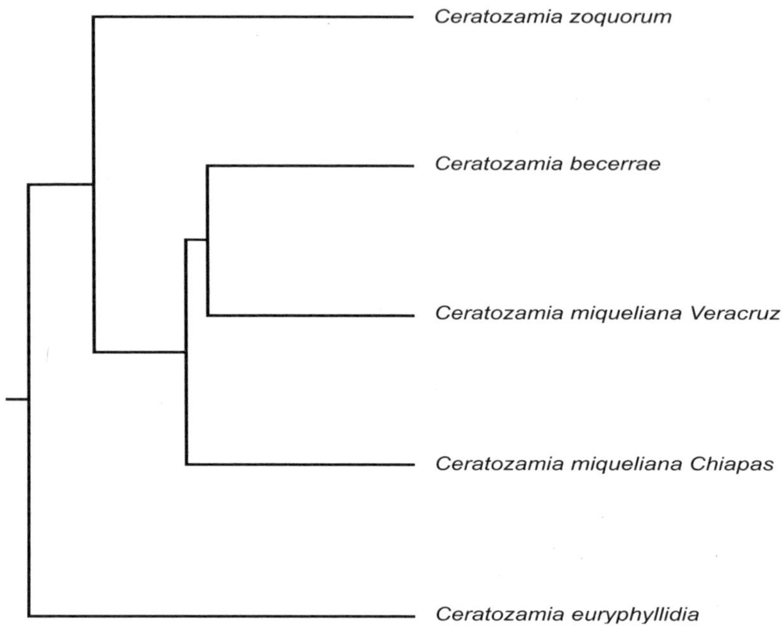

Figure 30-3. Fenograma de acuerdo a las distancia genéticas.

euryphyllidia y *C. zoquorum* (Nm=4.64) separados por unos 179 km y los más altos se encontraron entre *C. miqueliana* de Chiapas y *C. becerrae* (Nm=10.19) separados por 89.3 Km (Tabla 30-8). No se encontró correlación entre las matrices de Nm-distancia geográfica, (Mantel, r=− 0.08, p=0.81), Nm-diferencia de altitud (Mantel r=− 0.17 p=0.59) y distancia genética-distancia geográfica (Mantel r=0.38 p=0.28) (Tabla 30-8), pero si se encontró una fuerte correlación entre distancia genética y diferencia altitudinal (Mantel r=0.67, p=0.03) entre las poblaciones de *Ceratozamia,* lo cual nos indican que la diferenciación genética no siguen un patrón de aislamiento por distancia geográfica, sino por diferencia altitudinal. El modelo de regresión para la diferencia de altitud y la distancia genética estimada fue significativa y explica 56.66 % de la varianza en todos los casos e indica que mientras más aumenta la diferencia altitudinal entre las especies de *Ceratozamia* más aumenta su distancia genética (Tabla 30-8, Fig. 30-4).

En la Figura 30-5, se visualiza las relaciones espaciales entre las especies basado en el análisis de coordenadas principales a partir de datos de RAPD (presencia y ausencia de bandas). Una decisión para usar las dos primeras coordenadas, es porque contienen la clara

Tabla 30-8 Relación de identidad, distancia genética y flujo genético con respecto a su distancia geográfica y diferencia altitudinal.

	Distancia genética de acuerdo a Nei (1978)				
Identidad genética	*C. zoquorum*	*C. becerrae*	*C. miqueliana* Ver.	*C. miqueliana* Chis.	*C. euryphyllidia*
C. zoquorum	****	0.9791	0.96	0.9732	0.9662
C. becerrae	0.0211	****	0.9651	0.9911	0.9836
C. miqueliana Ver	0.0408	0.0356	****	0.9734	0.9614
C. miqueliana Chis	0.0272	0.009	0.0269	****	0.986
C. euryphyllidia	0.0344	0.0165	0.0394	0.0141	****
Distancia geográfica (Km)					
Flujo genético	*C. zoquorum*	*C. becerrae*	*C. miqueliana* Ver.	*C. miqueliana* Chis.	*C. euryphyllidia*
C. zoquorum	****	26.8	280	63.1	179
C. becerrae	8.02	****	293	89.3	202
C. miqueliana Ver.	5.76	10.14	****	253	146
C. miqueliana Chis	5.86	10.19	8.75	****	125
C. euryphyllidia	4.64	5.32	5.61	5.1	****
Distancias geográficas (Km)					
Distancia genéticas	*C. zoquorum*	*C. becerrae*	*C. miqueliana* Ver.	*C. miqueliana* Chis.	*C. euryphyllidia*
C. zoquorum	****	26.8	280	63.1	179
C. becerrae	0.0211	****	293	89.3	202
C. miqueliana Ver.	0.0408	0.0356	****	253	146
C. miqueliana Chis.	0.0272	0.009	0.0269	****	125
C. euryphyllidia	0.0344	0.0165	0.0394	0.0141	****
Diferencia altitudinal					
Distancia genéticas	*C. zoquorum*	*C. becerrae*	*C. miqueliana* Ver.	*C. miqueliana* Chis.	*C. euryphyllidia*
C. zoquorum	****	26.8	280	63.1	179
C. becerrae	0.0211	****	293	89.3	202
C. *miqueliana* Ver	0.0408	0.0356	****	253	146
C. *miqueliana* Chis	0.0272	0.009	0.0269	****	125
C. *euryphyllidia*	0.0344	0.0165	0.0394	0.0141	****

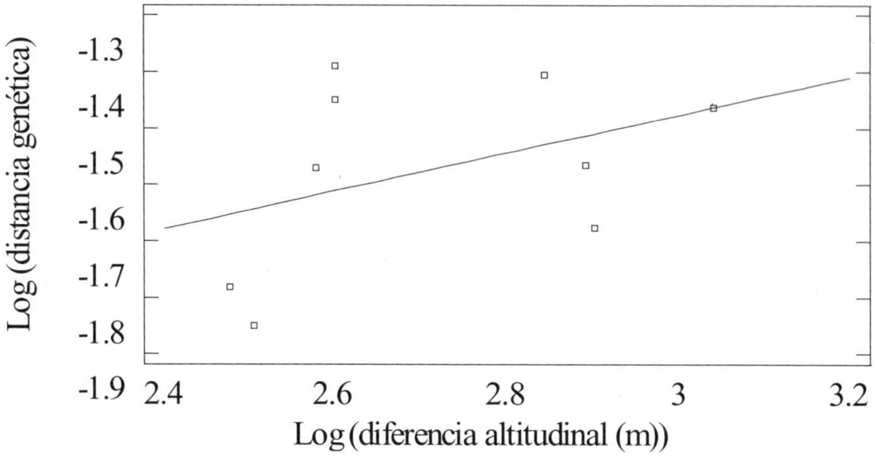

Figure 30-4. Relación entre la distancia genética y la diferencia altitudinal entre las especies del complejo *Ceratozamia miqueliana* (F = 10.46, g.l. = 9 $p = 0.0120$; Coeficiente de correlación = 0.75, $R^2 = 56.65$ %, Error estándar = 0.151392; Log [distancia genética] = –2.48185 + 0.334857* log [diferencia altitudinal]).

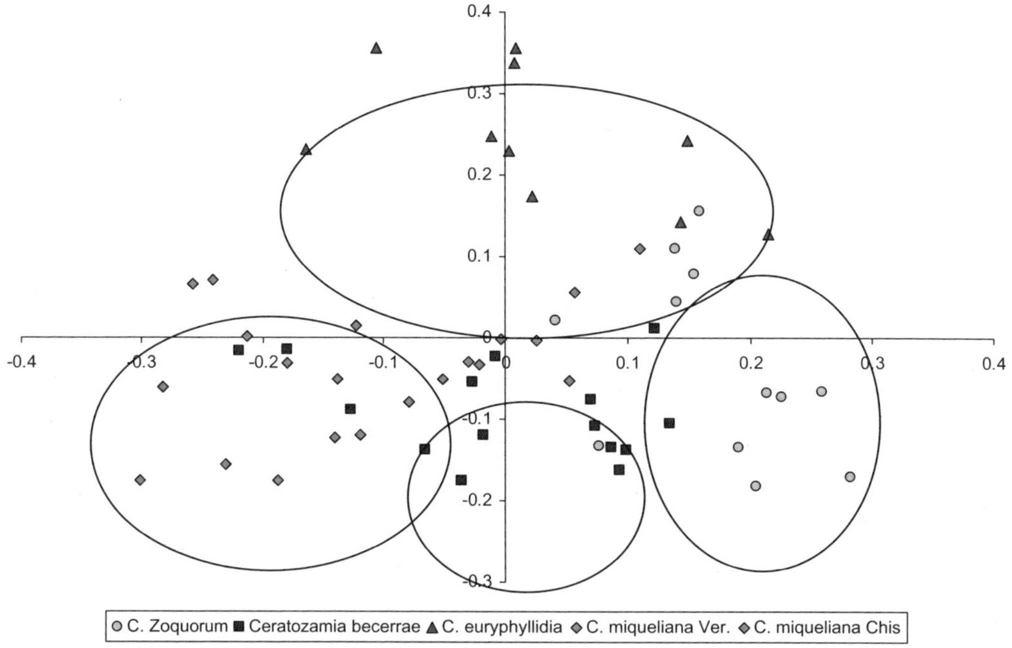

Figure 30-5. Análisis de coordenadas principales (PCO) de las especies de *Ceratozamia* del complejo *C. miqueliana* basado en la matriz de presencia y ausencia de bandas.

mayoría de la varianza (76 %) y divide éste en cuatro grupos. Este análisis demuestra que *C. miqueliana*, *C. zoquorum* y *C. becerrae* forman un grupo, los cuales concuerdan con los resultados morfométricos (Fig. 30-2).

Discusión y Conclusiónes

Variación Morfológica

Los caracteres morfológicos importantes en la diferenciación de estas especies son: diámetro del tronco; número, largo y ancho del foliolo; ancho de hoja y ancho de la semilla. *Ceratozamia euryphyllidia* sobresale dentro del complejo *C. miqueliana*, debido a que es la especie que posee los foliolos más anchos y con las hojas más largas (hasta 3 m), mientras que *C. becerrae* posee los troncos más pequeños inclusive dentro del género. La importancia de estos caracteres es evidenciada por su elevado peso en el análisis de discriminantes.

Es importante comentar que a pesar de que el diámetro de los troncos y el número de los foliolos fueron caracteres importantes para la discriminación de las especies dentro de este complejo. No obstante, esta información nunca está disponible en los ejemplares de herbario (tronco) o las hojas procesadas están incompletas. Igualmente sucede con los caracteres reproductivos tales como el largo del microsporofilo y megasporofilo y ancho de semilla que emergieron como variables reproductivas que ayudan a discriminar las diferencias entre las especies. Muchas veces esta información no está disponible en los ejemplares de herbario o están incompletas.

El análisis descrito anteriormente demuestra substancialmente las diferencias morfológicas dentro de las especies del complejo *C. miqueliana*. Mientras algunos resultados dentro del análisis univaridado demuestran sobrelape entre las especies; en el espacio multivariado las cuatro especies no se sobrelapan. El análisis provee evidencia para reconocer que al menos de las cuatro especies analizadas pertenecen cada uno a especies diferentes.

Análisis Genético

Los resultados encontrados sobre diversidad y diferenciación genética en este complejo fueron similares a los encontrados en el complejo *C. norstogii* (Pérez-Farrera dato, no publicado). Esta baja diversidad y diferenciación genética fue tanto dentro como entre poblaciones de las especies ($H_s=0.19$ y $G_{st}=0.11$) y son muy parecidos a registrados por Sharma et al. (1999) con el complejo de *Macrozamia heteromera* (H_t=0.165, H_s=0.149, G_{st}=0.10). Sin embargo el complejo *C. miqueliana* presento una mayor diversidad total (H_t=0.22) y un mayor porcentaje de loci polimórficos que la registrada para este mismo complejo de *Macrozamia*. Los valores de A=1.33, *p*=76.68 %, H_e=0.36 del complejo *C. miquelina,* fueron similares que el complejo *C. norstogii,* excepto que este complejo presento un porcentaje más elevado de polimorfismo. No obstante, éstos fueron más altos que los registrados por Sharma et al. (1999) para *Macrozamia heteromera* (A=1.3, *p*=26%, H_e=0.06) y Ellstrand et al. (1990) para *M. communis* (A= 1.61, *p*=50% y H_e=0.08), excepto por el número de álelos observados por Hamrick y Godt (1989) para otras plantas (A=1.51, P=34%, H_e=0.14), pero más bajo por los reportados por Byrne y Jame (1991) para *M rieddlei* (A=2.4, p=93 %, H_e=0.27). Sin embargo los valores de diversidad total fueron similares como los encontrados por *M. communis* (H_t=0.19) y *M. Riedlei* (H_t=0.29), lo cual sugiere una divergencia reciente entre las cuatro especies.

El resultado inesperado fueron los bajos valores de diversidad genética, sobre todo cuando se compara con otras especies de biología similar. Por ejemplo, especies con poca dispersión, tasas de crecimiento lento y perennes (Hamrick & Godt, 1989). La comparación de los valores de distancia genética pareadas por poblaciones dentro de las especies son usualmente 0.10 o menos (Crawford, 1983). No obstante el rango de los valores de distancia genética obtenida dentro de este complejo entre las especies indican un alto grado de similitud (rangos de 0.0090 y 0.0408), la cual se refleja en identidad genética (0.99) y contrasta significativamente con los valores usualmente encontrados con otras especies congenéricas (0.33) (Gottlieb, 1981).

Los resultados también indican un alto grado de endogamia a pesar de que las especies de *Ceratozamia* son dioicas y son polinizadas por insectos. Esto probablemente está relacionado con el hecho de que las semillas no son dispersadas a largas distancias, sino por gravedad (Pérez-Farrera et al., 2000; Pérez-Farrera & Vovides, 2004). Estas especies se establecen en laderas con pendientes pronunciadas (Vovides et al., 2004a), aunque se han registrado algunas observaciones de depredación de semillas por el jabalí *Tayassus tajacu* no se sabe con exactitud si intervienen en su dispersión, pero se cree que interviene en la colonización de nuevos hábitats (Pérez-Farrera et al. 2000).

El análisis de UPGMA de las distancias genéticas parece indicar la agrupación de tres grupos: el primero conformado por *C. zoquorum*, el segundo conformado por *C.becerrae* y *C. miqueliana* el tercero por *C. eurphyllidia*. Desde el punto de vista taxonómico esto tiene cierta consistencia debido a que *C. becerrae* es muy parecido a *C. miqueliana*. Ambos tienen foliolos oblanceolados y son coriáceos. Sin embargo, pueden diferenciarse debido a que los foliolos de *C. becerrae* son más anchos que en *C. miqueliana* y las articulaciones de *C. becerrae* son amarillentas, mientras que *C. miqueliana* son verdes. Este último carácter es muy importante en la separación de estas dos especies. Sin embargo este carácter se pierde al momento de procesar los ejemplares en el herbario.

Las bajas distancias genéticas encontradas entre las especies del complejo *Ceratozamia miqueliana* puede deberse a diversos factores tales como el rango geográfico, historia de vida, características ecológicas de las especies o a otros factores tales como: a) baja tasa de mutación, b) que la variabilidad no se ha acumulado o se han fijado completamente en el pool genético de las especies, c) que se hayan originado de un ancestro genéticamente deplorable con un rango geográfico restringido.

La morfología de este complejo está influida en gran parte por presión selectiva de "microambientes" (sustrato de crecimiento y altitud) en donde las diferencias genéticas no se han fijado completamente en el pool genético de cada una de las especies. Debido a que se encontró una correlación significativa entre distancias genéticas y altitud, mientras que las relaciones con otros variables tales como flujo genético y distancia geográfica; y distancia genética y distancia geográfica no fueron significativas, se alude a un proceso de especiación ecológica, por efecto de gradiente ambiental. Recientes estudios parecen demostrar que la selección natural opera a través de gradientes ecológicos y que puede ser más importante que el aislamiento geográfico (Schneider et al., 1999). Por ejemplo, Bruschi et al., 2003, encontraron relaciones entre la variación molecular y factores climáticos dados por efectos altitudinales con poblaciones de *Quercus petrea*. Es raro observar particularmente que todas las especies del complejo *C. miqueliana* crecen en el mismo hábitat (Bosque tropical perennifolio), sin embargo han colonizado "microhabitats" diferentes por ejemplo *C. becerrae* y *C. zoquorum*, crecen a diferentes altitudes y sobre rocas, mientras que *C. miqueliana* y *C. eurphyllidia* crecen en suelo arcilloso. Además, las distancias geográficas entre especies y poblaciones son relativamente cercanas por ejemplo *C. becerrae* y *C. zoquorum* solamente están separadas por escasos 26 km. Aunque morfológicamente son distintas, presentan poca variabilidad genética entre las especies, pero con rangos o gradientes altitudinales distintos, por lo que existe una influencia muy grande del factor ambiental.

Las especies del complejo *Ceratozamia miqueliana* pudieron haberse originado a través de la adaptación de diferentes zonas ecológicas y desarrollaron morfologías individuales en el proceso. Un caso similar es el reportado por Haufler et al. (2000) en donde encontraron valores altos de identidad genética entre las especies de *Pleopeltis.* Ellos sugieren un proceso reciente y rápido de especiación y proponen que el vector de especiación fue probablemente el disturbio climático a gran escala y el resultado del desplazamiento a nivel de especies. Aunque los rangos de especies de *Pleopeltis* se sobrelapan, estos parecen ocupar hábitats discretos ecológicamente diferentes. Lo mismo pasa con las especies del complejo *C. miqueliana.* Todas las especies habitan en bosque tropical perennifolio, pero a altitudes diferentes y habitando microambientes diferentes, por ejemplo tanto *C. zoquorum* como *C. becerrae* crecen sobre rocas cársticas, mientras que *C. miqueliana* y *C. euryphyllidia* lo hacen en suelo arcilloso. Sin embargo, algunas poblaciones de *C. miqueliana* crecen en bosque mesófilo y en suelo arcilloso con rocas basálticas. *C. eurphyllidia* aunque crece en suelo arcilloso, siempre está asociada a laderas con vientos fuertes y creciendo siempre en sotobosques muy oscuros.

La poca variabilidad genética presente dentro de este complejo, concuerda con los resultados por Gonzáles y Vovides (2002), con secuencia en ADN ribosomal y de ADN del cloroplasto *trn*L-F. Esto presupone un aislamiento relativamente reciente, quizás posterior a las fluctuaciones climáticas que prevalecieron en el Pleistoceno durante el Cenozoico en esta área, conocida comúnmente como refugio de la zona del arco (Wendt, 1989) o posterior a las migraciones de dispersores que prevalecieron durante y después del Pleistoceno de América central hacia México (Burnham & Graham, 1999). Los registros palinológicos y de glaciación en la zona del arco indican la presencia de polen de *Ceratozamia* del Mioceno inferior en Simojovel y Pichucalco, Chiapas (Graham, 1999; Palacios & Rzedowski, 1993).

Los resultados del análisis morfológico demuestran que las especies que comprende el complejo *C. miqueliana* está compuesto por entidades discretas, que por falta o escasez de información morfológica, mal procesamiento en los ejemplares de herbario y pérdida de sus caracteres al momento de procesarlas hacen complicado su delimitación taxonómica en el herbario, por lo cual han sido tratados como complejos. Sin embargo, los resultados genéticos indican que estas especies o entidades que comprende el complejo son de reciente especiación.

Agradecimientos

Los autores agradecen a: Nayely Martínez Meléndez, Rubén Martínez Camilo, Emerit Meléndez López y Claudia Virgen Montesinos. El autor principal agradece a Montgomery Botanical Center por el financiamiento otorgado a través del proyecto "Revisión taxonómica del género *Ceratozamia* para el Neotrópico" y el proyecto CONACYT-SEMARNAT 2002-C01-0183.

Literatura Citada

Bruschi, P., G. G.Vendramin, F. Bussoti & P. Grossoni. 2003. Morphological and molecular diversity among Italian population of *Quercus petrea* (Fagaceae). Ann. Bot. 91: 707–716.

Burnham, R. J. & A. Graham. 1999. The history of neotropical vegetation: new developments and status. Ann. Missouri Bot. Gard. 86: 546–589.

Byrne, M. & S. H. Jame 1991. Genetic diversity in the cycad *Macrozamia riedlei*. Heredity 67: 35–39.

Crawford, D. L. 1983. Phylogenetic and systematic inferences from electrophoretic studies. Pp 257–287 *in* S.D. Tanksley, & T.J. Orton (eds), Isozymes in plant genetics and breeding. Elsevier, Amsterdam.

Ellstrand, N. C., R. Ornduff & J. M. Clegg. 1990. Genetic structure of the Australian cycad *Macrozamia communis* (Zamiaceae). Amer. J. Bot. 77: 677–681.

Graham, A. 1999. Studies in neotropical paleobotany. XIII. An Oligo-Miocene palynoflora from Simojovel (Chiapas, México). Amer. J. Bot. 86: 17–31.

González, D. & A. P. Vovides. 2002. Low intralineage divergence in the genus *Ceratozamia* Brongn. (Zamiaceae) detected with nuclear ribosomal DNA ITS and chloroplast DNA *trn*L-F non-coding region. Syst. Bot. 27: 654–661.

Gottlieb, L. D. 1981. Electrophoretic evidence and plant populations. Prog. Phytochem. 7: 1–46.

Haufler, C. H., E. A. Hooper & J. P. Therrien. 2000. Modes and mechanisms of speciation in pteridophytes of contrasting patterns in ferns representing temperate and tropical habitats. Plant Species Biology 15: 223–236.

Hamrick, J. L. & M. J. W. Godt. 1989. Allozyme diversity in plant species. Pp. 46–55 *in* A. D. H. Brown, M. T. Clegg, A. L. Kahler & B. S. Weir (eds.), Population genetics breeding, and genetic resources. Sinauer Assiciates, Sunderland, MA.

Kimura, M. & Crow, J. F. 1964. The number of alleles that can be maintained in a finite population. Genetics 49:725–738.

Lewontin, R. C. 1972. The apportionment of human diversity. Evolutionary Biology 6: 381–398.

McDermott, J. M. & B. A. McDonald. 1993. Gene flow in plant pathosystems. Annual Review Phytopathology 31: 353–373.

McCune, B. & J. Mefford.1997. Multivariate analysis of ecology data V 3.17, JPM Software Gleneden Beach, Oregon, U.S.A.

Moretti, A., S. Sabato & M. Vázquez Torres. 1980. The distribution of *Ceratozamia* Brongn. (Zamiaceae). Delpinoa 21: 13–21.

Nei, M. 1973. Analysis of gene diversity in subdivided populations. Proc. Natl. Acad. Sci. USA 70: 3321–3323.

———. 1978. Estimation of average heterozygosity and genetic distance from a small number of individuals. Genetics 89: 583–590.

Palacios, C. R. & J. Rzedowski. 1993. Estudio palinológico de las floras fósiles del Mioceno inferior y principios del Mioceno Medio de la región de Pichucalco, Chiapas, México. Acta Botánica Mexicana 24: 1–96.

Pérez-Farrera, M. A., P. F. Quintana-Ascencio, B. Salvatierra & A. P. Vovides. 2000. Population dynamics of *Ceratozamia Matudai* Lundell (Zamiaceae) in El Triunfo Biosphere Reserve, Chiapas, México. J. Torrey Bot. Soc. 127: 291–299.

———, A. P. Vovides & C. Iglesias. 2001. A new species of *Ceratozamia* (Zamiaceae) from Chiapas, México. Bot. J. Linn. Soc. 137: 77–80.

——— & A. P. Vovides. 2004. Spatial distribution, population structure and fecundity of *Ceratozamia matudai* Lundell in The Triunfo Biosphere Reserve, Chiapas, México. Bot. Rev. 70: 299–311.

Schneider, C. J., T. B. Smith, B. Larison & C. Moritz. 1999. A test of alternative model of diversification in tropical forest: ecological gradients vs rainforest refugia. PNAS. 96: 13869–13873.

Sharma, I. K., D. L. Jones, P. I. Foster & A. G. Young. 1999. Low isozymic differentiation among five species of the *Macrozamia heteromera* group (Zamiaceae). Biochem. Syst. Ecol. 27: 67–77.

Sokal, R. R. & F. J. Rohlf. 1985. *Biometry*, 3rd ed. New York.

Stevenson, D., S. Sabato & M. Vázquez Torres. 1986. A new species of *Ceratozamia* (Zamiaceae) from Veracruz, México with comments on species relationships, habitats, and vegetative morphology in *Ceratozamia*. Brittonia 38:17–26.

Vovides, A. P., J. D. Rees & M. Vázquez Torres. 1983. Flora de Veracruz. Familia Zamiaceae. Fascículo 26. INIREB, Xalapa, Veracruz, México.

———, M. A. Pérez-Farrera & C. Iglesias. 2001. Another new species of *Ceratozamia* (Zamiaceae) from Chiapas, Mexico. Bot. J. Linn. Soc. 137: 81–85.

———, D. González, M. A. Pérez-Farrera, S. Avendaño & C. Bárcenas. 2004a. A review of research on the cycad genus *Ceratozamia* Brongn. (Zamiaceae) in México. Taxon 53: 291–297.

———, M. A. Pérez-Farrera, B. Schutzman, C. Iglesias, L. Hernández-Sandoval & M. Martínez. 2004b. A new Species of *Ceratozamia* (Zamiaceae) from Tabasco and Chiapas, México. Bot. J. Linn. Soc. 146: 1–6.

Wendt, T. 1989. Las selvas de Uxpanapa, Veracruz-Oaxaca, México: Evidencia de refugios florísticos Cenozoicos. Anales Inst. Biol. UNAM Ser. Bot. 58: 29–54.

Whitelock, L. M. 2004. Classification concepts in *Ceratozamia* (Zamiaceae). Pp. 95–108 *in* T. Walters & R. Osborne (eds), Cycad classification: concepts and recommendations. CABI Publishing, Wallingford.

Yeh, F. C. & T. Boyle. 1996. POPGENE Versión 1.11, Microsoft Windows-base software for population genetic analysis.

SYSTEMATICS

CHAPTER 31

Cycas annaikalensis from the Western Ghats of India with a Discussion on the Specific Limits of Cycas circinalis L.

Rita Singh, P. Radha, and Prabha Sharma

Abstract

The Western Ghats of India has been identified as one of the hotspots of the world and a center of biodiversity. Linnaeus established the genus *Cycas* and its type species *C. circinalis* from the illustrations of Van Rheede's Todda Panna. Rheede's locality of Todda Panna falls within the limits of Western Ghats. We extensively explored all the existing populations of *Cycas* in southern Western Ghats during 2001–2004 and identified several morphological, anatomical, reproductive, and cytological distinctions within the populations of these areas of the Western Ghats. Consequently, we identified and described a new species, *Cycas annaikalensis* R. Singh & P. Radha, from the Palaghat range. This chapter describes the characteristic features of *C. circinalis* L., its distribution, and its comparison with *C. annaikalensis* as described from the Malabar Coast. An attempt has been made to discuss the specific limitations of *C. circinalis*, to solve the Indian *Cycas* taxonomy.

Resumen

Los Ghats Occidentales de India han sido identificados como uno de los puntos críticos del mundo y un centro de biodiversidad. Linnaeus estableció el género *Cycas* y su especie típo *circinalis,* desde las ilustraciones de *Todda Panna* de Van Rheede. La localidad de *Todda Panna* de Rheede también se ubica dentro los límites de los Ghats Occidentales. Se realizó una exploración extensiva durante los últimos tres años de todas las poblaciones existentes de *Cycas* en el sur de los Ghats occidentales. Se ha identificado numerosas diferencias morfológicas, anatómicas, reproductivas y citológicas dentro de las poblaciones de estas dos áreas de los Ghats Occidentales. Una nueva especie *Cycas annaikalii* de la región Palaghat fue identificada y descrita por Singh y Radha (en prensa). Se describe los rasgos característicos de *Cycas circinalis*, su distribución y se compara con la nueva especie de la costa de Malabar. Se hizo un intento en discutir las limitaciones específicas de *C. circinalis* para resolver la taxonomía de las *Cycas* indias.

Introduction

Cycads of India are represented by a single genus, *Cycas*, with four species that occur in wild localities of the northeast (Orissa, Bihar, Sikkim, Assam, Manipur, Tripura, and areas adjacent to Myanmar). In the southern part, *Cycas* occurs in the Western Ghats starting from southern Maharashtra, Karnataka, Kerala, and Tamil Nadu and also in Andhra Pradesh and the Andaman and Nicobar Islands (Pant, 2002). *Cycas* L. is the type genus of the order Cycadales and *C. circinalis* L. is the type species, which was described by Linnaeus (1753, 1754) on the basis of Van Rheede's illustrations. Hendrik Adriaan Van Rheede described almost all the groups of plants from southern Kerala by their local names and also mentioned their medicinal values. In *Hortus Malabaricus*, he described *Cycas* by its local names of *Todda Panna* or *Mouta Panna* along with "the palms," probably owing to their common external appearance (Van Rheede, 1682). *Cycas circinalis* is the only species known so far to occur in the wild throughout the Western Ghats. According to Pant (2002), its distribution in the "Western part of peninsular India in deciduous forest of Western Ghats up to 1067 m (3500 ft) altitude from Malabar southwards to hills on the east side as far as Orissa. Hill (1995) mentioned that "*Cycas circinalis* L., as we now understand it, is an Indian endemic restricted to the Western Ghats and hills of the southern peninsula, as far north-east as Madras, in the states of Kerala, Karnataka, Tamil Nadu and Maharashtra." However, Pant (2002) had mentioned that *C. circinalis* of Mysore (Karnataka) described by Swamy (1948) could be a distinct variety owing to the peculiar mode of branching exhibited by plants of *Cycas* in populations in this area. Pant's assessment may be quite true regarding the possibility of variation within *C. circinalis*, but requires confirmation. All authors working on *Cycas* after Linnaeus (1753) have been of the opinion that the *Cycas* of Western Ghats is represented by a single endemic species, *C. circinalis* L., an opinion that is explored here.

Western Ghats comprises a vast mountain range that runs about 1600 km north–south between the river Tapti and Gujarat and Kanyakumari in Tamil Nadu, covering an area of approximately 160,000 km^2 that is one of the diversity hotspots of India, with a unique and rich assemblage of flora and fauna, in which many of the taxa have yet to be formally described (Biju & Bossuyt, 2005). In India, 38% of endemic flowering plants occur in the Western Ghats. *Cycas* is also one of them, but Indian taxonomists have given it little attention. We therefore decided to explore *C. circinalis* (*sensu* Todda Panna of Van Rheede) of southern Western Ghats, (Malabar Coast) starting from Wayanad in the north, crossing Kozhikode and Malapuram up to Trissur in the south, because in the *Hortus Malabaricus*, Rheede described and illustrated the plants of Todda Panna from Tirtsjour, which is now known as Trissur (Trichur) and Couronuada Provinces.

We collected data on all possible continuous, disjunct, or relictual *Cycas* populations. During this exploration, we found an isolated population of *Cycas* on Annaikal hills in Palaghat range, which is morphologically and anatomically distinct from *C. circinalis* L. and described it as a new species, *C. annaikalensis* Singh & Radha (2006), from the Malabar Coast of Western Ghats. In this chapter we also incorporate the description of *C. annaikalensis*, its comparison with *C. circinalis L.*, followed by a discussion on the specific limits of *Cycas circinalis.*

Materials and Methods

We extensively explored the wild localities of *Cycas* in the south Western Ghats crossing Kerala and studied more than 30 *Cycas* populations. We collected vegetative as well as reproductive parts at different time intervals. *Cycas* is dioecious, like other cycad genera. Its coning period is nonsynchronous; therefore repeated field visits were made over 2001–2004 to study and collect the reproductive parts at different stages of development. We prepared herbarium vouchers of leaves and reproductive parts from all the populations examined. Specimens are housed in the newly established herbarium (IPUH) in Guru Gobind Singh Indraprastha University, Delhi, India. Photographs of wild localities as well as plants were taken, and selected negatives have also been deposited in the herbarium for future reference. An average of at least 50 readings was taken to provide measurements of morphological and anatomical characters.

Results

Taxonomic Description

Cycas annaikalensis R. Singh & P. Radha, Brittonia 58: 119–123, 2006. (Figs. 31-1 to 31-11, 31-27B)

Typus: India, Kerala, Palaghat, Annaikal hills (940 m) Rita Singh, P. Radha and Prabha Sharma s.n., May 2003 (Holotype: –IPUH(0491) 0144; Isotype: DD).

Stem arborescent, smooth, steel grayish without any persistent armor of leaf bases, rarely branched, approximately 5 m tall, 19–61 cm diam. *Leaves* 1–2.5 m long with 100–120 pairs of pinnae, emergent foliage loosely tomentose; *petiole* 50–90 cm long, spinulose, prickles 0.9–3 mm long, attachment of median pinnae at 50–55 degrees to rachis, *pinnae* 26–35 × 0.8–1.2 cm, with entire margins, in cross section the extreme base shows three mucilage canals adjacent to the inner side of the bundle sheath toward the phloem side, with the number declining to two at the median level and up to one near the apical region. *Microsporangiate cone* subconical, yellowish orange, before dehiscence, 30–50 cm long 15–21 cm diam.; *microsporophylls* 6.3–8.3 × 1.8–2.6 cm (including the sterile and fertile portion), fertile zone length 2–3.2 cm, sterile zone 3.8–4.7 cm long (including upturned caudate apex). *Megasporophylls* 9–24.2 cm long (including fertile stalk with ovules and sterile lamina), ferruginous trichomes densely cover the megasporophylls when young, sterile apex of megasporophylls triangular-scutiform 1.9–5.5 × 1.3–4.0 cm (sterile tip width measured at the broad region of lamina), regularly dentate with 5–18 lateral teeth 0.05–0.8 mm long, apical spine distinct from lateral teeth, apical spine 3–25 mm long. *Ovules* 2–10 globose, 3.8–4.9 × 3.5–4.3 cm. Seeds platyspermic, sarcotesta green when young, becoming mango yellow

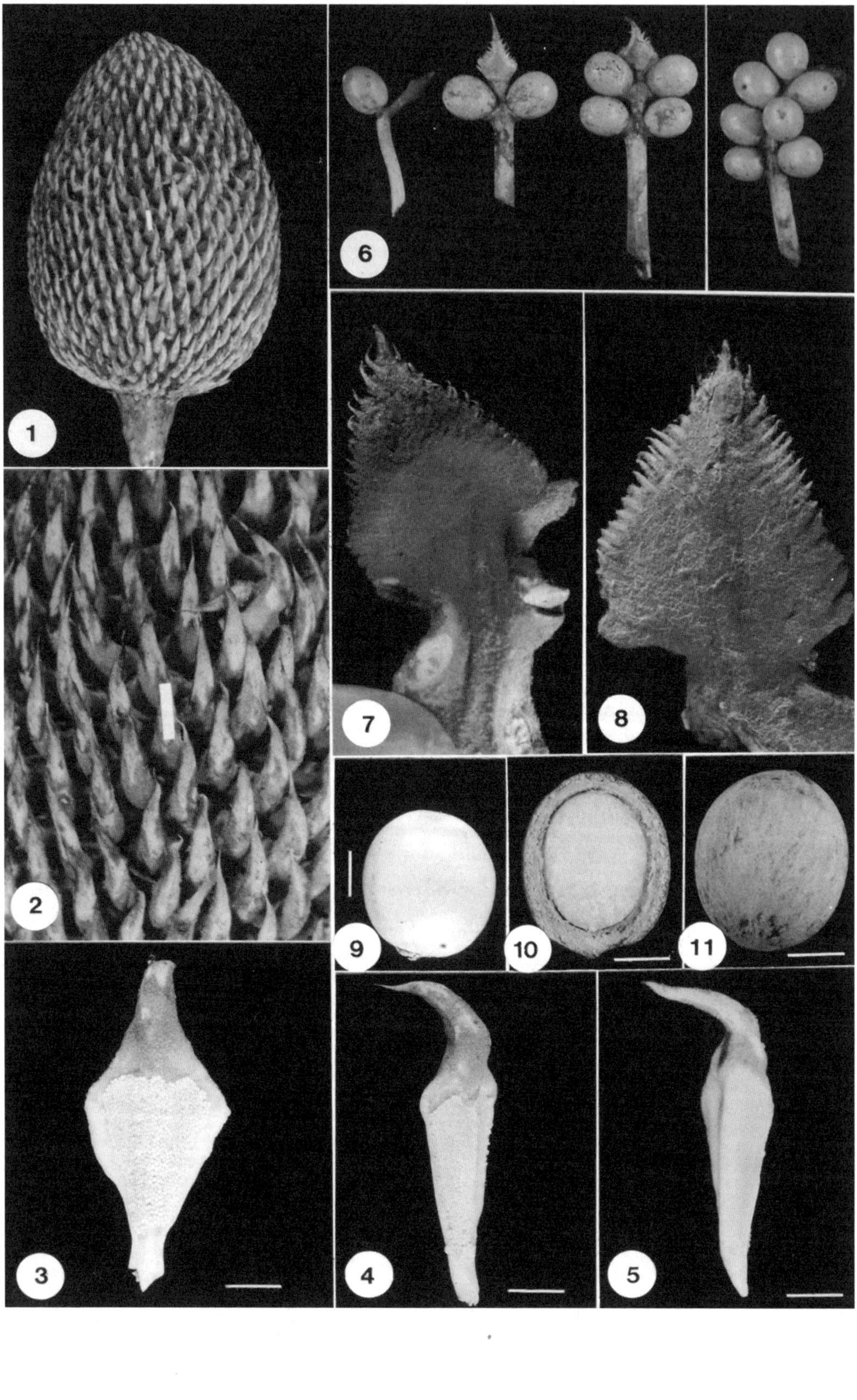
1
2
3
4
5
6
7
8
9
10
11

at maturity, 3 mm thick, sclerotesta stony, middle fibrous layer absent inner layer membranous, germination hypogeal.

Specimens examined: India: Kerala, Palaghat district, Annaikal, October 2001, 940 m *Rita Singh, P. Radha & Prabha Sharma, 0491*, 0041–0052; May 2003, *Rita Singh & P. Radha (0491) 0075,0076,0118,0119,0120* (all IPUH).

Etymology

The specific epithet refers to the name of the locality, Annaikal Hills, where this population is located.

Distribution and Habitat

Cycas annaikalensis is currently described from a single population of fewer than 100 individuals over an area of 100–250 m^2 on the Annaikal hills near Palaghat at an altitude of 940 m. Plants grow as an understory element on the steep slopes of coarse black humus soil with pH 4.7–5.7 covered by the thickest canopy of tropical flowering trees. Assessing the habitat, it is assumed that this species may also occur in adjacent Reserve Forest areas, but comprehensive exploration is needed to assess the number of populations occur in the region (Singh & Radha, 2006).

Conservation Status

The *Cycas annaikalensis* population occurs mainly within the hilly forest region of Palaghat district, and it has yet to be confirmed that there are other sporadic populations in the vicinity of the current locality. The presence of dangerous species of wild animals such as wild elephants, leopards, and boars adds to difficult terrain in the Reserve Forest

---

Figures 31-1 to 31-8. ***Cycas annaikalensis.*** **R. Singh & P. Radha. 1. Microsporangiate cone. 2. Close up showing details of apical spine. 3, 4, 5. Microsporophylls. 6. Megasporophylls with mature ovules. 7, 8. Sterile tips of megasporophylls.**

Figures 31-9 to 31-11. ***Cycas annaikalensis.*** **R. Singh & P. Radha. Ovules. 9. With sarcotesta. 10. Cross section. 11. Without sarcotesta.**

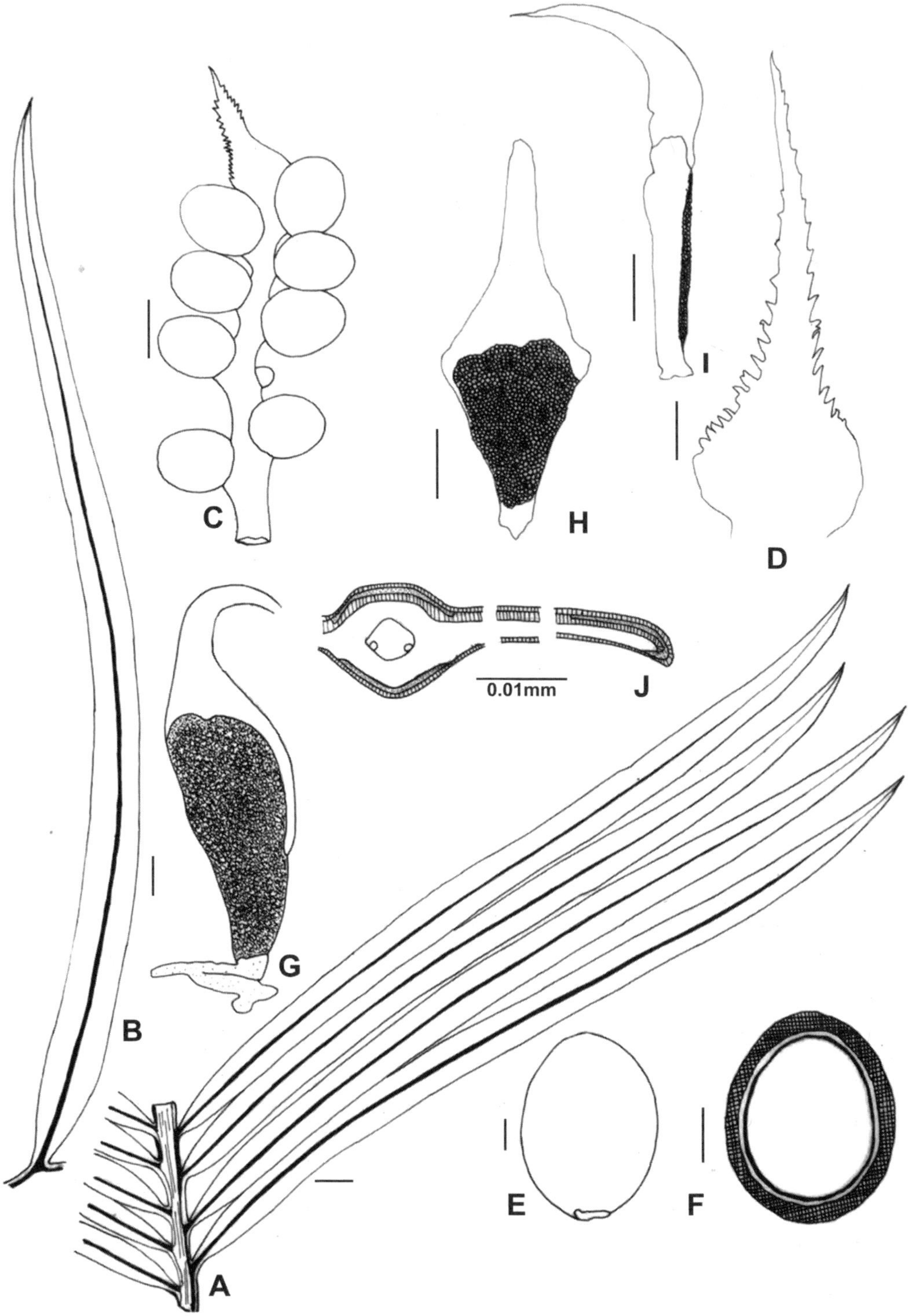

Figure 31-12. *Cycas circinalis* L. **A**. Part of leaf. **B**. Single pinna. **C**. Megasporophyll with ovules. **D**. Sterile tip of megasporophyll. **E**. Entire ovule. **F**. Section of ovule. **G**, **H**, **I**. Microsporophylls. **J**. Cross section of pinna. Bar = 1 cm.

Figures 31-13 to 31-17. *Cycas circinalis* L. **13, 14, 15.** Habit, showing branching and characteristic armour of leaf bases. **16.** Female plant with megasporophylls. **17.** Male plant with microsporangiate cone.

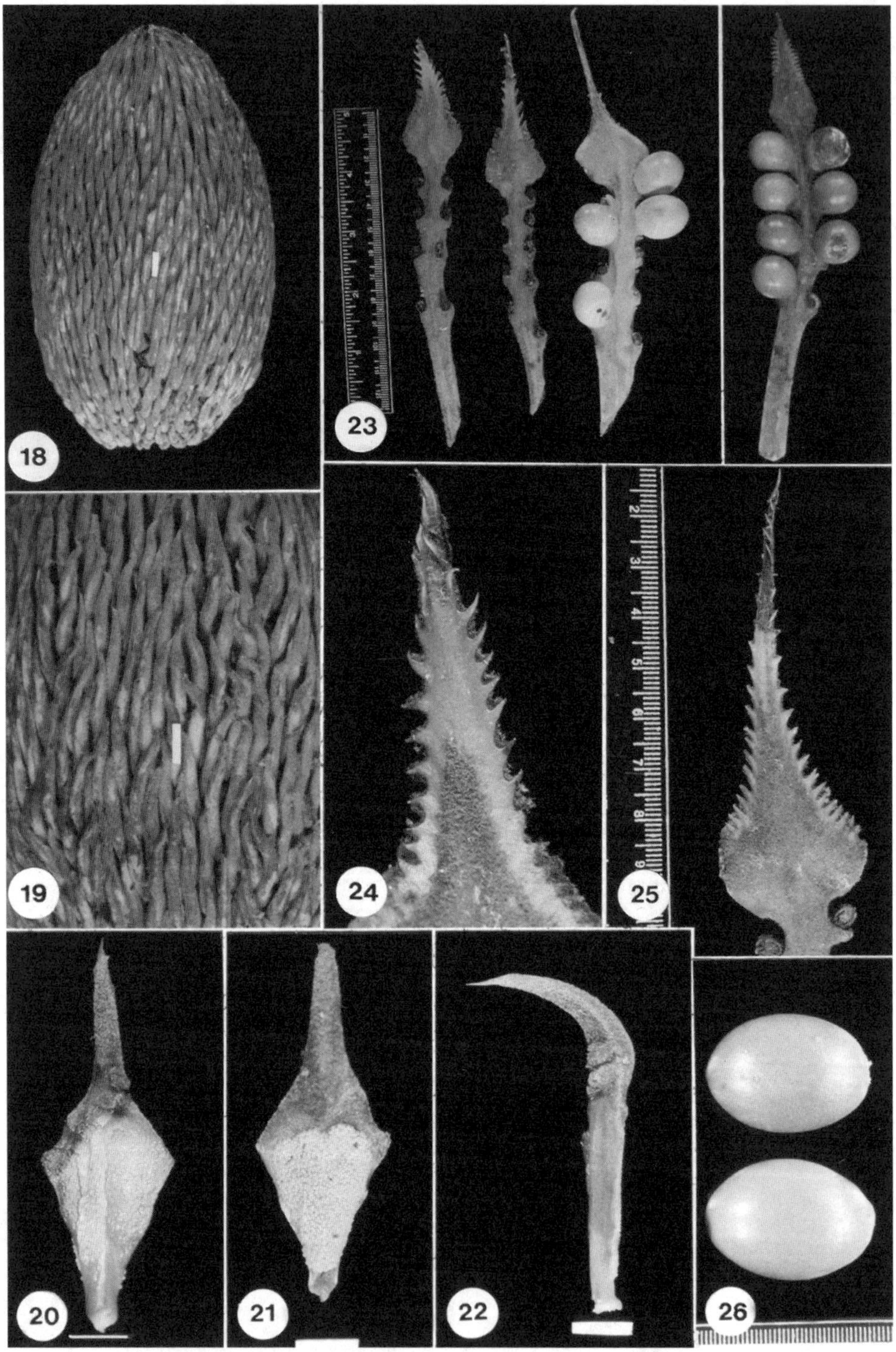

Figures 31-18 to 31-26. *Cycas circinalis* L. **18.** Microsporangiate cone. **19.** Closeup showing details of apical spine and arrangement of microsporophylls. **20, 21, 22.** Microsporophylls. **23.** Megasporophylls. **24, 25.** Sterile tip of megasporophylls. **26.** Ovules with sarcotesta.

areas of the Western Ghats (Malabar Coast), making field exploration slow and difficult. Therefore, *C. annaikalensis* may be currently categorized under the Data Deficient (DD) category of IUCN (2001).

Description of *Cycas circinalis* L.

Cycas circinalis L., Sp. Pl., 1188. 1753. (Figs. 31-12 to 31-26, 31-27B, C Table 31-1).

Type: Lectotype: the illustration: Rheede, Hortus Indicus Malabaricus Hortus III: Tab. 19, fide Stevenson in Jarvis et al. (1993).

Topotype: India, Kerala, Kozhikode, *Rita Singh & P.Radha– 0495, 0116,* (IPUH).

Stem arborescent, approximately 5 m tall, columnar with armor of leaf bases, sometimes branched, 18–83 cm diam., diameter of stem uniform from base to the distal region. Leaves 1–2.5 m long with 100–120 pairs of pinnae, new growth densely covered with ferruginous trichomes, *petiole* 50–70 cm long armed with spines approximately 2.5 mm long, median pinnae 40- 45 degrees to rachis, 30–35 cm long and 0.9–1.4 cm wide, margin entire, cross section of pinna shows presence of three mucilage canals at the either adjacent to or on the bundle sheath toward the phloem side of the vascular bundle, with the number declining to two at the median level and remains one near the tip region, columnar palisade cells with characteristic transverse partition. *Microsporangiate* cone oblong to oval, deep yellow to orange, 30–50 cm long, *microsporophylls* 3.1–8.1 cm long 1.4–2.5 cm wide, fertile zone 1.9–4.0 cm long, sterile zone length 1.5–4.2 cm (including the apical spine). *Megasporophylls* 15.3–33 cm long (including fertile stalk with ovule and apical sterile lamina), sterile apex of the megasporophylls 5–8.8 cm long and 2–4.1 cm wide (sterile tip width measured in the broad region of lamina) regularly dentate with caudate apex 1–4 cm long. *Ovules* 6–12 elliptic (3:2), 2.5–5.2 cm long 2–4.5 cm wide, green becoming orange to brown at maturity. Seeds platyspermic, sarcotesta approximately 3 mm thick, middle sclerotesta layer and membranous inner layer present, germination hypogeal.The additional white fibrous layer of the sarcotesta present in *Cycas beddomei* has not been seen in any of the *Cycas* seeds collected from the south Western Ghats.Measurements of various organs with ranges are given in Table 31-1.

Table 31-1 Comparison of Morphoanatomical Characters of *C. circinalis* and *C. annaikalensis.*

Characters (cm)	*C. circinalis*	*C. annaikalensis*
Plant height	70–500	52–500
Diameter	18– 83	19–61
Petiole length	50–70	50–90
Median pinnae length	30–35	26–35
Median pinnae width	0.9–1.4	0.8–1.2
Median pinnae angle to rachis (°)	40–50	50–55
Microsporophyll length (fertile portion)	1.9–4.0	2.0–3.2
Microsporophyll sterile portion length including spine	1.5–4.2	3.8–4.7
Megasporophyll length (fertile + sterile portion)	15.3–33.0	9.0–24.2
Megasporophyll sterile apex length + apical spine	5.0–8.8	1.9–5.5
Megasporophyll sterile apex width	2.0–4.1	1.3–4.0
Megasporophyll apical spine length	1.0–3.5	0.3–2.5
Ovule length × width	2.5–5.2 × 2.0–4.5	3.8–4.9 × 3.5–4.3
Number of ovules per megasporophyll	6–12	2–10
Pinnae anatomy (μm)		
Median pinnae thickness	240–320	290–360
Palisade cells length (two celled)	50–100	40–105
Stomata length × width	61.8 × 28.2	64.4 × 24.1
Stomatal index	5.2–5.9	4.6–5.2

Discussion

Superficially, in habit and foliage coloration, *C. annaikalensis* is similar to *C. circinalis*, but differs markedly from *C. circinalis* in having a steel grayish smooth trunk, in the angle of attachment of the median pinnae to the rachis, in the subconical microsporangiate cone with microsporophylls having straight upturned caudate apices, in the triangular-scutiform sterile apex of megasporophyll (1.9–5.5 × 1.3–4.0 cm) that has a prominently long-caudate apical median spine (0.3–2.5 cm long) and lateral teeth 0.05–0.8 mm long, and in the number of globose ovules per megasporophyll, which ranges from 2 to 10.

The confusion related to the specific limits of *C. circinalis* began with its inception, when it was established as the type of the genus *Cycas* by Linnaeus (1753) on the basis of a series of illustrations of "Todda Panna" or "Mouta Panna" (Rheede 1682) from India; no specimen relating to *Hortus Malabaricus* is known. Later, on identification and separation of *Cycas revoluta* Thunb. and *Cycas rumphii* Miq. out of the eight references cited by Linnaeus in the protologue of *Cycas circinalis,* there was a clear indication of the conglomeration and confusion related to the taxonomy of *C. circinalis*. However, amid all the confusion prevailing over inadequate knowledge of the true nature of *C. circinalis* L., one opinion that remained unchanged over the time in a world of profound changes was that Rheede's Todda Panna is the original *C. circinalis* (Miquel, 1839; De Candolle, 1868; Stapf. 1916; Schuster, 1932; Hill, 1995; Pant, 2002), and now one of Rheede's illustrations has been formally designated as its lectotype by Stevenson in Jarvis et al. (1993). Owing to the lack of a holotype of *C. circinalis*, the exact morphological nature of the species is not known, and as a result the majority of plants in cultivation have been known as *C. circinalis*, until Hill (1995) pointed out that they in fact belong to the *C. rumphii* complex with the exceptions of *Cycas beddomei* and *Cycas pectinata* in India, which is quite distinctive. Therefore, we perceived a need for exploration of Todda Panna localities in the Western Ghats. We described *C. circinalis* on the basis of extensive and intensive observations made during repeated field studies during 2001–2004, on the coastal and inland forest areas from Wayanad, Kozhikode, Malapuram, and Trissur districts of Malabar Coast. At each locality, the collected specimens/photographs were compared with Rheede's illustrations (Rheede, 1682, pp. 17–24). The comparisons included the shape of the columnar stem, leaf bases, leaf structure, megasporophylls, and shape of the sterile apex of the megasporophylls (Fig. 31-27).

The taxonomy of *C. circinalis* L. may be clearly understood in the context of geological formations and ecology of the Western Ghats. The Western Ghats or Sahayadri mountains run along the Western edge of Indian Deccan Plateau and separate the Plateau from the narrow coastal plain along the Arabian Sea. The range starts from Gujarat and runs approximately 1600 km almost to the southern tip of the Indian Peninsula. The only major gap or interruption in the range is the Palaghat Gap, which is approximately 30 km wide and joins the floral and faunal elements of Kerala to Tamil Nadu. Small ranges, including the Nilgiri Hills of northwestern Tamil Nadu and Biligirirangans southeast of Mysore, meet the Servarayan range (Shevaroys) and Thirumala range further east, linking

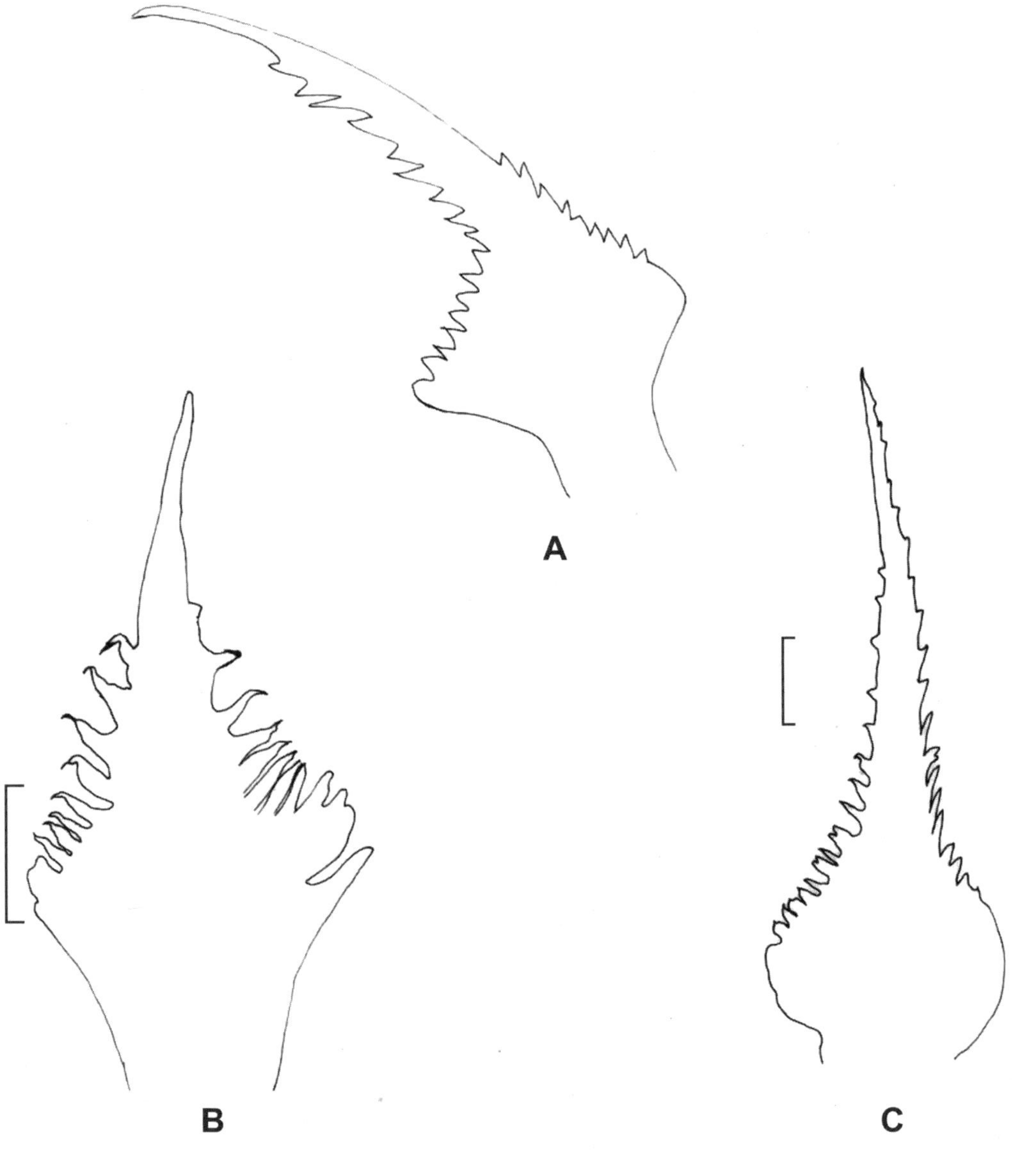

Figure 31-27. Comparison of sterile laminar tip of megasporophylls. **A.** A portion enlarged Todda Panna (Rheede, 1682, T 19). **B.** *C.annaikalenis* (Palaghat). **C.** *C.circinalis* (Kozhikode). Bar = 1 cm.

the Western Ghats to the Eastern Ghats. These hill ranges also serve as important wildlife corridors for the migration of both flora and fauna. The natural habitat of *C. circinalis* (Rheede's Todda Panna)occurs in the southern wet ecoregion of the Western Ghats (Malabar Coast), ranging from Wayanad as its northern limit and crossing Kozhikode, Malapuram, up to Trissur in the south. The suburban areas around Kozhikode still have intact mature populations of *C. circinalis* despite anthropogenic developmental disturbances. Therefore, we propose that this region be identified as the Topotype locality of *C. circinalis* L. (Todda panna). *Cycas circinalis* further extends into a drier ecoregion of Karnataka through Wayanad, where it shows variation in branching and other characters, as mentioned by Pant (2002).

Conclusions

The discovery of the new species *C. annaikalensis* from the hilly region of Palaghat Gap led us to confirm that *C. circinalis* (Todda Panna) is restricted to the region between Wayanad and Trissur of the south Western Ghats. It may or may not extend toward the eastern slopes of the Malabar Coast or it may be present on the other side as another taxon such as var. *swamyii* in southern Karnataka (Pant, 2002), since the *Cycas* of the eastern slopes (Alagarkoil–Mundanthurai range) lack characteristic mucilage canals present in the *Cycas* species of Western Ghats. Owing to lack of sufficient field data, the exact diversity in terms of species, subspecies, or variety may not be predicted now but the age-old assumption of most of the cycadologists throughout the world that Western Ghats of India is represented by a single species that is, *C. circinalis* is far from certain.

Acknowledgments

We thank the Ministry of Environment and the Forests, Govt. of India for financial support (D. O. No. J-22018/54/2000–CSC [BC], AICOPTAX Project) and Mr. P. Prakasam for helping in the field and collection of plant material. Rita Singh also thanks the Vice Chancellor, GGS Indraprastha University for approving full grants to attend the 7th International Conference on Cycad Biology CYCAD 2005 in Mexico.

Literature Cited

Biju, S. D. & F. Bossuyt. 2005. A new species of frog (Ranidae, Rhacophorinae, Philautus) from the rainforest canopy in the Western Ghats, India. Curr. Sci. 88(1): 175–178.

De Candolle, A. P. 1868. Prodromus systematis naturalis regni vegetabilis. 16(2). Paris.

Hill, K. D. 1995. The genus *Cycas* (Cycadaceae) in the Indian region, with notes on the application and typification of the name *Cycas circinalis*. Taxon 44: 1–9.

IUCN. 2001. IUCN Red List categories and criteria: version 3.1. IUCN Species Survival Commission. IUCN, Gland, Switzerland and Cambridge, UK.

Jarvis, C. E., Barrie, F. R., D. M. Allan & J. L. Reveal. 1993. A list of Linnean generic names and their types. Regnum Veg. 127: 1–100.

Linnaeus, C. 1753. Species plantarum. Stockholm.

———. 1754. Genera plantarum. Ed. 5. Stockholm.

Miquel, F.A.W. 1839. Melanges botaniques. Bull. Sci. Phys. Nat. Neerl. 2: 37–48.

Pant, D. D. 2002. An Introduction to gymnosperms, *Cycas* and Cycadales. BSIP Monograph No. 4 Birbal Sahni Institute of Palaeobotany, Lucknow, India.

Schuster, J. 1932. Cycadaceae.99 (4, 1) *in* Engler, A.(ed.), Das Pflanzenreich. W. Engelmann, Leipzig.

Singh. R. & P. Radha. 2006. A new species of *Cycas* from the Malabar Coast, Western Ghats, India. Brittonia 58: 119–123.

Stapf, O. 1916. *Cycas thourasii*. Bulletin of Miscellaneous Information 1–8.

Swamy, B.G.L. 1948. Contributions to the life history of a *Cycas* from Mysore (India). Am. J. Bot. 35(2): 77–88.

Van Rheede, H. A. 1682. Hortus indicus malabaricus, III: 17–24. (English translation).

SYSTEMATICS
CHAPTER 32

Variability of Natural Populations and Conservation Issues Facing Plicate-Leaved Zamia Species in Central and Western Panama

Alberto S. Taylor B, Jody L. Haynes, Gregory Holzman, and Jorge Mendieta

Abstract

The Isthmus of Panama is host to at least 12 described *Zamia* taxa, some of which occupy a very wide swath of western Panama. Many natural populations have been extensively and intensively studied for both vegetative and reproductive variability, taking into account such parameters as position, size, and relationship of stems, leaves, and cones; coning period; and reproductive mechanism. We obtained and analyzed data for seven population groups, and the results are given herein, with the aim of clarifying population dynamics, species plasticity, and the development of a coherent postulate of population structure of the species implied.

Resumen

El Istmo de Panamá es sede de por lo menos 12 taxa descritos del género *Zamia,* algunos de los cuales ocupan una extensión muy considerable del oeste de Panamá. Muchas poblaciones naturales han sido estudiadas en forma extensiva e intensiva tanto por sus características vegetativas como por su variabilidad reproductiva, tomando en cuenta parámetros tales como la posición, tamaño y la interrelación de tallos, hojas y conos; período de formación de conos; y mecanismo reproductor. Se han obtenido y analizado datos de siete grupos poblacionales y los resultados se presentan aquí, con la finalidad de esclarecer la dinámica poblacional, plasticidad específica, y el desarrollo de un postulado coherente sobre la estructura poblacional de las especies implicadas.

Introduction

The genus *Zamia* in the Isthmus of Panama (Stevenson, 1990; Schutzman et al., 1998; Jones, 2002; Whitelock, 2002; Hill et al., 2003) represents the highest concentration of species in the neotropics (Stevenson, 1993). The distribution of *Zamia* in Panama extends from the border with Colombia in the east to Costa Rica in the west (Stevenson, 1993, 2001). With at least 13 known species (Stevenson, 1993; Schutzman et al., 1998), there is considerable debate on species limits and relationships. There are notable works, both morphological and molecular, dealing with the subject (Hill et al., 2003; Caputo et al., 2004; Schutzman, 2004;). Populations of taxonomically misunderstood plicate-leaved species of *Zamia*, referable to both *Z. skinneri* and *Z.neurophyllidia* D. W.Stevenson, exist

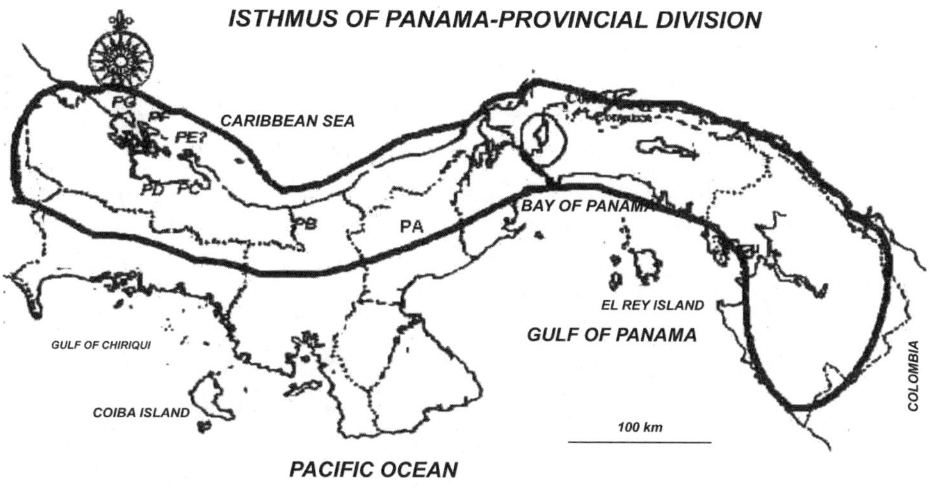

Figure 32-1. Distribution of plicate-leaved populations of Zamia in north–central to western Panama between 1982 and 2005.

mostly in western Panama, and are here considered as the *Zamia skinneri–Zamia neurophyllidia* complex.

The Isthmus of Panama was formed from a series of oceanic islands of volcanic origin, and the eastern part is only 3 million years old; hence, the introduction and colonization of plants and animals have been quite rapid (Coates, 1997). This process, of course, includes species of *Zamia* that are from northern or southern sources, as well as others that have evolved on the Isthmus itself, where they have formed localized endemics (Stevenson, 1993).

Population studies in cycads have addressed numerous issues, including herbivory and postfire events (Negrón-Ortiz et al., 2000), hurricane damage (Hirsch & Marier, 2002), gender performance in a cultivated cohort (Ornduff, 1996), among others discussed by Pérez-Farrera et al. (2004). The *Z. skinneri–Z. neurophyllidia* complex has been referred to before and attention has ensued, in both written (Stevenson, 1993; Whitelock, 2002; and Schutzman, 2004) and oral (D. Wm. Stevenson, pers. comm.) formats.

We conducted an extensive and intensive study on seven population of this species complex, which occupies a large swath of the isthmus from the central northeast, near the boundary of the Canal enclave, to the extreme northwest, impinging on the Costa Rican border (Fig. 32-1). The prime objective of this research was to amass a large database of vegetative and reproductive characteristics that could be used to compare populations within the complex to arrive at a better understanding of species delimitations. This research represents the most extensive population-level study conducted to date on this group, and the results presented herein will diminish some of the confusion surrounding this controversial species complex.

Materials and Methods

Every population represented in Table 32-1 and Figures 32-1 to 32-7 was studied by one or more of the authors, beginning with a general plant survey in early February 1982 by faculty and students of the Botany Department of the University of Panama. This initial survey benefited from the contribution of the systematist Dr. Robert Dressler. Dressler later wrote an unpublished manuscript, which he gave to Dr. Dennis Stevenson to carry

Table 32-1 Emergent leaf color and averages with standard deviation for variability of vegetative structures in populations of plicate-leaved *Zamia* species in central and western Panama.

Population	A	B	C	D	E	F	G
Plant structure							
Color of new leaf	Red	Red	Red	Green	Reddish	Green	Green
No. of leaves	3 1	3 2	5	5 2	12 5	11 5	10 4
Petiole length	101 (16)	96 (25)	67 (12)	68 (13)	75 (13)	62 (10)	77 (15)
Rachis length	51 (17)	62 (25)	71 (12)	49 (21)	80 (15)	84 (21)	74 (17)
Leaf length	150 (32)	156 (49)	137 (20)	117 (30)	154 (25)	146 (23)	151 (27)
Ratio P/R	2.1 (0.6)	2 (1)	1 (0.2)	2 (1)	1 (0.2)	0.8 (0.3)	1 (0.3)
No. of leaflets	9 (2)	10 (3)	11 (3)	10 (2)	16 (2)	19 (5)	16 (4)
Width apical l ft	14 (2)	12 (2)	13 (1)	12 (2)	10 (2)	9 (1)	10 (1)
Length apical l ft	44 (5)	48 (8)	45 (2)	35 (6)	39 (5)	29 (4)	30 (4)
Ratio W/L	0.3 (0.04)	0.26 (0.04)	0.3 (0.02)	0.3 (0.05)	0.3 (0.04)	0.3 (0.04)	0.3 (0.05)
Width median l ft	15 (2)	13 (2)	14 (2)	11 (2)	10 (1)	9 (1)	9 (1)
Length median l ft	47 (6)	50 (8)	46 (3)	35 (7)	41 (6)	29 (3)	30 (3)
Ratio W/L	0.3 (0.04)	0.25 (0.04)	0.3 (0.02)	0.3 (0.05)	0.23 (0.03)	0.28 (0.02)	0.3 (0.03)
Width basal l ft	14 (2)	11 (2)	10 (2)	10 (2)	9 (2)	7 (2)	8 (1)
Length basal l ft	45 (6)	47 (9)	45 (4)	32 (6)	37 (6)	27 (4)	27 (3)
Ratio W/L	0.3 (0.05)	0.5 (0.05)	0.2 (0.05)	0.3 (0.05)	0.2 (0.05)	0.3 (0.04)	0.3 (0.04)

Measurements are in cm and only healthy-looking leaves at least 100 cm in length are measured. Numbers in parentheses stand for 1 standard deviation.

on the work on the systematics of *Zamia* in Panama (Stevenson, 1993). The first author, fortunately, participated in field work with Stevenson, Russell Adams, and others, and embarked on investigating the Panamanian zamias, principally focused on ecology and reproductive biology.

We surveyed 7 populations of the *Zamia skinneri–Z. neurophyllidia* complex , and 31 morphological variables for vegetative and reproductive structures were recorded. Four of the populations had more than 50 mature individuals (coning specimens and non-coning plants).

Although the age of zamias cannot be accurately determined in the field, any plant with leaf size comparable to that of reproductive specimens in the population was considered an adult. Three populations (C, D, and F) were less intensively surveyed, due to highly dispersed individuals, or plants having been cut down, but measurements were taken in these populations for purposes of comparison.

Figures 32-2 to 32-7. Habit, leaf shape and reproduction of plicate-leaved *Zamia* species in Central and Western Panama. 2. Plants in population A. 3. Plant in population B. 4. Plants in population E. 5. Plant in population G. 6. Dehiscing pollen cone in island population E. 7. Clavicorn beetle *Pharaxonotha*, collected from dehiscing pollen cones in every population. Figure shown is from population E.

Vegetative data included the number of leaves and leaflets pairs, stem length and diameter, petiole length, rachis length, entire leaf length, and width and length of apical, median, and basal leaflets. Petiole to rachis length and leaflet width-to-length ratios were calculated. Annotations were made on the maturity and sex of cones when present. Dehiscing pollen cones were considered mature and so were seed cones with visibly ripe seeds.

Figures 32-8 to 32-13. Reproduction, growth and regeneration in plicate-leaved *Zamia* populations in western Panama. **8, 9.** Scanning electron micrograph of *Pharaxnotha* beetle. Proximal and distal, respectively. **10.** Close-up view of apparent root regeneration from detached petiole base. **11.** Old plant in population E, growing over and quite near the surface of the sea. **12.** Regeneration in population F. **13.** Giant plant in population F, destroyed between 2002 and 2004.

Other data included cone length and diameter, length of peduncle or stipe, and maximum diameter of the same near the base of the cone. Number of sporophyll columns and rows were counted, and the ratio of peduncle diameter to length for both seed and pollen cones was calculated. The data in Table 32.1 are simple statistics for values of average (Avg) and standard deviation (SD), and the color of emergent leaves, a very conspicuous and stable characteristic in four of the populations studied, was included. A statistical analysis between populations A, B, E, and G, making use of the SAS ANOVA procedure for Duncan's multiple range test, was made because more data based on adult size of individuals were available (Table 32-2). All populations, with more inclusive data including reproductive organs, will be addressed in a future publication dealing with the systematics of the of this species complex.

We collected pollinator beetles from pollen cones in all populations over the years, one population having been subjected to exclusion experiments by the last author (population G), as described elsewhere (Norstog and Nicholls, 1997; Stevenson, pers. comm.).

Scanning electron micrographs were made at the University of Panama (Universidad de Panamá) scanning microscope facility. Uncoated beetles were used (Figs. 32-8 and 32-9).

Results and Discussion

Population Structure and Variability

Population A occurs at about 600 or more m.a.s.l., while population B, considered to be conspecific, is found from sea level up to about 200 m. Populations C and D are also land forms near sea level and in shaded rain forest, in contrast with populations E, F, and G, which are found at sea level and near the sea, receiving on some occasions salt spray from the water, and being lightly shaded to barely shaded in many places.

Populations A and B are here considered as representatives of *Zamia skinneri*, and F and G are somewhat similar to the description of *Z. neurophyllidia* D. W. Stevenson, although some differences have been observed. Population E appears to be a new taxon with intermediate features between populations A/B and F/G. Adequate data for populations C and D were not available at the time of this writing, and neither were there enough samples of population F. These will be included in the future systematic assessment of the species complex.

Table 32-1 shows that there is much variability in every parameter measured for all populations, based on simple statistics of means and standard deviations. It should be noted that the color of emergent leaves in adult plants and in most plantlets of the same population is a reliable measure of variability.

Table 32-2 Results of the SAS ANOVA procedure for Duncan's multiple-range test for comparison of select vegetative variables of plicate-leaved populations of Zamia in central and western Panama.

	Variables									
	Nleaf	**Pet**	**rac**	**pet/rac**	**Tleaf**	**Nfol 1**	**nfol r**	**wmdfol**	**lmdfol**	**wmdfol/lmdfol**
F value	86.17	41.06	32.25	83.05	3.77	124.35	114.93	95.73	111.36	73.00
	A PopE	A PopB	A PopE	A PopA	A PopB	A PopE	A PopE	A PopA	A PopB	A PopG
Duncan	B PopG	A PopA	A PogG	B PopB	B PopE	A PopG	A PopG	B PopB	B PopA	A PopA
Grouping	C PopA	B PopG	B PopB	C PopG	B PopG	B PopB	B PopB	C PopE	C PopE	B PopB
	C PopB	B PopE	C PopA	C PopE	B PopA	C PopA	C PopA	C PopG	D PopG	C PopE
	Wapfol	**lapful**	**wapfol/ lapfol**	**wbslfol**	**Lbslfol**	**wbslfol/ lbslfol**				
F value	60.66	109.76	48.95	68.82	105.67	27.48				
	A PopA	A PopB	A PopG	A PopA	A PopB	A PopA				
Duncan	B PopB	B PopA	B PopA	B PopB	A PopA	A PopG				
Grouping	C PopG	C PopE	C PopB	C PopE	B PopE	B PopB				
	C PopE	D PopG	C PopE	C PopG	C PopG	B PopE				

nleaf = number of leaf on plant; pet = petiole length; rac = rachis length; tleaf = leaf length; nfoll = number of left side leaflets; nfolr = number of right side leaflets; wmdfol = width median leaflets; lmdfol = length median leaflet; wapfol = width apical leaflet; lapfol = length apical leaflet; wbslfol = width basal leaflet; lbslfol = length basal leaflet.pr > $F < 0.0001$. Populations with the same letter at the left in each column are not significantly different for the characteristic.

Table 32-2 gives the result of the SAS ANOVA of vegetative characteristics of populations A, B, E, and G. Populations A and B, although occurring in separate habitats, are similar in the number of leaves on trunk, petiole length, and length of basal leaflets. Populations E and G share a large number of size features of the leaves, including petiole length, rachis length, ratio of petiole length to rachis length, total leaf length, number of leaflets, and maximum widths of leaflets (apical, median, and basal). Populations A and E share same value for total leaf length; populations A and G also share similar values for total leaf length, and the width-to-length ratio of median and basal leaflets. Populations B and E share the same values for the width-to-length ratio of apical leaflets and basal leaflets. The data show quite a number of similarities in size of vegetative characteristics of leaves among the four populations analyzed, and some of this may be due to ecological differences. Although the results show many size differences of leaf features between populations A and B, both considered conspecific, the differences could be due to ecological and habitat differences, because both populations appear to be the same in leaf number, petiole size, emergent leaf color, and reproductive organs characteristics, the latter to be dealt with in a future publication on the systematics of the group.

Although it has been stressed that size and even form differences in leaf characteristics do not always mean specific differences among populations growing in different habitats (Newell, 1986), when these differences are many, the interpretation for considering species differences might vary among different workers. This has been the case in the research of Newell (1986), while studying the variations of leaflet morphology among populations of *Zamia* in Porto Rico and in which differences in leaflet morphology were found in three populations of the genus. The "narrow leaflet" form has been ascribed to *Zamia portoricensis* Urban, while the "wide leaflet" form has been considered to be *Zamia latifoliolata* Prenl. or *Zamia debilis* L. f. Some workers, such as Eckenwalder (1980), argue for synonymy based on a presumptive continuous variation or frequency distribution in leaflet widths throughout the Caribbean, but Stevenson and others argue for at least five or more valid species among Caribbean zamias (Norstog & Nicholls, 1997).

Much research on the value of leaf characteristic variations in cycads has been carried out by workers at La Selva Biological Station in Costa Rica, Central America (Clark & Clark, 1987, 1988; Clark et al., 1992) with populations of a species considered to be *Zamia skinneri* but which has since come to be considered by others as *Z. neurophyllidia* D. W. Stevenson, citing the same works (Norstog & Nicholls, 1997). However, there is some doubt as to the real specific entity of the tax on, and future research, including both the Costa Rica and Panama forms, will be undertaken to clear up matters. According to Clark and Clark (1987), the result of their 6-year research on the aforementioned species shows correlation between reproductive performance and plant size, given as number of leaves per plant or as number of leaflets in the largest leaf. This study was circumscribed to temporal and environmental patterns of reproduction. Light availability in primary and secondary forests also had an influence on reproductive growth. More light in open canopy of secondary forest meant a higher degree of sex organs formation and function. This result was the outcome of a 6-year study in a guarded biological station in a tropical rain forest. In contrast, all the populations in the present work are of free access to the public and are being continually degraded by human actions (building of fences for cattle, cutting down of plants to make pathways through the forest, etc.). Even population A, within a national park, is vulnerable to poaching and destruction, albeit on a smaller scale. This means that population values tend to change on a yearly basis. Even so, enough data have been gathered to make more than preliminary assumptions on a population scale, with four populations under analysis and two to study the effect of plant size on reproductive growth.

Tables 32-3 and 32-4 contain data correlating plant size, measured by number of leaves and number of leaflets on largest leaf, respectively, with cone production. Usually, more pollen plants than ovulate plants are present, although the highest percentage of coning was intermediate, considering the lower and the higher measures of plant size and coning activity. In other words, the highest values for plant size did not correlate with more coning activity than lower values. More data during additional years are warranted here to be comparable with the 6-year study at La Selva in Costa Rica for estimating the so-called operational sex ratio, and the "adult" or cumulative sex ratio. The former refers to pollen dehiscence correlated with ovule receptivity in seed cones, and the second to cumulative data on coning in the populations during many years of observation (Clark & Clark, 1987). However, coning sex ratios are given for the years of observation in many of the populations (Table 32-5), and it can be clearly seen that they change from year to year, that coning, per se, is very low in all populations and that in some years no seed cones were observed. In every case, when a ratio of unity was not obtained, the population was definitely male biased. This reinforces the fact that ovulate plants are more resource dependent than pollen plants, as has been seen elsewhere (Clark & Clark, 1987, 1988), although

Table 32-3 Maximum plant size by number of leaves and leaflets respectively correlated with reproductive activity in a *Zamia skinneri* population in northwestern Panama during 2004–2005.

Number of Leaves	Vegetative Plants	Pollen Plants	Seed Plants
2–4	(22) 67	(9) 27	(2) 6
5–7	(6) 35	(9) 53	(2) 12
8–10	(2) 50	(1) 25	(1) 25
Leaflets on largest leaf			
8–11	(13) 72	(3) 17	(2) 11
12–15	(10) 48	(9) 43	(2) 9
16–18	(5) 50	(4) 40	(1) 10

Sample size is given in parentheses with percentages in the population at right.

Table 32-4 Maximum plant size by number of leaves and leaflets respectively correlated with reproductive activity in a *Zamia* sp. nov. in western Panama in 2004.

Number of Leaves	Vegetative Plants	Pollen Plants	Seed Plants
6–12	(26) 81	(2) 6	(4) 13
13–19	(11) 69	(2) 12	(3) 19
20–28	(3) 67	(0) 0	(1) 33
Leaflets on largest leaf			
10–14	(18) 95	(0) 0	(1) 5
15–17	(10) 63	(4) 25	(2) 12
18–20	(12) 71	(0) 0	(5) 29

Sample size is given in parentheses with percentages in population at right.

Table 32-5 Estimated population size, time of cone dehiscence, and coning sex ratios of plicate-leaved *Zamia* populations in central and western Panama.

Population	*N*	DT	Sex Ratio (Male:Female)	*n*	Year
A	>100	Nov	2 : 0	(2)	2003
	>100	Nov	1 : 1	(4)	2004
	>100	Nov	1 : 3	(8)	2005
B	<100	Nov	1 : 1	(2)[a]	2003
	<100	Nov	1 : 1	(2)	2004
	<100	Oct	8 : 1***	(26)	2005
C	<50	Sept	1 : 0	(1)	2004
	<50	Sept	5 : 1	(6)	2005
D	17	Oct	1 : 0	(1)	2000
	> 40	Sept	7 : 1	(8)	2005
E	>200	Sept–Oct	4.3 : 1	(7)	2004
F	<100	Apr–May	1 : 1	(4)	2000
	>500	June	1 : 0	(1)	2001
	>1000[b]	Oct	2 : 1	(3)[c]	2004
	>1000[b]	Sept	2 : 1	(26)[d]	2005
G	>400	Apr–June,	2 : 1	(6)	2000
	>400	Sept–Dec			
	>400	Sept–Oct	1 : 1	(2)	2001
	<100	Sept	1 : 1	(15)	2005

N = number of plants seen or estimated; *n* = number of coning plants seen in a given year; DT = time of observed cone dehiscence. \

[a]Seed cone not seen but existence based on cone with ripe seeds in 2004.

[b]> 1000 with estimates > 5000.

[c]One male plant had 4 cones and another was 7-branched with 4 cones on each of four of the apices .

[d]More cones were seen but only a random selection was made. Sex ratios were tested for departure from 1:1 with significance by a binomial test. *** $p < 0.001$.

pollen plants can form up to seven cones on the same plant in a coning cycle. There was a higher tendency toward unity in coning sex ratio in population G, even with a higher output of cones. But it must be mentioned that population G has been severely destroyed and the remaining individuals are growing in minimal shade and very near the sea shore. This fact, perhaps, is reminiscent of a population of *Zamia neurophyllidia* (Clark & Clark, 1987), reported as *Z. skinneri* in the original publication, at La Selva Biological Station in Costa Rica, growing under higher canopy openness.

Reproductive Biology

Phenology

During the study of the various populations of the *Zamia skinneri–Zamia neurophyllidias*–complex, notes were made of certain aspects of reproductive biology. Pollen cone reproductive cycle for populations A and B is from about late July to early November, with dehiscing cones in late October to November. Data for the other populations were extrapolated from the time of cone dehiscence, and previous knowledge for almost all the known Panamanian Zamias. In these groups, with the exception of *Zamia elegantissima* Schutzman, Vovides and Adams, coning events take 3.5–4 months, from the time of first sight of cone emergence between the bracts, to time in which a pollen cone is seen dehiscing or an ovulate cone is seen receptive (usually later than the pollen cones) with dehiscence occurring over a 1- to 2-month period. Spent pollen cones were observed in population G in April, September, and November, and new cone formation was observed in June and September. In this case, it appears that pollen cones are produced many times during the year, instead of during a single coning cycle. This also has been observed in population F, while populations A, B, and E appear to maintain an annual cycle. In populations C and D, both receptive seed cones and dehiscing pollen cones have been found in September, and it appears that this maturing of cones would proceed into early or mid-October. The same was found for population E. Seed cones take more than 14 months to mature after pollination in population G, but only a year in population E. In both populations A and B seed mature in the cones about a year after pollination. In previous unpublished reports, it has been noted that certain *Zamia* populations do indeed form mature seed cones at different times, even within the same population.

Pollination Biology

Pollinators (Figs. 32-7 to 32-9) have been recovered from dehiscing pollen cones (Fig. 32-6) in all populations and, in every case, they have turned out to be one or more species of the clavicorn beetle *Pharaxonotha*, which has also been found in other zamias, such as *Zamia pumila* and *Z. furfuracea* (Norstog et al., 1986; Tang, 1987; Norstog & Fawcett, 1989; Stevenson et al., 1998). *Pharaxonotha*, accompanied sometimes by the weevil *Rhopalotria*, has also been found as the major pollinator in all the Panamanian zamias that have been studied in this respect (Taylor, unpubl. data). The importance of insect pollination for the survival of cycads cannot be underestimated, considering the wide-ranging data for such in different genera and species (Norstog & Fawcett, 1989; Ornduff, 1991, 1992; Donaldson, 1997; Vovides et al., 1997; Stevenson et al., 1998; Mound & Terry, 2001; Schneider, 2002; Wilson, 2002; Hall et al., 2004). For effective conservation of cycads in the wild, their pollinators must also be conserved (Norstog & Fawcett, 1989).

Considering the aforementioned factors, pollination by the same species of insect could be additional evidence for close phylogenetic relationships of this species complex in western Panama.

Conservation Status and Threats

Many of the *Zamia* populations of western Panama are threatened by agricultural expansion, and changes in land use and some sites have been under intense pressure (Fig. 32-12). Other localities under less pressure have had plants recover naturally, even after being exposed to seawater (Fig. 32-11). Many plants, or parts thereof, whether in the wild or under cultivation, have grown new apices from detached trunks, and, in two occasions, from what appears to be roots growing out of a petiole base. In both cases, the petioles with roots were found in population B. (Fig. 32-10). It has been suggested that this might have been a case of roots growing out of small pieces of trunk attached to the petiole (Vovides, pers. comm.) so the matter remains open to further research. Given the opportunity, new apices can reform new plants.

The threats to the cycads arise from a complex of circumstances based on many factors. The Atlantic side of Panama lacks an extensive road system and, because of this, populations are relatively difficult to access. This has been quite advantageous in allowing undisturbed cycad habitats to remain intact. Many of the populations are accessible only by boat, which is a major source of transportation in the area. Many people who live in the cycad sites do not know anything about them or have very little use for them in their daily living, yet others have medicinal uses for the cycads or employ the mucilage in the assembly of guitar or other local musical instruments. Where the cycads are used, a certain amount of destruction from harvesting can hamper a population's recruitment ability for a very long time. Because of this, all cycads in western Panama that are part of this study are vulnerable to habitat destruction, the inability to reproduce, or to the possible depletion from over-collecting.

Population A might survive for many years, unless an adequate road is made available to people living near the site; however, population B is practically decimated to the extent that almost half of the mature plants observed in November 2004 had been cut down by October 2005.

At the present time, habitat destruction is a threat to each of the populations studied, except population E. The plants in populations F and G live in beach strand habitat and are being exploited for their perceived medicinal value by the local people, resulting in a sustained harvest of trunks. Because of the very large cycad populations, the harvest of trunks at the sites has not had a pronounced adverse effect on regeneration, but perhaps the tallest *Zamia* in the region, measuring greater than 3 m (Fig. 32-13), is now gone. Regrown damaged trunks are commonplace, and the plants are typically in full reproductive capacity within a 10-year period. The fact that many plants were observed with multiple trunk scars in succession is evidence that the harvest is a long-standing tradition. Habitat destruction from development of beach house lots is more of a threat on the island sites in which these plants grow. Fortunately, a national park, created for the protection of sea turtle nesting sites on one of the islands, has, for the moment, provided the nearby cycad population a sanctuary. However, another large population in the area is under threat from ongoing development.

Population E is one of the few populations without large-scale habitat destruction, which appears to be due to its isolation and human population expansion. The real threat here comes in the form of future development of a resort or exploitation by collectors, because of its relatively small area. The uses of the trunks by humans in this area are unknown at this time, but are most likely of medicinal origin.

In population C, cycads are being wiped out by slash-and-burn agricultural practices. The plants grow on steep hillsides that are rocky and unsuitable for crops, but the forest has been largely cut down to make it easier to hunt a large rodent, similar to the capybara of South America. The animal burrows into the hills and the hunters find it easier to capture without the thick vegetation. People of the area know the plants but have no use for them. The cycads are seen as being beautiful, and, after the landowners were informed of their botanical value, they became interested in helping preserve what remained of the hillside population. It is our belief that education is very helpful in the preservation of these plants, but the habitat will be preserved only if alternative sources of revenue are secured for the local people. Aboriginal people living in the area have no uses for the plants. These were found far from human settlements along creeks in rain forest. Many of the creeks are used as water sources, and the water is piped to distant villages. The cycads are neither used nor appreciated as being special, and, because of the spines on the petioles, they are cut with machetes (as are all spiny plants growing along the pathways). In this area, more cycads were observed cut down than standing, and the main reason for this was their simply being in the way. Interestingly, the cut pieces readily regenerate new growth, even the decapitated apices. Because the population consisted of very few individuals of low reproductive rate, no seedlings were found. Undoubtedly, there are other undisturbed populations within the same general area, and further study might yield healthy populations. However, the lack of human uses and their spiny nature put these plants in peril. The aboriginal people in this area are very aware of the area's vulnerability and limited resources. Again, education will be a key factor in conservation, and the few people familiar with the plants immediately appreciated their beauty when informed of their botanical significance. In fact, following a brief "on-the-spot" history lesson, some were seen looking in botanical texts under candlelight with their children to learn more about the plants.

Conclusions and Recommendations

Summary of Morphometric and Pollination Biology Results

The populations assessed in this study can be distinguished based on general features of leaf size, petiole length-to-rachis length ratios, and emergent leaf color. Stem height and diameter were not taken into account, as most plants in almost every population had been repeatedly cut.

The data delimit at least three distinct species, two of which have been described and at least one that is new. Plants of populations A and B can be ascribed to *Z. skinneri*, and those of populations F and G have been ascribed by many workers to *Z. neurophyllidia*. Population C needs further study, because the plants appear to represent different taxa, some being similar to *Z skinneri* and some of dubious affinity. Population E is very interesting because the plants appear in some ways to be intermediate between populations A/B and F/G but in other ways are distinct from them both. The data herein suggest that population E might represent a species lacking description, and that will be studied in detail in an upcoming systematic treatment.

The data shed light on either an incomplete circumscription of *Z. neurophyllidia*, according to Stevenson or the existence of a distinct species represented by population F and possibly G. The future systematic publication will take up this issue in detail. .

In contrast to Stevenson's (1993) report of *Z. neurophyllidia* being somewhat small in stature (Fig. 32-5), when left undisturbed for decades these plants attain a very large size indeed (Fig. 32-13). In fact, Stevenson's (1993) description of *Z. neurophyllidia* and reassessment of *Z. skinneri* appear to include a mixed bag of representative specimens, including plants that are more like those in populations A/B and others that are more like those in populations F/G. Accordingly, it is thought that the taxa represented by populations E and F/G are in need of description and this study can aid in achieving that goal.

Suggestions for Conserving the *Z. Skinneri*–*Z. neurophyllidia* Complex

1. Information on populations, number of individuals, and present and future threats should be supplied to the country's environmental institutions.
2. The owners of lands on which the plants grow should be informed of their importance, made aware of the threats against their survival, and convinced to participate in their conservation within the framework of ongoing developments, so that certain areas are preserved for these plants.
3. In some cases, arrangements should be made for transplantation of a certain number of individual plants to educational institutions, especially to the University of Panama, where there is already a functional cycad garden, albeit very modest in scope and with hardly any financial assets of its own.
4. The environmental authorities, with the assistance of cycad workers at the university level and growers if these exist, should organize seminars at the grass-roots level and also at higher levels to make the lay public aware of the importance of

preserving these plants. This recommendation was put into practice by three of the authors in western Panama in a rural village elementary school, with the attendance of the environmental authorities, education representatives, teachers, and schoolchildren.

The National Environmental Authority (ANAM Spanish acronym) has been informed of potential and ongoing threats and destructions of plants, especially in populations E, F, and G, but the institution is understaffed and underfinanced, so the problem also needs to be addressed by the private sector (nongovernmental organizations) and international cooperative efforts.

Assignment/Reassignment of Population Groups (Taxa) to the IUCN Red List of Threatened Species

According to the 2004 IUCN Red List of Threatened Species (IUCN, 2004), only *Z. skinneri* among the plicate-leaved populations of central and western Panama is considered threatened, being assigned the following codes, according to the 2001 Red List Categories and Criteria (ver. 3.1): EN A2acd; C1.However, in view of the destruction of *Zamia* sites observed during this study, and taking into consideration the 2001 Red List Categories and Criteria (ver. 3.1), we propose a reassignment of this species (i.e., the mainland red-emergent leaved populations) to critically endangered status (CR B2abc). Our population E, at the moment considered as *Zamia* sp. nov, should be assigned as CR B1ab(ii,iii,v)+2ab(ii,iii,v). Our populations F and G, tentatively considered as being similar to *Z. neurophyllidia* D.W. Stevenson, until further treatment in an upcoming publication, should be assigned the codes CR B1ab(i,ii,iii,iv,v)+2ab(i,ii,iii,iv,v). For the present assessment, the following issues were considered: geographic range, comprising extent of occurrence of each population being less than 100 km^2, estimates indicating severely fragmented, continuing decline observed in extent of occurrence, area of occupancy and area /extent and / or quality of habitat (rapidly being destroyed by human activities).

Acknowledgments

Many thanks are due the Universidad de Panama (University of Panama) for academic and infrastructure support through the offices of the President, Dr. Gustavo García de

Paredes, and the Vice President for Research and Graduate Studies; Dr. Betty Ann Rowe de Catsambanis; Montgomery Botanical Center; the Autoridad Nacional de Ambiente (National Environmental Authority); Dr. Ivan Valdespino, who always gave material and moral support; and to the New York Botanical Garden through the supportive auspices of Dr. D. Wm. Stevenson, with whom we had the opportunity to find and inspect new sites. Thanks are also due Roberto Cambra, research assistant in the G. B. Fairchild Invertebrate Museum of the University of Panama, who identified the first *Pharaxonotha* specimens; Russell Adams, who accompanied us and gave valuable suggestions along the way; engineer Maritza Taylor (daughter of the first author), who helped with the statistical analyses and preparation of the plates; Professor Edilberto De León of the Statistics Department of the Universidad de Panama, for assistance with the SAS ANOVA program; Captain Louis Anceaux (US Ret.), who helped finding populations and worked with three of the authors in the field; the colleagues of the Departamento de Botánica (Botany Department) of the University of Panama, for continuing moral support; and to all colleagues, students (especially to Alex Espinoza); farmers, and people (special thanks due here to Santiago Quiroz) who helped in showing the way.

Literature Cited

Caputo, P., S. Cozozolino, P. De Luca, A. Moretti & D. Wm. Stevenson. 2004. Molecular phylogeny of *Zamia* (Zamiaceae). Pp. 149–157 *in* T. Walters & R. Osborne (eds.), Cycad classification: Concepts and recommendations. CABI, Cambridge, MA.

Clark, D. A. & D. B. Clark. 1987. Temporal and environmental patterns of of reproduction in *Zamia skinneri,* a tropical rain forest cycad. J. Ecol. 75: 135–149.

——— & ———. 1988. Leaf production and the cost of reproduction in the

neotropical rain forest cycad, *Zamia skinneri*. J. Ecol. 76: 1153–1163.

Clark, D. B., D. A. Clark & M. H. Grayum. 1992. Demography of a neotropical Rain forest cycad, *Zamia skinneri* (Zamiaceae). Am. J. Bot. 79: 28–33.

Coates, A. G. (ed.). 1997. Central America: a natural and cultural history. Yale University Press, New Haven, CT.

Donaldson, J. S. 1997. Is there a floral parasite mutualism in cycad pollination? The pollination biology of *Encephalartos villosus* (Zamiaceae). Am. J. Bot. 84: 1398–1406.

Eckenwalder, J. E. 1980. Taxonomy of West Indian cycads. J. Arnold Arb. 61: 701–722.

Hall, J. A., Walter, G. H., Bergstrom, D. M. & Machin, P. 2004. Pollination ecology of the Australian cycad *Lepidozamia peroffskyana* (Zamiaceae). Austral. J. Bot. 52(3): 333–343.

Hill, K. D., M. W. Chase, D. W. Stevenson, H. G. Hills & B. Schutzman. 2003. The families and genera of cycads: a molecular phylogenetic analysis of Cycadophyta based on nuclear and plastid DNA sequences. Int. J. Plant Sci. 164: 933–948.

Hirsh, H. & T. Marler. 2002. Damage and recovery of *Cycas micronesica* after typhoon Paka. Biotropica 34: 598–602.

IUCN. 2004. 2004 IUCN Red list of threatened species. IUCN, Gland Switzerland, and Cambridge, UK.

Jones, D. L. 2002. Cycads of the world. 2nd ed. Smithsonian Institution Press, Washington, DC.

Mound, L. A. & I. Terry. 2001. Thrips pollination of the central Australian cycad *Macrozamia macdonnellii* (Cycadales). Int. J. Plant Sci. 162: 147–154.

Negrón-Ortiz, V., D. L. Gorchov & G. J. Breckon. 2000. Effects of fire season and postfire herbivory on the cycad *Zamia pumila* (Zamiaceae) in Slash Pine Savanna, Everglades National Park, Florida. Int. J. Plant Sci. 161: 659–669.

Newell, S. J. 1986. Variation in leaflet morphology among three populations of *Zamia* L. in Puerto Rico. Taxon 35: 234–342.

Norstog, K. J. & P. Fawcett. 1989. Insect-cycad symbiosis and its relation to the pollination of *Zamia furfuracea* (Zamiaceae) by *Rhopalotria mollis* (Curculionidae). Am. J. Bot. 76: 1380–1394.

——— & T. J. Nicholls. 1997. The biology of the cycads. Cornell University Press, Ithaca, NY.

———, D. Stevenson & K. Niklas. 1986. The role of beetles in the pollination of *Zamia furfuracea* L. fil. (Zamiaceae). Biotropica 18: 300–306.

Ornduff, R. 1991. Size classes, reproductive behavior, and insect associates of *Cycas media* (Cycadaceae) in Australia. Bot. Gaz.152: 203–207.

———. 1992. Features of coning and foliar phenology, size classes, and insect associates of *Cycas armstrongii* (Cycadaceae) in the Northern Territory, Australia. Bull. Torr. Bot. Club 119: 39–43.

———. 1996. Gender performance in a cultivated cohort of the cycad *Zamia integrifolia* (Zamiaceae). Am. J. Bot. 83: 1006–1015.

Pérez-Farrera, M., A. P. Vovides, L. Hernández-Sandoval, D. González & M. Martínez. 2004. A morphometric analysis of the *Certatozamia norstogii* complex (Zamiaceae), Pp. 127–136 *in* T. Walters & R. Osborne (eds.), Cycad classification: concepts and recommendations. CABI, Cambridge, MA.

Schneider, D., M. Wink, F. Sporer & P. Lounibos. 2002. Cycads: their evolution, toxins, herbivores and insect pollinators. Naturwissenschaften 89: 281–294.

Schutzman, B. 2004. Systematics of Meso-American *Zamia* (Zamiaceae). Pp. 159–172 *in* T. Walters & R. Osborne (eds.), Cycad classification: concepts and recommendations. CABI, Cambridge, MA.

———, A. P. Vovides & R. S. Adams. 1998. A new *Zamia* (Zamiaceae, Cycadales) from Central Panama. Phytologia (September 1998) 85(3): 137–145.

Stevenson, D. Wm. 1990. Morphology and systematics of the Cycadales. *in* Stevenson, D. Wm. (ed.), The biology, structure and systematics of the Cycadales. Mem. NY Bot. Gard. 57: 8–55.

———. 1993. The Zamiaceae in Panama with comments on phytogeography and species relationships. Brittonia 45: 1–16.

———. 2001. Flora de Colombia: Monografía No. 21. Instituto de Ciencias Naturales , Universidad Nacional de Colombia, Bogotá, D.C.

———, K. J. Norstog & P. K. Fawcett. 1998. Pollination biology of cydads. Pp. 227–294 *in* S. J. Owens & P. J. Rudall (eds.), Reproductive biology in systematics, conservation, and economic botany. Royal Botanic Gardens, Kew, UK.

Tang, W. 1987. Insect pollination in the cycad *Zamia pumila* (Zamiaceae). Am. J. Bot. 74: 90–99.

Vovides, A. P., N. Ogata, V. Sosa & E. Peña-García. 1997. Pollination of endangered Cuban cycad *Microcycas calocoma* (Miq.) A.D.C. Bot. J. Linn. Soc. 125: 201–210.

Whitelock, L. M. 2002. The cycads. Timber Press, Portland, OR.

Wilson, G. W. 2002. Insect pollination in the cycad genus *Bowenia*, ex Hook. f (Stangeriaceae). Biotropica 34: 438–441.

SYSTEMATICS
CHAPTER 33

Morphological Variation of Dioon (Zamiaceae) from the Pacific Slope of Mexico using a Multivariate Analysis

Georgina Vargas Amado, Jorge Alberto Pérez de la Rosa, José Pablo Torres Morán, and Waldina Patricia Reyes Velázquez

Abstract

The descriptions of the species of the genus *Dioon* by taxonomists have shown variations that put in doubt the characterization of different taxa. Hence, we analyzed the populations distributed along the Pacific slope of Mexico to study the morphological variation of *Dioon* using a principal components analysis, a multivariate method. We sampled 12 localities, from Sonora to Chiapas states, and compared the species with *D. edule* and *D. spinulosum*, three localities from the Gulf of Mexico slope. The results showed five groups. The Pacific slope groups exhibited a clinal morphological variation pattern north–south, except for the Sonora populations, which showed a multivariate effect similar to that of the Gulf 's populations. With these results, the identity of *D. sonorense* was confirmed to be the northernmost population in Mexico. Principal components analysis supports the results of previous work on the morphological variation clinal of *D. edule*. In addition, a close relationship exists among populations from the southern and western Pacific, showing that individuals from the type localities of *D. holmgrenii* and *D. tomasellii* shared similar morphological characteristics in such a way that the two taxa established a morphological complex.

Resumen

La descripción de las especies del género *Dioon* por los taxónomos han mostrado variaciones que ponen en duda la caracterización de los diferentes taxa. Por lo anterior, el objetivo del presente estudio fue analizar las poblaciones de la vertiente pacífica para conocer la variación morfológica de las plantas del género mediante la aplicación del método multivariado conocido como Análisis de Componentes Principales. Se estudiaron 12 localidades ubicadas en la vertiente del océano Pacífico Mexicano del estado de Sonora al estado de Chiapas. Se contrastaron con las especies testigo (*D. edule* y *D. spinulosum*) correspondientes a tres localidades de la vertiente del Golfo de México. Los resultados mostraron cinco grupos; los grupos de la vertiente del Pacífico mostraron un patrón de variación morfológica clinal norte – sur, excepto para las poblaciones de Sonora, apreciándose un efecto multivariado semejante a los testigos, por lo que se confirmó la denominación de *D. sonorense*, al norte de México. El análisis de Componentes Principales confirmó los resultados de trabajos previos de variación morfológica clinal en *D. edule*. Además se observó una estrecha relación entre las poblaciones del sur de la vertiente del Pacífico con las del occidente, apreciándose que las localidades tipo de *D. holmgrenii y D. tomasellii* comparten características morfológicas semejantes de tal manera que ambos taxa constituyen un complejo morfológico.

Introduction

The cycads have been considered as living fossils, that is to say, they have remained practically immutable for millions of years. This condition seems to be exclusive to its phenotype, as some proposals suggested that its mutation and evolution rates have been high, although the species persists to the present time with few morphological changes during a long geological period (Soltis et al., 2002). However, other authors think that the genus *Dioon* had a recent speciation, influenced by the increment of habitats during the post Pleistocene, after the glaciations (González-Astorga et al., 2003a; Gregory & Chemnich, 2004). This discovery is quite recent; in fact that it has suggested that new taxa may be discovered.

The majority of the species of *Dioon* were established by traditional taxonomy and the species concept has been defined for just a few characters.

The different names that have been given during the process of determination of the plants of the genus, even from the same locality, have been a challenge for the floristic inventory lists and the evaluation of the species in its inclusion in the Mexican Official Norm (NOM) or in the IUCN Red List of Threatened Species. In other words, confusion exists regarding the morphological limits of the species.

The genus *Dioon* was established in 1843 by Lindley with *Dion edule*. Subsequently, Miquel (De Luca et al., 1982) corrected the spelling *Dion* to *Dioon*, which has been conserved for the International Code of Botanical Nomenclature to the present time (Whitelock, 2002).

Until 1979, four species were recognized from Mexico: *Dioon edule*, *D. angustifolium* Miquel (1848), *D. spinulosum* Dyer (1884), and *D. purpusii* Rose (1909). It is important to emphasize that *D. angustifolium* was under synonymy of *D. edule* var. *angustifolium* since 1851 by Miquel (De Luca et al., 1982); however, it was recently recognized as a species by González-Astorga et al. (2003b) in their population genetics survey. Currently, eight more species are known from Mexico. In the Pacific slope, De Luca et al. (1981a, b, 1984) described three new species and one variety. In the state of Chiapas, delimited by the Isthmus of Tehuantepec, they founded individuous with leaflets arranged in the rachis in an acute angle, overlapping and inserted obliquely on the same one. Such characteristics were ultimately be attributed to *D. merolae* De Luca, Sabato & Vázquez Torres. The same authors discovered in Oaxaca a population in which the individuals show flat leaves, spinulose, and spaced leaflets that was described as *D. holmgrenii* De Luca, Sabato & Vázquez Torres. Subsequently, they proposed the name of *D. tomasellii* De Luca, Sabato & Vázquez Torres for the plants distributed from Sonora to Guerrero. They defined this taxon by its falcate or subfalcate leaflet. In addition, they found that the plants of these populations show a continuous variation pattern, except for the ones in Sonora and N Sinaloa. In these localities, the plants show narrow and glaucous leaflets. Because of this discrete variation in the northern populations, an infraspecific taxon was described as *D. tomasellii* var. *sonorense* De Luca, Sabato & Vázquez Torres, which led automatically to the creation of the autonym *D. tomasellii* var. *tomasellii* De Luca, Sabato & Vázquez Torres.

McVaugh and Pérez de la Rosa (1992), in their revision of the genus *Dioon* for Nueva Galicia (Jalisco and adjacent areas), noted that De Luca et al. (1981b, 1984) inappropriately discussed the taxonomic value of the distinctions between *D. edule* and the other taxa. Therefore, they determined that *D. edule* is present in Nueva Galicia, having more similarities than differences with the species from the Gulf of Mexico slope. Moreover, they distinguished *D. edule* var. *edule* from the Pacific slope, with altitudinal ranks between 600 and 1000 (1200) m and the *D. edule var. sonorense* (De Luca et al.) McVaugh & Pérez de la Rosa, at higher elevations on the valleys and mountains at 1200–1900 m. Also, they proposed *D. holmgrenii* from Oaxaca in synonymy under *D. edule*.

Finally, Chemnick et al. (1997), taking into account the Moretti et al. (1993) database in their phylogenetic analysis, concluded that the two varieties of *D. tomasellii* have enough differences to be considered as species, *D. sonorense* (De Luca, Sabato & Vázquez Torres) J. Chemnick, T. Gregory and Salas-Morales. The morphological characteristics supporting the new status are the numerous, smallest and twisted leaves, rachis and petiole less tomentose and thinner than *D. tomasellii,* and narrower, linear-lanceolate and glabrous leaflets bearing one to three small spines.

For all the aforementioned reasons, we aimed to study the morphological variation of the genus *Dioon* from the Pacific slope, as well as to determine the taxonomic groups defined by natural limits via a multivariate method.

Methods

Study Area

Twelve localities were sampled in the states of Sonora, Sinaloa, Durango, Nayarit, Jalisco, Michoacán, Guerrero, Oaxaca, and Chiapas so as to include a great morphological variation for this study. To compare the differences of the groups generated for the analysis, we considered two species from three populations from the Gulf of Mexico slope: *Dioon edule*, collected in San Luis Potosí and Querétaro, and *D. spinulosum* from Oaxaca. The reasons for these selections were first that *D. edule* and *D. spinulosum* are well established species and second, they represent the extreme of the variation in their taxonomic categories, which corresponding to two of the natural groups of Sabato & De Luca (1985).

Sample Collection

Following Jones (1988) and Hair (1999), approximately 25 plants were sampled per locality to obtain a ratio of 10 individuals per variable. A total of 373 samples (cases or individuals) were obtained. The specimens consist of a complete leaf removed from the very base of the plant. One collection per locality will be deposited at IBUG and the duplicate sheets will be distributed to the main national and international herbaria.

Measurements

Because the vegetative characteristics are useful to define species and reproductive characters are utilized to distinguish natural groups (Sabato & De Luca, 1985) in this work, we considered 19 vegetative variables (Table 33-1). The variables are metric; no binary or categorical variable was considered.

Statistical Analysis

The multivariate analysis used in this study was the principal components analysis (PCA). According to Hair (1999) and Johnson (2000), this is an interdependent technique that transforms an assembly of variables correlated in a lesser assembly of intra-independent

Table 33-1 Leaf characteristic measures and the key via which they are identified.

No.	Part	Name (units)	Key
1	Leaf	Length (cm)	H1
2	Leaf	Width (cm)	H2
3	Leaf	Length/width	H3
4	Petiole	Length (cm)	hp
5	Leaflet	Length (cm)	f1
6	Leaflet	Width (mm)	f2
7	Leaflet	Thickness (mm)	f3
8	Leaflet	Base length (mm)	f4
9	Leaflet	Length/width	f5
10	Leaflet	Number	f6
11	Leaflet	Number of nervadures	f7
12	Leaflet	Distance between each one (mm)	f9
13	Leaflet	Base/width	f10
14	Leaflet	Imbrication (mm)	f11
15	Leaflet	Leaflet angle (°)	f12
16	Leaflet	Imbrication percentage	f13
17	Spine	Spine number on the leaflet upper margin	E1
18	Spine	Spine number on the leaflet lower margin	e2
19	Spine	Length (mm)	e3

variables called principal components (PC). The purposes of this study were to (1) generate new variables that can express the information contained in the original database matrix; (2) reduce the dimensions of the problem for future analysis; and (3) eliminate the variables that contribute with little information (Pla, 1986). We applied the statistical package SPSS Base 9.0 (1999), and analysis of factors without rotation was carried out to include these objects. According to Johnson (2000), this analysis is also useful to make of groupings of each one of the cases in subgroups of similar types for which the components here extracted were represented in scatterplots.

Results

Nineteen PCs were produced, from which we selected the first six. They contain the 84.5% of the variation of the original database matrix (Table 33-2). This selection was defined under the "latent root" criterion, which implies eigenvalues greater that 1.0, in agreement with Hair et al. (1999), as seen in Figure 33-1.

Table 33-2 Results of extraction of the first six principal components.

Principal component (CP)	Eigenvalue	Percentage of variation (%)	Cumulative percentage (%)
1	6.11984	32.2	32.2
2	3.87465	20.4	52.6
3	2.04314	10.8	63.4
4	1.53948	8.1	71.5
5	1.31817	6.9	78.4
6	1.16050	6.1	84.5

The morphological characteristics of PC 1, which contains 32.2% of the total variation, are shown in Figures 33-2 and 33-3.The variables having strong correlations with PC 1 are leaf length (h1), leaflet width (f2), leaflet base length (f4), number of veins (f7), number of spines on the leaflet upper margin (e1), number of spines on the leaflet lower margin (e2), and spine length (e3). These correlations are positive, and individuals with extremely positive PC 1 values (loads) have long leaves, wide leaflets, large bases, numerous veins and spines, with long spines (Fig. 33-2). Conversely, individuals with extremely negative values PC 1 have small leaves, narrow leaflets with small bases, and few veins and margins without spines (Fig. 33-3). Individuals close to zero loads show medium sized leaf length, wide leaflets, leaflet base and spines, and number of veins that are intermediate between negative and positive values.

PC 2, containing 20.4% of the total variation, is shown in Figures 33-4 and 33-5. The variables that have strong positive correlations are leaf width (h2), leaflet length (f1), leaflet length/wide index— leaflet form—(f5), distance between leaflets (f9), and leaflet base length/wide index (f10); the variables with strong negative correlations are leaf length/wide index—form of the leaf—(h3) and leaflet thickness (f3). PC 2 individuals with extremely positive values showed wide, blunt leaves; narrow and long, sharp leaflets; and slender and spaced leaflets (Fig. 33-4). Conversely, in individuals with extremely negative values, the leaf tends to be narrow and sharp; the leaflets are wide and short, that is, flattened, thick and close together to overlapping (Fig. 33-5). In individuals with a value

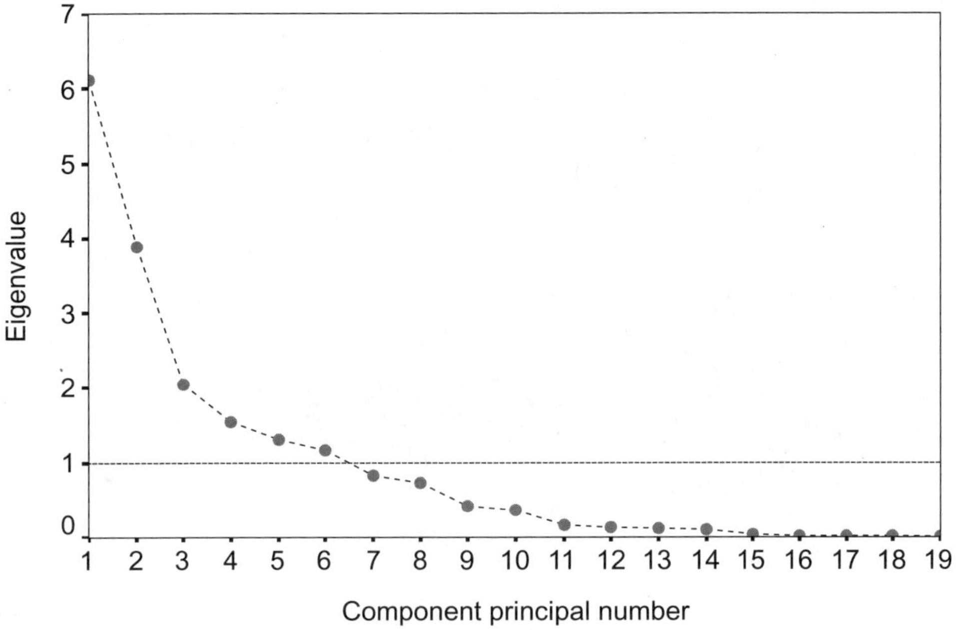

Figure 33-1. Factor plot showing the latent root criterion in the selection of the number of principal components. The dotted line divides the plot in two, for eigenvalues greater and smaller than 1.0.

close to the zero, the morphology of the leaves is intermediate between the two previous types.

PC 3 contains 10.8% of the variation, including the following variables: imbrication of leaflets (f11) and percentage of imbrication of leaflet (f13). Both had positive values. The individuals that have high loads are characterized by highly spaced leaflets, and the ones with possess low loads have overlapping leaflets.

The remaining PCs did not have a strong correlation with any variable, with the exception of PC 6, for which the petiole length (hp) and the leaflet insertion angle on the rachis (f12) are included. Because of the limited variation of the remaining PCs, none of them was utilized to construct scatterplots.

Once we defined the PC and its relationship with the variables studied, we constructed scatterplots to delimit the groups. PC 1 and PC 2 allowed the conformation of diverse groups, as seen in Figure 33-6. Because of its morphological characteristics they were defined like group I, *D. spinulosum*, characterized by having PC 1 with high positive values (Fig. 33-2); group II corresponds to *D. edule* from Querétaro and San Luis Potosí

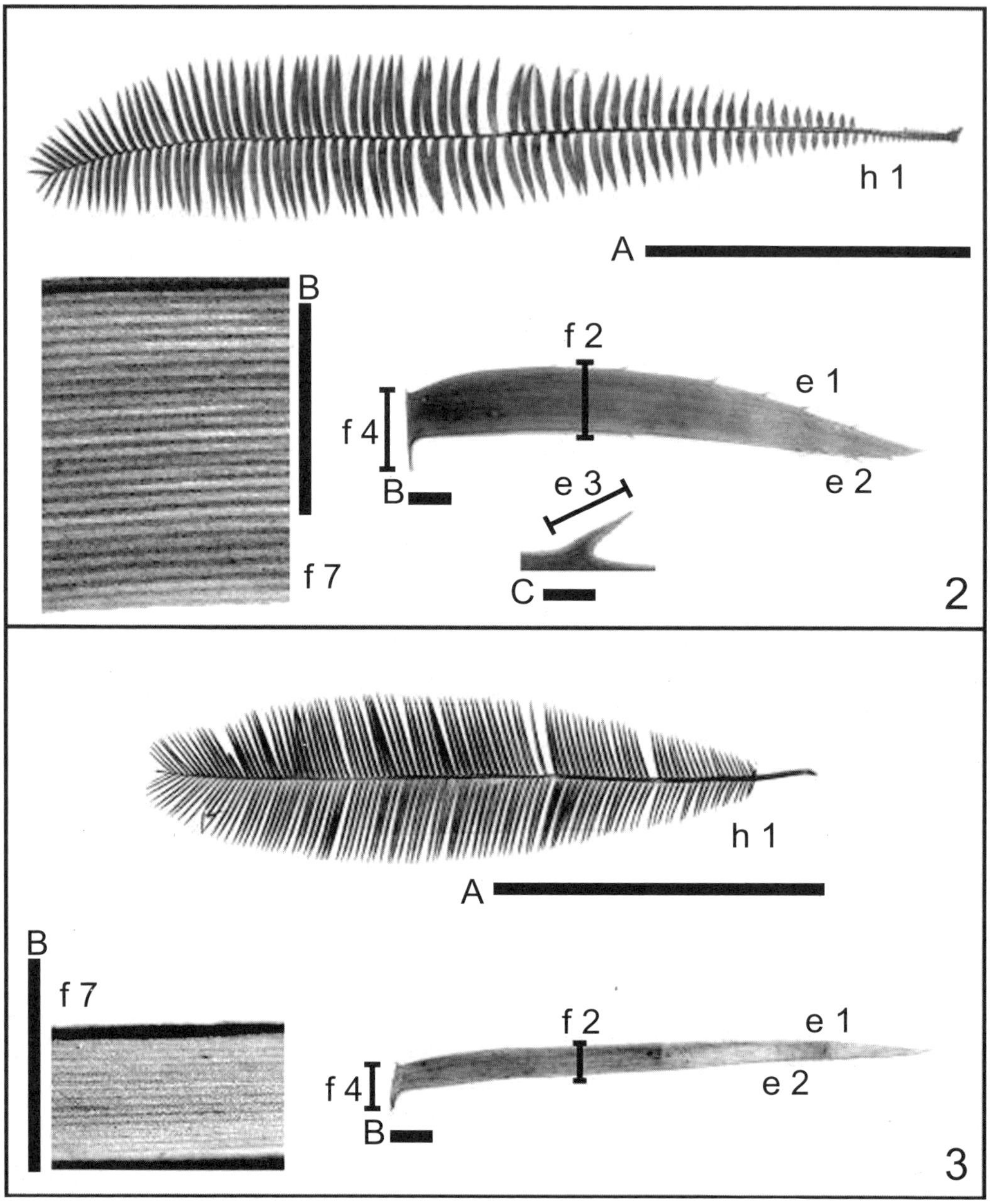

Figures 33-2 and 33-3. Phenotypes of principal component 1 (PC 1). 2. Phenotypes that correspond to individuals with extreme positive values. 3. Phenotypes that correspond to individuals with extremely negative values. Bars: A = 50 cm, B = 1 cm, C = 1 mm. The lowercase letters with numbers correspond to the variables of Table 33.1.

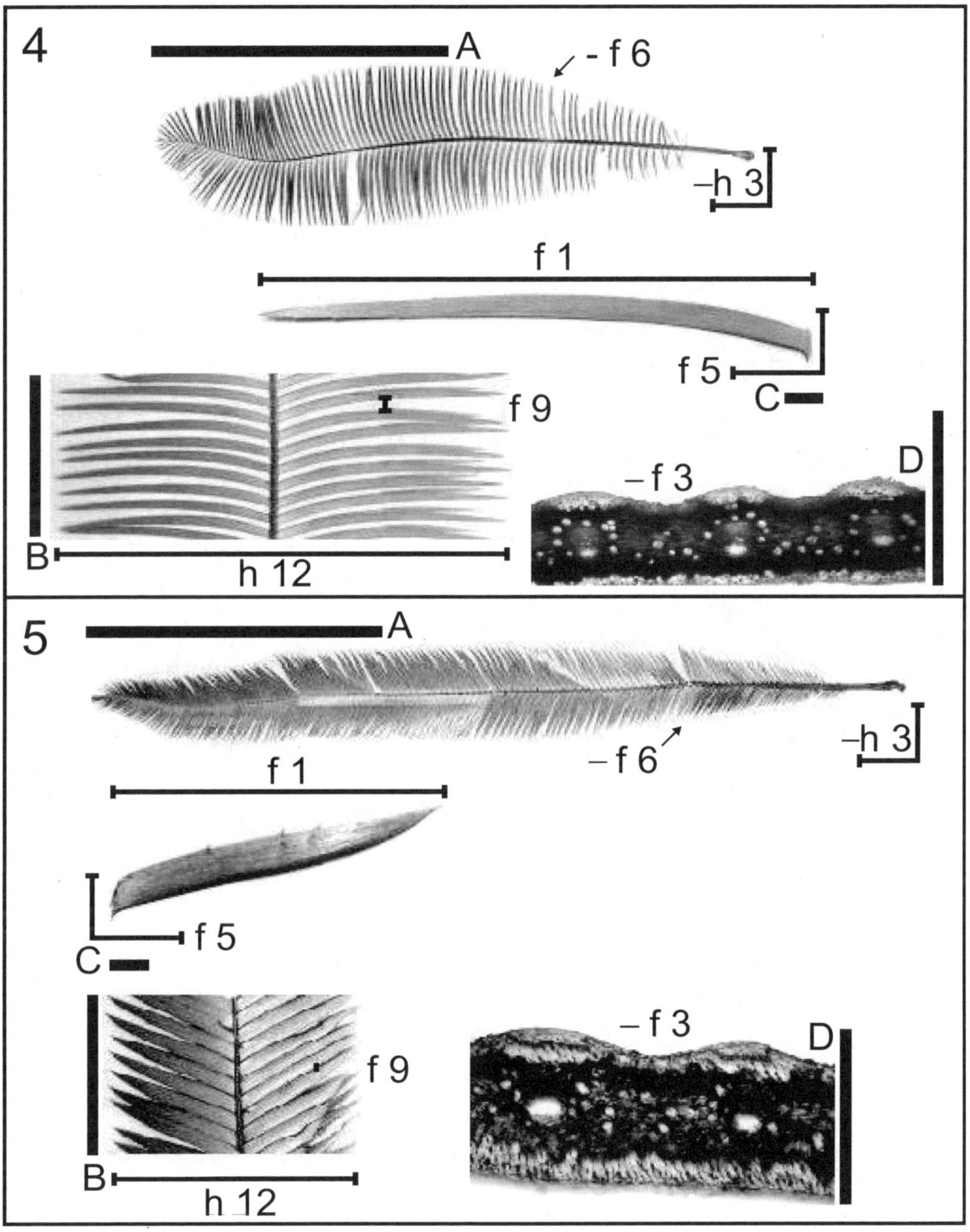

Figure 33-4 and 33-5. Phenotypes of principal component 2 (PC 2). **4.** Phenotypes that correspond to individuals with extremely positive values. **5.** Phenotypes that correspond to individuals with extremely negative values. Bars: A = 50 cm, B = 10 cm, C = 1 cm, D = 1 mm. The lowercase letters with numbers correspond to the variables of Table 33.1.

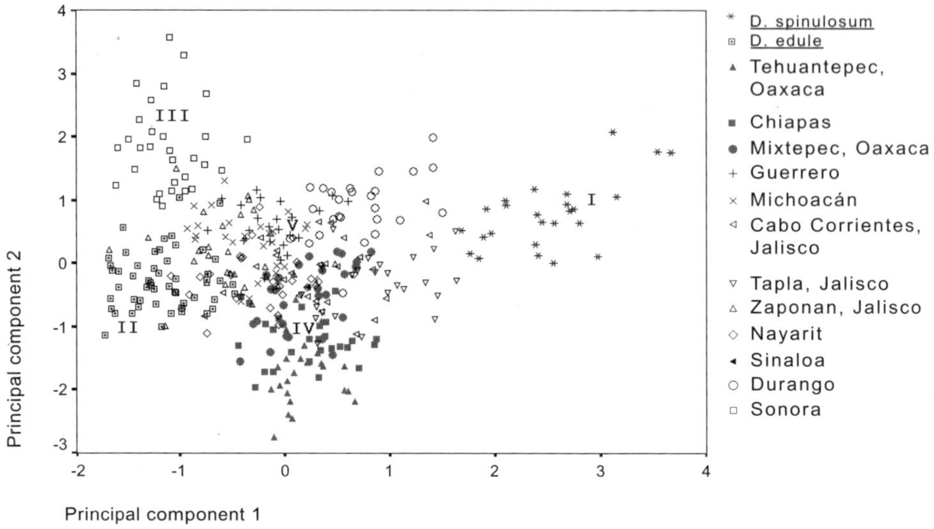

Figure 33-6. CP 1 and CP 2 scatterplot. The Roman numbers indicate the groups of the individuals of *Dioon* collected.

and was distinguished by having PC 1 with high negative values (Fig. 33-3); group III belonged to the Sonoran population and was differentiated by showing PC 2 with high positive values (Fig. 33-4); group IV, which corresponds to the individuals from Mixtepec (Oaxaca), Tehuantepec (Oaxaca), and Chiapas, was determined by their PC 2 with negative values (Fig. 33-5); and the group V, comprising individuals from Sinaloa, Durango, Nayarit, Jalisco (Zapopan, Talpa, Cabo Corrientes), Michoacán, and Guerrero, are characterized by having PC 1 and PC 2 close to zero loads.

The last two groups do not show a clear limitation based on PC 1 and PC 2, for which to improve the resolution of the little delimited groups we utilized the PC 2 vs. PC 3 scatterplot (Fig. 33-7). On both graphics an overlapping area is observed (indicated by two parallel lines). To locate the southern population of group IV that had more morphological similarities with the group V, the PC 1 vs. PC 2 scatterplot was utilized (Fig. 33-8). This plot contains only the individuals of groups IV and V. The overlapping area previously observed between groups IV and V allows recognition that the individuals from Mixtepec, Oaxaca are morphologically similar to the center of the Pacific slope (group V). This scatterplot also shows that the coastal populations of Jalisco (Talpa and Cabo Corrientes) and Nayarit show greater morphological similarity with the Mixtepec, Oaxaca population.

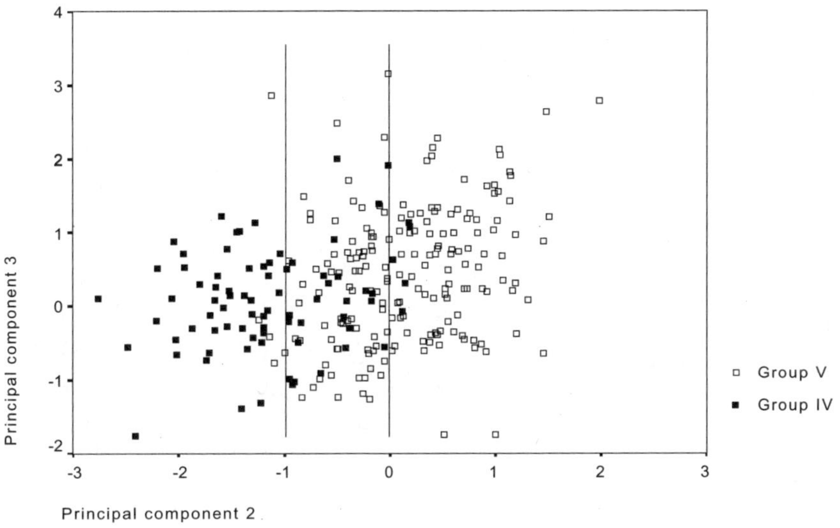

Figure 33-7. PC 2 vs. PC 3 scatterplot of groups IV and V. The overlapping area is found between −1 and 0 and is indicated by two parallel lines.

Discussion

In the present study, the application of a multivariate method permitted the creation of five groups and the discovery of patterns of variation in the genus *Dioon* of the Pacific slope of Mexico. The exploration of PC 1 was informative in defining the major groups, *D. spinulosum* (group I), *D. edule* (group II), the Sonoran population (group III), and the southern and center complex of the Pacific slope (groups IV and V). The first two were the reference groups for contrasting the morphological variation between the Pacific and the Gulf slopes, and the three last represent the study groups of the present work. The application of a multivariate method by Favela (1991) and Pérez de la Rosa (1993) has shown similar results in the genus *Pinus*; the authors mentioned that in the pine complexes study, the species were concentrated on different conglomerates and were morphologically distinguished. PC 2 defined an assembly of populations that respond to a continuous clinal morphological variation pattern north–south, except in the Sonora group, where there is no overlapping among individuals and the individuals of the center of the Pacific slope. This was also observed by De Luca et al. (1984); nevertheless, they did not

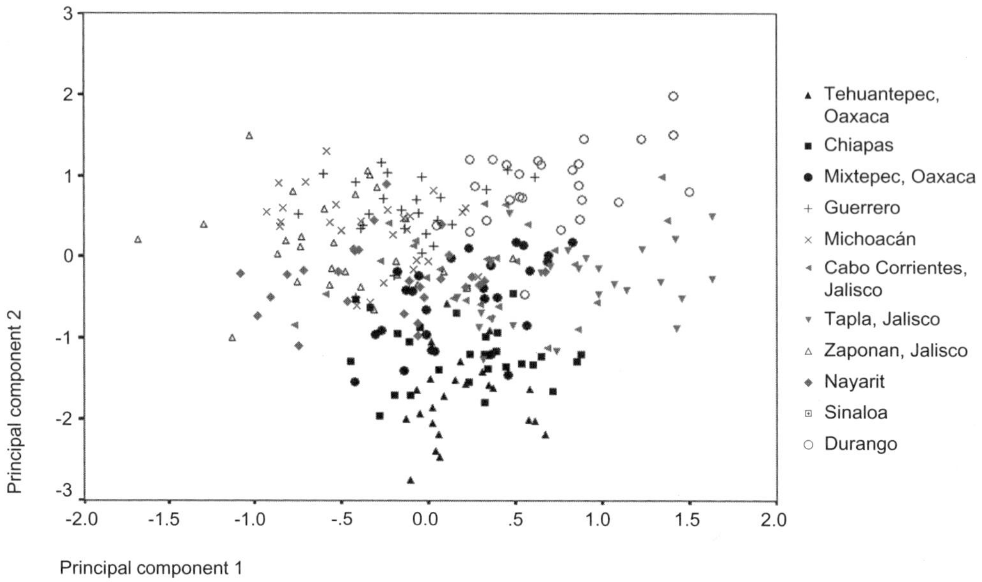

Figure 33-8. PC 1 vs. PC 2 scatterplot with localities of group IV and V.

find sufficient evidence to segregate the populations of Sonora and N. Sinaloa as different species. The morphological proximity of the *D. edule* populations of Querétaro and San Luis Potosí and, the Gulf of Mexico slope center (group II) with the Pacific slope center populations (grupo V) with respect to PC 2 provide evidence of the north–south variation pattern that both groups share but in different slopes; the previous assignation was also shown by González-Astorga (2003a) where he found a north–south variation pattern in the wide leaflet of *D. edule* and its geographical correspondence with the north and central populations and south of the Gulf 's slope. The PCA applied to morphological variation study in the pines group Oocarpa carried out by Pérez de la Rosa (2001) revealed a north–south variation pattern in *Pinus oocarpa* Schiede ex Schlechtendal, and the plants showed a relationship between their morphology and the geographical distance of their origin, a result observed in the present study. Previously it had been suggested that the effect of clinal variation in the populations of *D. edule* in the Gulf 's slope would be observed in the populations of the Pacific side and vice versa (De Luca et al., 1984); this correlation in the habitat and its morphology aims at an ecotopic differentiation among populations. The cause is the adaptation to the environmental factors, like light intensity and water stress, which could imply a process of speciation, influenced by the climatic change during the post Pleistocene (González Astorga et al., 2003a).

PC 3 was useful for the exploration of the overlapping area between the groups IV and V.

It was found that in the overlapping elements of group IV with V, some are from the type locality of *D. holmgrenii*, with morphologically similar individuals. McVaugh and Pérez de la Rosa (1992) used this similarity to propose *D. holmgrenii* as a synonym of *D. edule* in Nueva Galicia, which l is confirmed by the results of the present study because in the overlapping area, individuals from the type locality of *D. holmgrenii* overlap with the individuals of *Dioon* on the coast of Jalisco and Nayarit, which is the type locality of *D. tomasellii*, synonymous species under *D. edule,* according to McVaugh and Pérez de la Rosa (1992).

As the present survey did not find overlap between the group II, which corresponds to *D. edule*, and group V, we cannot support synonymy of *D. tomasellii* under *D. edule*.

The representation of the petiole length (hp) and the leaflet insertion angle on the rachis (f12) in PC 6, and the slight variation thus resulting, suggest that these characteristics were of little use for the groupings, although the angle of the leaflet has been utilized to distinguish the southern populations of the remainder of the Pacific slope (De Luca et al. 1981a, 1981b, 1984). Pérez de la Rosa (2001) found, via multivariate analysis, that the characters of the female cone, which are utilized in the classifications, they were not sufficient to make groupings, whereas in this survey they were fundamental for the constitution of groups PC 1, PC 2, and PC 3.

Dioon spinulosum (group I) was utilized as a morphologically distant reference species to the populations of the Pacific slope. It is observed that, although the individuals were dispersed, all of them constitute an independent morphological group. This idea agrees with that of Sabato and De Luca (1985), who proposed that *D. spinulosum* and *D. rzedowskii* De Luca, Moretti, Sabato & Vázquez Torres constitute a natural group of species, distinguishable from the remainder of *Dioon* in Mexico, which compose another natural group. On the other hand, Moretti et al. (1993) suggested that *D. spinulosum, D. rzedowskii,* and *D. mejiae* Standl. & L. O. Williams developed from a common ancestral form, different from the one from which the remainder of the species of the genus originated.

The *D. edule* reference group reveals the morphological separation of its members with relation to the others from the Pacific slope (groups III, IV, and V) and *D. spinulosum* of the Gulf 's slope.

The Sonoran population (group III) showed the same separation between *D. spinulosum* and *D. edule*; in other words, they served as a good reference. This result is consistent with the proposed name of *D. sonorense* by Chemnick et al. (1997), in the sense that the plants from Sonora and N. Sinaloa, described before as *D. tomasellii* var. *sonorense*, are morphologically and genetically different from those of the remainder of the Pacific slope. McVaugh and Pérez de la Rosa (1992) mentioned that the populations of Sonora and N. Sinaloa are similar to those of *D. edule*, with a slight differentiation that distinguishes them at the variety level. Nevertheless, the results of the present study do not exhibit great overlapping among the group III with groups II and V, which show morphological similarity.

Individuals of the taxa included in groups IV and V are morphologically similar and are not distinguishable in PC 1, yet they were morphologically continuous in PC 2. The discordance of the clinal variation pattern of the Pacific slope in the population of Tehuantepec suggests a moderate isolation and an incipient and fast process of morphological distancing. Little differentiation between the individuals of the population of Mixtepec, Oaxaca (the type locality of *D. holmgrenii*) and the individuals of the population of Chiapas, called *D. merolae* by De Luca et al. (1981a), both of the group V, was observed. Chemnick et al. (1997) found a single difference between *D. holmgrenii* and *D. merolae* in the chloroplast DNA restriction fragment length polymorphisms database matrix of the analysis of Moretti et al. (1993). Application of discriminant canonical analysis permitted Favela (1988) to find significant differences among *Pinus hartwegii*, which grows at medium and high elevations, and that approach could be used for this analysis to define appropriately *D. holmgrenii* and *D. merolae*.

Separation of the group V from group II indicates that the species occurring from Sinaloa to Oaxaca is not *D. edule,* as proposed by McVaugh and Pérez de la Rosa (1992), but indeed these groups show the same clinal effect. The distinction that these two authors made between the two varieties, *D. edule* var. *edule* in lower elevations and *D. edule* var. *sonorense* in Sonora, S. Sinaloa, and the mountains of Jalisco, based on morphological characteristics and altitudinal ranges themselves, is not confirmed in the present study, because the plants from the population of Zapopan, Jalisco, which are found in faraway mountainous regions of the coast, are similar to individuals of the populations of the remainder of the center of the Jalisco on the Pacific slope.

Soltis et al. (2002) mentioned that the mutation rates in cycads change quickly, with a minimum of morphological differentiation, which we here tested in the overlapping area of groups IV and V. The overlap of these two groups could indicate an incipient speciation process due to the decrease of genetic flow, a consequence of the natural division of the populations during the period post Pleistocene that occurred in the canyon systems of N. Oaxaca, as suggested by Gregory et al. (2003) and Gregory and Chemnick (2004).

As previously reported, the application of multivariate techniques with the purpose of testing the morphological differentiation in the taxonomic complex has produced successful results in diverse plant taxonomy studies. In the case of the cycadales, the use of this type of analysis was used to test the existence of three specific entities in *Ceratozamia alvarezii* Pérez-Farrera, Vovides & Iglesias, *C. mirandae* Vovides, Pérez-Farrera & Iglesias, and *C. norstogii* D. W. Stevenson taxonomic complex (Pérez-Farrera et al., 2004).

Conclusions

The majority of the variables used in the study had a strong relationship with PC 1, PC 2, and PC 3. The variables of petiole length and leaflet insertion angle on the rachis contributed less to the variation and are not adequate for the distinction of taxa.

PC 1 was useful for the separation of the major groups into the genus *Dioon*: the reference groups (*D. spinulosum* and *D. edule*) of the Gulf of Mexico slope, *D. sonorense,* and the southern and western complex of the Pacific slope. PC 2 turned was convenient for the definition of the morphological variation patterns in the populations of the Pacific slope. A north–south clinal variation pattern was interrupted in the north of Sinaloa and Sonora, which permitted us to confirm the name *D. sonorense* for these populations. PC 2 was also to corroborate prior work on clinal variation on *D. edule* from the Gulf of Mexico slope.

The populations of the west showed a tight relationship with those of southern Mexico. It was found that the individuals from the type locality of *D. holmgrenii* and some individuals from a close region to the type locality of *D. tomasellii* share similar morphological characteristics, in such a way that both taxa constitute a morphological complex that requires new studies.

The application of the principal components analysis is useful for the clarification of the taxonomic complex and offers a guideline for its possible solution.

Acknowledgments

This work was supported by a Universidad de Guadalajara grants CCB 03/04/085/2004 and Proyect No. 34961 of Cuerpo Académico CA-23. We thank Idea Wild for supplying the equipment for field work. We thank M. C. Luz María González Villarreal for facilitating bibliographical support and for help in the translation of this manuscript. We also thank biologist Arturo Castro Castro for his unconditional aid in measuring botanical samples and all the guides who helped us in the field work.

Literature Cited

Chemnick, J., T. J. Gregory & S. Salas-Morales. 1997. A revision of *Dioon tomaselli* (Zamiaceae) from western Mexico, a range extension of *D. merolae*, and clarification of *D. purpusii*. Phytologia 83: 1–6.

De Luca, P., S. Sabato & M. Vázquez Torres. 1981a. *Dioon merolae* (Zamiaceae), a new species from Mexico. Brittonia 33: 179–185.

———, ——— & ———. 1981b. *Dioon holmgrenii* (Zamiaceae), a new species from Mexico. Brittonia 33: 552–555.

———, ——— & ———. 1982. Distribution and variation of *Dioon edule* (Zamiaceae). Brittonia 34: 355–362.

———, ——— & ———. 1984. *Dion tomasellii* (Zamiaceae), a new species with two varieties from western Mexico. Brittonia 36: 223–227.

Favela L. S. 1988. Taxonomy of some mexican pines. Master of Science Thesis. University of Oxford, Oxford, UK.

———. 1991. Taxonomía de *Pinus pseudostrobus* Lindl., *P. montezumae* Lamb. y *P. hartwegii* Endl. Reporte Científico No. 26. Facultad de Ciencias Forestales. Universidad Autónoma de Nuevo León. Linares, Nuevo León.

González-Astorga, J., A. P. Vovides & C. Iglesias. 2003a. Morphological and geographic variation of the cycad *Dioon edule* Lindl. (Zamiaceae): ecological and evolutionary implications. Bot. J. Linn. Soc. 141: 465–470.

———, ———, M. M. Ferrer & C. Iglesias. 2003b. Population genetics of *Dioon edule* Lindl. (Zamiaceae, Cycadales): biogeographical and evolutionary implications. Biol. J. Linn. Soc. 80: 457–467.

Gregory, T. J. & Chemnick J. 2004. Hypotheses on the relationship between biogeography and speciation in *Dioon* (Zamiaceae). Pp. 137–148 *in* T. Walters, & R. Osborne (eds.), Cycad classification: concepts and recommendations. CABI Publishing, Wallingford, UK.

———, ———, S. Salas M. & A. P. Vovides. 2003. A new species in the genus *Dioon* (Zamiaceae) from North-Central Oaxaca, Mexico. Bot. J. Linn. Soc. 141: 471–476.

Hair, J. F. Jr., R. E. Anderson, R. L. Tatham & W. C. Black. 1999. Análisis multivariante. . Ed. 5. Prentice-Hall, Madrid.

Johnson, D. E. 2000. Métodos multivariados aplicados al análisis de datos. International Thomson Editores. México, D.F.

Jones, S. B., Jr. 1988. Sistemática vegetal. Ed. 2. McGraw-Hill, México, D.F.

Lindley, J. 1843. *Dion edule*. Edward's Bot. Reg. 29: 59–60.

McVaugh, R. & J. A. Pérez de la Rosa. 1992. Zamiaceae. Pp. 107–119 *in* R. McVaugh (ed.), Flora Novogaliciana v. 17. Gymnosperms and pteridophytes. University of Michigan Herbarium.

Moretti, A., P. Caputo, S. Cozzolino, P. De Luca, L. Gaudio, G. Siniscalco G. & D. W. Stevenson. 1993. A phylogenetic analysis of *Dioon* (Zamiaceae). Am. J. Bot. 80: 204–214.

Pérez de la Rosa, J. A. 1993. Taxonomía de *Pinus ayacahuite* y *Pinus strobiformis*. Master of Science Thesis. Colegio de Postgraduados. Montecillo, México.

———. 2001. Variación morfológica y taxonomía de *Pinus* Grupo "*Oocarpa*" (Martínez, 1948), Pinaceae. Ph D. Thesis. Facultad de Ciencias, Universidad Nacional Autónoma de México. México, D.F.

Pérez-Farrera, M. A., A. P. Vovides, L. Hernández S., D. González & M. Martínez. 2004. A morphometric analysis of the *Ceratozamia norstogii* complex (Zamiaceae). Pp. 127–136 *in* T. Walters and R. Osborne (eds.), Cycad classification concepts and recommendations. CABI, Cambridge, MA.

Pla, L. E. 1986. Análisis multivariado: método de componentes principales. Secretaría General de la Organización de los Estados Americanos. Programa Regional de Desarrollo Científico y Tecnológico. Washington, DC.

Sabato, S. & P. De Luca. 1985. Evolutionary trends in Dion (Zamiaceae). Am. J. Bot. 72: 1353–1363.

Soltis, P. S., D. E. Soltis, V. Savolainen, P. R. Crane & T. G. Barraclough. 2002. Rate heterogeneity among lineages of tracheophytes: Integration of molecular and fossil data and evidence for molecular living fossils. Proc. Natl. Acad. Sci. USA 99: 4430–4435.

SPSS Base 9.0. 1999. Application guide. SPSS, Inc., Chicago, IL.

Whitelock, L. M. 2002. The cycads. Timber Press, Portland, OR.

SYSTEMATICS
CHAPTER 34

Subdivision of Encephalartos

Piet Vorster

Abstract

The subdivision of the genus *Encephalartos* into several genera (*Acanthozamia, Congostrobus, Dracostrobus, Dyerstrobus, Glaucostrobus, Inezamia, Pyrrhostrobus, Rugostrobus, Tanzamia, Viridestrobus,* and *Xanthostrobus*), as published by Cooper and Goode in *The Cycads and Cycad Moths of KwaZulu-Natal* (2004), is discussed in view of the authors' motivation as well as available morphological, geographical, and molecular data.

Resumen

Se discute la subdivisión de *Encephalartos* en varios géneros (según: Cooper & Goode en *The Cycads and Cycad Moths of KwaZulu-Natal*, 2004), i. e. *Acanthozamia, Congostrobus, Dracostrobus, Dyerstrobus, Glaucostrobus, Inezamia, Pyrrhostrobus, Rugostrobus, Tanzamia, Viridestrobus* y *Xanthostrobus*. Con los criterios propuestos por los autores, así como con caracteres morfológicos, geográficos y moleculares presentados en este trabajo.

Introduction

During 2004 a little book with the innocuous title of *The Cycads and Cycad Moths of KwaZulu-Natal* appeared under the authorship of Michael Cooper and Douglas Goode. It was reviewed by Vorster (2004b). What makes this book unusual among several recent books on cycads is that it described five new species and subdivided the genus *Encephalartos* into 12 genera.

Aim

The purpose of this paper is to make a closer scrutiny of these genera, to see to what extent their acknowledgment is justified. This will be done in terms of morphological and geographical evidence as well as the molecular evidence reported by Treutlein et al. (2005), who used nuclear ITS 1 and 2 and *rbc*L sequence data and genomic ISSR fingerprinting to compare species throughout the range of the genus. They produced cladograms based on ITS 1 and 2 sequences, on 55 morphological and geographical characteristics, on *rbc*L, and on ISSR fingerprint data. For comparative purposes, I use only the most informative of these, namely the ITS 1 and 2 sequence data-based cladogram (their Fig. 1, my Fig. 34-1).

Discussion

It has to be said right at the beginning that these genera are generally poorly circumscribed so that it is not at all always clear how the authors distinguish them. I here assess the species groups included in each of the genera against available evidence.

The genus *Encephalartos* sensu Cooper and Goode (2004) is typified, correctly so, by *Cycas caffra* Thunb. [*Encephalartos caffer* (Thunb.) Lehm.] (see Vorster, 2004c), and includes *E. aplanatus, E. cerinus, E. ngoyanus, E. umbeluziensis,* and *E. villosus*. The description is not diagnostic, but the drooping and sometimes fringed proximal "lip" of the megasporophyll face and the poor expression of facets on the scale faces, as well as the external similarity of male and female cones, are diagnostic. These species occupy different parts of a distinct geographical area along the eastern coast of South Africa (Fig. 34-2). For these reasons I have always considered these species to belong together. In Treutlein et al.'s

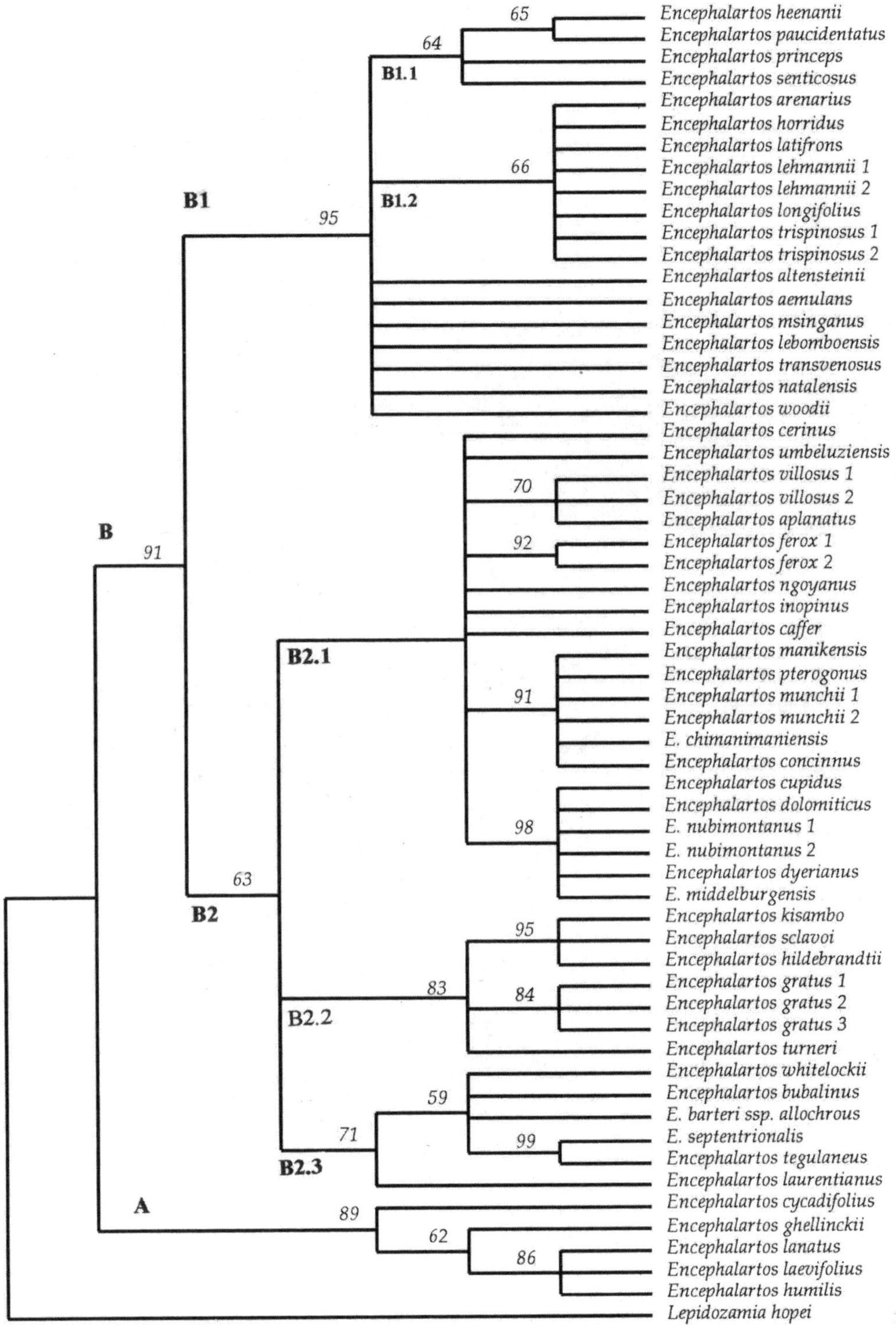

Figure 34-1. Cladogram based on ITS 1 & 2 sequence data. From Treutlein et al. (2005).

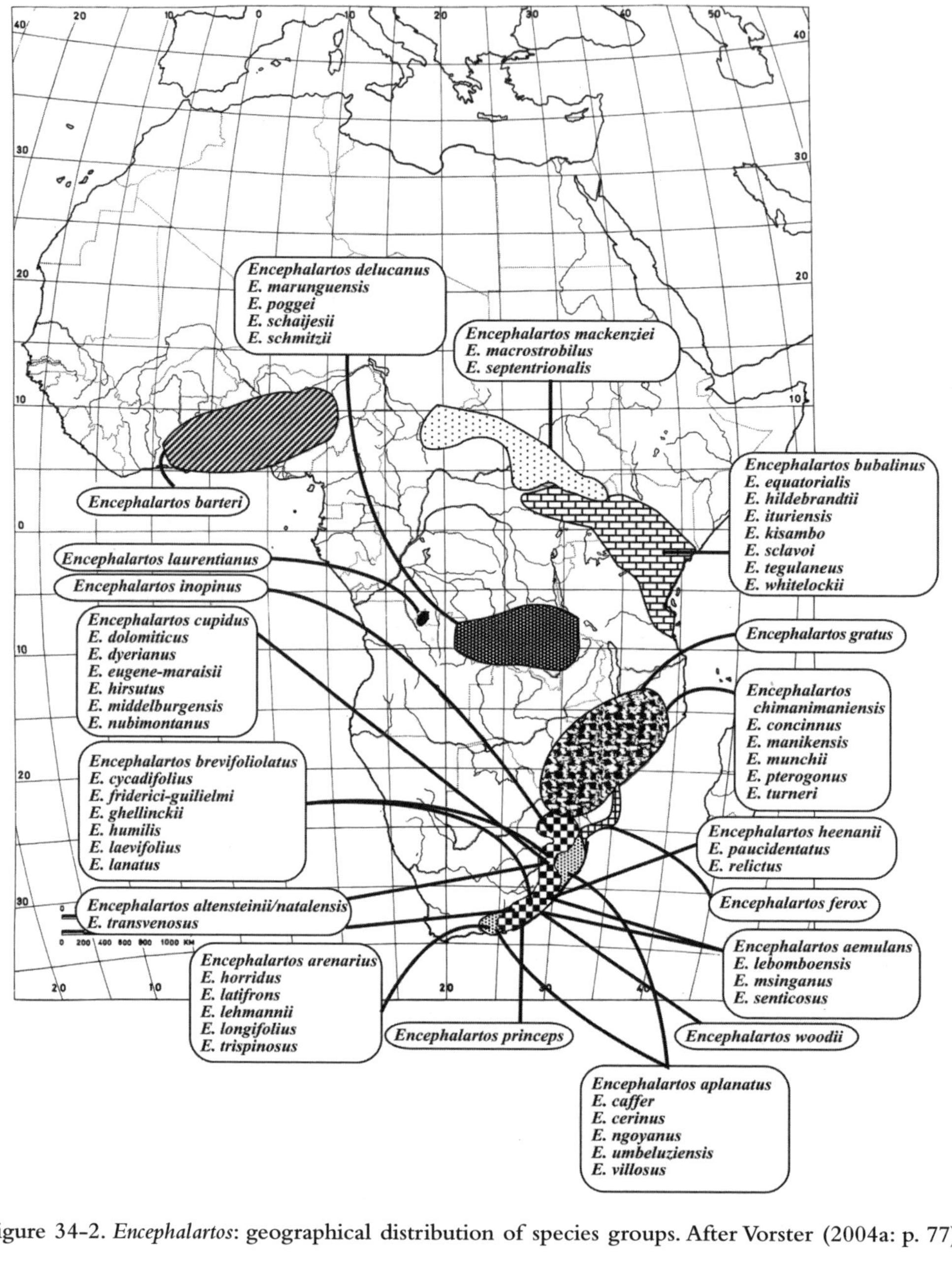

Figure 34-2. *Encephalartos*: geographical distribution of species groups. After Vorster (2004a: p. 77).

(2005) molecular cladogram, based on ITS 1 and 2 data, they are grouped more or less together, that is, as nonmonophyletic species (Fig. 34-3) in an unresolved polytomy (except for the clustering of *E. aplanatus* with *E. villosus* with moderate bootstrap support of 70%). Cooper and Goode (2004) stated that "the genus seems to have affinities with *Tanzamia* . . ." (p. xi), and "it seems to us, *Tanzamia* gen. nov. and *Encephalartos* are closely related" (p. xiv), but no supporting evidence is provided, and neither is it supported by Treutlein et al.'s cladogram (Fig. 34-1).

The genus *Dracostrobus* contains *D. brevifoliolatus* (*Encephalartos brevifoliolatus*), *D. cycadifolius* (*E. cycadifolius*), *D. dedekindii* (a newly described species), *D. ghellinckii* (*E. ghellinckii*), *D. humilis* (*E. humilis*), D. *laevifolius* (*E. laevifolius*), *D. lanatus* (*E. lanatus*), and *D. mkomaasiana* (another newly described species). Again the description is uninformative and undiagnostic, but these species have in common very narrow leaflets (up to 8 or rarely 10 mm wide), which are entire at maturity. Reproductively, they are well isolated, producing cones six months out of phase with all other species of *Encephalartos,* sensu lato, known to me. Furthermore, in this group the female cones mature and disintegrate within about three months, compared to in six to 12 months in most other species. It occurs in the eastern mountainous parts of South Africa (Fig. 34-2). This grouping is well supported by Treutlein et al.'s (2005) molecular cladogram (see Fig. 34-4), where they are grouped together in a monophyletic clade with bootstrap support of 89%.

The genus *Rugostrobus* comprises *R. arenarius* (*Encephalartos arenarius*), *R. latifrons* (*E. latifrons*), and *R. longifolius* (*E. longifolius*), "which seem to form a simple evolutionary series." The authors did not substantiate. None of the features listed in the description are diagnostic, that is, exclusive to this group of species. This grouping is not supported by Treutlein et al.'s (2005) ITS cladogram, where these three species form part of a larger unresolved group (Fig. 34-5). Neither does a more resolved allozyme-based cladogram by van der Bank et al. (2001) (Fig. 34-6) support this grouping.

The genus *Acanthozamia* consists of *A. horrida* (*Encephalartos horridus*), *A. lehmannii* (*E. lehmannii*), *A. princeps* (*E. princeps*), and *A. trispinosa* (*E. trispinosus*). Again none of the elements in the description are exclusive to this group of species, and neither are they convincing in combination. In Treutlein et al.'s (2005) cladogram they group together with *E. arenarius, E. latifrons,* and *E. longifolius* in an unresolved cluster with moderate bootstrap support of 66% (Fig. 34-5). However, van der Bank et al.'s (2001) isozyme cladogram (Fig. 34-6) shows more detail, very different from Cooper and Goode's recognition of the two

3

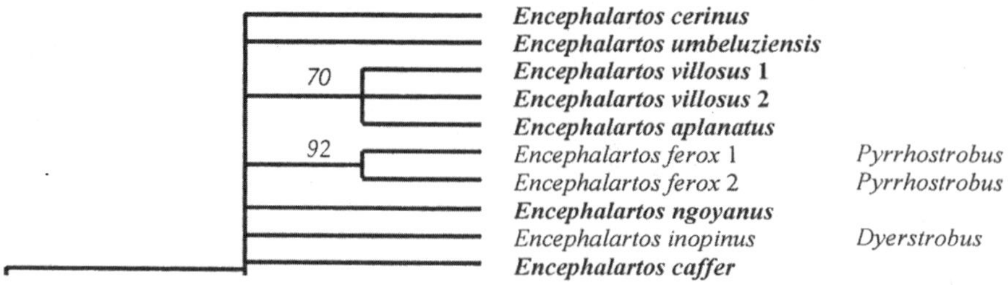

4

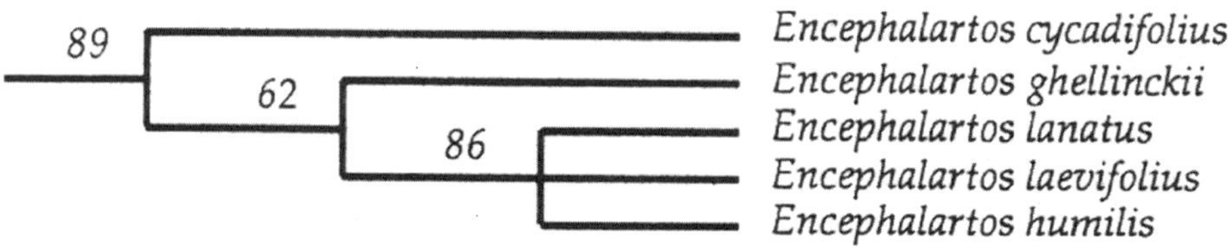

5

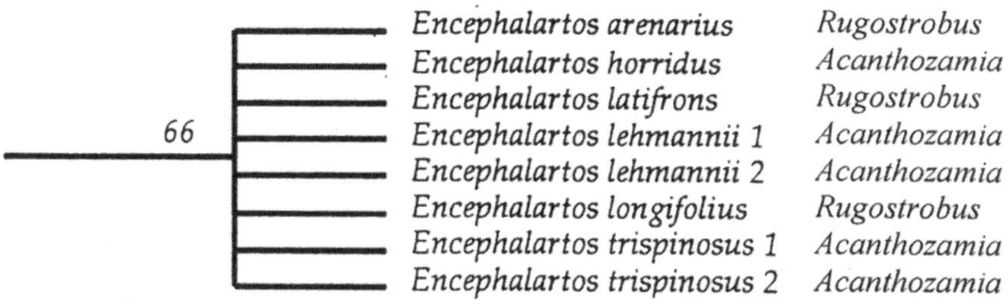

6

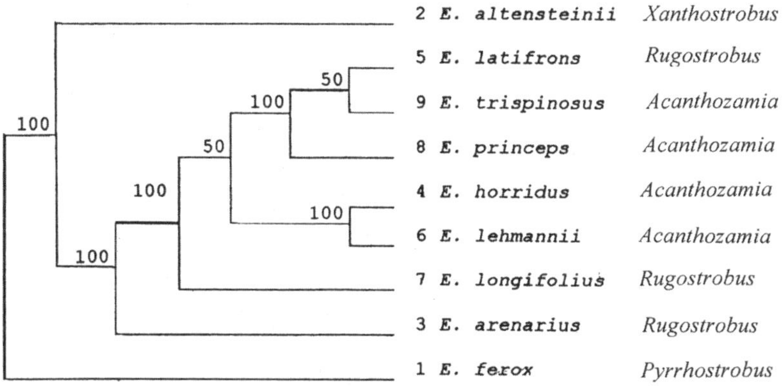

Figure 34-3. Clade from Treutlein et al.'s (2005) cladogram based on ITS 1 & 2 sequence data, showing clustering of the species in the genus *Encephalartos* in the sense of Cooper and Goode (2004).

Figure 34-4. Clade from Treutlein et al.'s (2005) cladogram based on ITS 1 & 2 sequence data, showing clustering of the species in Cooper and Goode's (2004) genus *Dracostrobus*.

Figure 34-5. Clade from Treutlein et al.'s (2005) cladogram based on ITS 1 & 2 sequence data, showing clustering of the species in Cooper and Goode's (2004) genera *Rugostrobus* and *Acanthozamia*.

genera *Rugostrobus* and *Acanthozamia*. The species of these two genera occur in the Eastern Cape Province of South Africa (Fig. 34-2) and have in common hard foliage with a tendency toward glaucousness, sometimes heavily dentate-lobed leaflets, and a tendency toward bearing solitary cones. While conceding that the available molecular results may not present the whole truth, there is in my opinion no evidence to support the recognition of two separate groupings, and I believe that these seven species form a single, closely knit group.

The genus *Pyrrhostrobus* consists of a single species, *P. ferox* (*E. ferox*). Though not emphasized by the authors, this species is unique in the genus *Encephalartos* in the broad sense by virtue of its wide and blunt-tipped leaflets with prominent teeth (not lobes) along the whole margin, its salmon red cones, and its reluctance to hybridize with any other species in breeding experiments. In Treutlein et al.'s (2005) ITS cladogram it forms part of an unsupported clade, but, significantly, the two accessions that were studied form a subclade with bootstrap support of 92% (Fig. 34-1). Cooper and Goode (2004), without any supporting evidence, stated that "its origin may lie in *Rugostrobus* . . . , close to *R. arenarius.*" This is contrary to the morphological, geographical (Fig. 34-2), and molecular evidence. It was used as outgroup for van der Bank et al.'s allozyme cladogram (Fig. 34-6), which shows no affinity of this species for *E. arenarius*.

The genus *Xanthostrobus* was created for *X. aemulans* (*Encephalartos aemulans*), *X. altensteinii* (*E. altensteinii*), *X. curachii* (a newly described species, similar if not identical to *E. lebomboensis*), *X. lebomboensis* (*E. lebomboensis*), *X. msinganus* (*E. msinganus*), *X. natalensis* (*E. natalensis*), *X. pietretiefii* (another newly described species similar if not identical to *E. lebomboensis*), *X. senticosus* (*E. senticosus*), and *X. woodii* (*E. woodii*). Yet again, none of the characteristics listed in the description are unique to this group, although among species with relatively wide and bright green leaflets it can be recognized by a variably expressed tomentum on the cones. They also mention verrucose leaflets, but this is not obvious and I would not read taxonomical significance in this trait. Strangely, *E. transvenosus*, with cones that are indistinguishable from those of *E. natalensis* and which in the vegetative stage of some forms is easily confused with *E. natalensis*, is not included here but placed in

Figure 34-6. Cladogram, based on allozyme data, showing molecular relationship between 8 Eastern Cape species of *Encephalartos* and containing the species in Cooper and Goode's (2004) genera *Rugostrobus* and *Acanthozamia*. From van der Bank et al. (2001).

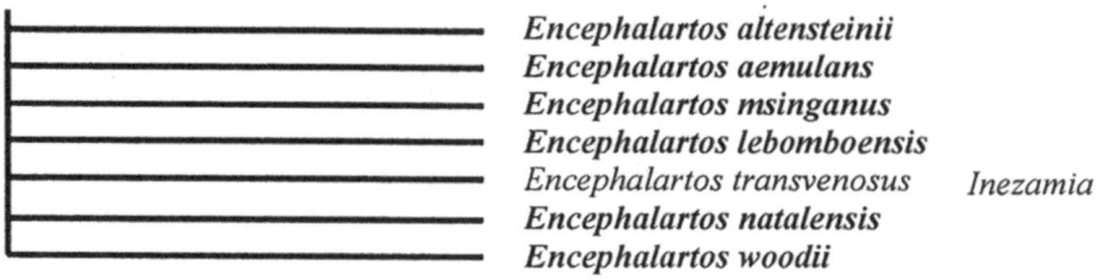

7

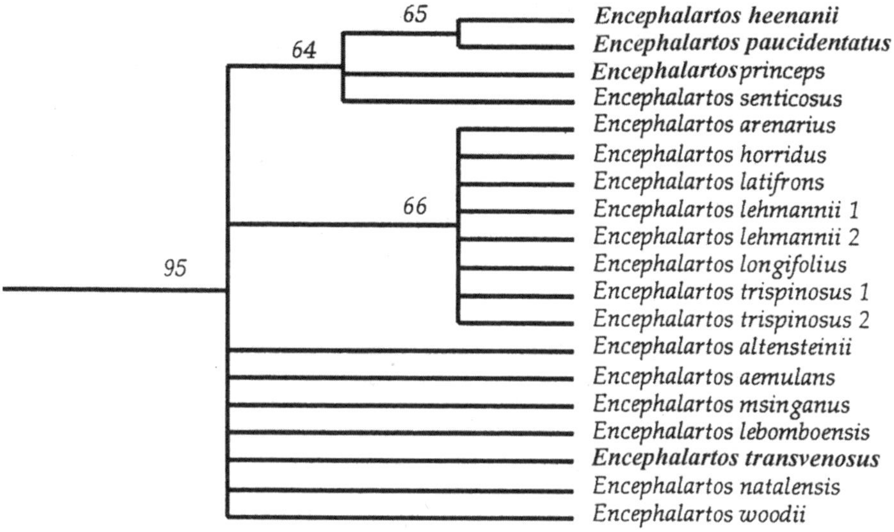

8

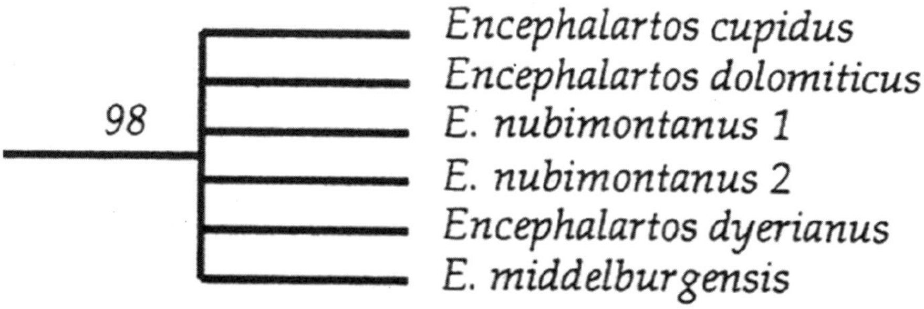

9

Figure 34-7. Cluster from Treutlein et al.'s (2005) cladogram based on ITS 1 & 2 sequence data, showing clustering of the species in Cooper and Goode's (2004) genus *Xanthostrobus*.
Figure 34-8. Clade from Treutlein et al.'s (2005) cladogram based on ITS 1 & 2 sequence data, showing clustering of the species in Cooper and Goode's (2004) genus *Inezamia*.
Figure 34-9. Clade from Treutlein et al.'s (2005) cladogram based on ITS 1 & 2 sequence data, showing clustering of the species in Cooper and Goode's (2004) genus *Glaucostrobus*.

the next genus, *Inezamia*. Perhaps Cooper and Goode (2004) were influenced by a rather skeletal isozyme cladogram of Van der Bank et al. (1998), which placed *E. transvenosus* basal to a group of mostly unrelated species, including *E. altensteinii* and *E. natalensis.* All these species occur along the eastern and northern escarpment of South Africa (Fig. 34-2). In Treutlein et al.'s (2005) ITS cladogram, it forms part of an unresolved clade that excludes *E. senticosus* but includes *E. transvenosus* (Fig. 34-7). The available information therefore supports this grouping, but poorly.

The genus *Inezamia* contains *I. heenanii* (*Encephalartos heenanii*), *I. paucidentata* (*E. paucidentatus*), *I. relicta* (*E. relictus*), and *I. transvenosa* (*E. transvenosus*). Cooper and Goode (2004) attempted to separate it from *Xanthostrobus* on account of the orientation of the leaflets and the longer indumentum of the cones, but there is really more morphological difference between *I. transvenosa* and the other three than between *I. transvenosa* and *Xanthostrobus natalensis*. Their grouping together of these species is clearly wrong, and all indications are that *I. transvenosa* should have been included in their genus *Xanthostrobus.* Moreover, the type species of *Inezamia* is *I. transvenosa*, so that the name *Inezamia* cannot be used for the remaining species. These remaining species (*E. heenanii, E. paucidentatus* and *E. relictus*) are recognizable by having prominently raised vascular bundles on the abaxial leaflet surfaces, incurved leaves, and more or less drooping leaflets. They occur in the eastern and western mountains in Swaziland (Fig. 34-2). Interestingly, the ITS cladogram of Treutlein et al. (2005) does not group *E. heenanii* and *E. paucidentatus* together with *E. transvenosus* or the species in Cooper and Goode's (2004) genus *Xanthostrobus*, but (with modest bootstrap support of 65%) with *E. princeps* and *E. senticosus,* for which there is no morphological support (Fig. 34-8). Finally, Cooper and Goode (2004), supposedly quoting van der Bank et al. (1998), stated about *Xanthostrobus*: "Allozyme data suggest closer affinities . . .'with species north of the Limpopo.'" That is not so: van der Bank et al. (1998) merely wrote ". . . the possibility cannot be discounted that the affinities of *E. transvenosus* are with species north of the Limpopo River rather than with *E. altensteinii* and *E. natalensis.*"

The genus *Glaucostrobus* comprises *G. cupidus* (*Encephalartos cupidus*), *G. dolomiticus* (*E. dolomiticus*), *G. dyerianus* (*E. dyerianus*), *G. eugene-maraisii* (*E. eugene-maraisii*), *G. hirsutus* (*E. hirsutus*), *G. middelburgensis* (*E. middelburgensis*), and *G. nubimontanus* (*E. nubimontanus*). These species have in common a glaucous foliage and occur along the northern escarpment of South Africa (Fig. 34-2). Otherwise, they exhibit a bewildering variety in the architecture

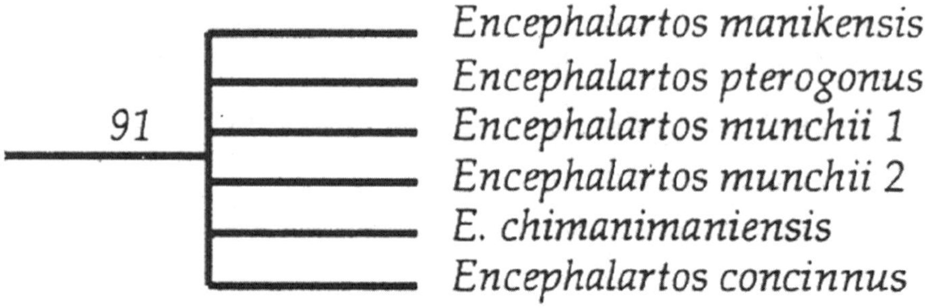

10

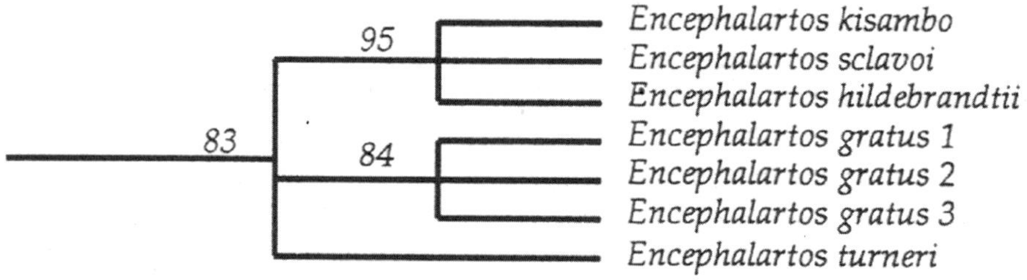

11

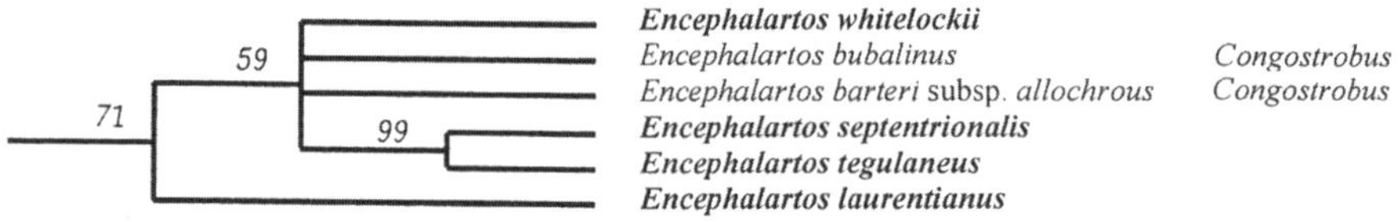

12

Figure 34-10. Clade from Treutlein et al.'s (2005) cladogram based on ITS 1 & 2 sequence data, showing the position of Cooper and Goode's (2004) genus *Viridestrobus*.
Figure 34-11. Clade from Treutlein et al.'s (2005) cladogram based on ITS 1 & 2 sequence data, showing clustering of some of the species in Cooper and Goode's (2004) genus *Tanzamia*.
Figure 34-12. Clade from Treutlein et al.'s (2005) cladogram based on ITS 1 & 2 sequence data, showing clustering of some of the species in Cooper and Goode's (2004) genus *Tanzamia*.

of their foliage and cones. Confusingly, *G. eugene-maraisii* looks very similar to *E. lehmannii* from the Eastern Cape and in Cooper and Goode's (2004) genus *Acanthozamia*. The geographical area of this group is, however, well isolated from that of the other glaucous-leaved South African species, notably *E. lehmannii* and *E. princeps*, so that one has reason to suspect that they are of different lineages. And indeed, Treutlein et al.'s (2005) ITS cladogram placed these species in a separate clade with very strong bootstrap support of 98% (Figs. 34-1 and 34-9).

The genus *Dyerstrobus* contains a single species, *D. inopinus* (*E. inopinus*). Again, the description fails to diagnose it. Yet this is a species with an arresting appearance. It has incurved leaves of a glaucous soapy green color, and drooping and entire leaflets, at a first glance not unlike *Microcycas calocoma*. The female cones are, however, almost indistinguishable from those of *E. dyerianus* and *E. nubimontanus,* which Cooper and Goode (2004) placed in their genus *Glaucostrobus*. Moreover, all attempts at artificial hybridization have failed, suggesting genetic isolation. Treutlein et al.'s (2005) ITS cladogram placed it in their second, unsupported clade, where it appears in a totally unresolved position (Fig. 34-3).

The genus *Viridestrobus* comprises *V. chimanimaniensis* (*E. chimanimaniensis*), *V. concinnus* (*E. concinnus*), *V. manikensis* (*E. manikensis*), *V. munchii* (*E. munchii*), and *V. pterogonus* (*E. pterogonus*). These species lack distinctive common features apart from the proximal leaflets being reduced to a series of prickles (commonly so in tropical species), emerging leaves that have a conspicuously white indumentum, and green cones. These species occur on the escarpment between Zimbabwe and Mozambique as well as in hilly country and island peaks in the vicinity (Fig. 34-2). *Viridestrobus chimanimaniensis* and *V. munchii* are morphologically distinct, but the other described species can be confusingly similar and may be variants of an evolving macrospecies. Treutlein et al.'s (2005) ITS cladogram placed all these species in a clade with bootstrap support of 91%, yet the terminal branches (species) are unresolved, thus supporting the notion that it is an evolving macrospecies (Fig. 34-10).

The genus *Tanzamia* comprises a wide spectrum of tropical species, namely *T. equatorialis* (*E. equatorialis*), *T. hildebrandtii* (*E. hildebrandtii*), *T. ituriensis* (*E. ituriensis*), *T. kisambo* (*E. kisambo*), *T. laurentiana* (*E. laurentianus*), *T. mackenziei* (*E. mackenziei*), "*T. powysii*" (presumably *E. tegulaneus* subsp. *powysii*), *T. sclavoi* (*E. sclavoi*), *T. septentrionalis* (*E. septentrionalis*), *T. tegulanea* (*E. tegulaneus*), and *T. whitelockii* (*E. whitelockii*). Provisionally, Cooper and Goode (2004) also included *T. grata* (*E. gratus*), *T. macrostrobilus* (*E. macrostrobilus*), and *T. turneri* (*E. turneri*) here. These are mostly tall and robust species, the proximal leaflets are reduced to a series of prickles, the male cones emerge in succession rather than simultaneously, and the cones are mostly glabrous (finely rust brown pubescent in *T. laurentiana*) and green to turning yellow. In general, there is a lack of species-distinctive characteristics, and the wide and regular geographical spacing of species (Fig. 34-2) provides few geographical clues. The ITS-based cladogram of Treutlein et al. (2005) placed these species in the same clade but in two separate subclades thereof, with bootstrap support of 83% and 71% respectively and several further subclades each (Figs. 34-11 and 34-12). *Encephalartos laurentianus* is basal in one of the subclades and thus stands alone.

Lastly, the genus *Congostrobus* contains *C. poggei* (*E. poggei*) as well as a number of similar species, namely, *C. delucans* (*E. delucanus*?), *C. marunguensis* (*E. marunguensis*), *C. schaijesii* (*E. schaijesii*), and *C. schmitzii* (*E. schmitzii*), as well as questionably (*fide* Cooper & Goode) *C. allochrous* (*E. barteri* subsp. *allochrous*?), *C. barteri* (*E. barteri*), and *c. bubalinus* (*E. bubalinus*). It is distinguished from the genus *Tanzamia* by "its very short to acaulescent stem [*sic*] [which often suckers profusely], sessile megastroboli [*sic*, not diagnostic], and [soft-textured] bluish-green leaves". These little plants lack very distinctive morphological characteristics and occur over a wide area in central Africa (Fig. 34-2) on sand in fire-prone savannah. They probably constitute a natural group, but no molecular data are available. However, to my mind the inclusion of *E. barteri* (both subspecies) and *E. bubalinus* is doubtful: these are far more robust plants with well-exposed if short stems, and the cones resemble those in the genus *Tanzamia*. Cooper and Goode (2004) wrote that "The affinities of the genus seem to be with *Tanzamia* . . .", without supporting evidence Unfortunately no molecular data are available for the *E. poggei*-like species, but Treutlein et al.'s (2005) ITS-based cladogram placed *E. barteri* subsp. *allochrous* and *E. bubalinus* in a subclade together with *E. septentrionalis* and *E. tegulaneus*, albeit with pretty low bootstrap support of 59% (Fig. 34-12).

Conclusions

A number of Cooper and Goode's (2004) genera appear to be natural groupings and are supported by morphological, geographical, and molecular evidence. These are *Dracostrobus, Dyerstrobus, Encephalartos* sensu Cooper & Goode, *Glaucostrobus, Pyrrhostrobus, Viridestrobus,* and *Xanthostrobus*. The genera *Congostrobus* and *Inezamia* contain discordant elements. The genera *Acanthozamia* and *Rugostrobus* are doubtfully distinct, and the genus *Tanzamia* is a rather amorphous conglomeration of species.

Should these genera, especially those that comprise apparently natural groupings be upheld? Whatever the phylogeny and genetic distance between taxa may be, an eminently practical consideration is that taxa, especially genera, should be recognizably distinct, especially in respect of their reproductive structures. In my opinion, this is not the case with these genera. In spite of certain trends in some groups, it is in many cases impossible to identify material as belonging to any of these genera if presented with a vegetative part or portion thereof, a cone, or even a complete plant of unknown provenance. Added to that the conclusion, rightly or wrongly, of Sharma et al. (1998, 1999), Treutlein and Wink (2002), Hill (2004), and Caputo et al. (2004), that the extant lineages of cycads, including *Encephalartos*, are of recent origin, it does not seem justified to recognize these groupings as taxonomically distinct at generic level. It should also be noted that the new genera are in their present form invalid, for lacking both Latin descriptions and diagnoses. The short English descriptions are not diagnostic.

Literature Cited

Caputo, P., S. Cozzolino, P. de Luca, A. Moretti & D.W. Stevenson. 2004. Molecular phylogeny of *Zamia* (Zamiaceae). Pp 149–157 *in* T. Walters & R. Osborne (eds.), Cycad classification, concepts and recommendations. CABI Publishing, Wallingford, UK and Cambridge, MA.

Cooper, M. R. & D. Goode. 2004. The cycads and cycad moths of KwaZulu-Natal. Peroniceras Press, New Germany, South Africa.

Hill, K. D. 2004. Character evolution, species recognition and classification concepts in the Cycadaceae. Pp. 13–44 *in* T. Walters & R. Osborne (eds.), Cycad classification, concepts and recommendations. CABI Publishing, Wallingford, UK and Cambridge, MA.

Sharma, I. K., D. L. Jones, P. I. Forster & A. G. Young. 1998. The extent and structure of genetic variation in the *Macrozamia pauli-guilielmii* complex (Zamiaceae). Biochem. Syst. Ecol. 26: 45–54.

———, ———, ——— & ———. 1999. Low isozymic differentiation among five species of the *Macrozamia heteromera* group (Zamiaceae). Biochem. Syst. Ecol. 27: 67–77.

Treutlein, J. & M. Wink. 2002. Molecular phylogeny of cycads inferred from *rbc*L sequences. Naturwissenschaften 89: 221–225.

———, P. Vorster & M. Wink. 2005. Molecular relationships in *Encephalartos* (Zamiaceae, Cycadales) based on nucleotide sequences of nuclear ITS 1 & 2, *rbc*L, and genomic ISSR fingerprinting. Pl. Biol. (New York) 7: 1–12.

Van der Bank, F. H., P. Vorster & M. van der Bank. 1998. Phylogenetic relationships, based on allozyme data, between six cycad taxa indigenous to South Africa. S. African J. Bot. 64: 182–188.

———, M. Wink, P. Vorster, J. Treutlein, L. Brand, M. van der Bank & J. Hurter. 2001. Allozyme and DNA sequence comparisons of nine species of *Encephalartos* (Zamiaceae). Biochem. Syst. Ecol. 29: 241–266.

Vorster, P. 2004a. Classification concepts in *Encephalartos*. Pp. 69–83 *in* T. Walters & R. Osborne (eds.), Cycad classification, concepts and recommendations. CABI Publishing, Wallingford, UK and Cambridge, MA.

———. 2004b. Book review. Cooper, Michael Robert & Goode Douglas. The cycads and cycad moths of KwaZulu-Natal. Encephalartos 80: 10–14.

———. 2004c. Typification of *Encephalartos*. Bothalia 34: 112–113.

SYSTEMATICS
CHAPTER 35

The Cycads of Mexico: 25 Years of Research and Conservation

Andrew P. Vovides, Jorge González-Astorga, Miguel Angel Pérez-Farrera, Dolores González, Christina Bárcenas, and Carlos Iglesias

Abstract

During the botanical explorations of the nineteenth century there was a proliferation of cycad species descriptions from the New World by botanists and plant hunters alike, and the first Mexican cycad to be cultivated in Europe was *Zamia furfuracea*, grown in the Royal Garden at Hampton Court, England, in 1691. The early twentieth century was marked by the extensive work of Charles Chamberlain, followed by a relative vacuum up until the late 1970s, when renewed interest blossomed among various workers at home and abroad. There has since been an increase of more than 300% in the number of named species, compared with the nine or so species known in the late 1960s and early 1970s. This has placed Mexico as second worldwide in the number of cycad species, with more than 80% endemics. An outline is presented of species, distribution, vegetation types, and current basic and applied research being done on Mexican cycads. Molecular and genetic research has shown that cycad speciation in Mexico is relatively recent, and we now have a better understanding of distribution patterns of *Dioon* and *Ceratozamia*. The geographical distribution of many cycad species in Mexico appears to be correlated with floristic refugia of the Cenozoic. As a result of this research, as well as of demographic and ecological studies on some species, projects aimed at sustainable management have been implemented and follow-up studies in cultivation improvement are underway.

Resumen

Durante las exploraciones botánicas del siglo 20 tardío hubo una proliferación de descripciones de especies de cícadas del Nuevo Mundo por botánicos y colectores de plantas y la primera cícada mexicana cultivada en Europa fue *Zamia furfuracea*, en el Jardín Real de Hampton Court, Inglaterra en el año 1691. El siglo 20 temprano estuvo marcado por los trabajos extensivos de Charles Chamberlain, seguido por un vacío relativo hasta la década de los 1970s, cuando surgió un interés renovado entre varios investigadores nacionales y extranjeros. Desde entonces hubo un incremento de más del 300% en el número de especies descritas de las aproximadamente nueve especies conocidas en los años 1960s y principios de los 1970s. Esto ha colocado México en el segundo lugar a nivel mundial en el número de especies con más del 80% endémicas al país. En este artículo presentamos un bosquejo de las especies, su distribución, típos de vegetación así como la investigación básica y aplicada que se está llevando acabo en las cícadas mexicnas. Investigaciones moleculares y genéticas han mostrado que la especiación en México es relativamente reciente y tenemos un mejor conocimento de los patrones de distribución de los géneros *Dioon* y *Ceratozamia*. La distribución geográfica de algunas especies de cícadas en México aparentemente se correlaciona con los refugios florísticos del Cenozoico. Como resultados de estas investigaciones así como estudios ecológicos y demográficos sobre algunas de las especies se desarrollaron proyectos enfocados hacía el manejo sustentable de las cícadas y estudios de seguimiento para la mejora de los métodos de cultivo están en marcha.

Introduction

Cycads in Mexico are distributed along the Sierra Madre Oriental, facing the Gulf of Mexico, and along the Sierra Madre Occidental, Sierra Madre del Sur, Sierra Madre de Chiapas, and the Sierra del Soconusco, all facing the Pacific and southeastern Mexico, as well as on the Isthmus of Tehuantepec and the Yucatan Peninsula. There are currently 48 named species within three genera, and we believe the number may reach over 50 species when all potential cycad habitats have been investigated and taxonomic studies are completed.

Mexico is particularly rich in cycads, with a high endemism, and we consider Mexico to be a center of diversity for the group in the neotropics (Vovides et al., 2003); it occupies second place worldwide for cycad diversity (Vovides, 2000). Much of our knowledge of the distribution of cycads in Mexico is due to the early exploratory work of Chamberlain (1919), and in recent years information has been greatly expanded by workers from Mexico, the United States, and Italy (Balduzzi et al., 1981–82; Moretti et al., 1982; Stevenson et al., 1986; Norstog & Nicholls, 1997; Pérez-Farrera et al., 2001a, 2001b). Three genera are present: *Ceratozamia* Brongn., *Dioon* Lindl., and *Zamia* L. *Ceratozamia* ranges from Mexico into Central America, and is present in Guatemala, Belize,

Figure 35–1 (left). Distribution of the genus *Ceratozamia* in Mexico.
Figure 35–2 (right). Distribution of the genus *Dioon* in Mexico.

and Honduras (Fig. 35-1), whereas the largely Mexican *Dioon* is vicariant only in Honduras and possibly Nicaragua, which are the southern limits of "Mega-Mexico 2" (sensu Rzedowski, 1993) for reasons of biotic affinity (Fig. 35-2).

The genus *Zamia* ranges widely, from Florida in the United States, through Mexico, all of Central America, the West Indies, and into South America, where the genus is particularly diverse in Colombia (Stevenson, 2001) (Fig. 35-3).

The genera *Zamia* and *Ceratozamia* are rife with nomenclatural problems that are hampering taxonomic research to some extent. A workshop on cycad classification concepts (Walters & Osborne, 2004) has set out guidelines for future work on cycads, especially the species concept within the group. In Mexico and the rest of the neotropics, cycad habitats vary from evergreen tropical rainforests, through seasonally dry tropical forests, cloud forests, oak and pine-oak forests, grasslands, savannahs, and rarely riverine/mangrove communities, such as that of the Colombian and Ecuadorian *Z. roezlii* Linden.

Cycad Research

As a result of renewed interest in cycads during the last 25 years or so, the known world cycad flora has increased in number to over 300 species (Hill et al.,2007) after a long period of relative stagnation following the pioneering work of Charles Joseph Chamberlain during the early twentieth century (Norstog & Nicholls, 1997). Work has been done in

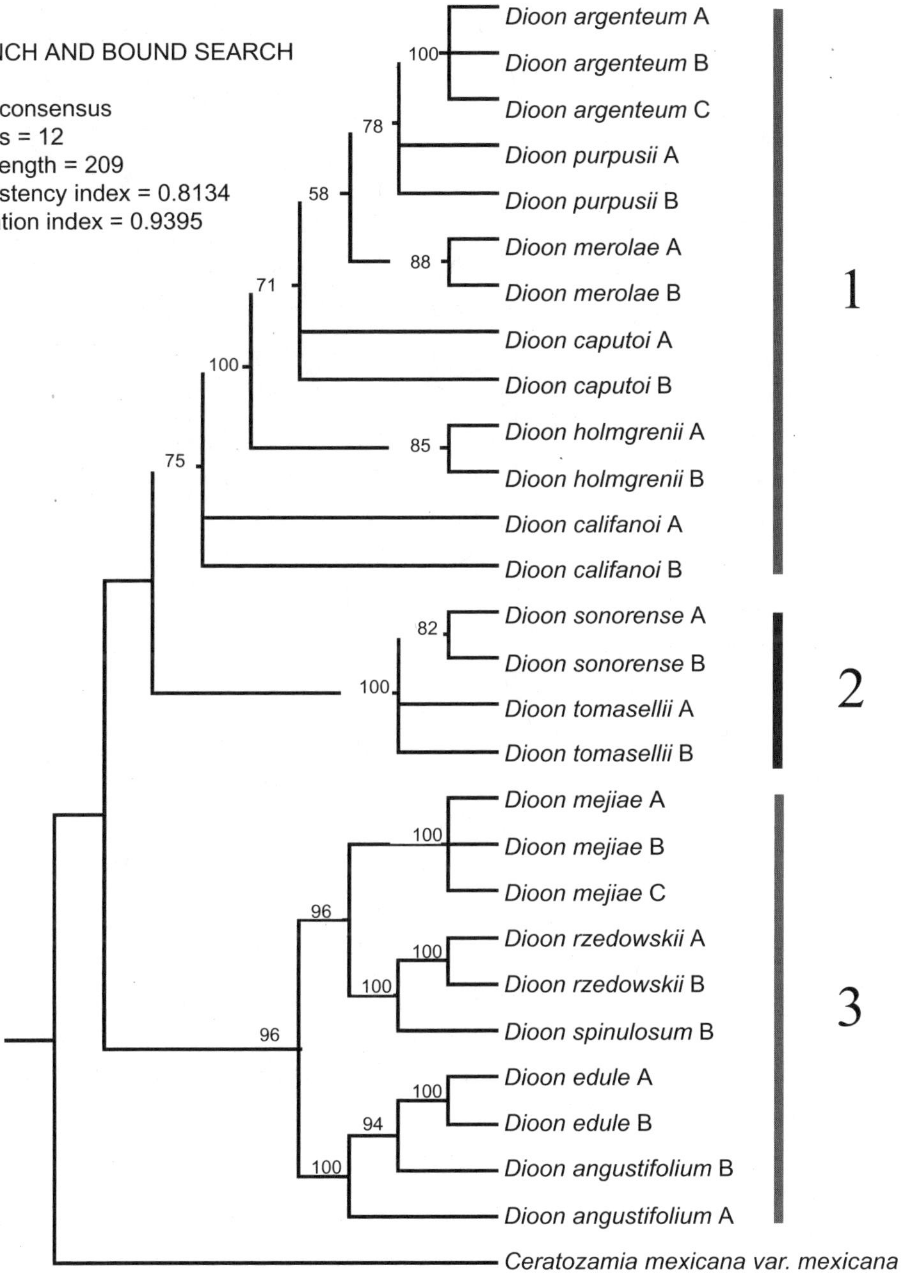

Figure 35-3. Distribution of the genus *Zamia* in Mexico.

the fields of toxicity (Whiting, 1963; Charlton et al., 1992; and more recently Cox et al, 2007; Shaw et al., 2007; and Brenner et al., 2007); population ecology of *Zamia* (Newell, 1983; Clark & Clark, 1987, 1988), of *Encephalartos* (Donaldson, 1995), and in Mexico of *Dioon* and *Ceratozamia* (Vovides, 1990 and Pérez-Farrera et al., 2000, 2007 respectively); genetics and evolution of Pacific *Cycas* (Keppel, 2002; Keppel et al., 2002) and of *Dioon* in Mexico (González-Astorga et al., 2003a, 2003b, 2005); cytotaxonomy (Norstog, 1980, 1981; Vovides, 1983, 1985; Moretti & Sabato, 1984; Moretti, 1990; Vovides & Olivares, 1996; Kokubugata et al., 2004); morphology and systematics (Stevenson, 1990); and pollination biology (Norstog et al., 1986; Tang, 1987; Norstog & Fawcett, 1989; Vovides, 1991a, 1991b; Vovides et al., 1993), to mention a few. On the whole, there is growing interest and research on the neotropical cycad flora (Castilo-Guevara, 2007; Chávez-González, 2007; Flores-Vázquez et al., 2007; González-Astorga & Vovides, 2007; Lazcano-Lara, 2007; Nicolalde, 2007; Sánchez-Tinoco et al., 2007).

Wind pollination was thought to be the mechanism of pollination in cycads until the work of Norstog et al. (1986), Norstog (1987), Tang (1987), and Norstog and Fawcett (1989) established pollination by insects (Curculionidae and Languriidae) for two species (*Zamia pumila* and *Z. furfuracea*), and also wind pollination of female cycad cones was found to be aerodynamically improbable (Niklas & Norstog, 1984). Entomophily has been postulated for other cycad species by evidence such as volatile emissions from male cones, differential starch content between male and female cones, and idioblast breakdown in female cones prior to receptivity (Pellmyr et al., 1991; Vovides, 1991b; Vovides et al., 1993; Stevenson et al., 1998), as well as by phylogenetic inference, such as in the case of *Microcycas calocoma* (Miq.) A.DC. (Vovides et al., 1997).

Though it may be thought that the insect/cycad relationship is the result of Mesozoic coevolution, molecular evidence suggests that the extant cycad species have evolved more recently (Treutlin & Wink, 2002), and this may also be the case for their insect pollinators (Schneider et al., 2002).

The Genus *Ceratozamia*

The genus *Ceratozamia* or "horned" *Zamia* (κερασ = ceras, Greek for horn) is distinguished from *Zamia* in having two horn-like projections on the distal faces of both

Table 35-1 Species list of *Ceratozamia* and distribution in Mexico.

Species	Distribution
C. alvarezii Pérez-Farr., Vovides & Iglesias	Chiapas
C. becerrae Pérez-Farr., Vovides & Schutzman	Chiapas, Tabasco
C. euryphyllidia Vázq. Torres, Sabato & D.W. Stev.	Oaxaca, Veracruz
C. hildae G.P. Landry & M.C. Wilson	Querétaro, San Luis Potosí
C. huastecorum S. Avendaño, Vovides & Cast.-Campos	Veracruz
C. kuesteriana Regel	Tamaulipas
C. latifolia Miq.	Hidalgo, Querétaro, San Luis Potosí
C. matudae Lundell	Chiapas
C. mexicana Brongn.	Hidalgo, Puebla, Veracruz
C. microstrobila Vovides & J.D. Rees	San Luis Potosí
C. miqueliana H. Wendl.	Chiapas, Veracruz
C. mirandae Vovides, Pérez-Farr. & Iglesias	Chiapas
C. mixeorum Chemnick, T.J. Greg. & S. Salas-Mor.	Oaxaca
C. morettii Vázq. Torres & Vovides	Veracruz
C. norstogii D.W. Stev.	Chiapas, Oaxaca
C. robusta Miq.	Chiapas, Oaxaca, Veracruz
C. sabatoi Vovides, Vázq. Torres, Schutzman & Iglesias	Hidalgo, Querétaro
C. whitelockiana Chemnick & T.J. Greg.	Oaxaca
C. zaragozae Medellin-Leal	San Luis Potosí
C. zoquorum Pérez-Farr., Vovides & Iglesias	Chiapas

mega- and microsporophylls. Adolphe T. Brongniart described the genus in 1846 with the type species *C. mexicana* Brongn. By 1980, only six species were known, but now there are over 20 species described for Mexico and this figure may increase in the future (Table 35-1).

Ceratozamia is largely distributed in the warm temperate and tropical regions of eastern and southern Mexico, including mid-elevation pine-oak forests in the state of Tamaulipas at its extreme northern limit in Mexico, to Belize and Guatemala in Central America (Vovides et al., 2004). Recently a taxon related to the *C. miqueliana* Wendl. complex of species has been found in northern Honduras; hence, the range of the genus has been extended. *Ceratozamia* is much like some robust species of *Zamia,* and according to Crane (1988), *Ceratozamia, Zamia,* and the Cuban endemic *Microcycas* are sister genera. There are 20 currently known species of *Ceratozamia* in Mexico (Table 35-1), and four taxa are undergoing study.

Morphology and Distribution

Some *Ceratozamia* species are basically arborescent but with stems rarely more than about 1 m tall. They are often leaning or curved but rarely branching, while others are semihypogeous and often branching. The leaves form an ascending or open to descending crown. Leaflets vary from linear lanceolate to widely obovate with margins mostly entire, asymmetrical in obovate types, and sometimes with irregular clefting toward the apex. The indumentum and color of the cones, especially the female cones, are often used as diagnostic features. Other species, especially those from northeastern Mexico, are generally characterized as having small and often branching trunks with few leaves per crown and generally small cones. Leaflets are usually narrow (less than 2 cm wide), as in *C. kuesteriana* Regel and *C. zaragozae* Medellín-Leal, with the exception of *C. microstrobila* Vovides & Rees and another two undetermined isolated species. Plants from south and southeast Mexico are generally large, single-trunked plants, and usually have obovate leaflets usually more than 2 cm wide, the widest being found in *C. euryphyllidia* Vázq. Torres, Sabato & Stevenson, though there are exceptions, such as *C. norstogii* Stevenson, *C. mirandae* Vovides, Pérez-Farrera & Iglesias, and *C. alvarezii* Vovides, Pérez-Farrera & Iglesias.

Throughout the distribution of *Ceratozamia*, about seven species complexes are found (Moretti et al., 1980; Vovides et al., 2004). This genus presents a number of distinct populations in Mexico, with the *C. robusta* Miq. group forming the most widely distributed complex, the majority of populations of which are found in southeast Veracruz, northwest Chiapas, and the southern Sierra of Oaxaca, Belize, and Guatemala. Stevenson et al. (1986) mention vicariance as a possible explanation for the present distribution of *Ceratozamia* and note that the genus is monophyletic. Pérez-Farrera et al. (2000) remark that at least some species may be dispersed by mammals such as the peccary. Some of the present odd distribution patterns may be the result of past dispersal agents that are now extinct, as suggested by Jansen and Martin (1982) for tropical trees bearing large fruits.

The spatial distribution patterns of both *Ceratozamia* spp. so far studied (*C. mirandae* and *C. matudae*) were found to be grouped or aggregate and have a reverse 'J' or Devey type III population structure curve indicating a higher frequency of seedlings and juvenile plants than adults, though gravity appears to be the major distribution agent (Pérez-Farrera et al., 2000; Pérez-Farrera & Vovides, 2004).

Systematics

Cytotaxonomic studies of the genus has revealed chromosome counts of $2n = 16$ with a stable karyotype of 12 mesocentric chromosomes, two acrocentric chromosomes, and two telocentric chromosomes (Vovides, 1983, 1985; Moretti, 1990), which contrasts with that reported for *Zamia* (Vovides et al., 2003). However, there are satellites in *Ceratozamia* that vary in number and position. The satellite number appears to be correlated with latitude, and the reason for this is not yet fully understood (Table 35-2). Fluorochrome banding (CMA/DAPI) and fluorescence in situ hybridization of rDNA appear to be promising at the genus level, since banding showed that *C. mexicana* Brongn. did not have any simple Robertsonian relationship to *Microcycas calocoma* (Miq.) A.DC. (Kokubugata & Kondo, 1998) and that chromosome evolution in the genus may have occurred through paracentric inversions (Kokubugata et al., 2004).

Table 35-2 Number of chromosome satellites registered in *Ceratozamia* karyotypes and the species distribution relative to the trans-Mexican mountain range.

Species	No. of satellites	Distribution
C. microstrobila	4	North
C. huastecorum	5	North
C. kuesteriana	2; 4	North
C. hildae	6	North
C. huastecorum	5	North
C. sabatoi	5	North
C. zaragozae	7	North
C. mexicana	*3?*	On the mountain range
C. sp. nov.	6	On the mountain range
C. robusta	6	On the mountain range
C. becerrae	1	South
C. euryphyllidia	2–4?	South
C. miqueliana	1?	South
C. matudae	2	South
C. norstogii	1?	South
C. robusta	1?	South
C. zoquorum	1?	South

Morphological and molecular characters have shown that *Ceratozamia* is a sister genus to both *Zamia* and *Microcycas* (Crane, 1988; Stevenson, 1990; Bogler & Francisco-Ortega, 2004), and the group is monophyletic. However, at the species level, DNA sequencing of noncoding regions from the chloroplast and nuclear genomes ([ITS] and *trn*L-F) of *Ceratozamia* have revealed very little variation. In spite of this, three main clades were obtained from a consensus tree, which is consistent with the distribution ranges of the species although only limited conclusions concerning phylogenetic relationships can be inferred. Two clades contain the species distributed in southern and southeastern Mexico, both at and south of the Trans-Mexican Neovolcanic Mountain Range of Pliocene-Quaternary (Pleistocene) age, suggesting a probable ancestral geographic area for *Ceratozamia* in southeast Mexico (González & Vovides, 2002). Another implication is that speciation within the genus appears to be associated with the post-Pleistocene spread of floristic communities from proposed Pleistocene tropical refugia in southeast Mexico to south of the Trans-Mexican Neovolcanic Mountain Range. This is consistent with the findings of Marshall and Liebherr (2000), who identified two biogeographic assemblages, one north of the Neovolcanic range and another to the south. We believe that *Ceratozamia* species are of recent speciation, which is in agreement with Treutlein and Wink (2002), who studied *rbc*L sequences in cycads and suggested that the emergence of the extant species may have occurred recently, perhaps during the Tertiary.

Morphological and Genetic Variation

The morphological variation of species within the *C. norstogii* complex (*C. alvarezii, C. mirandae, C. norstogii,* and *Ceratozamia* sp.) and the *C. miqueliana* complex (*C. zoquorum, C. miqueliana, C. becerrae,* and *C. euryphyllidia*) were evaluated using analysis of variance and discriminant analysis based on 17 vegetative and six reproductive characters, and the genetic variation by application of the random amplification of polymorphic DNA (RAPD) method with five primers to a total of 147 individuals among eight populations located in adjacent areas of Chiapas, Oaxaca, Veracruz, and Tabasco. Although notable morphological differences were observed among the species that were separated by the discriminant analysis, the high values of genetic identity (0.95–0.97) and the low values of genetic differentiation among the species of the complexes suggest that they have evolved recently, in agreement with González and Vovides (2002), very probably at the end of the Miocene (Pleistocene), and that allele mobility among the species has not been completely fixed, which suggests that each species analyzed within each complex forms a discrete and valid unit (see Pérez-Farrera et al.,2007).

The Genus *Dioon*

This genus was described by John Lindley (1843) with the type species *Dion edule* Lindl. Lindley derived the generic name from the Greek δισ=two and οων=egg or ova (dis+oön), and Miquel (1848) regarded *Dion* as an orthographic error and revised the spelling to *Dioon,* which has been formally conserved (Brummitt, 1987).

Our early knowledge of this cycad is attributed largely to Charles J. Chamberlain, who speaks of the "great abundance" of *D. edule* near Chavarillo during the early twentieth century. Chamberlain (1919) also estimated the age of an approximately 1.5-m-tall *D. edule* individual growing in a garden in Xalapa to be about 970 years, and mentions that *D. spinulosum* Dyer & Eichler has a much faster growth rate. Vovides and Peters (1987) and Vovides (1990), using field data and observations from over four years, confirmed the great age attained by *D. edule* plants in the wild, with some individuals reaching over 2000 years.

Dioon spp. generally grow in seasonally dry forest and oak forest. However, *D. spinulosum* and *D. rzedowskii* are found in humid tropical rain forests in karst topography. The state of Oaxaca is particularly rich in *Dioon* spp. and may be considered a center of diversity for the genus. There are currently 12 *Dioon* spp. known from Mexico (Table 35-3).

Morphology and Distribution

The genus consists of arborescent cycads, up to 15 m tall in the case of *D. spinulosum*, trunks usually solitary and erect, becoming decumbent and sometimes branching with age. Trunks are cylindrical and armed with persistent petiole bases forming a light to dark gray bark. Leaves are pinnate, up to 30 forming a tight apical crown. The large pendant woolly female cones of *D. spinulosum* can weigh 25 kg or more and contain up to 300 seeds, but the numbers are less spectacular in other *Dioon* species (Vovides et al., 1983).

Table 35-3 Species list of *Dioon* and distribution in Mexico.

Species	Distribution
D. angustifolium Miq.	Nuevo Leon, Tamaulipas
D. argenteum T.J. Greg, Chemnick, S. Salas-Mor. & Vovides	Oaxaca
D. califanoi De Luca & Sabato	Puebla
D. caputoi De Luca, Sabato & Vázq. Torres	Puebla
D. edule Lindl.	Hidalgo, Querétaro, San Luis Potosi, Tamaulipas, Veracruz
D. holmgrenii De Luca, Sabato & Vázq. Torres	Oaxaca
D. merolae De Luca, Sabato & Vázq. Torres	Chiapas, Oaxaca
D. purpusii Rose	Oaxaca
D. rzedowskii De Luca, A. Moretti, Sabato & Vázq. Torres	Oaxaca
D. sonorense (De Luca, Sabato & Vázq. Torres) Chemnick, T.J. Greg. & S. Salas-Mor.	Sinaloa, Sonora
D. spinulosum Dyer & Eichler	Oaxaca, Veracruz
D. tomasellii De Luca, Sabato & Vázq. Torres	Durango, Guerrero, Jalisco, Nayarit, Michoacán

Dioon is largely endemic to Mexico with vicariance in northern Honduras (*D. mejiae* Standley & Williams), and the genus can be considered endemic to Mega-Mexico 2, after Rzedowski (1993) (Fig. 35-2). The genus ranges from the state of Sonora in the northwestern Mexico, where *D. sonorense* (De Luca, Sabato & Vázq.Torres) Gregory, Chemnick & S. Salas-Morales is found at latitude 29°04′N, to latitude 15°32′N in Honduras, where *D. mejiae* is found. The southernmost distribution of the genus in Mexico is in the state of Chiapas, where *D. merolae* De Luca, Sabato & Vázq.Torres is found. The state of Oaxaca is particularly rich in *Dioon* spp., and we consider it a center of diversity for the genus with eight species, and probably more to be described as new explorations proceed. New populations of *D. caputoi* De Luca, Sabato & Vázq.Torres have already been reported (Flores-Vázquez et al., 2007). We believe that there are at least three species complexes in Oaxaca alone, and outside of Oaxaca we believe that three other species complexes can be recognized. In the Pacific coast states of Sinaloa, Nayarit, and Jalisco, and into Guerrero, *D. tomasellii* De Luca, Sabato & Vázq.Torres comprises a series of widely scattered populations, and several of these populations appear to have distinct morphological characteristics and may prove to be distinct species upon further study. Gregory and Chemnick (2004) discuss the hypotheses regarding relationships and distribution of *Dioon* in detail.

Genetic Variation

Dioon edule and *D. angustifolium* Miq. mirror morphologically the *D. tomasellii* and *D. sonorense* distribution in the west, in that *D. sonorense* and *D. angustifolium* have narrower leaflets than *D. tomasellii* and *D. edule* and are at the extreme northwest and northeast range of the distribution, respectively. *Dioon edule* is distributed in the states of Tamaulipas, Nuevo León, Hidalgo, Queretaro, San Luis Potosí, and Veracruz. *Dioon angustifolium* (previously considered *D. edule* var. *angustifolium* (Miq.) Miq.) is found in the northern part of the range and is morphologically, genetically, and geographically distinct from *D. edule,* which occurs further south (González-Astorga et al., 2003a, 2003b). Recent genetic variation studies confirm this distinction, suggesting that *D. edule* consists of disjunct populations with semidistinct morphology (González-Astorga et al., 2005). *Dioon angustifolium* shows the least genetic variation compared with the southernmost populations of *D. edule* in Veracruz (González-Astorga et al., 2003a, 2003b) and has the smallest, and possibly the most easily dispersed, seeds in the genus, and some of the largest populations. It might be considered an opportunistic colonizer of new habitats made available by postglacial warming, which appears consistent with the findings of González-Astorga et al. (2003a, 2003b, 2005). This again supports a south to north species migration during the post-Pleistocene climate warming, as suggested for *Ceratozamia* by González and Vovides (2002). Both *D. edule* and *D. angustifolium* show very high genetic variation compared with other cycads, probably because of recent genetic bottleneck effects. For further details see González-Astorga et al. (2005) and González-Astorga and Vovides (2007).

Ecology and Conservation

Dioon inhabits seasonally dry tropical forests and oak and pine-oak forests, including transitional areas between these vegetation types, and to a limited extent tropical evergreen and subdeciduous rain forests (*D. spinulosum* and *D. rzedowskii* De Luca, A. Moretti, Sabato & Vázq.Torres) but never cloud forests (Gregory & Chemnick, 2004). Species of the genus occur as primary vegetation on rocky substrates, steep slopes, and poor soils (Vovides et al., 1983; Sabato & De Luca, 1985).

Dioon edule has been shown to present a variant of crassulacean acid metabolism (CAM) cycling, a photosynthetic metabolic sequence present in plants that occur in water-stressed habitats (Vovides et al., 2002a). It is the first cycad to be reported with a variant of CAM physiology, and *Welwitschia* is the only other gymnosperm reported to show CAM cycling (Ting & Burk, 1983). That CAM is an ancient metabolic sequence is supported by its presence in *Isoëtes* (Lycophyta). The Lycopsids are a group of ancient plants whose origins can be traced to the Carboniferous. Cycads also appear to be of Paleozoic origin, and gymnosperms were already numerous during the Carboniferous (Norstog & Nicholls, 1997). This is particularly interesting from an evolutionary point of view since the drying of climates and desertification during the Permo-Triassic transition caused mass extinctions well into the Triassic (Wilson, 1992) and may have favored survival of those plant groups that had already developed a precursory variant of CAM by late Paleozoic times, which may have contributed to the success and abundance of cycads during the Mesozoic (Vovides et al., 2002a).

Demographic studies have shown a population structure (*D. edule*) with a reverse J or Devey type III curve and a spatial distribution pattern with a preference for shallow rocky soils and an extremely slow growth rate. The great ages (over 2000 years) attained by individuals of this cycad are remarkable (Vovides & Peters, 1987; Vovides, 1990). The longevity of *D. edule* and the presence of arborescent branches in old individuals enables production of multiple cones per plant, and this agrees with the type III cycad life cycle of Donaldson (1995). This life-cycle type is characterized by the persistence of adults and a high reproductive potential. Mathematical modeling and projection matrices for three populations of *D. edule* in Veracruz have shown that the reproductive adults occupy the most important stage in the life cycle, and it is therefore of utmost importance to conserve these rather than reintroduce artificially propagated seedlings (Pablo-Aguilar et al., unpublished data).

Systematics

Moretti et al. (1993) used cpDNA restriction fragment length polymorphism (RFLP) analysis for a phylogenetic analysis of the genus *Dioon,* and a consensus tree showed two well-defined major clades. The phylogeny was congruent with the biogeography of the genus but only partly congruent with morphological data. Speciation by vicariance was proposed, with a center of diversification of one of the major clades in the mountain system in the states of Oaxaca, Puebla, and Veracruz. Bogler and Francisco-Ortega (2004),

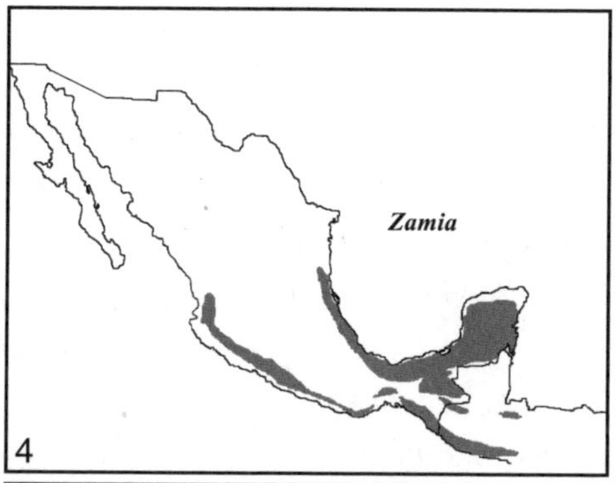

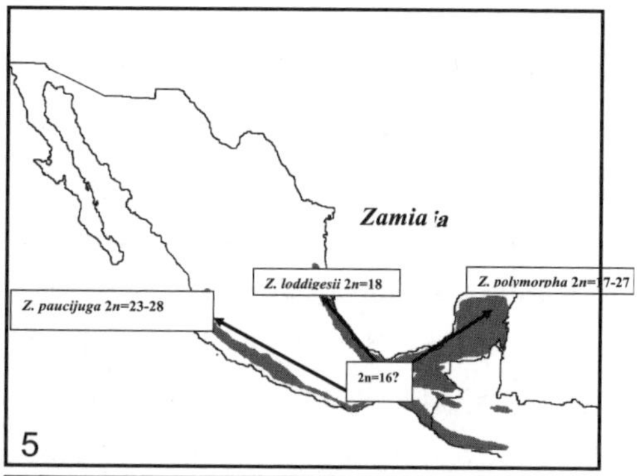

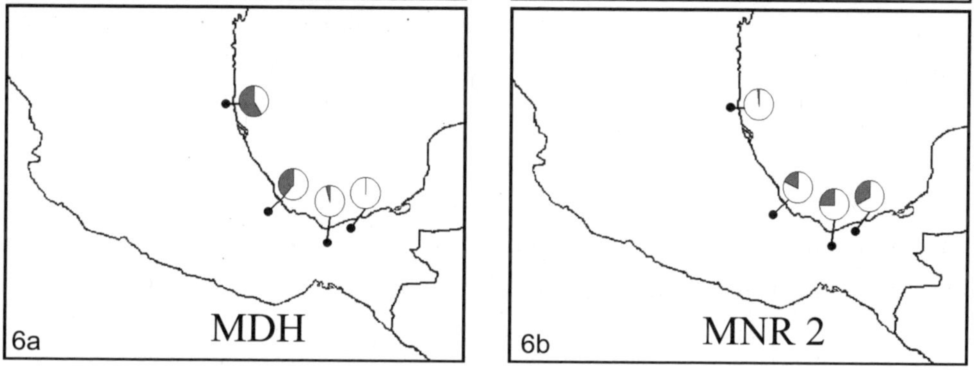

Figure 35-4. Molecular phylogeny for the genus *Dioon* based on nuclear ITS 1 and 2 regions.
Figure 35-5. Probable migration of cytotypes of *Zamia polymorpha, Z. paucijuga,* and *Z. loddigesii* from a putative ancestor of somatic genome of $2n=16$ on the Isthmus of Tehuantepec.
Figure 35-6. Clinal variation of the allele frequencies at two loci in four populations of *Zamia loddigesii* in N.E. and S.E. Mexico: **a.** locus *MDH*; **b.** locus *MNR2.*

using *trn*L intron and ITS 2 rDNA sequences of ten cycad genera, found that the *trn*L intron sequence did not resolve the *Dioon* clade (five spp.) but that ITS 2 rDNA resolved the five *Dioon* spp. into two groups, in accord with Moretti et al. (1993).

Our preliminary sequences of the ITS 1 and 2 regions of 13 species of *Dioon* appear well suited for elucidating most relationships within *Dioon*. A strict consensus tree (from 12 trees) obtained from a branch and bound search showed good resolution in terminal branches among species (Fig. 35-4). Sequences from putative outgroup taxa such as *Cycas* and *Zamia* were of limited utility for polarizing the tree because of ambiguous alignment of the ITS sequence data. The inclusion of *Ceratozamia* to polarize the tree was based only on its higher nucleotide similarity in the alignment compared with other taxa. The topology obtained in the branch and bound search corresponds partially to previous relationships previously hypothesized in this genus.

Three major clades were obtained that appear to follow a biogeographical pattern. The relationships from this preliminary analysis suggest that *D. rzedowskii*, *D. spinulosum*, and *D. mejiae* form a clade, as previously hypothesized from morphology, cpDNA RFLP, and habitat conditions. The major groups are in accord with those suggested by Moretti et al. (1993) and Bogler and Francisco-Ortega (2004) and reflect biogeography. Further resolution has been obtained in that *D. tomasellii* and *D. sonorense* form a separate (bar 2) clade that is congruent with their distribution in western (*D. tomasellii*) to northwestern Mexico (*D. tomasellii* and *D. sonorense*). *Dioon califanoi, D. holmgrenii, D. caputoi, D. merolae, D. purpusii,* and *D. argenteum* form a clade with species distributed in southern Mexico (Oaxaca and Chiapas, bar 1). The Atlantic clade (bar 3) can be further subdivided into two smaller clades, reflecting the distribution of *D. edule* in eastern to northeastern Mexico and *D. angustifolium* in northeastern Mexico, and a clade with *D. spinulosum* and *D. rzedowskii* in southeastern Mexico, as well as *D. mejiae* in Honduras. Work is still in progress to fully solve phylogenetic relationships within this genus.

The Genus *Zamia*

The derivation of the generic name *Zamia* L. appears to be based on the Greek *azaniae*, which refers to pine cones (Jones, 1993), as described by Carl von Linnaeus in 1763. Like the previous genera, *Zamia* is a totally New World genus. At its northernmost range is *Z. integrifolia* L. fil. in southeastern Georgia and Florida in the United States, and *Z. boliviana* (Brongn.) A. DC. in Bolivia occupies its southernmost range. The genus therefore comprises the most geographically wide-ranging and morphologically diverse group of neotropical cycads, occupying a wide array of habitats, comparable only to the old world genus *Cycas* L. (Norstog & Nicholls, 1997).

Morphology, Distribution, and Ecology

The genus consists of plants with tuber-like or cylindrical trunks that may be unbranched or branching with age and either subterranean or partly so, to arborescent, with dehiscent petiole bases, or partly persistent leaf bases near the leaf crown. Leaves are bright green or

Table 35-4 Species list of *Zamia* and distribution in Mexico (state or region):

Species	Distribution
Z. cremnophila Vovides, Schutzman & Dehgan	Tabasco
Z. fischeri Miq.	Querétaro, San Luis Potosí, Nuevo León
Z. furfuracea L. f.	Veracruz
Z. herrerae Calderón & Standl.	Chiapas
Z. inermis Vovides, J.D. Rees & Vázq. Torres	Veracruz
Z. katzeriana (Regel) Rettig (= *Z. splendens* Schutzman)	Chiapas, Tabasco, Veracruz
Z. lacandona Schutzman & Vovides	Chiapas
Z. loddigesii Miq.	Several states
Z. paucijuga Wieland	Oaxaca, Guerrero, Michoacan, Jalisco Nayarit
Z. polymorpha D.W. Stev., A. Moretti & Vázq. Torres.	Campeche, Yucatan peninsula
Z. purpurea Vovides, J.D. Rees & Vázq. Torres	Veracruz, Oaxaca
Z. soconuscensis Schutzman, Vovides & Dehgan	Chiapas
Z. spartea A. DC.	Oaxaca, Chiapas
Z. variegata Warsz.	Chiapas
Z. vazquezii D.W. Stev., Sabato, A. Moretti & De Luca	Veracruz
Z. verschaffeltii Miq.	Extinct

reddish brown at emergence, pinnate, ascending, spreading or descendent, flat or keeled. The petiole and rhachis are armed with prickles or unarmed, swollen at the base, glabrous or pubescent. Leaflets are articulate at the base, evenly arranged along the rhachis, opposite to subopposite, linear lanceolate to widely oblanceolate, margin dentate to entire, veins immersed and not prominent to raised and highly prominent. Female cones are ovoid to barrel-shaped, puberulent and becoming glabrous at maturity. Seeds are ovoid to subglobose with a brightly colored sarcotesta that is often red, yellow, or orange, or sometimes brown.

Zamia habitats range from seasonal tropical dry forests and secondary vegetation, to tropical oak forests, evergreen tropical rainforest, cloud forests, to riverine and coastal habitats within seawater splash zones, as with the Mexican species *Z. furfuracea* L. fil., to epiphytes, as with the Panamanian *Z. pseudoparasitica* Yates. Most Mexican and all West Indian species of *Zamia* have subterranean trunks with only the leaf crowns visible, and a high number of Central and South American species are arborescent. Within Mexico the genus is distributed in lowland habitats under 1000 m in elevation (Table 35-4, Fig. 35-3) along the Sierra Madre Oriental facing the Gulf of Mexico (*Z. loddigesii* Miq., *Z. inermis* Vovides, Rees & Vázq. Torres, *Z. cremnophila* Vovides, Schutzman & Dehgan, *Z. vazquezii*

D.W. Stev., Sabato, Moretti & De Luca, and *Z. fischeri* Miq.), in coastal habitats (*Z. furfuracea*) and the Yucatan peninsula (*Z. polymorpha* D.W. Stev., Moretti & Vázq.Torres), on the Isthmus of Tehuantepec into Chiapas (*Z. spartea* A.DC., *Z. purpurea* Vovides, Rees & Vázq.Torres, *Z. loddigesii, Z. lacandona* Schutzman & Vovides, *Z. variegata* Warsz. *Z. splendens* Schutzman, *Z. soconuscensis* Schutzman, Vovides & Dehgan, and *Z. herrerae* Calderón & Standl.), and in coastal Oaxaca and along the Sierra Madre Occidental facing the Pacific as far to the northwest as Nayarit and Islas Marias (*Z. paucijuga* Wieland).

Systematics and Genetic Variation

There is a *Zamia* species complex in southeastern Mexico, found in moist rain forests that constitute the Pleistocene "arc" floristic refuge of Wendt (1987), undergoing study. These plants are characterized by reddish brown leaves when emergent, wide leaflets, and underground stems. This species complex includes *Z. purpurea, Z. cremnophila, Z. lacandona, Z. splendens,* and the enigmatic holotypes *Z. verschaffeltii* Miq. and *Z. katzeriana* (Regel) Ritter. Nomenclatural problems involving the identity of the latter three taxa have been solved by both field studies and morphometric techniques using discriminant analysis (Nicolalde et al., in revision).

Cytotaxonomic studies of the genus have revealed varying chromosome numbers and karyotypes; $2n = 16, 17, 18, 22, 23, 24, 25, 26, 27$, or 28 (Norstog, 1980, 1981; Vovides, 1983; Moretti & Sabato, 1984; Moretti, 1990; Vovides & Olivares, 1996). Karyotype evolution in *Zamia* is thought to have proceeded in the direction of increasing symmetry produced by fusion of many telocentric chromosomes to give rise to fewer metacentric chromosomes (Norstog, 1980, 1981). However, others consider asymmetrical karyotypes to be the more highly evolved, with evolution thus proceeding in the other direction by centromeric splitting or fission of the larger metacentric chromosomes to produce a higher number of telocentric chromosomes (Khoshoo, 1969; Stebbins, 1971). Notwithstanding, the genus *Cycas,* phylogenetically the most basal of the cycads (Stevenson, 1990), has an asymmetrical karyotype, and *Z. pumila* L., considered one of the more highly evolved species, has a fairly symmetrical karyotype (Norstog & Nicholls, 1997). Jones (1977) mentions that cycles of symmetry and asymmetry through Robertsonian changes may have occurred during the long course of plant evolution, proceeding in one direction by fusion and then in the other by fission, thus making it difficult to decide which is the more derived. Vovides and Olivares (1996) comment that the karyotype evolution of *Z. polymorpha* in the Yucatan peninsula with $2n = 17, 24, 25, 26$, or 27 appears to be in response to increasing habitat dryness. This appears to reflect that of *Z. paucijuga* of the Pacific region ($2n = 23, 25, 26, 27, 28$), with the most asymmetrical cytotypes found toward the extreme northwest of its range (Moretti & Sabato, 1984), where seasonal dry tropical forest increases. From this cytological evidence, we hypothesize the existence of a putative ancestor of *Z. loddigesii, Z. paucijuga,* and *Z. polymorpha* somewhere on the isthmus of Tehuantepec with a somatic chromosome number of $2n = 16$ and a probably symmetric karyotype, which gave rise through Robertsonian changes to the present cytotypes observed in these species, very likely by hybridization, in response to climatic factors (Fig. 35-5).

The diversity and genetic structure of four populations of the cycad *Z. loddigesii* were studied throughout its range in Mexico, and patterns of genetic diversity within and among populations of plants were obtained. The results of allozyme electrophoresis conducted on 15 loci indicated that *Z. loddigesii* has significant levels of genetic differentiation among populations, with 18% of the total variation due to differences among populations. The mean F_{st} value for *Z. loddigesii* is similar to those reported for 16 tree species in tropical zones, 19.9%, and 13 cycad species, 21.3% (González-Astorga et al., unpublished data). These results and comparisons illustrate the effect of habitat fragmentation, and therefore isolation, on populations of *Z. loddigesii*, a process typical of tropical zone plant speciation (Alvarez-Buylla et al., 1996). The geographical isolation among populations of *Z. loddigesii* generated allele loss, as well as a clinal variation of the frequencies of two loci (*MDH* and *MNR2*) in relationship to the latitudinal distribution of populations (Fig. 35-6). The populations have become fragmented by the ever increasing pressure of habitat conversion and disturbance (González-Astorga et al., unpublished data).

Cycad Conservation Through Sustainable Utilization

Cycads are much sought after as ornamental commodities, and the illegal trade in wild-collected specimens along with habitat destruction has lead to the decimation of their populations in the wild. They are considered threatened and endangered and are protected by Mexican and international laws (Anonymous, 1994; IUCN, 2003).

Since demographic studies of some Mexican cycad species have been ongoing for a number of years (Vovides, 1990; Pérez-Farrera et al., 2000; Pérez-Farrera & Vovides, 2004), an obvious spin-off to these studies is assessment of in situ management projects with the aim of attaining sustainable utilization on the part of rural farming communities (Vovides & Iglesias, 1994; Pérez-Farrera & Vovides, 1997; Vovides et al., 2002b). The first of these nurseries was begun in 1990 at Monte Oscuro near Chavarillo, Veracruz, for the management of *D. edule*. The producers collect seed from a nearby conserved population, germinate the seeds, and cultivate the plants until they reach a saleable size after 4–5 years. Approximately 80 ha of cycad habitat has been conserved as a community reserve, and now the nursery has plants of a range of ages for sale.

Five additional cycad nurseries have been established in Chiapas for *C. mirandae* Vovides, Pérez-Farr. & Iglesias, *C. matudae* Lundell, *Z. soconuscensis* Schutzman, Vovides & Dehgan, and *D. merolae* in two biosphere reserves.

In a reintroduction experiment started during September 1997, about 300 *D. edule* seedlings 2, 4, or 7 years of age were planted in a forest near Monte Oscuro and monitored yearly for establishment, growth, and deaths; the total death rate in the field was not more than 20% by 2004. Comparison of the growth rate with plants of the same cohorts that remained in the nursery showed that nursery plants developed five times faster (leaf production and caudex diameter) than those reintroduced, in which no appreciable growth was observed. Sexual maturity (male cones) of *D. edule* was attained after 15 years from seed in the nursery.

Though incentive to conserve cycad habitats has been established among the farmers, sales and exports have been sporadic in all of the nurseries. This situation can be improved through marketing assessment and promotion (Vovides et al., 2002b). The experience of these nurseries is optimistic since a number have been created since 1990, and new ones are being assessed. These activities are in accord with Article 9 of the Convention on Biological Diversity (CBD) on ex situ conservation, which states that "*ex situ* conservation should preferably be in the countries of origin." This facilitates research, rescue, and propagation of threatened species, regulates and names ex situ collections that do not threaten ecosystems and in situ populations, and provides financing for ex situ conservation, especially in developing countries. The CBD also refers to public education and awareness under Article 13, and technical and scientific cooperation is covered by Article 18 (Given, 1997). The role of botanic gardens in both ex situ and in situ conservation of threatened species cannot be overstressed (Vovides et al., 1995), and detailed outlines of their role are described by Wyse Jackson and Sutherland (2000).

Discussion and Conclusions

Floristic Refuges and Hot Spots

There is general agreement that floristic and faunistic refuges of great age exist in southern Mexico (Brown, 1976; Toledo, 1982; Wendt, 1987). This refugium hypothesis indicates that tropical species were restricted to a few areas within the geographical tropics, where climatic changes during periods of glaciation were not as extreme as elsewhere. Lowland forest was replaced by savannah/scrub vegetation at several sites during the last glacial maximum (Leyden, 1984). During the past 40,000 years, tropical forests in Mexico have been disrupted and displaced to lower latitudes by climatic changes, with cycles of cold-dry, cold-wet, and warm-dry climates. It appears that *Ceratozamia* (*Cycadopodites*) was present in the Miocene flora of the southern Mexican regions of Pichucalco and Chiapas (Palacios & Rzedowski, 1993), and in Cenozoic *Engelhardtia* forests in Oaxaca, where the fossil pollen spectrum is remarkably similar to the modern spectrum (Rzedowski & Palacios, 1977). Our evidence from population genetics, morphology, and distribution studies on cycads supports the floristic refuge hypothesis (González & Vovides, 2002; Gonzalez-Astorga et al., 2003a, 2003b, 2005). We therefore put forward the hypothesis that cycad speciation took off from these refugia when post-Pleistocene climatic amelioration took place. The south and southeastern regions in Mexico are particularly rich in these refuges, which is paralleled by cycad species richness in these areas as well as other endemic flora. These are obvious hot spots where in situ conservation efforts should be concentrated.

Systematics, Population Genetics, and Demography

Recent molecular systematic studies have helped to throw light on relationships at the species level and to a greater extent on phytogeographical patterns (Moretti et al., 1993; González & Vovides, 2002; Hill et al., 2003; Bogler & Francisco-Ortega, 2004). However, much work is still needed to solve relationships within species complexes, which are difficult to elucidate using general morphology alone, and even more difficult when only herbarium vouchers are used. Furthermore, the genera *Zamia* and *Ceratozamia* are rife with nomenclatural problems, largely because of incomplete, missing, or lost type specimens and incomplete twentieth century descriptions with nonexistent or at best ambiguous locality information. The work of Stevenson and Sabato (1986a, 1986b) has helped to clear the stage for further work in this respect, and only thorough fieldwork will bring to light new information.

Population genetics has shown that there is an unexpectedly high genetic variation in the cycad populations thus far studied, suggesting much larger populations in the past, which have since been fragmented, accompanied by genetic drift associated with bottlenecks generated by vicariance or founder effects, which may have driven or accompanied rapid speciation (González-Astorga et al., 2003a; 2005). Indeed, fragmentation of some *D. edule* populations have occurred over the past three decades: The first author has seen the reduction of once large populations of this species into isolated groups of a few dozen plants as a result of cattle ranching activities in central Veracruz. Small or declining populations are exposed to a higher extinction risk, since inbreeding, loss of genetic diversity, and fixation of new deleterious mutations are unavoidable (Lande, 1999; Frankham, 2005).

The cycads studied are long-lived and adult plants can attain over 1000 years (Vovides, 1990); since adult plants were sampled for the genetic variation study, the variation observed is therefore historic. However, when seeds and seedlings of one population were analyzed, they were found to be less diverse, owing to the reduction in effective population size (Octavio-Aguilar, unpublished data). Demographic studies on both *Dioon* and *Ceratozamia* have shown a high mortality in seeds and seedlings, with dispersal being mainly by gravity, though animal dispersion has been observed (Pérez-Farrera et al., 2000; Vovides, 1990). Elasticity values of the life cycle of *D. edule* indicate the highest values for reproductive adults, and the lowest for seeds and seedlings (Octavio-Aguilar et al., in review). A very high germination rate has been observed during germination trials, and this has enabled us to set goals for rural nursery management projects. It is recommended that conservation of reproductive adults should take priority over seedling reintroduction.

Sustainable Management of Cycads

The concept of conservation by means of sustainable utilization has been highlighted by many conservationists and plays a mayor role in the framework of the CBD. It has been postulated (and questioned by others) that putting an economic value on populations can create an incentive for their conservation by local people. However, positive examples where this has actually happened are few, especially in the area of over-utilized plant resources. The sustainable management of cycads can be regarded as an encouraging case study in which "Conservation through utilization" has indeed taken place. The prospect of being able to sell propagated plants has created an incentive for the farmers to protect the wild populations. However, this kind of project is typically multidisciplinary and requires mid- to long-term financing (Global Environment Facility, 1998), especially for marketing.

Acknowledgments

The authors are grateful for support by SEMARNAT-CONACyT grant No. 2002-C01-0183, which financed much of the present investigation, and CONACyT research grants No. 29379 N and 34077-N, and grant No. B2-00/013 of FMCN, A. C.

Literature Cited

Alvarez-Buylla, E. R., A. Chaos, D. Piñero & A. A. Garay. 1996. Demographic genetics of a pioneer tropical tree species: patch dynamics, seed dispersal, and seed banks. Evolution 50: 1155–1166.

Anonymous. 1994. Norma Oficial Mexicana NOM-CRN-001.ECOL/1994 que determina las especies y subespecies de flora y fauna silvestres, terrestres y acuáticas en peligro de extinción, amenazadas, raras y las sujetas a protección especial y que establece especificaciones para su protección. Diario Oficial 16 Mayo, México D.F., 1–25.

Balduzzi, A., P. De Luca & S. Sabato. 1981–1982. A phytogeographical approach to the New World cycads. Delpinoa n.s. 23–24: 185–202.

Bogler, D. J. & J. Francisco-Ortega. 2004. Molecular systematic studies in cycads: evidence from *trn*L intron and ITS2 rDNA sequences. Bot. Rev. 70: 260–273.

Brenner, E. D., D. W. Stevenson, J. Chiu, S. J. Runko, P. M. Palenchar & G. M. Coruzzi. 2007. Defining the role of BMAA and its potential Glutamate Receptor targets in *Arabidopsis thaliana* and *Cycas rumphii*. Mem. New York Bot. Gard. 97: 236–252.

Brown. K. S. 1976. Centros de evolução, refugios Quaternários, e conservação de patrimônios genéticos na região neotropical: padrões de diferenciação em Ithomiinae (Lepidoptera: Nymphlidae). Acta Amazon. 7: 75–137.

Brummitt, R. K. 1987. Proposal 708. Taxon. 36: 74–75.

Castillo-Guevara, C. 2007. Herbivores and chemical defenses in cycads—an ecological and evolutionary approach. Mem. New York Bot. Gard. 97: 87–103.

Chamberlain C. J. 1919. The living cycads. Hafner, New York.

Charlton T. S., A. M. Marini, S. P. Markey, K. Norstog & M. W. Duncan. 1992. Quantification of the neurotoxin 2-amino-3-(methylamino)-propanoic acid (BMAA) in Cycadales. Phytochemistry 31: 3429–3432.

Chávez González, A. D., B. Rendón Aguilar & A. Luna José. 2007. Ecología de *Ceratozamia aff. robusta* en la Sierra Madre del Sur de Oaxaca. Mem. New York Bot. Gard. 97: 73–86.

Clark, D. A. & D. B. Clark. 1987. Temporal and environmental patterns of reproduction in *Zamia skinneri*, a tropical rain forest cycad. J. Ecol. 75: 135–149.

———. 1988. Leaf production and the cost of reproduction in the neotropical rain forest cycad, *Zamia skinneri*. J. Ecol. 76: 1153–1163.

Cox, P. A., S. A. Banack & S. J. Murch. 2007. Cyanobacteria, Cycads, and Neurodegenerative Disease among the Chamorro People of Guam. Mem. New York Bot. Gard. 97: 253–285.

Crane, P. R. 1988. Major clades and relationships in the "higher" gymnosperms. Pp. 218–272 *in* C. B. Beck (ed.), Origin and evolution of gymnosperms. Columbia University Press, New York.

Donaldson, J. S. 1995. Understanding cycad life histories: an essential basis for successful conservation. Pp. 8–13 *in* J. S. Donaldson (ed.), Cycad conservation in South Africa. Cycad Society of South Africa, Matieland, South Africa.

Flores-Vázquez, J. C., S. Salas-Morales & A. Saynes-Vásquez. 2007. The populations of *Dioon caputoi* in the Tehuacán-Cuicatlán Biosphere Reserve. Mem. New York Bot. Gard. 97: 388–399.

Frankham, R. 2005. Genetics and extinction. Biol. Conservation 126: 131–140.

Global Environment Facility (GEF). 1998. Global Environment Facility: GEF evaluation of experience with conservation trust funds. GEF Council, Washington, DC.

Given, D. R. 1997. Practical plant conservation—what can botanic gardens hope to achieve? Pp. 87–98 *in* D. H. Touchell & K. W. Dixon (eds.), Conservation into the 21st century. Botanic Gardens Conservation Internacional, Perth, Australia.

González, D. & A. P. Vovides. 2002. Low intralineage divergence in *Ceratozamia* (Zamiaceae) detected with nuclear ribosomal DNA ITS and chloroplast DNA trnL-F non-coding region. Syst. Bot. 27: 654–661.

González-Astorga, J. & A. P. Vovides. 2007. Genetic structure and diversity of *Dioon edule*: A comparison with other cycads. Mem. New York Bot. Gard. 97: 400–417.

———, ———, A. Cruz-Angón, P. Octavio-Aguilar & C. Iglesias. 2005. Allozyme variation in three extant populations of the narrowly endemic cycad *Dioon angustifolium* Miq. (Zamiaceae) from north-eastern Mexico. Ann. Bot. 95: 999–1007.

———, ———, M. M. Ferrer & C. Iglesias. 2003a. Population genetics of *Dioon edule* Lindl. (Zamiaceae, Cycadales): biogeographical and evolutionary implications. Biol. J. Linn. Soc. 80: 457–467.

———, ——— & C. Iglesias. 2003b. Morphological and geographical variation of the cycad *Dioon edule* Lindl. (Zamiaceae): ecological and evolutionary implications. Bot. J. Linn. Soc. 141: 465–470.

Gregory, T. J. & Chemnick J. 2004. Hypotheses on the relationship between biogeography and speciation in *Dioon* (Zamiaceae). Pp. 137–148 *in* T. Walters, & R. Osborne (eds.), Cycad classification: concepts and recommendations. CABI Publishing, Wallingford, UK.

Hill, K. D., M. W. Chase, D. W. Stevenson, H .G. Hills & B. Schutzman. 2003. The families and genera of cycads: a molecular phylogenetic analysis of Cycadophyta based on nuclear and plastid DNA sequences. Int. J. Pl. Sci. 164: 933–948.

Hill, K. D., D. W. Stevenson & R. Osborne. 2007. The world list of cycads. /[La lista mundial de cícades). Mem New York Bot. Gard. 97: 454–483.

IUCN. 2003. IUCN red list of threatened species. www.iucnredlist.org.

Jansen, D. H. & P. S. Martin. 1982. Neotropical anachronisms: the fruits the Gomphotheres ate. Science 215: 19–27.

Jones, D. L. 1993. Cycads of the world. Reed, Chatswood, Australia.

Jones, K. 1977. The role of Robertsonian change in karyotype evolution in higher plants. Chromosomes Today 6: 121–129.

Keppel, G. 2002. Low genetic variation in a Pacific cycad: conservation concerns for *Cycas seemannii* (Cycadaceae). Oryx 36: 41–49.

———, S. W. Lee & P. D. Hodgskiss. 2002. Evidence for long isolation among populations of a Pacific cycad: genetic diversity and differentiation in *Cycas seemannii* A. Br. (Cycadaceae). J. Heredity 93: 133–139.

Khoshoo, T. N. 1969. Chromosome evolution in cycads. Chromosomes Today 2: 236–240.

Kokubugata, G. & K. Kondo. 1998. Comparative karyotype analysis of *Ceratozamia mexicana* and *Microcycas calocoma* (Zamiaceae) using fluorochrome banding (CMA/DAPI) and fluorescence in situ hybridization of ribosomal DNA. Pl. Syst. Evol. 210: 41–50.

———, A. P. Vovides & K. Kondo. 2004. Mapping 5s ribosomal DNA on somatic chromosomes of four species of *Ceratozamia* and *Stangeria eriopus* (Cycadales). Bot. J. Linn. Soc. 145: 499–504.

Lande, R. 1999. Extinction risks from anthropogenic, ecological, and genetic factors. Pp. 1–22 *in* L. F., Landweber, & A. P. Dobson (eds.), Genetics and the extinction of species. Princeton University Press, Princeton, NJ.

Lazcano, J. C. 2007. *Ex situ* conservation of *Microcycas calocoma, Zamia amblyphyllidia, Z. integrifolia, Z. ottonis,* and *Z. pygmaea* germplasm at the National Botanic Garden, Cuba. Mem. New York Bot. Gard. 97: 32–44.

Leyden, B. W. 1984. Guatemalan forest synthesis after Pleistocene aridity. Proc. Natl. Acad. Sci. U.S.A. 81: 4856–4859.

Lindley, J. 1843. *Dion edule*. Edward's Bot. Reg. 29: 59–60.

Marshall, C. J. & J. K. Liebherr. 2000. Cladistic biogeography of the Mexican transition zone. J. Biogeogr. 27: 203–216.

Miquel, F. A. W. 1848. Over eenige niewe of zeldzame Cycadeën in den Hortus Botanicus te Amsterdam. Tijdschr. Wetenschap. 1: 197–206.

Moretti, A. 1990. Karyotypic data on north and central American Zamiaceae, (Cycadales) and their phylogenetic implications. Amer. J. Bot. 77: 1016–1029.

——— & S. Sabato. 1984. Karyotype evolution by centromeric fission in *Zamia* (Cycadales). Pl. Syst. Evol. 146: 215–223.

———, ——— & M. Vázquez-Torres. 1980. The distribution of *Ceratozamia* (Zamiaceae). Delpinoa 20: 13–21.

———, ——— & ———. 1982. The rediscovery of *Ceratozamia kuesteriana* (Zamiaceae) in Mexico. Brittonia 34: 185–188.

———, P. Caputo, S. Cozzolino, P. De Luca, L. Gaudio, G. Siniscalco-Gigliano & D. W. Stevenson. 1993. A phylogenetic analysis of *Dioon* (Zamiaceae). Amer. J. Bot. 80: 204–214.

Newell, S. J. 1983. Reproduction in a natural population of cycads *Zamia pumila* L. in Puerto Rico. Bull. Torrey Bot. Club 110: 464–473.

Nicolalde-Morejón, F. 2007. Taxonomía, distribución y estado de conservación de *Zamia* en Ecuador. Mem. New York Bot. Gard. 97: 45–63.

Niklas, K. J. & K. J. Norstog. 1984. Aerodynamics and pollen grain depositional patterns on cycad megastrobili: implications on the reproduction of three cycad genera (*Cycas, Dioon* and *Zamia*). Bot. Gaz. 145: 92–104.

Norstog, K. 1980. Chromosome numbers in *Zamia* (Cycadales). Caryologia 33: 419–428.

———. 1981. Karyotypes of *Zamia chigua* (Cycadales). Caryologia 34: 255–260.

———. 1987. Cycads and the origin of insect pollination. Amer. Sci. 75: 270–279.

——— & P. K. S. Fawcett. 1989. Insect-cycad symbiosis and its relation to the pollination of *Zamia furfuracea* (Zamiaceae) by *Rhopalotria mollis* (Curculionidae). Amer. J. Bot. 76: 1380–1394.

——— & T. J. Nicholls. 1997. *The biology of the cycads.* Cornell University Press Ithaca, NY.

———, D. W. Stevenson & K. J. Niklas. 1986. The role of beetles in the pollination of *Zamia furfuracea* L.fil. (Zamiaceae). Biotropica 18: 300–306.

Palacios, C. R. & J. Rzedowski. 1993. Estudio palinológico de las floras fósiles del Mioceno inferior y principios del Mioceno Medio de la región de Pichucalco, Chiapas, México. Acta Bot. Mex. 24: 1–96.

Pellmyr, O., W. Tang, I. Groth, G. Bergstrom & L. Thien. 1991. Cycad cone and angiosperm floral volatiles: inferences for the evolution of insect pollination. Biochem. Syst. Ecol. 19: 623–627.

Pérez-Farrera, M. A., P. F. Quintana-Ascencio, B. Salvatierra-Izaba & A. P. Vovides. 2000. Population dynamics of *Ceratozamia matudai* Lundell (Zamiaceae) in El Triunfo Biosphere Reserve, Chiapas, Mexico. J. Torrey Bot. Club 127: 291–299.

——— & A. P. Vovides. 1997 Manual para el cultivo y propagación de cycadas. INE-SEMARNAP, México, D.F.

——— & ———. 2004. Spatial distribution, population structure, and fecundity of *Ceratozamia matudai* Lundell (Zamiaceae) in El Triunfo Biosphere Reserve, Chiapas, Mexico. Bot. Rev. 70: 299–311.

———, ———, D. González, L. H. Sandoval & M. Martínez. 2007. Variación morfológica y genética del complejo *Ceratozamia miqueliana* H. A. Wendl. (Cycadales, Zamiaceae). Mem. New York Bot. Gard. 97: 519–541.

———, ——— & C. Iglesias. 2001a. The cycad *Ceratozamia norstogii* Stevenson (Zamiaceae) from southern Mexico: new information on distribution, habitat and vegetative morphology. Bot. J. Linn. Soc. 137: 71–76.

———, ——— & ———. 2001b. A new species of *Ceratozamia* (Zamiaceae) from Chiapas, Mexico. Bot. J. Linn. Soc. 137: 77–80.

Rzedowski, J. 1993. Diversity and origins of the phanerogamic flora of Mexico. Pp. 129–146 *in* T.P. Ramamoorthy, R. Bye, A. Lot, & J. Fa (eds.), Biological diversity of Mexico: origins and distribution. Oxford University Press, New York.

——— & C. R. Palacios. 1977. El bosque de *Engelhardtia* (*Oreomunnea*) *mexicana* en la región de la Chinantla (Oaxaca, México). Una reliquia del Cenozoico. Bol. Soc. Bot. México 36: 93–123.

Sabato, S. & P. De Luca. 1985. Evolutionary trends in *Dion* (Zamiaceae). Amer. J. Bot. 72: 1353–1363.

Sánchez-Tinoco, M.Y., E. M. Engleman & S. D. Koch. 2007. The vascularization of the seed of *Ceratozamia mexicana* (Zamiaceae). Mem. New York Bot. Gard. 97: 223–235.

Schneider, D., M. Wink, F. Sporer & P. Lounibos. 2002. Cycads: their evolution, toxins, herbivores and insect pollinators. Naturwissenschaften 89: 281–294.

Shaw, C. A., J. M. B. Wilson, R. Cruz-Aguado, S. Singh, E. L. Hawkes, V. Lee & T. Marler. 2007. Cycad-Induced neurodegeneration in a mouse model of ALS-PDC: Is the culprit really BMAA or is a novel toxin to blame? Mem. New York Bot. Gard. 97: 286–307.

Stebbins, G. L. 1971. Chromosomal evolution in higher plants. E. Arnold, London.

Stevenson, D. 1990. Morphology and systematics of Cycadales. Mem. New York Bot. Gard. 57: 8–55.

———. 2001. Orden Cycadales. Pp. 1–90 *in* R. Bernal, & E. Forero (eds), Flora de Colombia. Universidad Nacional de Colombia, Bogotá. Colombia.

———, K. J. Norstog & P. K. S. Fawcett. 1998. Pollination biology of cycads. Pp. 277–294 *in* S. J. Owens, & P. J. Rudall (eds.), Reproductive biology. Royal Botanic Gardens, Kew, London.

——— & S. Sabato. 1986a. Typification of names in *Ceratozamia* Brongn, *Dion* Lindl. and *Microcycas* A.DC. (Zamiaceae). Taxon 35: 578–584.

——— & ———. 1986b. Typification of names in *Zamia* L. and *Aulacophyllum* Regel (Zamiaceae). Taxon 35: 134–144.

———, ——— & M. Vazquez-Torres. 1986. A new species of *Ceratozamia* (Zamiaceae) from Veracruz, Mexico with comments on species relationships, habitats, and vegetative morphology in *Ceratozamia*. Brittonia 38: 17–26.

Tang, W. 1987. Insect pollination in the cycad *Zamia pumila* (Zamiaceae). Amer. J. Bot. 74: 90–99.

Ting, I. P. & J. H. Burk. 1983. Aspects of carbon metabolism in *Welwitschia*. Pl. Sci. Lett. 32: 279–285.

Toledo, V. M. 1982. Pleistocene changes of vegetation in tropical Mexico. Pp. 93–111 *in* G. T. Prance (ed.), Biological diversification in the tropics. Columbia University Press, New York.

Treutlein, J. & M. Wink. 2002. Molecular phylogeny of cycads inferred from *rbc*L sequences. Naturwissenschaften 89: 221–225.

Vovides, A. P. 1983. Systematic studies on the Mexican Zamiaceae I. Chromosome numbers and karyotypes. Amer. J. Bot. 70: 1002–1006.

———. 1985. Systematic studies on Mexican Zamiaceae II. Additional notes on *Ceratozamia kuesteriana* from Tamaulipas, Mexico. Brittonia 37: 226–231.

———. 1990. Spatial distribution, survival and fecundity of *Dioon edule* (Zamiaceae) in a tropical deciduous forest in Veracruz, Mexico, with notes on its habitat. Amer. J. Bot. 77: 1532–1543.

———. 1991a. Insect symbionts of some Mexican cycads in their natural habitat. Biotropica 23: 102–104.

———. 1991b. Cone idioblasts of eleven cycad genera: morphology, distribution, and significance. Bot. Gaz. 152: 91–99.

———. 2000. México: segundo lugar mundial en diversidad de cícadas. Biodiversitas 6: 6–10.

———, M. E. Cortéz, C. G. Iglesias & M. Lascurain. 1995. The Jardín Botánico Francisco Javier Clavijero in Xalapa, Veracruz, Mexico. Bot. Gard. Conservation News 2: 32–39.

———, J. R. Etherington, P. Quentin Dresser, A. Groenhof, C. Iglesias & J. Flores-Ramirez. 2002a. CAM-cycling in the cycad *Dioon edule* Lindl. in its natural tropical deciduous forest habitat in central Veracruz, Mexico. Bot. J. Linn. Soc. 138: 155–162.

——— & C. G. Iglesias. 1994. An integrated conservation strategy for the cycad *Dioon edule* Lindl. Biodivers. & Conservation 3: 137–141.

———, C. Iglesias, M. A. Pérez-Farrera, M. Vázquez Torres & U. Schippmann. 2002b. Peasant nurseries: a concept for an integrated conservation strategy for cycads in Mexico. Pp. 421–444 *in* M. Maunder, C. Clubbe, C. Hankamer & M. Groves (eds.), Plant conservation in the tropics: perspectives and practice. Royal Botanic Gardens, Kew, UK.

———, K. J. Norstog, P. K. S. Fawcett, M. W. Duncan, R. J. Nash & D.V. Molsen. 1993. Histological changes during maturation in male and female cones of the cycad *Zamia furfuracea* and their significance in relation to pollination biology. Bot. J. Linn. Soc. 111: 241–252.

———, N. Ogata, V. Sosa & E. Peña-García. 1997. Pollination of the endangered Cuban cycad *Microcycas calocoma* (Miq.) A.DC. Bot. J. Linn. Soc. 125: 201–210.

——— & M. Olivares. 1996. Karyotype polymorphism in the cycad *Zamia loddigesii* (Zamiaceae) of the Yucatan Peninsula, Mexico. Bot. J. Linn. Soc. 120: 77–83.

———, M. A. Pérez-Farrera, J. González-Astorga, D. Gonález, T. Gregory, J. Chemnick, C. Iglesias, P. Octavio-Aguilar, S. Avendaño, C. Bárcenas & S. Salas-Morales. 2003. An outline of our current knowledge on Mexican Cycads (Zamiaceae, Cycadales). Curr. Topics Pl. Biol. 4: 159–174.

———, ———, D. Gonzalez & S. Avendaño. 2004. Relationships and phytogeography in *Ceratozamia* (Zamiaceae). Pp. 109–125 *in* T. Walters & R. Osborne (eds), Cycad classification: concepts and recommendations. CABI Publishing, Wallingford, UK.

——— & C. M. Peters. 1987. *Dioon edule*: la planta más Antigua de México. Ci. Desarr. 13(73): 19–24.

———, J. D. Rees & M. Vázquez-Torres. 1983. Zamiaceae. Pp. 1–31 *in* A. Gómez-Pompa & V. Sosa (eds.), Flora de Veracruz Fasc. 26, INIREB, Xalapa, Mexico.

Walters T. & R. Osborne (eds.). 2004. Cycad classification: concepts and recommendations. CABI Publishing, Wallingford, UK.

Wendt, T. 1987. Las selvas de Uxpanapa, Veracruz-Oaxaca, México: evidencia de refugios florísticos Cenozoicos. Anales Inst. Biol. Univ. Nac. México, Ser. Bot. 58: 29–54.

Whiting M. G. 1963. Toxicity of cycads. Econ. Bot. 17: 270–302.

Wilson, E. O. 1992. The diversity of life. W.W. Norton, New York.

Wyse Jackson, P. & L. A. Sutherland. 2000. International agenda for botanic gardens in conservation. Botanic Gardens International, United Kingdom.

Index

Andrew P. Vovides (right) was born in London, England, in 1944. His parents, both Cypriots, emigrated to London in the mid 1930s. His early education took place in London and he took engineering diplomas whilst working as an optical instruments engineer for a London based firm. He later took a position as a laboratory technician at the Botany department at Birkbeck College, London University.

He obtained an Honors Botany Degree at the University of Wales Cardiff and in 1975 he emigrated with his Mexican wife, Victoria, to Mexico where he took up a research position at the Instituto de Investigaciones Sobre Recursos Bióticos (INIREB) under Arturo Gómez-Pompa at Xalapa. His first Project with Arturo was the foundation of the Jardín Botánico Francisco Javier Clavijero that was inaugurated on 17 February 1977. In 1978 Andrew had a three month internship at the Royal Botanic Gardens, Kew where he worked with the Deputy Curator of Living Collections, Ian Beyer, and his staff on the intricacies of running a botanic garden. Andrew was also a founding member and first secretary of the Mexican Association of Botanic Gardens.

Andrew became interested in cycads when Arturo Gómez-Pompa invited him to collaborate with John Rees on the Flora de Veracruz treatment for Zamiaceae. At this time Andrew was becoming acquainted with Mexican cycads and studying for a Ph.D. on the biology of Mexican cycads as an external part-time student at the Botany Dept. at Cardiff. During his course of graduate studies he has built up the Mexican National Cycad Collection at the Garden and obtained his Ph.D. in 1988 at a time when INIREB was closed and the Garden passed under the administration of the present Instituto de Ecología, A.C. He served a one-year post-doctoral internship at Fairchild Tropical Botanic Garden with Knut Norstog working on aspects of cycad pollination.

Andrew has held scientific positions as a research botanist, botanic garden director and curator at Instituto de Ecología, and serves as a member IUCN Cycad Specialist Group. He has produced over 100 scientific articles and book chapters and is well known for his collaborative work with Arturo Gómez-Pompa on two interactive CD ROMs, the "Cycads of Mexico" and "Cycads: the Endangered Living Fossils," and a compilation about endangered plant species of Mexico, a precursor to the Official Mexican Norm list for plants. He has also pioneered rural nurseries aimed at the sustainable management of cycads and habitat conservation in Mexico. These nurseries are now operating in the states of Veracruz, Puebla, Querétaro, and Chiapas. For these efforts, he received the State of Veracruz Academic Prize for the Environment in June 2003. *(Photograph of Andrew P. Vovides [right] and Knut Norstog [left].)*

Dennis Wm. Stevenson was born in 1942 in rural Ohio where he grew up on a farm. He received his BS and MS degrees from Ohio State University in 1970 and 1971, respectively. It was during this time, at an Ohio Academy of Sciences meeting, that he met Knut Norstog and developed his interest in cycads. Stevenson completed his Ph.D. in 1975 at the University of California at Davis. From there he spent two years at Fairchild Tropical Garden as a Harvard University Fellow, one year at the Royal Botanic Gardens, Kew as a NATO Postdoctoral Fellow, followed by two more years at Fairchild Tropical Garden as a Harvard University Fellow. From 1980 to 1987, he was on the faculty at Barnard College of Columbia University. In 1987, he joined the staff at The New York Botanical Garden.

His activities have focused upon the evolution and classification of the cycads and their placement in seed plant phylogeny. He has conducted research on various facets of the biology of the cycads ranging among paleobotany, reproductive biology, anatomy, cytology, molecular systematics, as well as taxonomic monographs and treatments of these groups for various floras being prepared for the neotropics.

These research interests have been pursued by an active field program, particularly in the neotropics, to survey population sizes and demographics as part of a program to develop conservation strategies. As part of this field program, live plants are also collected for *ex situ* conservation research and for breeding purposes in efforts to save the cycads from extinction. He has served as a member of the IUCN Cycad Specialist Group since its founding and served as its Chairman from 1990–1996.

Roy Osborne was born in London, England, during WW II. After his family emigrated to South Africa in 1948, he attended Grey High School at Port Elizabeth and then studied botany and chemistry at Rhodes University, Grahamstown. It was during this period that he first became acquainted with cycads, many species of which are endemic to southern Africa. Roy held various scientific positions in Zimbabwe and South Africa and later continued postgraduate studies in botany and chemistry at the University of Natal, obtaining a Ph.D. for cycad-oriented research in 1988. He continued his dual interest in the subjects, and was later appointed Professor of Chemistry at the University of Natal.

Concerned about longer-term prospects in southern Africa, Roy, his wife Angela, and their four children, relocated to Australia in 1996. They settled in Brisbane, Queensland, where Roy is currently employed as Senior Chemist by Queensland Government.

Roy was founder and first President of the Cycad Society of South Africa, and has been a long-serving representative to the IUCN Cycad Specialist Group. He has attended most of the International Cycad Conferences and has been involved, as author, co-author, editor and publisher, in the production of some 175 books and scientific papers. He is best known for his collaborative effort in the popular texts *Cycads of Australia* (2001) and *Cycads of Vietnam* (2007). Roy maintains eclectic interests in botany, horticulture, music, and foreign languages.

Memoirs of The New York Botanical Garden
(ISSN 0077-8931)

The Memoirs are published at irregular intervals in volumes of various sizes and are designed to include results of original botanical research by members of The New York Botanical Garden's staff, or by botanists who have collaborated in one or more of The New York Botanical Garden's research programs. Ordinarily only manuscripts of 100 or more typewritten pages will be considered for publication.

Manuscripts should be submitted to the Editor. For further information regarding editorial policy and instructions for the preparation of manuscripts, contact the Editor.

Orders for published and forthcoming issues and volumes should be placed with:

The New York Botanical Garden Press
200th Street and Kazimiroff Boulevard
Bronx, New York 10458-5126 USA
Telephone: 718.817.8721 Fax: 718.817.8842
E-mail: nybgpress@nybg.org
Web site: http://www.nybgpress.org